Congratulations!

As a reader of ***Structure and Interpretation of Signals and Systems*** by Edward A. Lee and Pravin Varaiya, you are entitled to prepaid access to the book's **Companion Web site!** This Web site will be a key resource in helping you succeed in your study of signals and systems.

This prepaid subscription gives you full access to all student support areas of this Web site, including comprehensive signals and systems tutorials, many of which use Java applets to illustrate key concepts.

To Activate Your Prepaid Subscription:

1. Go to your book's Companion Web Site at http://www.aw.com/lee_varaiya
2. Click on "Protected Student Resources"
3. Select the "New Users Please Register First" link
4. Enter your preassigned Access Code, exactly as it appears below
5. Select "Submit"
6. Complete the online registration form to establish your personal User ID and Password
7. Once your personal User ID and Password are confirmed, you can log in to and begin using the Lee/Varaiya Companion Web site!

This Access Code can only be used once to establish a subscription. This subscription to the Lee/Varaiya Companion Web site is valid for six months upon activation and is not transferrable.

WARNING

If the activation code above is showing, it may no longer be valid. To obtain access to the Lee/Varaiya Companion Web site, please purchase a new copy of this textbook or visit http://www.aw.com/lee_varaiya for information on purchasing a subscription online.

STRUCTURE AND INTERPRETATION OF

Signals and Systems

STRUCTURE AND INTERPRETATION OF *Signals and Systems*

Edward A. Lee
Pravin Varaiya

UNIVERSITY OF CALIFORNIA AT BERKELEY

Boston San Francisco New York
London Toronto Sydney Tokyo Singapore Madrid
Mexico City Munich Paris Cape Town Hong Kong Montreal

Executive Editor	Susan Hartman Sullivan
Assistant Editor	Galia Shokry
Executive Marketing Manager	Michael Hirsch
Production Supervisor	Kim Ellwood
Project Management	P. M. Gordon Associates, Inc.
Compositor	Windfall Software, using ZzTEX
Cover Design	Gina Hagen Kolenda
Prepress and Manufacturing	Caroline Fell

Access the latest information about Addison-Wesley titles from our World Wide Web site: www.aw.com/cs

Library of Congress Cataloging-in-Publication Data

Lee, Edward A., 1957–
Structure and interpretation of signals and systems / Edward A. Lee, Pravin Varaiya.
p. cm.
Includes bibliographical references and index.
ISBN 0-201-74551-8
1. Signal detection. I. Varaiya, P. P. (Pravin Pratap) II. Title
TK5102.9.L43 2002
621.38′2—dc21 12420592

1 2 3 4 5 6 7 8 9 10—CRW—05 04 03 02

TO *RHONDA* AND *HELEN*,
AND TO *RUTH* AS ALWAYS

Contents

14 *Composition and feedback control* 549

A *Sets and functions* 589

B *Complex numbers* 619

Preface

This textbook is about signals and systems, a discipline rooted in the intellectual tradition of electrical engineering (EE). This tradition, however, has evolved in unexpected ways. EE has lost its tight coupling with the "electrical." Electricity provides the impetus, the potential, but not the body of the subject. How else could microelectromechanical systems (MEMS) become so important in EE? Is this not mechanical engineering? Or signal processing? Is this not mathematics? Or digital networking? Is this not computer science? How is it that control system techniques are profitably applied to aeronautical systems, structural mechanics, electrical systems, and options pricing?

This book approaches signals and systems from a computational point of view. It is intended for students interested in the modern, highly digital problems of electrical engineering, computer science, and computer engineering. In particular, the approach is applicable to problems in computer networking, wireless communication systems, embedded control, audio and video signal processing, and, of course, circuits.

A more traditional introduction to signals and systems would be biased toward the latter application, circuits. It would focus almost exclusively on linear time-invariant systems, and would develop continuous-time models first, with discrete-time models then treated as an advanced topic. The discipline, after all, grew out of the context of circuit analysis. But it has changed. Even pure EE graduates are more likely to write software than to push electrons, and yet we still recognize them as electrical engineers.

The approach in this book benefits students by showing from the start that the methods of signals and systems are applicable to software systems, and most interestingly, to systems that mix computers with physical devices such as circuits, mechanical control systems, and physical media. Such systems have become pervasive, and profoundly affect our daily lives.

The shift away from circuits implies some changes in the way the methodology of signals and systems is presented. While it is still true that a voltage that varies over time is a signal, so is a packet sequence on a network. This text defines *signals* to cover both. While it is still true that an RLC circuit is a system, so is a computer program for decoding Internet audio. This text defines *systems* to cover both. While for some systems the state is still captured adequately by variables in a differential equation, for many it is now the values in registers and memory of a computer. This text defines *state* to cover both.

The fundamental limits also change. Although we still face thermal noise and the speed of light, we are likely to encounter other limits—such as complexity, computability, chaos, and, most commonly, limits imposed by other human constructions—before we get to these. A voiceband data modem, for example, uses the telephone network, which was designed to carry voice, and offers as immutable limits such nonphysical constraints as its 3 kHz bandwidth. This has no intrinsic origin in the physics of the network; it is put there by engineers. Similarly, computer-based audio systems face latency and jitter imposed by the operating system. This text focuses on composition of systems so that the limits imposed by one system on another can be understood.

The mathematical basis for the discipline also changes. Although we still use calculus and differential equations, we frequently need discrete math, set theory, and mathematical logic. Whereas the mathematics of calculus and differential equations evolved to describe the physical world, the world we face as system designers often has nonphysical properties that are not such a good match for this mathematics. This text bases the entire study on a highly adaptable formalism rooted in elementary set theory.

Despite these fundamental changes in the medium with which we operate, the methodology of signals and systems remains robust and powerful. It is the methodology, not the medium, that defines the field.

The book is based on a course at Berkeley taught over the past four years to more than 2,000 students in electrical engineering and computer sciences. That experience is reflected in certain distinguished features of this book. First, no background in electrical engineering or computer science is assumed. Readers should have some exposure to calculus, elementary set theory, series, first-order linear differential equations, trigonometry, and elementary complex numbers. The appendices review set theory and complex numbers, so this background is less essential.

Approach

This book is about mathematical modeling and analysis of signals and systems, applications of these methods, and the connection between mathematical models and computational realizations. We develop three themes. The first theme is the use of sets and functions as a universal language to describe diverse signals and systems. Signals—voice, images, bit sequences—are represented as functions with an appropriate domain and range. Systems are represented as functions whose domain and range are themselves sets of signals. Thus, for example, a modem is represented as a function that maps bit sequences into voice-like signals.

The second theme is that complex systems are constructed by connecting simpler subsystems in standard ways—cascade, parallel, and feedback. The connections detennine the behavior of the interconnected system from the behaviors of component subsystems. The connections place consistency requirements on the input and output signals of the systems being connected.

Our third theme is to relate the declarative view (mathematical, "what is") with the imperative view (procedural, "how to"). That is, we associate mathematical analysis of systems with realizations of these systems. This is the heart of engineering. When EE was entirely about circuits, this was relatively easy, because it was the physics of the circuits that was being described by the mathematics. Today we have to somehow associate the mathematical analysis with very different realizations of the systems, most especially software. We make this association through the study of state machines, and through the consideration of many real-world signals, which, unlike their mathematical abstractions, have little discernable declarative structure. Speech signals, for instance, are far more interesting than sinusoids, and yet many signals and systems textbooks talk only about sinusoids.

Content

We begin in chapter 1 by describing signals as functions, focusing on characterizing the domain and the range for familiar signals that humans perceive, such as sound, images, video, trajectories of vehicles, as well as signals typically used by machines to store or manipulate information, such as sequences of words or bits.

In chapter 2, systems are described as functions, but now the domain and the range are themselves sets of signals. The telephone handset converts voice into an analog electrical signal, and the line card in the telephone central office converts the latter into a stream of bits. Systems can be connected to form a more complex system, and the function describing these more complex systems is a composition of functions describing the component systems.

Characterizing concretely the functions that describe signals and systems is the content of the book. We begin to characterize systems in chapter 3 using the notion of state, the state transition function, and the output function, all in the context of finite-state machines. In chapter 4, state machines are composed in various ways (cascade, parallel, and feedback) to make more interesting systems. Applications to feedback control illustrate the power of the state machine model.

In chapter 5, time-based systems are studied in more depth, first with discrete-time systems (which have simpler mathematics), and then with continuous-time systems. We define linear time-invariant (LTI) systems as infinite state machines with linear state transition and output functions and zero initial state. The input–output behavior of these systems is now fully characterized by their impulse response.

Chapter 6 bridges the finite-state machines of chapters 3 and 4 with the time-based systems of chapter 5, showing that they can be combined in useful ways to get hybrid systems. This greatly extends the applicability of LTI systems, because, although most systems are not LTI, many have modes of operation that are approximately LTI. The concept of modal models is illustrated with supervisory control systems. This chapter alone would justify the unified modeling approach in this text, because it offers a glimpse of a far more powerful conceptual framework than either state machines or LTI methods can offer alone.

Chapter 7 introduces frequency decomposition of signals; chapter 8 introduces frequency response of LTI systems; and chapter 9 brings the two together by discussing filtering. The approach is to present frequency domain concepts as a complementary toolset, different from that of state machines, and much more powerful when applicable. Frequency decomposition of signals is motivated first using psychoacoustics, and gradually developed until all four Fourier transforms (the Fourier series, the Fourier transform, the discrete-time Fourier transform, and the discrete Fourier transform) have been described. We linger on the first of these, the Fourier series, since it is conceptually the easiest, and then more quickly present the others as generalizations of the Fourier series. LTI systems yield best to frequency-domain analysis because of the property that complex exponentials are eigenfunctions. Consequently, they are fully characterized by their frequency response—the main reason that frequency domain methods are important in the analysis of filters and feedback control.

Chapter 10 covers classical Fourier transform material such as properties of the four Fourier transforms and transforms of basic signals. Chapter 11 applies frequency domain methods to a study of sampling and aliasing.

Chapters 12, 13, and 14 extend frequency-domain techniques to include the Z transform and the Laplace transform. Applications in signal processing and feedback control illustrate the concepts and the utility of the techniques. Mathematically, the Z transform and the Laplace transform are introduced as extensions of the discrete-time and continuous-time Fourier transforms to signals that are not absolutely summable or integrable.

The unified modeling approach in this text is rich enough to describe a wide range of signals and systems, including those based on discrete events and those based on signals in time, both continuous and discrete. The complementary tools of state machines and frequency-domain methods permit analysis and implementation of concrete signals and systems. Hybrid systems and modal models offer systematic ways to combine these complementary toolsets. The framework and the tools of this text provide a foundation on which to build later courses on digital systems, embedded software, communications, signal processing, hybrid systems, and control.

The Web site

The book has an extensive companion Web site, **http://www.aw.com/lee_varaiya**. It includes:

- **The laboratory component.** A suite of exercises based on MATLAB and Simulink®* help reconcile the declarative and imperative points of view. MATLAB is an imperative programming language. Simulink is a block diagram language, in which one connects blocks implementing simpler subsystems to construct more interesting systems. It is much easier to quickly construct interesting signals and systems using the extensive built-in libraries of MATLAB and Simulink than using a conventional programming language like C++, Java, or Scheme. These laboratory exercises involve audio, video, and images, which are much more interesting signals than sinusoids.
- **The applets.** An extensive set of interactive applets brings out the imperative view and illustrates concepts of frequency analysis. These include speech, music, and image examples, interactive applets showing finite Fourier series approximations, and illustrations of complex exponentials and phasors.
- **Instructor and student aids.** A large set of Web pages, arranged by topic, can be used by the instructor in class and by students to review the material. These pages integrate many of the applets, and thus offer more interactive and dynamic presentation material than what is possible with more conventional presentation material. At Berkeley, we use them in the classroom, as a supplement to the blackboard. Qualified instructors can download a snapshot of the Web pages, including the applets, so a network connection is not required in the classroom.
- **Additional sidebars.** The Web site includes additional topics in sidebar form, beyond those in the text. For example, there is a discussion of image encoding methods that are commonly used on the Web.
- **Solutions.** Solutions to exercises are available from the publisher to qualified instructors.

* MATLAB and Simulink are registered trademarks of The MathWorks, Inc.

Pedagogical features

This book has a number of highlights that make it well suited as a textbook for an introductory course.

1. "Probing Further" sidebars briefly introduce the reader to interesting extensions of the subject, to applications, and to more advanced material. They serve to indicate directions in which the subject can be explored.
2. "Basics" sidebars offer readers with less mathematical background some basic tools and methods.
3. Appendix A reviews basic set theory and helps establish the notation used throughout the book.
4. Appendix B reviews complex variables, making it unnecessary for students to have much background in this area.
5. Key equations are boxed to emphasize their importance. They can serve as the places to pause in a quick reading. In the index, the page numbers where key terms are defined are shown in bold.
6. The exercises at the end of each chapter are annotated with the letters E, T, or C to distinguish those exercises that are mechanical (E for excercise) from those requiring a plan of attack (T for thought) and from those that generally have more than one reasonable answer (C for conceptualization).

Notation

The notation in this text is unusual when compared to standard texts on signals and systems. We explain our reasons for this as follows:

Domains and ranges. It is common in signals and systems texts to use the form of the argument of a function to define its domain. For example, $x(n)$ is a discrete-time signal, while $x(t)$ is a continuous-time signal; $X(j\omega)$ is the continuous-time Fourier transform and $X(e^{j\omega})$ is the discrete-time Fourier transform. This leads to apparent nonsense like $x(n) = x(nT)$ to define sampling, or to confusion like $X(j\omega) \neq X(e^{j\omega})$ even when $j\omega = e^{j\omega}$.

We treat the domain of a function as part of its definition. Thus, a discrete-time, real-valued signal is a function $x : Integers \to Reals$, and its discrete-time Fourier transform is a function $x : Reals \to Complex$. The DTFT itself is a function whose domain and range are sets of functions,

$$DTFT\colon [Integers \to Reals] \to [Reals \to Complex].$$

Then we can unambiguously write $X = DTFT(x)$.

Functions as values. Most texts call the expression $x(t)$ a function. A better interpretation is that $x(t)$ is an element in the range of the function x. The difficulty with the former interpretation becomes obvious when talking about systems. Many texts pay lip service to the notion that a system is a function by introducing a notation like $y(t) = T(x(t))$. This makes it seem that T acts on the value $x(t)$ rather than on the entire function x.

Our notation includes set of functions, allowing systems to be defined as functions with such sets as the domain and range. Continuous-time convolution, for example, becomes

$$\mathit{Convolution}\colon [\mathit{Reals} \to \mathit{Reals}] \times [\mathit{Reals} \to \mathit{Reals}] \to [\mathit{Reals} \to \mathit{Reals}].$$

We then introduce the notation $*$ as a shorthand,

$$y = x * h = \mathit{Convolution}(x, h),$$

and define the convolution function by

$$\forall\, t \in \mathit{Reals}, \quad y(t) = (x * h)(t) = \int_{-\infty}^{\infty} X(\tau) y(t-\tau)\, d\tau.$$

Note the careful parenthesization. The more traditional notation, $y(t) = x(t) * h(t)$, would seem to imply that $y(t-T) = x(t-T) * h(t-T)$. But it is not so!

A major advantage of our notation is that it easily extends beyond LTI systems to the sorts of systems that inevitably arise in any real world application, such as mixtures of discrete event and continuous-time systems.

Names of functions. We use long names for functions and variables when they have a concrete interpretation. Thus, instead of x we might use *Sound*. This follows a long-standing tradition in software, where readability is considerably improved by long names. By giving us a much richer set of names to use, this helps us avoid some of the preceding pitfalls. For example, to define sampling of an audio signal, we might write

$$\mathit{SampledSound} = \mathit{Sampler}_T(\mathit{Sound}).$$

It also helps bridge the gap between realizations of systems (which are often software) and their mathematical models. How to manage and understand this gap is a major theme of our approach.

How to use this book

At Berkeley, the first 11 chapters of this book are covered in a 15-week, one-semester course. Even though it leaves Laplace transforms, Z transforms, and feedback control systems to a follow-up course, it remains a fairly intense experience. Each week consists of three 50-minute lectures, a one-hour problem session, and one three-hour laboratory. The lectures and problem sessions are conducted by a faculty member while the laboratory is led by teaching assistants, who are usually graduate students, but are also often talented juniors or seniors.

The laboratory component is based on MATLAB and Simulink, and is closely coordinated with the lectures. The text does not offer a tutorial on MATLAB, although the labs include enough material so that, combined with on-line help, they are sufficient. Some examples in the text and some exercises at the ends of the chapters depend on MATLAB.

At Berkeley, this course is taken by all electrical engineering and computer science students, and is followed by a more traditional signals and systems course. That course covers the material in the last three chapters plus applications offrequency-domain methods to colllnunications systems. The follow-up course is not taken by most computer science students. In a program that is more purely electrical and computer engineering than ours, a better approach might be to spend two quarters or two semesters on the material in this text, since the unity of notation and approach would be better than having two disjoint courses, the introductory one using a modern approach, and the follow-up course using a traditional one.

Acknowledgments

Many people have contributed to the content of this book. Dave Messerschmitt conceptualized the first version of the course on which the book is based, and later committed considerable departmental resources to the development of the course while he was chair of the EECS department at Berkeley. Randy Katz, Richard Newton, and Shankar Sastry continued to invest considerable resources in the course when they each took over as chair, and backed our efforts to establish the course as a cornerstone of our undergraduate curriculum. This took considerable courage, since the conceptual approach of the course was largely unproven.

Tom Henzinger probably had more intellectual influence over the approach than any other individual, and to this day we still argue in the halls about details of the approach. The view of state machines, of composition of systems, and of hybrid systems owe much to Tom. Gerard Berry also contributed a great deal to our way of presenting synchronous composition.

We were impressed by the approach of Harold Abelson and and Gerald Jay Sussman, in *Structure and Interpretation of Computer Programs* (MIT Press, 1996), who confronted a similar transition in their discipline. The title of our book shows their influence. Jim McLellan, Ron Shafer, and Mark Yoder influenced this book through their pioneering departure from tradition in signals and systems, *DSP First—A Multimedia Approach* (Prentice-Hall, 1998). Ken Steiglitz greatly influenced the labs with his inspirational book, *A DSP Primer: With Applications to Digital Audio and Computer Music* (Addison-Wesley, 1996).

A number of people have been involved in the media applications, examples, the laboratory development, and the Web content associated with the book.

These include Brian Evans and Ferenc Kovac. We also owe gratitude for the superb technical support from Christopher Hylands. Jie Liu contributed sticky masses example to the hybrid systems chapter, and Yuhong Xiong contributed the technical stock trading example. Other examples and ideas were contributed by Steve Neuendorffer, Cory Sharp, and Tunc Simsek.

For each of the past four years, about 500 students at Berkeley have taken the course that provided the impetus for this book. They used successive versions of the book and the Web content. Their varied response to the course helped us define the structure of the book and the level of discussion. The course is taught with the help of undergraduate teaching assistants. Their comments helped shape the laboratory material.

Several colleagues kindly consented to be interviewed: Panos Antsaklis, University of Notre Dame; Gerard Berry, Esterel Technologies; P. R. Kumar, University of Illinois, Urbana–Champaign; Dawn Tilbury, University of Michigan, Ann Arbor; Jeff Bier, BDTI; and Xavier Rodet, IRCAM, France. We thank them for sharing the experience that encouraged them toward a career in electrical and computer engineering.

Parts of this book were reviewed by more than 30 faculty members around the country. Their criticisms helped us correct defects and inconsistencies in earlier versions. Of course, we alone are responsible for the opinions expressed in the book, and the errors that remain. We especially thank: Jack Kurzweil, San Jose State University; Lee Swindlehurst, Brigham Young University; Malur K. Sundareshan, University of Arizona; Stéphane Lafortune, University of Michigan; Ronald E. Nelson, Arkansas Tech University; Ravi Mazumdar, Purdue University; Ratnesh Kumar, University of Kentucky; Rahul Singh, San Diego State University; Paul Neudorfer, Seattle University; R. Mark Nelms, Auburn University; Chen-Ching Liu, University of Washington; John H. Painter, Texas A&M University; T. Kirubarajan, University of Connecticut; James Harris, California Polytechnic State University in San Luis Obispo; Frank B. Gross, Florida A&M University; Donald L. Snyder, Washington University in St. Louis; Theodore E. Djaferis, University of Massachusetts in Amherst; Soura Dasgupta, University of Iowa; Maurice Felix Aburdene, Bucknell University; and Don H. Johnson, Rice University.

These reviews were solicited by Heather Shelstad of Brooks/Cole, Denise Penrose of Morgan-Kaufmann, and Susan Hartman and Galia Shokry of Addison-Wesley. We are grateful to these editors for their interest and encouragement. To Susan Hartman, Galia Shokry and Nancy Lombardi we owe a special thanks; their enthusiasm and managerial skills helped us and others keep the deadlines in bringing the book to print.

It has taken much longer to write this book than we expected when we embarked on this project five years ago. It has been a worthwhile effort nonetheless. Our friendship has deepened, and our mutual respect has grown as we learned from each other. Rhonda Lee Righter and Ruth Varaiya have been remarkably sympathetic and encouraging through the many hours at nights and on weekends that this project has demanded. To them we owe our immense gratitude.

CHAPTER 1

Signals and systems

Signals convey information. Systems transform signals. This book is about developing an understanding of both. We gain this understanding by dissecting their structure and by examining their interpretation. For systems, we look at the relationship between the input and output signals (this relationship is a **declarative** description of the system) and the procedure for converting an input signal into an output signal (this procedure is an **imperative** description of the system).

A sound is a signal. We leave a description of the physics of sound to texts on physics and instead show how a sound can be usefully decomposed into components that themselves have meaning. For example, a musical chord can be decomposed into a set of notes.

An image is a signal. We do not discuss the biophysics of visual perception; instead we show that an image can be usefully decomposed. We can use such decomposition, for instance, to examine what it means for an image to be sharp or blurred and thus to determine how to blur or sharpen an image.

Signals can be more abstract (less physical) than sound or images. A signal can be, for example, a sequence of commands or a list of names. We develop models for such signals and the systems that operate on them, such as a system that interprets a sequence of commands from a musician and produces a sound.

One way to get a deeper understanding of a subject is to formalize it by developing mathematical models. Such models admit manipulation with a level of confidence not achievable with less formal models. We know that if we follow the rules of mathematics, a transformed model still relates strongly to the original model. There is a sense in which mathematical manipulation preserves "truth" in a way that is elusive with almost any other intellectual manipulation of a subject. We can leverage this truth preservation to gain confidence in the design of a system, to extract hidden information from a signal, or simply to gain insight.

Mathematically, we model both signals and systems as functions. A **signal** is a function that maps a domain, often time or space, into a range, often a physical measure such as air pressure or light intensity. A **system** is a function that maps signals from its domain—its input signals—into signals in its range—its output signals. Both the domain and the range are sets of signals (**signal spaces**). Thus, systems are functions that operate on functions.

We use the mathematical language of sets and functions to make our models unambiguous, precise, and manipulable. This language has its own notation and rules, which are reviewed in appendix A. We begin to use this language in this chapter. Depending on the situation, we represent physical quantities such as time, voltage, current, light intensity, air pressure, or the content of a memory location by variables that range over appropriate sets. For example, discrete time may be represented by a variable $n \in Naturals$ (the natural numbers) or $n \in Integers$ (the integers); continuous time may be represented by a variable $t \in Reals_+$ (the nonnegative real numbers) or $t \in Reals$ (the real numbers). Light intensity may be represented by a continuous variable $x \in [0, I]$, a range of real numbers from zero to I, where I is some maximum value of the intensity; a variable in a logic circuit may be represented as $x \in Binary$ (the binary digits). A binary file is an element of $Binary^*$, the set of sequences of binary digits. A computer name such as `cory.eecs.Berkeley.edu` may be assigned to a variable in $Char^*$, the set of sequences of characters.

1.1 Signals

Signals are functions that carry information, often in the form of temporal and spatial patterns. These patterns may be embodied in various media; radio and broadcast TV signals are electromagnetic waves, and images are spatial patterns of light intensities of different colors. In digital form, images and video are bit strings. Sensors of physical quantities (such as speed, temperature, or pressure) often convert those quantities into electrical voltages, which are then often converted into digital numbers for processing by a computer. In this text, we study systems that store, manipulate, and transmit signals.

In this section, we study signals that occur in human perception, mechanical and electronic sensors, radio and television, and telephone and computer networks, and the description of physical quantities that change over time or over space. The most common feature of these signals is that their domains are sets representing time and space. However, we also study signals that are represented as sequences of symbols, in which position within the sequence has no particular association with the physical notions of time or space. Such signals are often used to represent sequences of commands or sequences of events.

We will model signals as functions that map a domain (a set) into a range (another set). Our interest for now is to understand through examples how to select the domain and range of signals and how to visualize them. To fully describe a signal, we need to specify not only its domain and range, but also the rules by which the function assigns values. Because the domain and range are often infinite sets, specification of the rules is rarely trivial. Much of the emphasis of subsequent chapters is on how to characterize these rules.

1.1.1 *Audio signals*

Our ears are sensitive to sound, which is physically just rapid variations in air pressure. Thus sound can be represented as a function,

$$\boxed{Sound\colon Time \to Pressure,}$$

where *Pressure* is a set consisting of possible values of air pressure, and *Time* is a set representing the time interval over which the signal lasts.*

Example 1.1: A one-second segment of a voice signal is a function of the form

$$Voice\colon [0, 1] \to Pressure,$$

where [0, 1] represents one second of time. An example of such a function is plotted in figure 1.1. Such a plot is often called a **waveform**.

The signal in figure 1.1 varies over positive and negative values, averaging approximately zero. But air pressure cannot be negative, and so the vertical axis does not literally represent air pressure. It is customary to normalize the representation of sound by subtracting the ambient air pressure (about 100,000 newtons per square meter) from the range. Our ears are, after all, not sensitive to constant ambient air pressure. Thus, we assume *Pressure* = *Reals* (the real numbers), in which negative pressure means a drop in pressure in relation to ambient air pressure.

In fact, the possible values of the function *Voice* as shown in figure 1.1 are 16-bit integers, suitable for storage in a computer. We can call the set of 16-bit integers $Integers16 = \{-32768, \ldots, 32767\}$. The audio hardware of the computer is responsible for converting members of the set *Integers16* into air pressure.

* For a review of the notation of sets and functions, see appendix A.

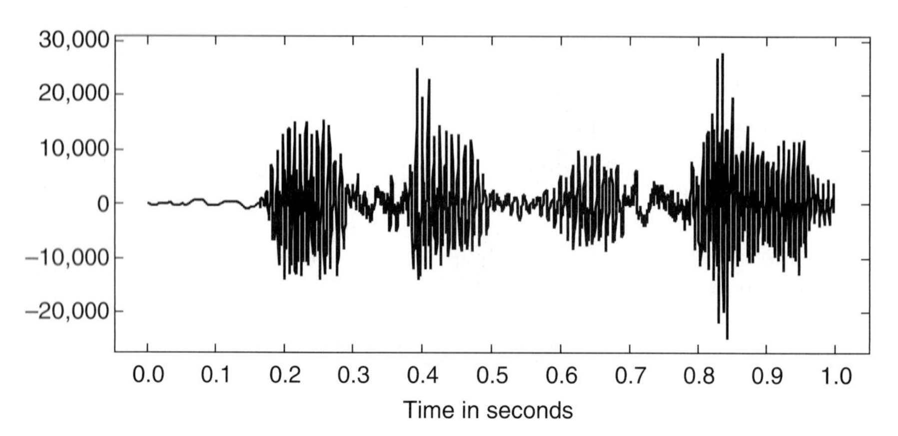

FIGURE 1.1: Waveform of a speech fragment.

The numbers in the computer representation *Integers16* are a subset of *Reals*. The units of air pressure in this representation are arbitrary; therefore, to convert to units of newtons per square meter, we need to multiply these numbers by some constant. The value of this constant depends on the audio hardware of the computer and on its volume setting. ❒

The horizontal axis in figure 1.1 suggests that time varies continuously from 0 to 1.0. However, a computer cannot directly handle such a continuum. The sound is represented not as a continuous waveform but rather as a list of numbers (for voice-quality audio, 8,000 numbers for every second of speech).* A close-up of a section of the speech waveform is shown in figure 1.2. That plot shows 100 data points (called **samples**). For emphasis, that plot shows a dot for each sample, rather than a continuous curve, and a stem connecting the dot to the horizontal axis. Such a plot is called a **stem plot**. Because there are 8,000 samples per second, the 100 points in figure 1.2 represent 100/8,000 seconds, or 12.5 milliseconds of speech.

Such signals are said to be **discrete-time signals** because they are defined only at discrete points in time. A discrete-time one-second voice signal in a computer is a function,

$$\boxed{ComputerVoice\colon DiscreteTime \to Integers16,}$$

where $DiscreteTime = \{0, 1/8{,}000, 2/8{,}000, \ldots, 8{,}000/8{,}000\}$ is the set of sampling times.

* In a compact disc (CD), there are 44,100 numbers per second of sound per stereo channel.

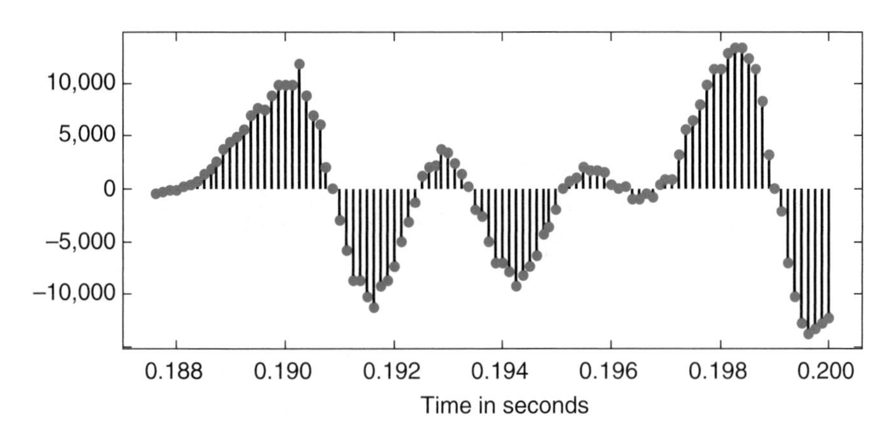

FIGURE 1.2: Discrete-time representation of a speech fragment.

In contrast, **continuous-time signals** are functions defined over a continuous interval of time (technically, a continuum in the set *Reals*). The audio hardware of the computer is responsible for converting the *ComputerVoice* function into a function of the form *Sound*: *Time* → *Pressure*. That hardware, which converts an input signal into a different output signal, is a system.

Example 1.2: We cannot represent the function *Voice* of Example 1.1 by a mathematical expression. We now consider an example in which there is such an expression. The sound emitted by a precisely tuned and idealized 440-Hz tuning fork over the infinite time interval $Reals = (-\infty, \infty)$ is the function

$$PureTone: Reals \to Reals,$$

where the time-to-(normalized) pressure assignment is*

$$\forall\, t \in Reals, \quad PureTone(t) = P \sin(2\pi \times 440t).$$

Here, P is the **amplitude** of the sinusoidal signal *PureTone*. It is a real-valued constant.

Figure 1.3 is a graph of a portion of this pure tone (showing only a subset of the domain *Reals*). In the figure, $P = 1$. ❒

* If the notation here is unfamiliar, see appendix A.

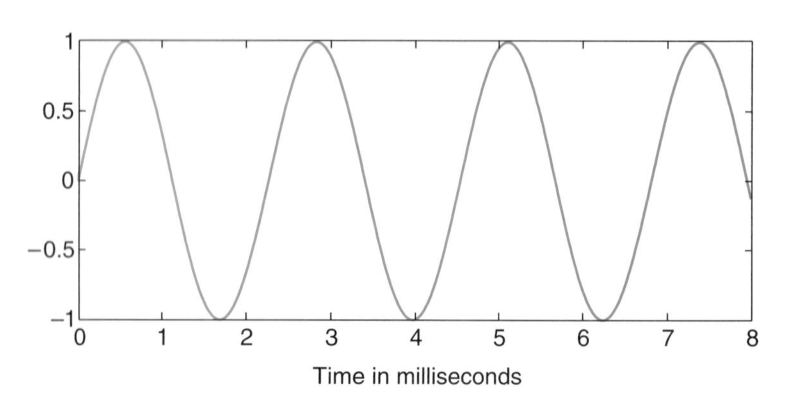

FIGURE 1.3: Portion of the graph of a pure tone with frequency 440 Hz.

The number 440 in this example is the **frequency** of the sinusoidal signal shown in figure 1.3, in units of **cycles per second** or **Hertz (Hz)**.* It simply means that the sinusoid completes 440 cycles per second. Alternatively, the sinusoid completes one cycle in 1/440 seconds, or about 2.3 milliseconds. The time to complete one cycle, 2.3 milliseconds, is called the **period**.

The *Voice* signal in figure 1.1 is much more irregular than *PureTone* in figure 1.3. According to an important theorem that we study in subsequent chapters, a function like *Voice*, despite its irregularity, is a sum of signals of the form of *PureTone* but with different frequencies. A sum of two pure tones of frequencies—say, 440 Hz and 660 Hz—is the function *SumOfTones*: *Reals* → *Reals*, given by

$$\forall\, t \in Reals, \quad SumOfTones(t) = P_1 \sin(2\pi \times 440t) + P_2 \sin(2\pi \times 660t).$$

Notice that summing two signals amounts to adding the values of their functions at each point in the domain. Of course, two signals can be added only if they have the same domain and compatible ranges (so that addition is defined between an element of one range and an element of the other). The two components are shown in figure 1.4. At any point on the horizontal axis, the value of the sum is simply the addition of the values of the two components.

*The unit of frequency called Hertz is named after physicist Heinrich Rudolf Hertz (1857–1894) for his research in electromagnetic waves.

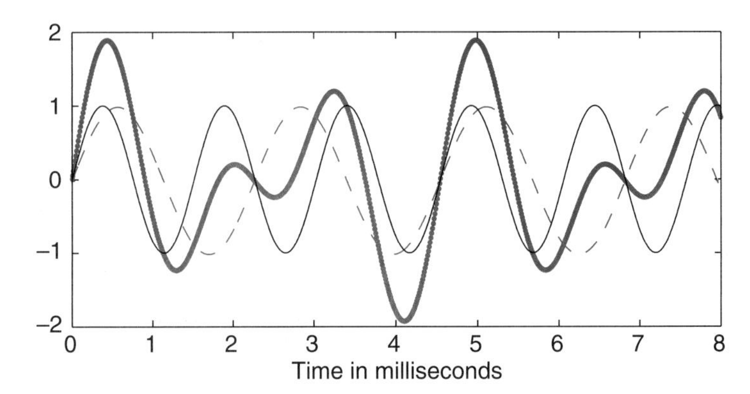

FIGURE 1.4: Sum (color line) of two pure tones, one at 440 Hz (dashed line) and the other at 660 Hz (black line).

PROBING FURTHER

Household electrical power

In the United States, household current is delivered on three wires: a **neutral wire** and two **hot wires**. The voltage between either hot wire and the neutral wire is around 110 to 120 volts, RMS (root mean square, the square root of the average of the voltage squared). The voltage between the two hot wires is around 220 to 240 volts, RMS. The higher voltage is used for appliances that need more power, such as air conditioners. Here, we examine exactly how this works.

The voltage between the hot wires and the neutral wire is sinusoidal, with a frequency of 60 Hz. Thus, for one of the hot wires, it is a function x: *Reals* → *Reals*, where the domain represents time and the range represents voltage, and

$$\forall\, t \in Reals, \quad x(t) = 170\cos(60 \times 2\pi t).$$

This 60-Hz sinusoidal waveform completes one cycle in a period of $T = 1/60$ seconds. Why is the amplitude 170 volts, rather than 120? Because the 120 voltage is **RMS** (**root mean square**); that is,

$$voltage_{RMS} = \sqrt{\frac{1}{T}\int_0^T x^2(t)dt} \text{ volts},$$

the square root of the average of the square of the voltage. This is calculated to be 120.

continued on next page

PROBING FURTHER

The voltage between the second hot wire and the neutral wire is a function $y\colon Reals \to Reals$, where

$$\forall\, t \in Reals, \quad y(t) = -170\cos(60 \times 2\pi t) = -x(t).$$

It is the negative of the other voltage at any time t. This sinusoidal signal is said to have a **phase shift** of 180 degrees, or π radians, in comparison with the first sinusoid. Equivalently, it is said to be 180 degrees **out of phase**.

We can now see how to get the higher voltage for power-hungry appliances. We simply use the two hot wires, rather than one hot wire and the neutral wire. The voltage between the two hot wires is the difference, a function $z\colon Reals \to Reals$, where

$$\forall\, t \in Reals, \quad z(t) = x(t) - y(t) = 340\cos(60 \times 2\pi t).$$

This corresponds to 240 volts, RMS. A plot is shown in figure 1.5.

The neutral wire should not be confused with the ground wire in a three-prong plug. The ground wire is a safety feature to allow current to flow into the earth rather than, say, through a person.

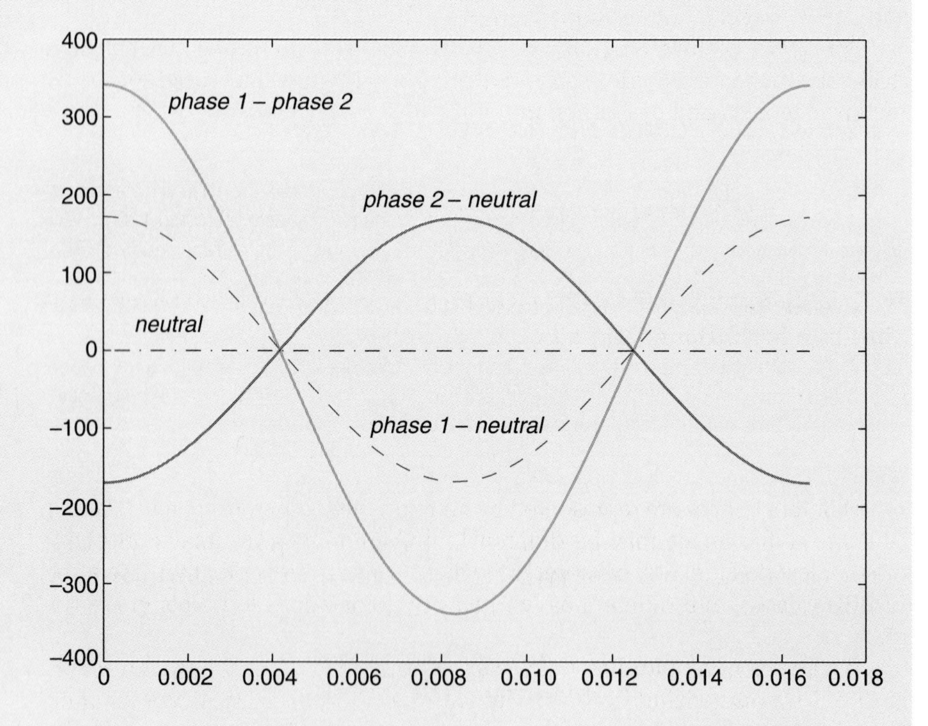

FIGURE 1.5: The voltages between the two hot wires and the neutral wire (*phase 1–neutral* and *phase 2–neutral*) and between the two hot wires (*phase 1–phase 2*) in household electrical power in the United States.

1.1.2 *Images*

If an image is a grayscale picture on an 11×8.5 inch sheet of paper, the picture is represented by a function

$$Image: [0, 11] \times [0, 8.5] \to [0, B_{max}], \tag{1.1}$$

where B_{max} is the maximum grayscale value (0 is black and B_{max} is white). The set $[0, 11] \times [0, 8.5]$ defines the space of the sheet of paper. More generally, a grayscale image is a function

$$\boxed{Image: VerticalSpace \times HorizontalSpace \to Intensity,}$$

where $Intensity = [black, white]$ is the intensity range from *black* to *white* measured in some scale. An example is shown in figure 1.6.

For a color picture, the reflected light is sometimes measured in terms of its RGB values (the magnitudes of the red, green, and blue colors), and so a color picture is represented by a function

$$\boxed{ColorImage: VerticalSpace \times HorizontalSpace \to Intensity^3.}$$

The RGB values assigned by *ColorImage* at any point (x, y) in its domain is the triple $(r, g, b) \in Intensity^3$, given by

$$(r, g, b) = ColorImage(x, y).$$

Different images are represented by functions with different spatial domains (the size of the image may be different), different ranges (we may consider a more or less detailed way of representing light intensity and color than grayscale or RGB values), and different assignments of color values to points in the domain.

Because a computer has finite memory and finite word length, an image is stored by discretizing both the domain and the range, as in the function *ComputerVoice*. Thus, for example, your computer may represent an image by storing a function of the form

$$ComputerImage: DiscreteVerticalSpace \times DiscreteHorizontalSpace \to Integers8,$$

FIGURE 1.6: Grayscale image (left) and its enlarged pixels (right).

where

$$\textit{DiscreteVerticalSpace} = \{1, 2, \ldots, 300\},$$

$$\textit{DiscreteHorizontalSpace} = \{1, 2, \ldots, 200\}, \text{ and}$$

$$\textit{Integers8} = \{0, 1, \ldots, 255\}.$$

It is customary to say that *ComputerImage* stores 300 × 200 pixels, whereby a **pixel** is an individual picture element. The value of a pixel is *ComputerImage* (*row*, *column*) ∈ *Integers8*, where *row* ∈ *DiscreteVerticalSpace* and *column* ∈ *DiscreteHorizontalSpace*. In this example, the range *Integers8* has 256 elements; therefore, in the computer these elements can be represented by an eight-bit integer (hence the name of the range, *Integers8*). An example of such an image is shown in figure 1.6, in which the right image is part of the left image magnified to show the discretization implied by the individual pixels.

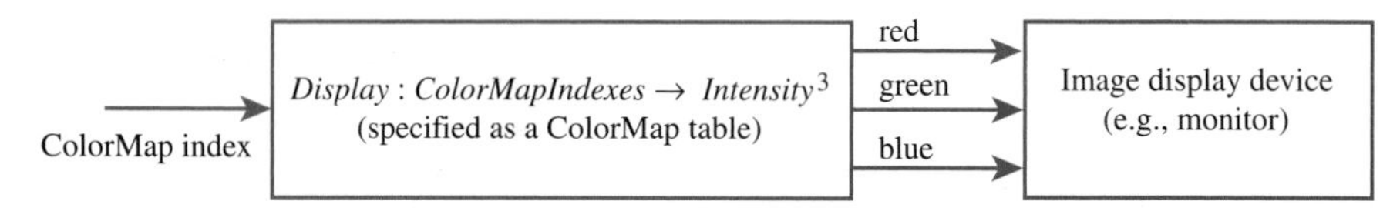

FIGURE 1.7: In a computer representation of a color image that uses a colormap, pixel values are elements of the set *ColorMapIndexes*. The function *Display* converts these indexes to an RGB (red, green, blue) representation.

A computer could store a color image in one of two ways. One way is to represent it as a function,

$$\begin{aligned} &\mathit{ColorComputerImage}\colon \mathit{DiscreteVerticalSpace} \times \\ &\qquad \mathit{DiscreteHorizontalSpace} \to \mathit{Integers8}^3, \end{aligned} \tag{1.2}$$

so that each pixel value is an element of $\{0, 1, \ldots, 255\}^3$. Such a pixel can be represented as three eight-bit integers. A common method that saves memory is to use a **colormap**. Define the set $\mathit{ColorMapIndexes} = \{0, \ldots, 255\}$, together with a *Display* function,

$$\mathit{Display}\colon \mathit{ColorMapIndexes} \to \mathit{Intensity}^3. \tag{1.3}$$

Display assigns to each element of *ColorMapIndexes* the three (r, g, b) color intensities. This is depicted in the block diagram in figure 1.7. Use of a colormap reduces the memory required to store an image by a factor of three because each pixel is now represented by a single eight-bit number. But only 256 colors can be represented in any given image. The function *Display* is typically represented in the computer as a lookup table (see lab 2 in the Lab Manual at **www.aw.com/lee_varaiya**).

1.1.3 *Video signals*

A **video** is a sequence of images displayed at a certain rate (frequency). National Television System Committee (NTSC) video (the standard for analog video in the United States) displays 30 images (called **frames**) per second. The eye is unable to distinguish between successive images displayed at this frequency, and so a television broadcast appears to be continuously varying in time.

PROBING FURTHER

Color and light

The human eye is sensitive to electromagnetic waves of certain frequency, which we call light. The frequency f in Hertz of a purely sinusoidal electromagnetic wave is related to its *wavelength* λ in meters by the formula $f = c/\lambda$, where c is the speed of light (about 3×10^8 meters/second). The wavelengths of visible light range from about 350–400 nanometers, or 10^{-9} meters, to 750–800 nanometers (nm). We experience light of different wavelengths as having different colors: violet (350 nm), indigo, blue, green, yellow, orange, and red (800 nm).

The retina has three groups of cones, each sensitive to one of the three primary colors: red, green, and blue. Other colors are perceived when these three groups are stimulated in different combinations, as shown in the diagram. By combining its red, green, and blue light sources in different amounts, a computer monitor can create the perception of all colors. The color white is obtained by adding all three primary colors, and the absence of any light is perceived as black. This "additive" model of color perception was proposed in 1802 by Thomas Young and H. L. F. Helmholz.

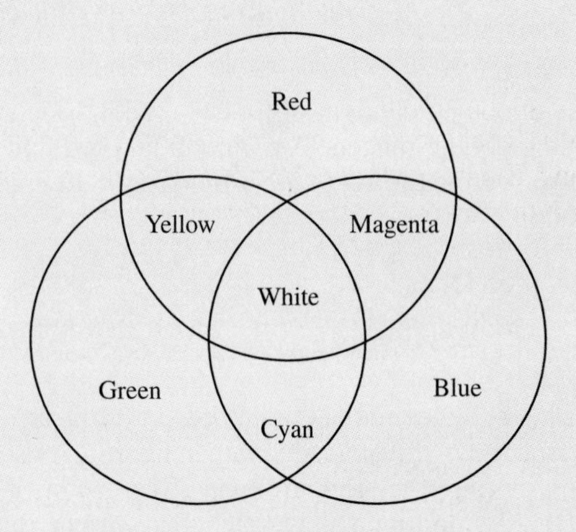

In a computer, if the amount of each primary color is represented by an eight-bit word and each color is represented by three eight-bit words, the computer could generate $2^8 \times 2^8 \times 2^8 = 2^{24} = 16,777,216$ colors. An eight-bit color map, in contrast, can generate only 256 colors.

Painting works by subtraction: different pigments of color absorb (subtract) light of different wavelengths. The primary subtractive colors are magenta, yellow, and cyan. Similarly, if you wear rose-tinted glasses, things look pink because the glasses filter out the nonred wavelengths.

The ear and the eye are very different perceptual systems. If we listen to a sound consisting of the sum of two pure tones, we can distinguish the two tones. However, we cannot perceive the difference between, say, a yellow light source and an appropriate combination of red and green sources. The ear can be modeled as a linear time-invariant system (see chapter 5); the eye cannot.

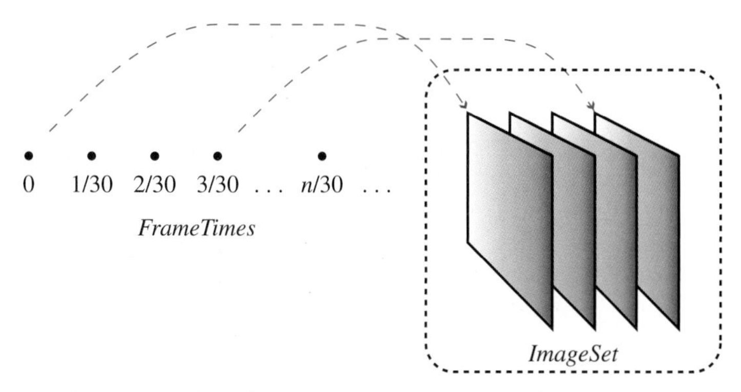

FIGURE 1.8: Illustration of the function *Video*.

Thus, the domain of a video signal is discrete time, $FrameTimes = \{0, 1/30, 2/30, \ldots\}$, and its range is a set of images, *ImageSet*. For **analog video**, each image in *ImageSet* is a function of the form

$$VideoFrame\colon DiscreteVerticalSpace \times HorizontalSpace \to Intensity^3.$$

An analog video signal is discretized in the vertical direction but not in the horizontal direction.* The image is composed of a set of horizontal lines called **scan lines**, in which the intensity varies along the line. The horizontal lines are stacked vertically to form an image.

A video signal is therefore a function

$$Video\colon FrameTimes \to ImageSet. \tag{1.4}$$

For any time $t \in FrameTimes$, the image $Video(t) \in ImageSet$ is displayed. The signal *Video* is illustrated in figure 1.8.

An alternative way of specifying a video signal is by the function $Video'$, whose domain is a product set as follows:

$$Video'\colon FrameTimes \times DiscreteVerticalSpace \times HorizontalSpace \to Intensity^3.$$

Like *Video* in figure 1.8, $Video'$ is depicted in figure 1.9. The RGB value assigned to a point (x, y) at time t is

$$(r, g, b) = Video'(t, x, y). \tag{1.5}$$

*This is a simplification. Most analog video images are **interlaced**, which means that successive frames use different sets for *DiscreteVerticalSpace* so that scan lines in one frame lie in between the scan lines of the previous frame. Also, the range $Intensity^3$ has an interesting structure that ensures compatibility between black-and-white and color television sets.

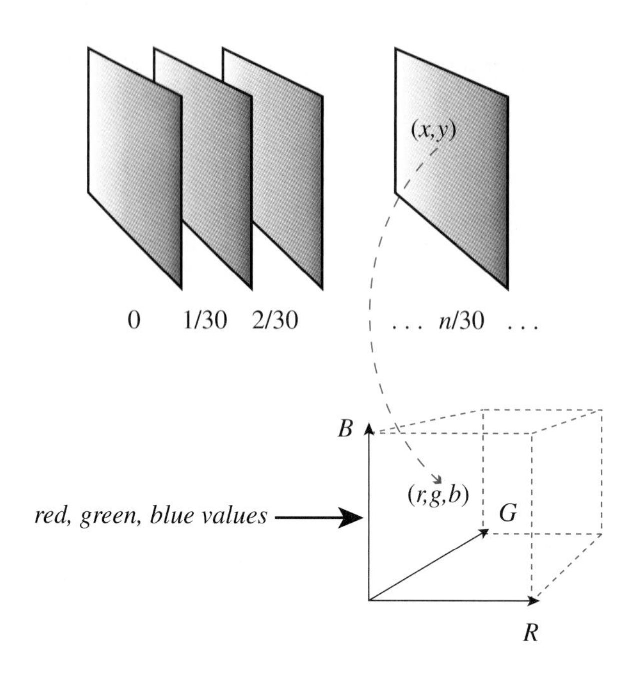

FIGURE 1.9: Illustration of the function *Video*′.

If the signals specified in (1.4) and (1.5) represent the same video, then for all $t \in \mathit{FrameTimes}$ and $(x, y) \in \mathit{DiscreteVerticalSpace} \times \mathit{HorizontalSpace}$,

$$(\mathit{Video}(t))(x, y) = \mathit{Video}'(t, x, y). \tag{1.6}$$

It is worth pausing to understand the notation used in (1.6). *Video* is a function of t, and so $\mathit{Video}(t)$ is an element in its range *ImageSet*. Because elements in *ImageSet* themselves are functions, $\mathit{Video}(t)$ is a function. The domain of $\mathit{Video}(t)$ is the product set $\mathit{DiscreteVerticalSpace} \times \mathit{HorizontalSpace}$, and so $(\mathit{Video}(t))(x, y)$ is the value of this function at the point (x, y) in its domain. This value is an element of $\mathit{Intensity}^3$. On the right side of (1.6), Video' is a function of (t, x, y), and so $\mathit{Video}'(t, x, y)$ is an element in its range, $\mathit{Intensity}^3$. The equality (1.6) asserts that for all values of t, x, y, the two sides are the same. On the left side of (1.6), the parentheses enclosing $\mathit{Video}(t)$ are not necessary; we could equally well write $\mathit{Video}(t)(x, y)$. However, the parentheses improve readability.

1.1.4 *Signals representing physical attributes*

The changes over time in the attributes of a physical object or device can be represented as functions of time or space.

Example 1.3: The position of an airplane can be expressed as

$$Position\colon Time \to Reals^3,$$

where

$$\forall\, t \in Time, \quad Position(t) = (x(t), y(t), z(t))$$

is its position in three-dimensional space at time t. The position and velocity of the airplane is a function

$$s\colon Time \to Reals^6, \tag{1.7}$$

where

$$s(t) = (x(t),\ y(t),\ z(t),\ v_x(t),\ v_y(t),\ v_z(t)) \tag{1.8}$$

gives its position and velocity at $t \in Time$.

The position of the pendulum shown on the left side of figure 1.10 is represented by the function

$$\theta\colon Time \to [-\pi, \pi],$$

where $\theta(t)$ is the angle with the vertical made by the pendulum at time t.

The position of the upper and lower arms of a robot arm depicted on the right side of figure 1.10 can be represented by the function

$$(\theta_u, \theta_l)\colon Time \to [-\pi, \pi]^2,$$

where $\theta_u(t)$ is the angle at the shoulder made by the upper arm with the vertical at time t, and $\theta_l(t)$ is the angle made by the lower arm at the elbow at time t. Note that we can regard (θ_u, θ_l) as a single function with range as the product set $[-\pi, \pi]^2$ or as two functions, θ_u and θ_l, each with range $[-\pi, \pi]$. Similarly, we can regard s in (1.7) as a single function with range $Reals^6$ or as a collection of six functions, each with range *Reals*, as suggested by (1.8).
❒

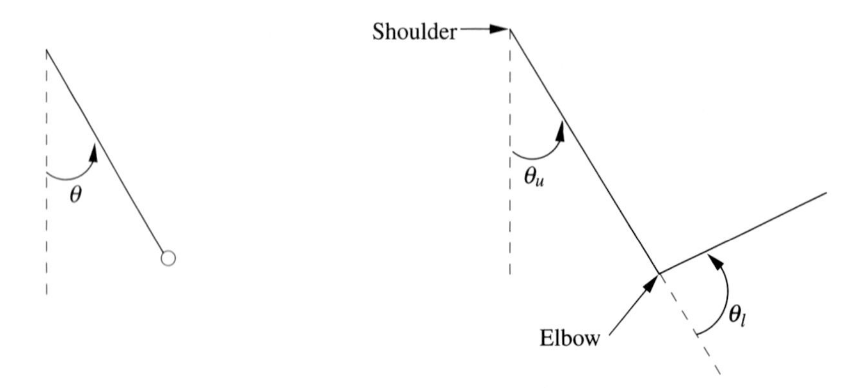

FIGURE 1.10: Position of a pendulum (left) and upper and lower arms of a robot (right).

Example 1.4: The spatial variation of temperature over some volume of space can be represented by a function

$$AirTemp: X \times Y \times Z \to Reals,$$

where $X \times Y \times Z \subset Reals^3$ is the volume of interest, and $AirTemp(x, y, z)$ is the temperature at the point (x, y, z). ❐

1.1.5 *Sequences*

We have studied examples in which temporal or spatial information is represented by functions of a variable representing time or space. The domain of time or space may be continuous, as in *Voice* and *Image*, or discrete, as in *ComputerVoice* and *ComputerImage*.

In many situations, information is represented as **sequences** of symbols rather than as functions of time or space. These sequences occur in two ways: as a representation of **data** or as a representation of an **event stream**. Sequences, in fact, are special sorts of functions.

Examples of data represented by sequences are common. A file stored in a computer in binary form is a sequence of bits, or binary symbols—that is, a sequence of 0s and 1s. A text, like that in this book, is a sequence of words. A sheet of music represents a sequence of notes.

Example 1.5: Consider an N-bit-long binary file,

$$b_1, b_2, \cdots, b_N,$$

where each $b_i \in Binary = \{0, 1\}$. We can regard this file as a function

$$File: \{1, 2, \ldots, N\} \to Binary,$$

with the assignment $File(n) = b_n$ for every $n \in \{1, \ldots, N\}$.

Sometimes we can give a mathematical expression for a binary signal. For instance, the N-bit-long binary file *Alt*, consisting of an alternating sequence of 0s and 1s, is given as follows:

$$\forall\, n \in \{1, 2, \ldots, N\}, \quad Alt(n) = \begin{cases} 0, & n \text{ even} \\ 1, & n \text{ odd} \end{cases}$$

If instead of *Binary* we take the range to be *EnglishWords*, then an N-word-long English text is a function

$$EnglishText\colon \{1, 2, \ldots, N\} \to EnglishWords. \;\square$$

In general, data sequences are functions of the form

$$\boxed{Data\colon Indices \to Symbols,} \qquad (1.9)$$

where $Indices \subset Naturals$—in which *Naturals* is the set of natural numbers—is an appropriate index set such as $\{1, 2, \ldots, N\}$, and *Symbols* is an appropriate set of symbols such as *Binary* or *EnglishWords*.

One advantage of the representation (1.9) is that we can then interpret *Data* as a discrete-time signal, and so some of the techniques that we develop in later chapters for those signals will automatically apply to data sequences. However, the domain *Indices* in (1.9) does not represent uniformly spaced instances of time. All we can say is that if m and n are in *Indices* with $m < n$, then the mth symbol $Data(m)$ occurs in the data sequence *before* the nth symbol $Data(n)$, but we cannot say how much time elapses between the occurrence of those two symbols.

The second way in which sequences arise is as representations of event streams. An **event stream**, or **trace**, is a record or log of the significant events that occur in a system of interest. Here are some everyday examples.

Example 1.6: When you call someone by phone, the normal sequence of events is

$$LiftHandset, HearDialTone, DialDigits, HearTelephoneRing,$$
$$HearCalleeAnswer, \ldots,$$

but if the other phone is busy, the event trace is

$$LiftHandset, HearDialTone, DialDigits, HearBusyTone, \ldots.$$

When you send a file to be printed, the normal trace of events is

$$CommandPrintFile, FilePrinting, PrintingComplete,$$

but if the printer has run out of paper, the trace might be

$$CommandPrintFile, FilePrinting, MessageOutofPaper, InsertPaper, \ldots. \quad \square$$

Example 1.7: When you enter your car, the starting trace of events might be

$$StartEngine, SeatbeltSignOn, BuckleSeatbelt, SeatbeltSignOff, \ldots.$$

Thus, event streams are functions of the form

$$\boxed{EventStream: Indices \to Symbols.} \quad \square$$

We show in chapter 3 that the behavior of finite state machines is best described in terms of event traces, and that systems that operate on event streams are often best described as finite-state machines.

1.1.6 *Discrete signals and sampling*

Voice and *PureTone* are said to be continuous-time signals because their domain *Time* is a continuous interval of the form $[\alpha, \beta] \subset Reals$. The domain of *Image*, similarly, is a continuous two-dimensional rectangle of the form $[a, b] \times [c, d] \subset Reals^2$. The signals *ComputerVoice* and *ComputerImage* have domains of time and space that are discrete sets. *Video* is also a discrete-time signal, but in principle it could be a function of a space continuum. We can define a function *ComputerVideo* in which all three sets that are composed to form the domain are discrete.

Discrete signals often arise from signals with continuous domains by **sampling**. We briefly explain the purpose of sampling here; a detailed discussion is taken up later. Continuous domains have an infinite number of elements. Even the domain $[0, 1] \subset Time$, representing a finite time interval, has an infinite number of elements. The signal assigns a value in its range to each of these infinitely many elements. Such a signal cannot be stored in a finite digital memory device such as a computer or CD-ROM. If we wish to store, say, *Voice*, we must approximate it by a signal with a finite domain.

A common way to approximate a function with a continuous domain, such as *Voice* and *Image*, by a function with a finite domain is by uniformly sampling its continuous domain.

Example 1.8: If we sample a 10-second-long domain of *Voice*,

$$Voice\colon [0, 10] \to Pressure,$$

10,000 times a second (i.e., at a frequency of 10 kHz), we get the signal

$$SampledVoice\colon \{0,\ 0.0001,\ 0.0002,\ \ldots,\ 9.9998,\ 9.9999,\ 10\} \to Pressure, \tag{1.10}$$

with the assignment

$$SampledVoice(t) = Voice(t), \forall\, t \in \{0,\ 0.0001,\ 0.0002, \ldots,\ 9.9999,\ 10\}. \tag{1.11}$$

Notice from (1.10) that uniform sampling means picking a uniformly spaced subset of points of the continuous domain [0, 10]. ❐

In the example, the **sampling interval**, or **sampling period**, is 0.0001 second, which corresponds to a **sampling frequency**, or **sampling rate**, of 10,000 Hz. Because the continuous domain is 10 seconds long, the domain of *SampledVoice* has 100,000 points. A sampling frequency of 5,000 Hz would yield the domain $\{0, 0.0002, \ldots, 9.9998, 10\}$, which has half as many points. The sampled domain is finite, and its elements are discrete values of time.

Notice also from (1.11) that the pressure assigned by *SampledVoice* to each time in its domain is the same as that assigned by *Voice* to the same time; that is, *SampledVoice* is indeed obtained by sampling the *Voice* signal at discrete values of time.

Figure 1.11 shows an exponential function $Exp\colon [-1, 1] \to Reals$ defined by

$$Exp(x) = e^x.$$

SampledExp is obtained by sampling with a sampling interval of 0.2. Thus, its domain is

$$\{-1,\ -0.8,\ \ldots,\ 0.8,\ 1.0\}.$$

The continuous domain of *Image* given by (1.1), which describes a grayscale image on an 8.5 × 11 inch sheet of paper, is the rectangle $[0, 11] \times [0, 8.5]$, representing the space of the page. In this case, too, a common way to approximate *Image* by a signal with a finite domain is to sample the rectangle. Uniform sampling with a **spatial resolution** of, say, 100 dots per inch in each dimension

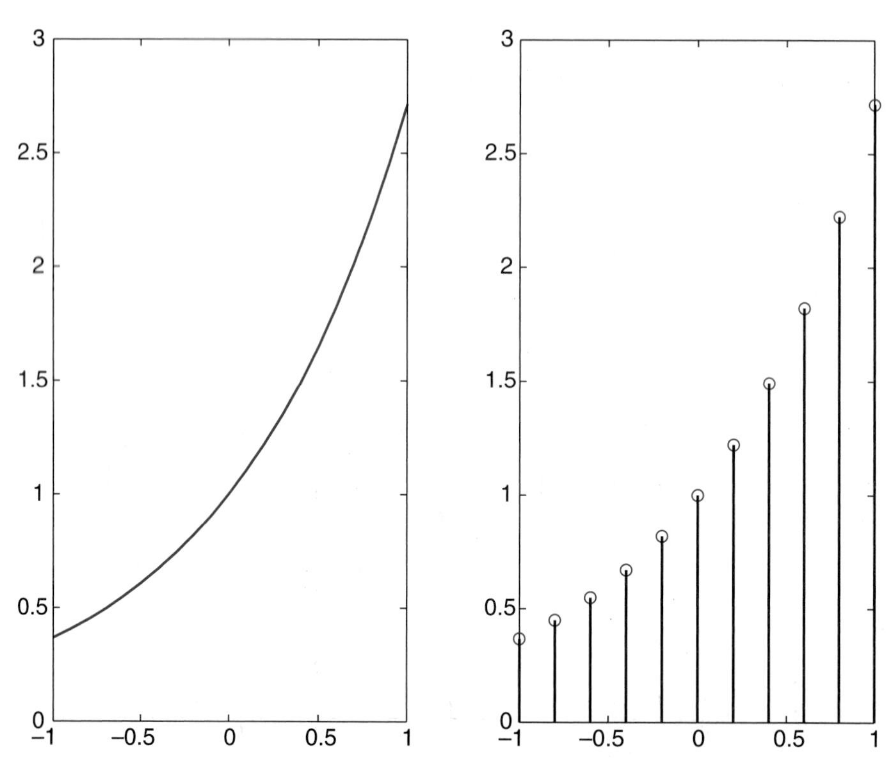

FIGURE 1.11: The exponential functions *Exp* (left) and *SampledExp* (right), obtained by sampling with a sampling interval of 0.2.

yields the finite domain $D = \{0, 0.01, \ldots, 8.49, 8.5\} \times \{0, 0.01, \ldots, 10.99, 11.0\}$. Therefore, the sampled grayscale picture is

$$SampledImage: D \to [0, B_{max}]$$

with

$$SampledImage(x, y) = Image(x, y), \forall\ (x, y) \in D.$$

As mentioned before, each sample of the image is called a pixel, and the size of the image is often given in pixels. The size of a computer screen display, for example, may be 600 × 800 or 768 × 1,024 pixels.

Sampling and approximation

Let f be a continuous-time function, and let *Sampledf* be the discrete-time function obtained by sampling f. Suppose we are given *Sampledf*, as, for example, in the left graph of figure 1.12. Can we reconstruct or recover f from *Sampledf*? This question lies at the heart of digital storage and communication technologies. The general answer to this question tells us, for example, what audio quality we can

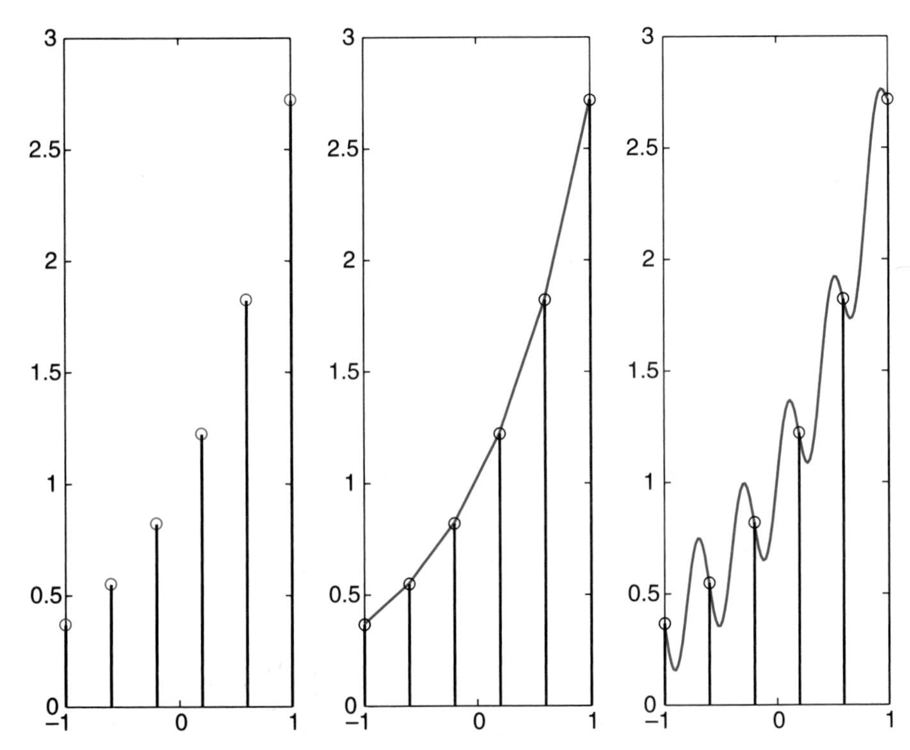

FIGURE 1.12: The discrete-time signal (left) is obtained by sampling the continuous-time signal (middle or right).

obtain from a given discrete representation of a sound. The format for a compact disc is based on the answer to this question. We discuss it in much detail in later chapters.

For the moment, note that the short answer to the question above is no. For example, we cannot tell whether the discrete-time function in the left graph of figure 1.12 was obtained by sampling the continuous-time function in the middle graph or the function in the right graph. Indeed, there are infinitely many such functions, and a choice must be made. One option is to connect the sampled values by straight line segments, as shown in the middle graph. Another choice is shown in the right graph. The choice made by a compact disc player is different from both of these, as explored further in chapter 11.

Similarly, an image like *Image* cannot be uniquely recovered from its sampled version *SampledImage*. Several different choices are commonly used.

Digital signals and quantization

Even though *SampledVoice* in example 1.8 has a finite domain, we may nonetheless be unable to store it in a finite amount of memory. To see why, suppose that the range *Pressure* of the function *SampledVoice* is the continuous interval $[a, b]$.

To represent every value in $[a, b]$ requires infinite **precision**. In a computer, in which data are represented digitally as finite collections of bits, such precision would require an infinite number of bits for just one sample. But a finite digital memory has a finite word length in which we can store only a finite number of values. For instance, if a word is eight bits long, it can have 2^8, or 256, different values. So we must approximate each number in the range $[a, b]$ by one of 256 values. The most common approximation method is to **quantize** the signal. A common approach is to choose 256 uniformly spaced values in the range $[a, b]$ and to approximate each value in $[a, b]$ by the one of the 256 values that is closest. An alternative approximation, called **truncation**, is to choose the largest of the 256 values that is less than or equal to the desired value.

Example 1.9: Figure 1.13 shows a *PureTone* signal, a *SampledPureTone* obtained after sampling, and a quantized *DigitalPureTone* obtained by means of four-bit, or 16-level, truncation. *PureTone* has continuous domain and continuous range, whereas *SampledPureTone* (depicted with circles) has discrete

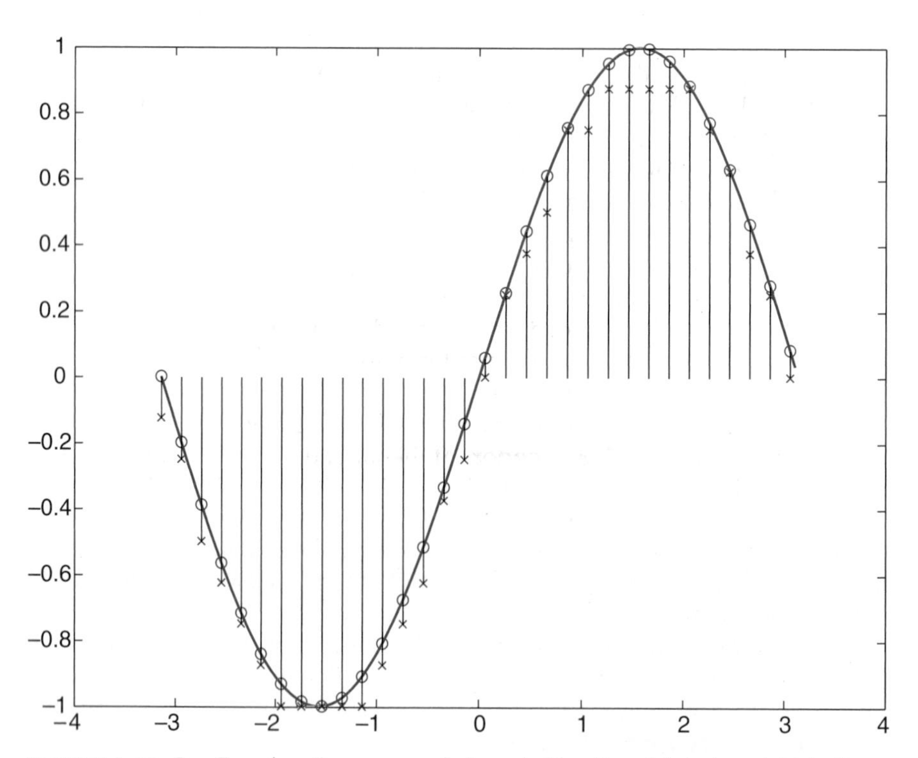

FIGURE 1.13: *PureTone* (continuous curve), *SampledPureTone* (circles), and *DigitalPureTone* signals (×s).

domain and continuous range, and *DigitalPureTone* (depicted with ×s) has discrete domain and discrete range. Only the last of these can be precisely represented on a computer. ❐

It is customary to call a signal with continuous domain and continuous range, such as *PureTone*, an **analog signal** and to call a signal with discrete domain and range, such as *DigitalPureTone*, a **digital signal**.

Example 1.10: In digital telephony, voice is sampled every 125 microseconds (μsec) or at a sampling frequency of 8,000 Hz. Each sample is quantized into an eight-bit word, or 256 levels. This yields an overall rate of $8{,}000 \times 8 = 64{,}000$ bits per second. The worldwide digital telephony network is therefore composed primarily of channels capable of carrying 64,000 bits per second or multiples of this (so that multiple telephone channels can be carried together). In cellular phones, voice samples are further compressed to bit rates of 8,000 to 32,000 bits per second. ❐

1.2 Systems

Systems are functions that transform signals. There are many reasons for transforming signals. A signal carries information. A transformed signal may carry the same information in a different way. For example, in a live concert, music is represented as sound. A recording system may convert that sound into a pattern of magnetic fields on a magnetic tape. The original signal, the sound, is difficult to preserve for posterity. The magnetic tape has a more persistent representation of the same information. Thus, **storage** is one of the tasks accomplished by systems.

A system may transform a signal into a form that is more convenient for **transmission**. Sound signals cannot be carried by the Internet. There is simply no physical mechanism in the Internet for transporting rapid variations in air pressure. The Internet provides instead a mechanism for transporting sequences of bits. A system must convert a sound signal into a sequence of bits. Such a system is called an **encoder** or **coder**. At the far end, of course, a **decoder** is needed to convert the sequence back into sound. When a coder and a decoder are combined into the same physical device, the device is often called a **codec**.

A system may transform a signal to hide its information so that snoops do not have access to it. This is called **encryption**. For encryption to be useful, matching **decryption** is needed.

A system may **enhance** a signal by emphasizing some of the information it carries and deemphasizing some other information. For example, an **audio equalizer** may compensate for poor room acoustics by reducing the magnitude of certain low frequencies that happen to resonate in the room. In transmission, signals are often degraded by **noise** or distorted by physical effects in the transmission medium. A system may attempt to reduce the noise or reverse the distortion. When the signal is carrying digital information over a physical channel, the extraction of the digital information from the degraded signal is called **detection**.

Systems are also designed to **control** physical processes such as the heating in a room, the ignition in an automobile engine, and the flight of an aircraft. The state of the physical process (room temperature, cylinder pressure, aircraft speed) is sensed. The sensed signal is processed to generate signals that drive actuators, such as motors or switches. Engineers design a system called the **controller** that, on the basis of the processed sensor signal, determines the signals that control the physical process (turn the heater on or off, adjust the ignition timing, change the aircraft flaps) so that the process has the desired behavior (room temperature adjusts to the desired setting, engine delivers more torque, aircraft descends smoothly).

Systems are also designed for **translation** from one format to another. For example, a command sequence from a musician may be transformed into musical sounds. Or the detection of risk of collision in an aircraft might be translated into control signals that trigger evasive maneuvers.

1.2.1 *Systems as functions*

Consider a system S that transforms input signal x into output signal y. The system is a function, so $y = S(x)$. Suppose $x: D \to R$ is a signal with domain D and range R. For example, x might be a sound, x: *Reals* $\to$ *Pressure*. The domain of S is the set X of all such sounds, which we write

$$\boxed{X = [D \to R] = \{x \mid x: D \to R\}.} \tag{1.12}$$

This notation reads "X, also written $[D \to R]$, is the set of all x such that x is a function from D to R." This set is called a **signal space** or **function space**. A signal or function space is a set of all functions with a given domain and range.

Example 1.11: The set of all sound segments with duration $[0, 1]$ and range *Pressure* is written

$$[[0, 1] \to \mathit{Pressure}].$$

Notice that square brackets are used for both a subset of reals, as in $[0, 1]$, and for a function space, as in $[D \to R]$, although the meanings of these two notations are obviously very different.

The set *ImageSet* considered in section 1.1.3 is the function space

$$\mathit{ImageSet} = [\mathit{DiscreteVerticalSpace} \times \mathit{HorizontalSpace} \to \mathit{Intensity}^3].$$

Because this is a set, we can define functions that use it as a domain or range, as we have done earlier with

$$Video\colon FrameTimes \to ImageSet.$$

Similarly, the set of all binary files of length N is

$$BinaryFiles = [Indices \to Binary],$$

where $Indices = \{1, \ldots, N\}$. ❐

A system S is a function mapping a signal space into a signal space,

$$\boxed{S\colon [D \to R] \to [D' \to R'].}$$

Systems are therefore much like signals, except that both their domain and their range are signal spaces. Thus, if $x \in [D \to R]$ and $y = S(x)$, then it must be that $y \in [D' \to R']$. Furthermore, if z is an element of D', $z \in D'$, then

$$y(z) = S(x)(z) = (S(x))(z) \in R'.$$

The parentheses around $S(x)$ in $(S(x))(z)$ are not necessary but may improve readability.

1.2.2 *Telecommunications systems*

We give some examples of systems that occur in or interact with the global telecommunications network. This network is unquestionably one of the most remarkable accomplishments of humankind. It is astonishingly complex, composed of hundreds of distinct corporations, and linking billions of people. We often think of it in terms of its basic service, plain old telephone service (**POTS**). POTS is a voice service, but the telephone network is in fact a global, high-speed digital network that carries not just voice but also video, images, and computer data, including much of the traffic in the Internet.

Figure 1.14 depicts a small portion of the global telecommunications network. In POTS, represented at the upper right, a **twisted pair** of copper wires connects a central office to a home telephone. The twisted pair is called the **local loop** or **subscriber line**. At the central office, the twisted pair is connected to a **line card**, which usually converts the signal from the telephone immediately into digital form. The line card, in turn, is connected to a **switch**, which routes incoming and outgoing telephone connections.

The digital representation of a voice signal, a sequence of bits, is routed through the telephone network. It is usually combined with other bit sequences,

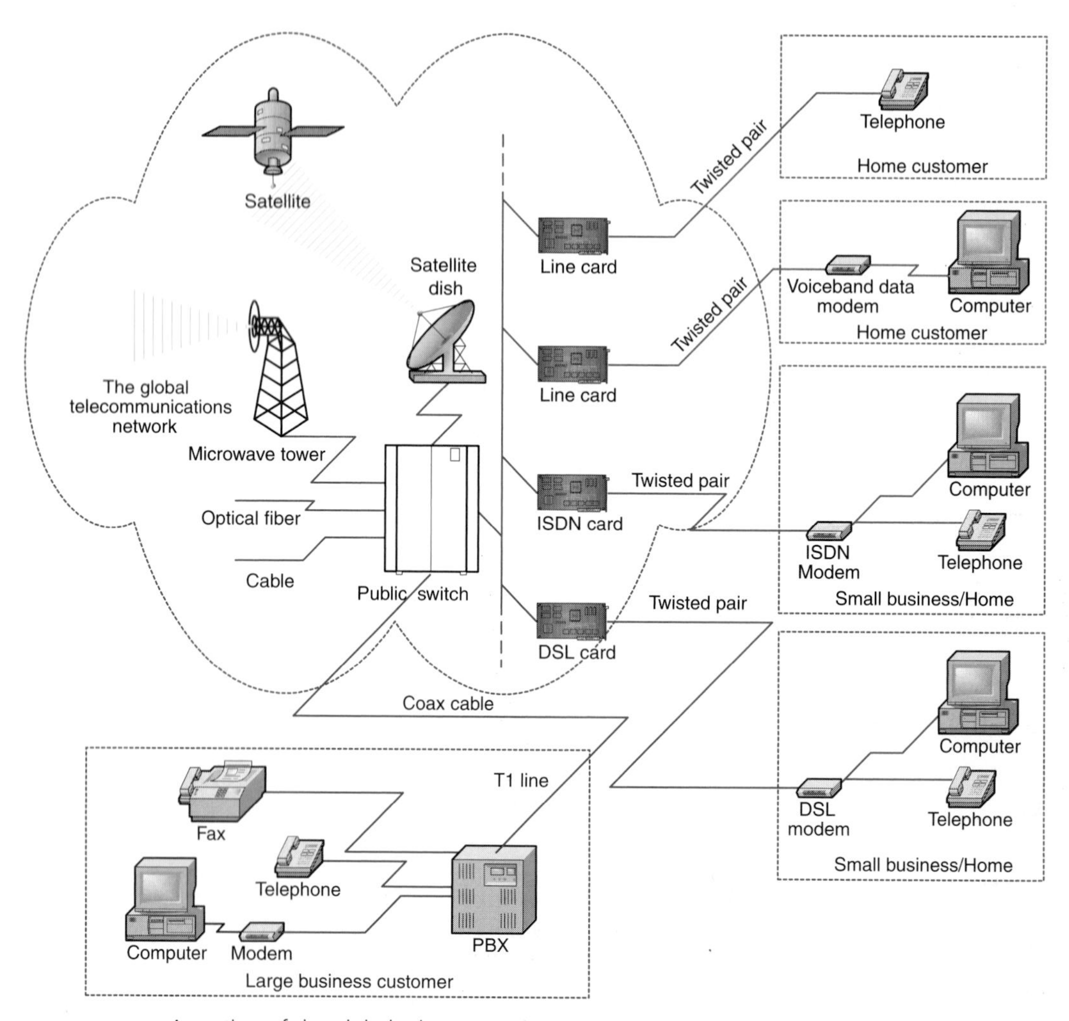

FIGURE 1.14: A portion of the global telecommunications network.

which are other voices or computer data, and sent over high-speed links implemented with optical fiber, microwave radio, coaxial cable, or satellites.

Of course, a telephone conversation usually involves two parties, so the network delivers to the same line card a digital sequence representing the far-end speaker's voice. That digital sequence is decoded and delivered to the telephone via the twisted pair. The line card therefore includes a codec.

The telephone itself, of course, is a system. It transforms the electrical signal that propagates down the twisted pair into a sound signal and transforms a local sound signal into an electrical signal that can propagate down the twisted pair.

POTS can be abstracted as shown in figure 1.15. The entire network is reduced to a model that accepts an electrical representation of a voice signal

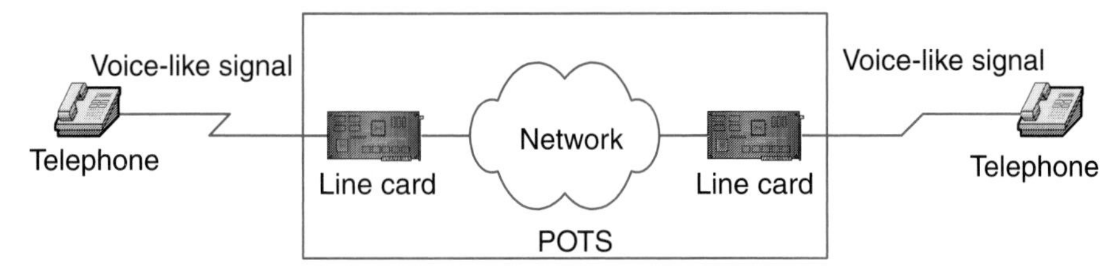

FIGURE 1.15: Abstraction of plain old telephone service (POTS).

and transports it to a remote telephone. In this abstraction, the digital nature of the telephone network is irrelevant. The system simply transports (and degrades somewhat) a voice signal.

PROBING FURTHER

Wireless communication

The telephone network has been freeing itself of its dependence on wires. **Cellular telephones**, which came into widespread use in the 1990s, use radio waves to connect a small, hand-held telephone to a nearby **base station**. The base station connects directly to the telephone network.

There are three major challenges in the design of cellular networks. First, radio spectrum is scarce. Frequencies are allocated by regulatory bodies, often constrained by international treaties. Finding frequencies for new technologies is difficult. Thus, wireless communication devices have to be efficient in their use of the available frequencies. Second, the power available to drive a cellular phone is limited. Cellular phones must operate for reasonably long periods of time with the aid of only small batteries that fit easily within the handset. Although battery technology has been improving, the low power that these batteries can deliver severely limits the range of a cellular phone (how far it can be from a base station) and the processing complexity (the microprocessors in a cellular phone consume considerable power). Third, networking is complicated. In order to be able to route telephone calls to a cellular phone, the network needs to know where the phone is (or, more specifically, which base station is closest). Moreover, the network needs to support phone calls in moving vehicles, which implies that a phone may move out of range of one base station and into the range of another during the course of a telephone call. The network must **hand off** the call seamlessly.

Although "radio telephones" have existed for a long time, particularly for maritime applications in which wireline telephony is impossible, it was the cellular concept that made it possible to offer radio telephony to large numbers of users. The concept is simple. Radio waves propagating along the surface of the earth lose power approximately proportionally to the inverse of the fourth power of distance. In other words, if at distance d meters from a transmitter you receive w watts of radio power, then at distance $2d$ you will receive approximately $w/2^4 = w/16$ watts of radio power. This fourth-power propagation loss was traditionally considered to be only a hindrance to wireless communication. It had to be overcome by greatly boosting the transmitted power. The cellular concept turns this

continued on next page

PROBING FURTHER

hindrance into an advantage: Because the loss is so high, beyond a certain distance the same frequencies can be reused without significant interference. Thus, the service area is divided into cells. A second benefit of the cellular concept is that, at least in urban areas, a cellular phone is never far from a base station. Thus, it does not need to transmit a high-power radio signal to reach a base station. This makes it possible to operate on a small battery.

PROBING FURTHER

LEO telephony

Ideally, a cellular phone, with its one phone number, could be called anywhere in the world, wherever it happens to be, without the caller's needing to know where it is. The technological and organizational infrastructure is evolving to make this possible. When a phone "roams" out of its primary service area, it negotiates with the local service provider in a new area for service. If that service provider has a business agreement with the customer's main service provider, then it provides service. This requires complex networking so that telephone calls to the customer are routed to the correct locale.

However, digital cellular service, particularly in the United States, remains spotty; many rural areas are not served. Providing such service by installing base stations is expensive. Moreover, maritime service away from coastlines is technically impossible with cellular technology.

One candidate technology for truly global telephony services is based on low-earth-orbit (**LEO**) satellites. One such project (now bankrupt) is the Iridium project, spearheaded by Motorola, and so named because in the initial conception, there would be 77 satellites. The iridium atom has 77 electrons. The idea is that enough satellites are put into orbit so that one is always near enough to communicate with a hand-held telephone. When the orbit is low enough so that a hand-held telephone can reach the satellite (a few hundred kilometers above the surface of the earth), the satellites move by fairly quickly. As a consequence, during the course of a telephone conversation, the connection may have to be handed off from one satellite to another. In addition, in order to be able to serve enough users simultaneously, each satellite has to reuse frequencies according to the cellular concept. To do that, it focuses multiple beams on the surface of the earth, using multielement antenna arrays.

As of this writing, this approach has not yet proved economically viable. The investment already has been huge, with at least one high-profile bankruptcy to date, and so the risks are high. Better networking of terrestrial cellular services may provide formidable competition, particularly as service improves to rural areas. The LEO approach, however, has one advantage that terrestrial services cannot hope to match anytime soon: truly worldwide service. The satellites provide service essentially everywhere, even in remote wilderness areas and at sea.

Dual-tone, multifrequency

Even in POTS, not all the information transported is voice. At a minimum, the telephone needs to be able to convey to the central office a telephone number in order to establish a connection. A telephone number is not a voice signal. It is intrinsically discrete. Because the system is designed to carry voice signals, one option is to convert the telephone number into a voice-like signal. A system is needed with the structure shown in figure 1.16. The block labeled dual-tone multifrequency (DTMF) is a system that transforms a sequence of numbers (coming from the keypad on the left) into a voice-like signal.

The DTMF standard provides precisely such a mechanism. As indicated at the left in figure 1.16, when the customer pushes one of the buttons on the telephone keypad, a sound that is the sum of two sinusoidal signals is generated. The frequencies of the two sinusoids are given by the row and column of the key. For example, a "0" is represented as a sum of two sinusoids with frequencies 941 Hz and 1,336 Hz. The waveform for such a sound is shown in figure 1.17. The line card in the central office measures these frequencies to determine which digit was dialed.

Modems

Because POTS is ubiquitous, it is attractive to find a way for it to carry computer data. Like the numbers on a keypad, computer data are intrinsically discrete. Computer data are represented by bit sequences, which are functions of the form

$$\boxed{BitSequence\colon Indices \to Binary,}$$

where $Indices \subset Naturals$, the natural numbers, and $Binary = \{0, 1\}$. Like keypad numbers, in order for a bit sequence to traverse a POTS phone line, it has to be transformed into something that resembles a voice signal. Furthermore, a

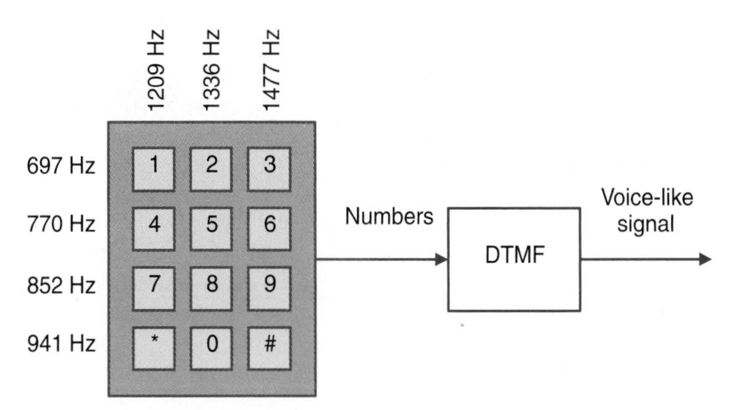

FIGURE 1.16: DTMF system converts numbers from a keypad into a voice-like signal.

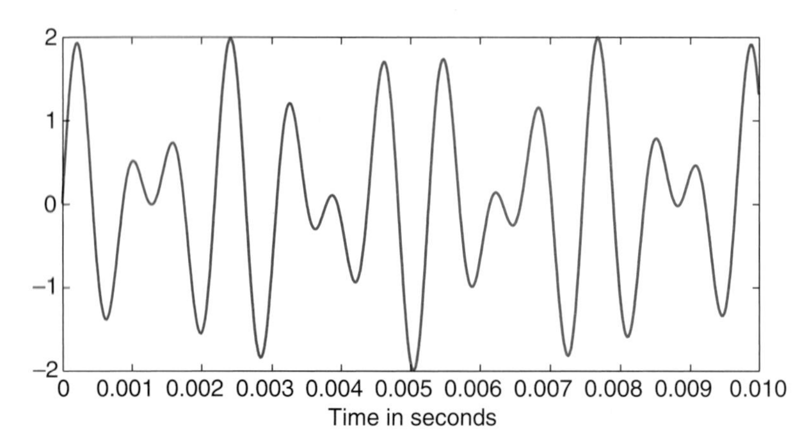

FIGURE 1.17: Waveform representing the "0" key in DTMF system.

system is needed to transform the voice-like signal back into a bit sequence. A system that does that is called a **voiceband data modem**, shown just below the upper right in figure 1.14. The word **modem** is a contraction of "modulator" and "demodulator." A typical arrangement is shown in figure 1.18.

The voice-like signal created by modern modems does not sound like the discrete tones of DTMF; rather it sounds more like hissing. This is a direct conse-

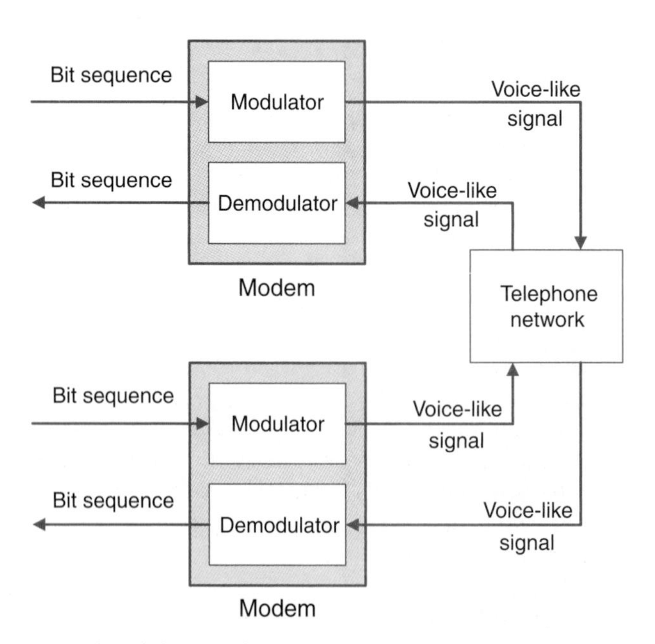

FIGURE 1.18: Voiceband data modems.

quence of the fact that modern modems carry much more data per second than DTMF can (up to 54,000 bits per second rather than just a few digits per second).

Most line cards involved in POTS convert the voice signal into a digital bit stream at the rate of 64,000 bits per second. This bit stream is then transported by the digital network. A voiceband data modem gains access to the digital network rather indirectly, by first constructing a voice-like signal to send to the line card. This gives the voiceband data modem its universality. It works anywhere because the telephone network is designed to carry voice, and it is making the digital data masquerade as voice.

Digital networks

The first widely available service that gave direct access to the global digital telephone network was the integrated services digital network (**ISDN**). The ISDN service required that a different line card be installed at the central office; it was therefore not as universally available as POTS. In fact, after it was developed in the early 1980s, it took nearly 10 years in the United States for ISDN to become widely installed.

The configuration for ISDN is shown below the voiceband data modem in figure 1.14. It requires a special modem on the customer side as well as a special line card in the central office. ISDN typically provides two channels at rates of 64,000 bits per second plus a third control channel with a rate of 16,000 bits per second. One of the 64 kilobytes per second (kbps) channels can be used for voice while, simultaneously, the other two channels are used for data.

A more modern service is the digital subscriber line (**DSL**). As shown at the lower right in figure 1.14, the configuration is similar to that of ISDN. Specialized modems and line cards are required. Asymmetric DSL (**ADSL**) is a variant that provides an asymmetric bit rate, whereby the rate in the direction from the central office to the customer is much higher than the rate from the customer to the central office. This asymmetry reflects the reality of most Internet applications, in which relatively few data flow from the client and torrents of data (including images and video) flow from the server.

Modems are used for many other channels besides the voiceband channel. Digital transmission over radio, for example, requires that the bit sequence be transformed into a radio signal that conforms with regulatory constraints on that radio channel. Digital transmission over electrical cable requires transforming the bit sequence into a form that propagates well over such cable and that does not radiate too much radiofrequency interference. Digital transmission over optical fiber requires transforming the bit sequence into a light signal, whose intensity is usually modulated at very high rates. At each stage in the telephone network, therefore, a voice signal has a different physical form with properties that are well suited to the medium through which the signal propagates. For example, voice, which in the form of sound travels well only over short distances, is converted to an electrical signal that carries well over copper wires for a few kilometers. Copper wires, however, are not as well suited for long distances as optical fiber. Most long-distance communication channels today use optical fiber, although satellites still have certain advantages.

PROBING FURTHER

Encrypted speech

Pairs of modems are used at opposite ends of a telephone connection, each with a transmitter and a receiver, to achieve bidirectional (**full duplex**) communication. Once such modems are in place, and once they have been connected via the telephone network, they function as a bidirectional "bit pipe." That bit pipe is then usable by other systems, such as a computer.

One of the strangest uses is to transmit digitally represented and encrypted voice signals. Here is a depiction of this relatively complicated arrangement:

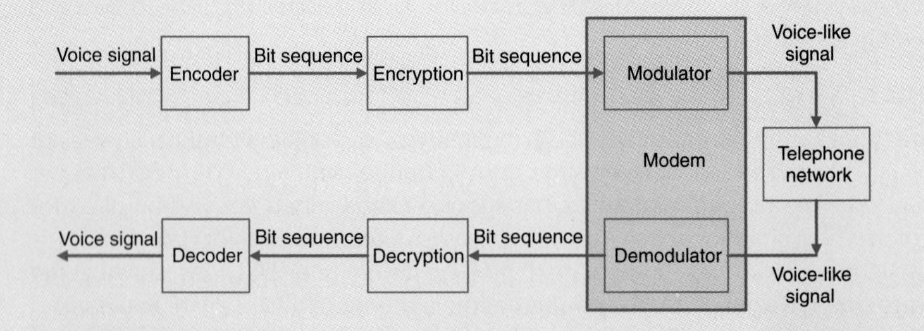

What is actually sent through the telephone network sounds like hissing, which by itself provides a modicum of privacy. Casual eavesdroppers are unable to understand the encoded speech. However, this configuration also provides protection against sophisticated listeners. A listener who is able to extract the bit sequence from this sound is nonetheless unable to reconstruct the voice signal because the bit sequence is encrypted.

Only one end is shown. The encoder and decoder, which convert voice signals to and from bit sequences, are fairly sophisticated systems, as are the encryption and decryption systems. The fact that such an approach is cost effective has more to do with economics and sociology than with technology.

Signal degradation

A voice received via the telephone network is different from the original in several ways. These differences can be modeled by a system that degrades the voice signal.

First, there is a loss of information because of sampling and quantization in the encoder, as discussed in section 1.1.6. Moreover, the media that carry the signal, such as the twisted pair, are not perfect; they distort the signal. One cause of distortion is addition of **noise** to the signal. Noise, by definition, is any undesired component in the signal. Noise in the telephone network is sometimes audible as background hissing, or as **crosstalk**—that is, leakage from other telephone channels into your own. Another degradation results because the medium attenuates the signal, and this attenuation depends on the signal frequency. The line card, in particular, usually contains a **bandlimiting filter** that discards the high-frequency components of the signal. This is why telephone channels do not transport music well. Finally, the signal propagates over a physical medium at a finite speed, bounded by the speed of light, and so there is a delay between the time you say something and the time when the person at the other end hears what you say. Light travels through 1 kilometer (km) of optical fiber in approximately 5 μs, and so the 5,000 km between Berkeley and New York causes a delay of about 25 milliseconds (ms); this delay, however, is not easily perceptible.*

Communications engineering is concerned with how to minimize the degradation for all kinds of communication systems, including radio, television, cellular phones, and computer networks (such as the Internet).

1.2.3 *Audio storage and retrieval*

We have seen how audio signals can be represented as sequences of numbers. Digital audio storage and retrieval are all about finding a physical and persistent representation for these numbers. These numbers can be converted into a single sequence of bits (binary digits) and then "printed" onto some physical medium from which they can later be read back. The transformation of sound into its persistent representation can be modeled as a system, as can the reverse or playback process.

> **Example 1.12:** In the case of compact discs (CDs), the physical medium is a layer of aluminum on a platter into which tiny pits are etched. In the playback device, a laser aimed at the platter uses an interference pattern to determine whether a pit exists at a particular point in the platter. These pits thus naturally represent binary digits, inasmuch as they can exist in two states (present or not present).

* A phone conversation relayed by satellite has a much larger delay. Most satellites traditionally used in the telecommunications network are **geosynchronous**, which means that they hover at the same point over the surface of the earth. To do that, they have to orbit at a height of 22,300 miles, or about 35,900 km. A radio signal takes about 120 ms to traverse that distance; because a signal has to go up and back, there is an end-to-end delay of at least 240 ms (which does not include delays in the electronics). If you are using this channel for a telephone conversation, the round-trip time from when you say something to when you get a reaction is a minimum of 480 ms. This delay can be quite annoying, impeding your ability to converse until you get used to it. If you use Internet telephony, the delays are even longer, and can be irregular, depending on how congested the Internet is when you call.

Whereas a voiceband data modem converts bit sequences into voice-like signals, a musical recording studio does the reverse, creating a representation of the sound that is a bit sequence:

$$RecordingStudio\colon Sounds \to BitStreams.$$

There is a great deal of engineering in the details, however. For instance, CDs are vulnerable to surface defects, which may arise in manufacturing or from hands-on use. These defects may obscure some of the pits or fool the reading laser into detecting a pit where there is none. To guard against this, a very clever error-correcting code called a Reed-Solomon code is used. The coding process can be viewed as a function,

$$Encoding\colon BitStreams \to RedundantBitStreams,$$

where $RedundantBitStreams \subset BitStreams$ is the set of all possible encoded bit sequences. These bit sequences are redundant, in that they contain more bits than are necessary to represent the original bit sequence. The extra bits are used to detect errors and, sometimes, to correct them. Of course, if the surface of the CD is too badly damaged, even this clever scheme fails, and the audio data are not recoverable.

CDs also contain **metadata**, which is extra information about the audio signal. This information allows the CD player to identify the start of a musical number, its length, and sometimes the title and the artist.

The CD format can also be used to contain purely digital data. Such a CD is called a **CD-ROM** (read-only memory). It is called this because, like a computer memory, it contains digital information. Unlike the data in a computer memory, however, that information cannot be modified.

Digital video discs (**DVDs**) take this concept much further, including much more metadata. They may eventually replace CDs. They are entirely compatible, in that they can contain exactly the same audio data that a CD can. DVD players can play CDs; however, CD players cannot play DVDs. DVDs can also contain digital video information and, in fact, any other digital data. **Digital audio tape** (**DAT**) is also a competitor of CDs but has not captured much of a market. ❐

1.2.4 *Modem negotiation*

A very different kind of system is the one that manages the establishment of a connection between two voiceband data modems. These two modems are at physically different locations, are probably produced by different manufacturers, and possibly use different communication standards. Both modems convert bit streams to and from voice-like signals, but other than that, they do not have much in common.

When a connection is established through the telephone network, the answering modem emits a tone that announces, "I am a modem." The initiating modem listens for this tone and, if it fails to detect it, assumes that no connection can be established and hangs up. If it does detect the tone, it answers with a voice-like signal that announces, "I am a modem that can communicate according to ITU standard x," x being one of the many modem standards published by the **International Telecommunication Union** (**ITU**).

The answering modem may or may not recognize the signal from the initiating modem. The initiating modem, for example, may be a newer modem that operates under a standard that was established after the answering modem was manufactured. If the answering modem does recognize the signal, it responds with a signal that says, "Good, I too can communicate using standard x, so let's get started." Otherwise, it remains silent. If the initiating modem fails to get a response, it tries another signal, announcing, "I am a modem that can communicate according to ITU standard y," y being typically now an older (and slower) standard. This process continues until the two modems agree on a standard.

Once agreement is reached, the modems need to make measurements of the telephone channel to compensate for its distortion. They do this by sending each other preagreed signals called **training signals**, defined by the standard. The training signal is distorted by the channel, and, because the receiving modem knows the signal, it can measure the distortion. It uses this measurement to set up a device called an **adaptive equalizer**. Once both modems have completed their setup, they begin to send data to one another.

As systems go, modem negotiation is fairly complex. It involves both event sequences and voice-like signals. The voice like signals need to be analyzed in fairly sophisticated ways, sometimes producing events in the event sequences. It takes this entire book to analyze all parts of this system. The handling of the event sequences is treated through the use of finite-state machines, and the handling of the voice-like signals is treated through the use of frequency-domain concepts and filtering.

1.2.5 *Feedback control systems*

Feedback control systems are composite systems in which a **plant** embodies a physical process whose behavior is guided by a control signal. A plant may be, for example, a mechanical device, such as the power train of a car; a chemical process; or an aircraft with certain inertial and aerodynamic properties. Sensors attached to the plant produce signals that are fed to the controller, which then generates the control signal. This arrangement, whereby the plant feeds the controller and the controller feeds the plant, is a complicated sort of composite system called a **feedback control system**. It has extremely interesting properties, which we explore in much more depth in subsequent chapters.

In this section, we construct a model of a feedback control system by using the syntax of block diagrams. The model consists of several interconnected components. We identify the input and output signals of each component and how the components are interconnected, and we argue on the basis of common-sense physics how the overall system will behave.

Example 1.13: Consider a forced-air heating system, which heats a room in a home or office to a desired temperature. Our first task is to identify the individual components of the heating system. These are

- a furnace/blower unit (which we will simply call the heater) that heats air and blows the hot air through vents into a room,
- a temperature sensor that measures the temperature in a room, and
- the control system that compares the specified desired temperature with the sensed temperature and turns the furnace/blower unit on or off, depending on whether the sensed temperature is below or above the demanded temperature.

The interconnection of these components is shown in figure 1.19.

Our second task is to specify the input and output signals of each component system (the domain and range of the function), ensuring the input–output matching conditions. The heater produces hot air if it is turned on. Therefore, its input signal is simply a function of time that takes one of two values, *On* or *Off*. We call input to the heater (a signal) *OnOff*,

$$OnOff\colon Time \to \{On, Off\},$$

and we take $Time = Reals_+$, the nonnegative real numbers. Thus, the input signal space is

$$OnOffProfiles = [Reals_+ \to \{On, Off\}].$$

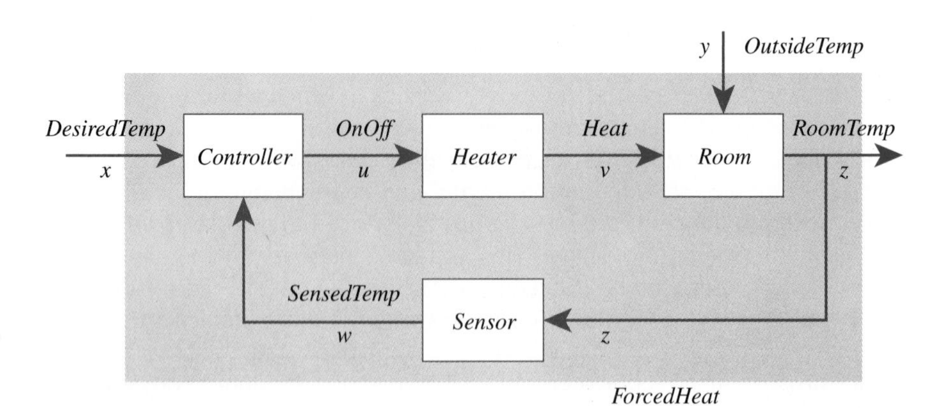

FIGURE 1.19: The interconnected components of a forced-air heating system.

(Recall that the notation $[D \to R]$ defines a function space, as explained in section 1.2.1.) When the heater is turned on, it produces heat at some rate that depends on the capacity of the furnace and blower. We measure this heating rate in British thermal units (BTUs) per hour. Thus, the output signal of the heater, which we name *Heat*, is of the form

$$Heat: Reals_+ \to \{0, B_c\},$$

where B_c is the heater capacity measured in BTUs per hour. If we name the output signal space *HeatProfiles*, then

$$HeatProfiles = [Reals_+ \to \{0, B_c\}].$$

Thus the *Heater* system is described by a function,

$$Heater: OnOffProfiles \to HeatProfiles. \quad (1.13)$$

Commonsense physics tells us that when the heater is turned on, the room will begin to warm up, and when the heater is turned off, the room temperature will fall until it reaches the outside temperature. The room temperature therefore depends on both the heat delivered by the heater and the outside temperature. Thus the input signal to the room is the pair $(Heat, OutsideTemp)$. We can take *OutsideTemp* to be of the form

$$OutsideTemp: Reals_+ \to [min, max],$$

where $[min, max]$ is the range of possible outside temperatures, measured, say, in degrees Celsius. The output signal of the room is of course the room temperature,

$$RoomTemp: Reals_+ \to [min, max].$$

If we denote

$$OutsideTempProfiles = [Reals_+ \to [min, max]]$$

and

$$RoomTempProfiles = [Reals_+ \to [min, max]],$$

then the behavior of the *Room* system is described by the function

$$Room: HeatProfiles \times OutsideTempProfiles \to RoomTempProfiles. \quad (1.14)$$

In a similar manner, the *Sensor* system is described by the function

$$Sensor\colon RoomTempProfiles \to SensedTempProfiles, \tag{1.15}$$

with input signal space *RoomTempProfiles* and output signal space

$$SensedTempProfiles = [Reals_+ \to [min, max]].$$

The *Controller* is described by the function

$$\begin{aligned} Controller\colon & DesiredTempProfile \times \\ & SensedTempProfile \to OnOffProfile, \end{aligned} \tag{1.16}$$

where

$$DesiredTempProfiles = [Reals_+ \to [min, max]].$$

We have constructed a model in which every output drives a compatible input.

The overall forced-air heating system (the shaded part of figure 1.19) has a pair of input signals (desired temperature and outside temperature) and one output signal (room temperature). Therefore, it is described by the function

$$\begin{aligned} ForcedHeat\colon & DesiredTempProfiles \times \\ & OutsideTempProfiles \to RoomTempProfiles. \end{aligned}$$

If we are given the input signal value x of desired temperature and the value y of outside temperature, we can compute the value $z = ForcedHeat(x, y)$ by solving the following four simultaneous equations:

$$\begin{aligned} u &= Controller(x, w) \\ v &= Heater(u) \\ z &= Room(y, v) \\ w &= Sensor(z) \end{aligned} \tag{1.17}$$

Given x and y, we must solve these four equations to determine the four unknown functions u, v, w, and z of which u, v, and w are the internal signals and z is the output signal. Of course, to solve these simultaneous equations, we need to specify the four system functions. So far we have simply given names to those functions and identified their domain and

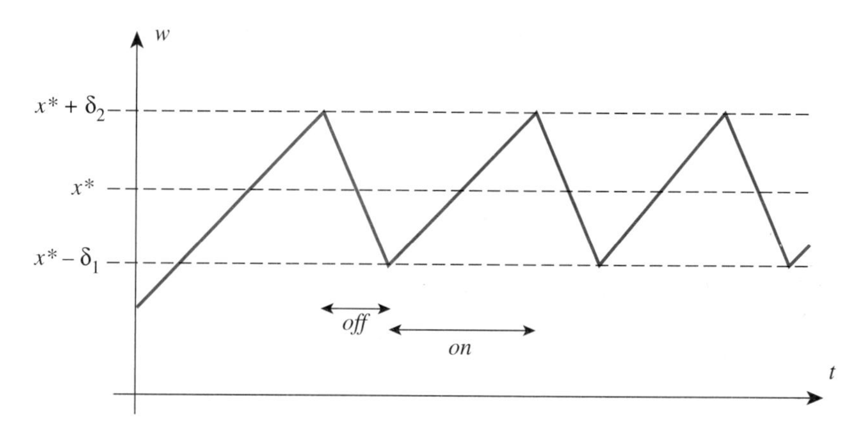

FIGURE 1.20: With a thermostatic controller, the room temperature fluctuates around the desired temperature setting, x^*.

range. To complete the specification, we must describe how those functions assign output signals to input signals.

If the sensor is functioning properly, we expect *Sensor's* output signal to be the room temperature; that is, for all z and for all $t \in Reals_+$,

$$w(t) = Sensor(z)(t) = z(t).$$

A thermostatic controller has a simple behavior: It turns the heater on if the sensed temperature falls below the desired temperature by a certain amount, say δ_1, and it turns the heater off if the sensed temperature rises above the desired temperature by, say, δ_2. Thus, for all x, w and for all $t \in Reals_+$,

$$u(t) = Controller(x, w)(t) = \begin{cases} On, & \text{if } w(t) - x(t) \leq -\delta_1 \\ Off, & \text{if } w(t) - x(t) \geq \delta_2. \end{cases}$$

Suppose, finally, that the desired temperature is set to some constant—say, x^*; therefore, for all $t \in Reals_+$,

$$x(t) = x^*.$$

We can expect the behavior depicted in figure 1.20. When $x^* - w(t)$ drops below $-\delta_1$, the controller turns on the heater, the room temperature increases until $x^* - w(t)$ rises above δ_2, and then the controller turns off the heater. Thus, the room temperature fluctuates around the desired temperature, x^*. ❒

1.3 *Summary*

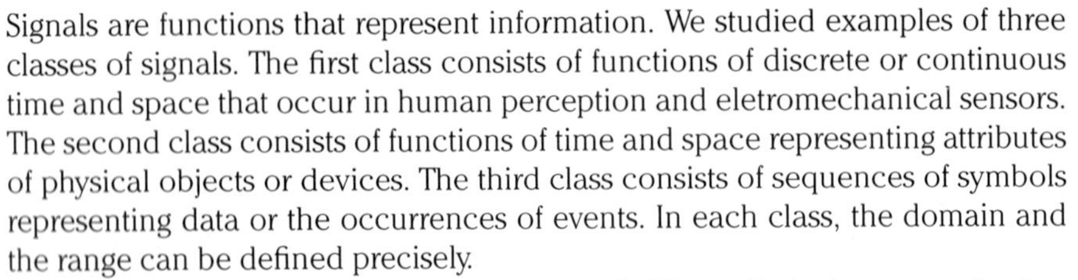

Signals are functions that represent information. We studied examples of three classes of signals. The first class consists of functions of discrete or continuous time and space that occur in human perception and eletromechanical sensors. The second class consists of functions of time and space representing attributes of physical objects or devices. The third class consists of sequences of symbols representing data or the occurrences of events. In each class, the domain and the range can be defined precisely.

Systems are functions that transform signals. We studied telecommunication systems, in which a network that was originally designed for carrying voice signals is used for many other kinds of signals today. One way to accomplish this is to design systems such as modems that transform signals so that they masquerade as voice-like signals. We also examined system models for signal degradation and for storage of signals. We studied systems that are concerned primarily with discrete events and command sequences, and we examined a feedback control system. Both the telephone system and the forced-air heating system were described with the use of block diagrams as interconnections of simpler component systems. In all cases, systems were given as functions in which the domain and the range are function spaces, or sets of functions.

EXERCISES

Each problem is annotated with the letter **E** (exercise), **T** (requires some thought), or **C** (requires some conceptualization). Problems labeled **E** are usually mechanical; those labeled **T** require a plan of attack; and those labeled **C** usually have more than one defensible answer.

E 1. The function $x : Reals \to Reals$ given by

$$\forall\, t \in Reals, \quad x(t) = \sin(2\pi \times 440t)$$

is a mathematical example of a signal in the signal space $[Reals \to Reals]$. Give a mathematical example of a signal x in each of the following signal spaces.

(a) $[Integers \to Reals]$

(b) $[Reals \to Reals^2]$

(c) $[\{0, 1, \ldots, 600\} \times \{0, 1, \ldots, 400\} \to \{0, 1, \ldots, 255\}]$

(d) Describe a practical application for the signal space $[\{0, 1, \ldots, 600\} \times \{0, 1, \ldots, 400\} \to \{0, 1, \ldots, 255\}]$; that is, what might a function in this space represent?

C 2. For each of the continuous-time signals below, represent the signal in the form of $f: X \to Y$ and as a sketch like figure 1.1. Carefully identify the range and domain in each case.

(a) The voltage across the terminals of a car battery.

(b) The closing prices on each day of a share of a company.

(c) The position of a moving vehicle on a straight one-lane road of length L.

(d) The simultaneous position of two moving vehicles on the same straight one-lane road of length L.

(e) The sound heard in both of your ears.

E 3. In digital telephony, voice is sampled at a rate of 8,000 samples/second; therefore, the sampling period is $1/8{,}000 = 125$ microseconds. What are the sampling period and the sampling frequency of sound in a compact disc (CD)?

E 4. Figure 1.4 displays the plots of two sinusoidal 3ignals and their sum. Sketch by hand the plots of the four functions *Step*, *Triangle*, *Sum*, and *Diff*, all with domain $[-1, 1]$ and range *Reals*, defined by, $\forall\, t \in [-1, 1]$,

$$Triangle(t) = 1 - |t|,$$

$$Step(t) = \begin{cases} 0 & \text{if } t < 0 \\ 1 & \text{if } t \geq 0, \end{cases}$$

$$Sum(t) = Triangle(t) + Step(t),$$

$$Diff(t) = Triangle(t) - Step(t).$$

C 5. The following examples of spatial information can be represented as a signal in the form of $f: X \to Y$. Specify a reasonable choice for the range and domain in each case.

(a) An image impressed on photographic paper.

(b) An image stored in computer memory.

(c) The height of points on the surface of the earth.

(d) The location of the chairs in a room.

(e) The household voltage in Europe, which has frequency 50 Hz and is 210 volts, RMS.

C 6. The image called *Albers* consists of an eight-inch yellow square in the center of a white 12-inch square background. Express *Albers* as a function by choosing the domain, range, and function assignment.

E 7. How many bits are there in a $768 \times 1{,}024$ pixel image in which each pixel is represented as a 16-bit word? How long would it take to download this image over a 28-kbps voiceband modem, a 384-kbps DSL modem, and a 10 Megabits per second (Mbps) Ethernet local area network?

C 8. Represent these examples as data or event sequences. Specify reasonable choices for the range and domain in each case.

(a) The result of 100 tosses of a coin.

(b) The sequence of button presses inside an elevator.

(c) The sequence of main events in a soda vending machine.

(d) Your response to a motorist who is asking directions.

(e) A play-by-play account of a game of chess.

C 9. Formulate the following items of information as functions. Specify reasonable choices for the domain and range in each case.

(a) The population of U.S. cities.

(b) The white pages in a phone book (careful: the white pages may list two identical names and may list the same phone number under two different names).

(c) The birth dates of students in class.

(d) The broadcast frequencies of AM radio stations.

(e) The broadcast frequencies of FM radio stations (look at your radio dial or at the web page:

`http://www.eecs.berkeley.edu/~eal/eecs20/sidebars/radio/index.html`).

E 10. Use MATLAB to plot the graph of the following continuous-time functions defined over $[-1, 1]$, and on the same plot display 11 uniformly spaced samples (0.2 seconds apart) of these functions. Are these samples good representations of the waveforms?

(a) $f\colon [-1, 1] \to \mathit{Reals}$, where for all $x \in [-1, 1]$, $f(x) = e^{-x}\sin(10\pi x)$.

(b) $\mathit{Chirp}\colon [-1, 1] \to \mathit{Reals}$, where for all $t \in [-1, 1]$, $\mathit{Chirp}(t) = \cos(10\pi t^2)$.

E 11. Suppose the pendulum of figure 1.10 is rotating counterclockwise at a speed of one revolution per second over the five-second interval $[0, 5]$. Sketch a plot of the resulting function: $\theta\colon [0, 5] \to [-\pi, \pi]$. Assume $\theta(0) = 0$. Also specify this function mathematically. Your plot is discontinuous, but the pendulum's motion is continuous. Explain this apparent inconsistency.

T 12. There is a large difference between the sets X, Y, and $[X \to Y]$. This exercise explores some of that difference.

(a) Suppose $X = \{a, b, c\}$ and $Y = \{0, 1\}$. List all the functions from X to Y—that is, all the elements of $[X \to Y]$. Note that part of the problem here is to figure out *how* to list all the functions.

(b) If X has m elements and Y has n elements, how many elements does $[X \to Y]$ have?

(c) Suppose

$$ColorMapImages = [DiscreteVerticalSpace \times DiscreteHorizontalSpace \to ColorMapIndexes].$$

Suppose the domain of each image in this set has 6,000 pixels and the range has 256 values. How many distinct images are there? Give an approximate answer in the form of 10^n. Hint: $a^b = 10^{b \log_{10}(a)}$.

CHAPTER 2

Defining signals and systems

The previous chapter describes the representation of signals and systems as functions, concentrating on how to select the domain and range. This chapter is concerned with how to give more complete definitions of these functions. In particular, we need an **assignment rule**, which specifies how to assign an element in the range to each element in the domain.

There are many ways to give an assignment rule. A theme of this chapter is that these different ways have complementary uses. Procedural descriptions of the assignment rule, for example, are more convenient for synthesizing signals or constructing implementations of a system in software or hardware. Mathematical descriptions are more convenient for analyzing signals and systems and for determining their properties.

In practice, it is often necessary to use several descriptions of assignment rules in combination, because of their complementary uses. In designing systems, a practicing engineer is often reconciling these diverse views to ensure, for instance, that a particular hardware device or piece of software indeed implements a system that is specified mathematically. We begin with a discussion of functions in general and then specialize to signals and systems.

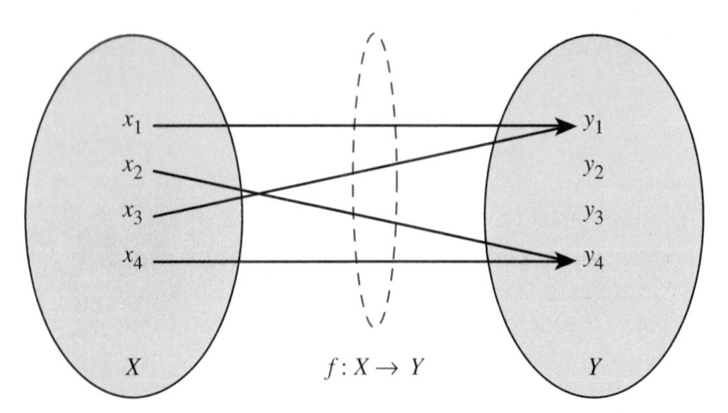

FIGURE 2.1: A function $f: X \to Y$ assigns to each element in X a single element in Y.

2.1 Defining functions

A function $f: X \to Y$ assigns to each element in X a single element in Y, as illustrated in figure 2.1. This assignment can be defined by declaring the mathematical relationship between the value in X and the value in Y, by graphing or enumerating the possible assignments, by giving a procedure for determining the value in Y given a value in X, or by composing simpler functions. We describe each of these methods in more detail in this section.

Example 2.1: In section 1.1.5, we mentioned that sequences are a special kind of function. An infinite sequence s is a function that maps the natural numbers into some set Y, as illustrated in figure 2.2. This function fully defines any infinite sequence of elements in Y. ❐

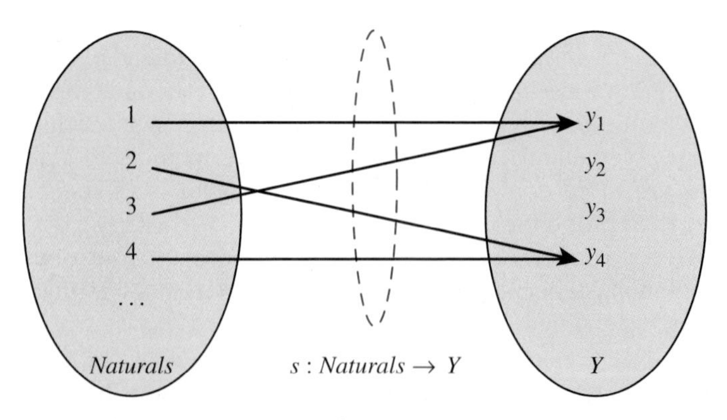

FIGURE 2.2: An infinite sequence s is a function s: *Naturals* $\to Y$ that assigns to each element in *Naturals* a single element in Y.

2.1.1 *Declarative assignment*

Consider the function *Square*: *Reals* → *Reals* given by

$$\forall\, x \in Reals, \quad Square(x) = x^2. \tag{2.1}$$

In (2.1), we have used the universal quantifier symbol ∀, which means "for all" or "for every" to declare the relationship between values in the domain of the function and values in the range. Statement (2.1) is read "for every value of x in *Reals*, the function *Square* evaluated at x is assigned the value x^2." The expression "$Square(x) = x^2$" in (2.1) is an assignment.*

Expression (2.1) is an instance of the following prototype for defining functions. Define $f: X \to Y$ by

$$\boxed{\forall\, x \in X, \quad f(x) = \text{expression in } x.} \tag{2.2}$$

In this prototype, f is the name of the function to be defined, such as *Square*; X is the domain of f; Y is the range of f; and "expression in x" specifies the value in Y assigned to $f(x)$.

The prototype (2.2) does not say how the "expression in x" is to be evaluated. In the *Square* example, it was specified by the algebraic expression $Square(x) = x^2$. Such a definition of a function is said to be **declarative**, because it declares properties of the function without directly explaining how to construct the function.

Example 2.2: Here are some examples of functions of complex variables.†

The magnitude of a complex number is given by *abs*: *Complex* → *Reals*$_+$, where *Complex* is the set of complex numbers, *Reals*$_+$ is the set of set of nonnegative real numbers, and

$$\forall\, z = x + iy \in Complex, \quad abs(z) = \sqrt{(x^2 + y^2)}.$$

The complex conjugate of a number, *conjugate*: *Complex* → *Complex*, is given by

$$\forall\, z = x + iy \in Complex, \quad conjugate(z) = x - iy.$$

* See appendix A for a discussion of the use of "=" as an assignment, as opposed to its use as an assertion.

† See appendix B for a review of complex variables.

The exponential of a complex number, *exp*: *Complex* → *Complex*, is given by*

$$\forall\, z \in Complex, \quad exp(z) = \sum_{n=0}^{\infty} \frac{z^n}{n!}.$$

It is worth emphasizing that the last definition is declarative: It does not give a procedure for calculating the exponential function, inasmuch as the sum is infinite. Such a calculation would never terminate. ❐

Example 2.3: The signum function gives the sign of a real number, *signum*: *Reals* → $\{-1, 0, 1\}$:

$$\forall\, x \in Reals, \quad signum(x) = \begin{cases} -1 & \text{if } x < 0 \\ 0 & \text{if } x = 0 \\ 1 & \text{if } x > 0. \end{cases} \tag{2.3}$$

The right side of this assignment is a tabulation of three expressions for three different subsets of the domain. In section 2.1.3, we consider a more extreme case of this in which every value in the domain is tabulated with a value in the range. ❐

Example 2.4: The size of a matrix, *size*: *Matrices* → *Naturals* × *Naturals*, is given by

$$\forall\, M \in Matrices, \quad size(M) = (m, n),$$

where m is the number of rows of the matrix M, n is the number of columns of M, and *Matrices* is the set of all matrices.

This definition relies not only on formal mathematics but also on the English sentence that defines m and n. Without that sentence, the assignment would be meaningless. ❐

2.1.2 *Graphs*

Consider a function $f: X \to Y$. To each $x \in X$, f assigns the value $f(x)$ in Y. The pair $(x, f(x))$ is an element of the product set $X \times Y$. The set of all such pairs is called the **graph** of f, written *graph*(f). Using the syntax of sets, *graph*(f) is the subset of $X \times Y$ defined by

$$graph(f) = \{(x, y) \mid x \in X \text{ and } y = f(x)\}, \tag{2.4}$$

or, slightly more simply,

$$graph(f) = \{(x, f(x)) \mid x \in X\}.$$

* If this notation is unfamiliar, see Basics: Summations box on page 69.

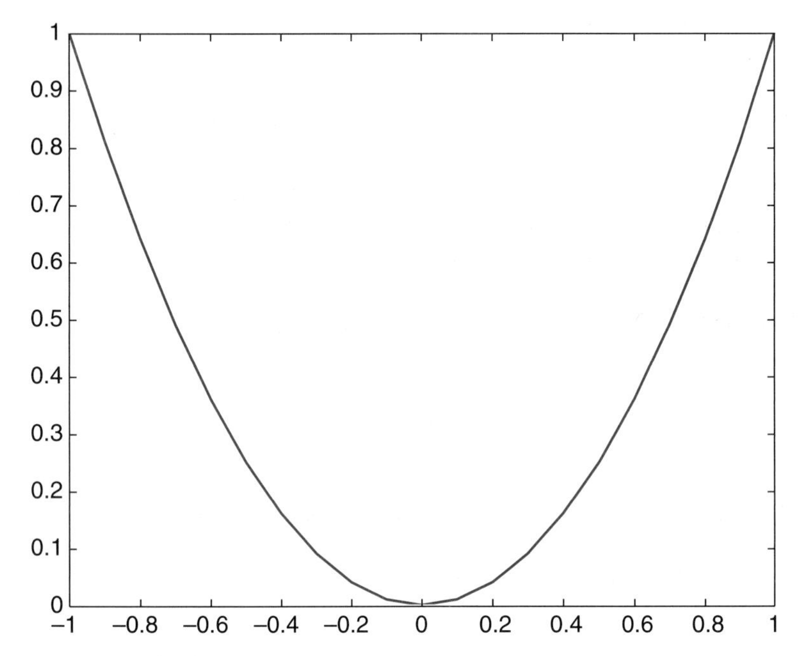

FIGURE 2.3: Graph of *Square*.

The vertical bar | is read "such that," and the expression after it is a **predicate** that defines the set.*

When $X \subset Reals$ and $Y \subset Reals$, we can plot *graph(f)* on a page.

Example 2.5: Consider the graph of the function *Square*,

$$graph(Square) = \{(x, x^2) \mid x \in Reals\},$$

which is plotted in figure 2.3. In that figure, the horizontal and vertical axes represent the domain and the range, respectively (more precisely, a subset of the domain and a subset of the range). The rectangular region enclosed by these axes represents the product of the domain and the range; every point in that region is a member of $(Reals \times Reals)$. The graph is visually rendered by placing a black dot at every point in that region that is a member of *graph(Square)*. The resulting picture is the familiar plot of the *Square* function. ❒

* See appendix A for a review of this notation.

Although the graph of $f: X \to Y$ is a subset of $X \times Y$, it is a very particular sort of subset. For each element $x \in X$, there is exactly one element $y \in Y$ such that $(x, y) \in graph(f)$. In particular, there cannot be more than one such $y \in Y$, and there must exist such a $y \in Y$. This is, in fact, what we mean when we say that f is a function.

Example 2.6: Let $X = \{1, 2\}$ and $Y = \{a, b\}$. Then

$$\{(1, a), (2, a)\}$$

is the graph of a function, but

$$\{(1, a), (1, b)\}$$

is not, because two points a and b are assigned to the same point, 1, in the domain. Neither is

$$\{(1, a)\},$$

because no point in the range is assigned to the point 2 in the domain. ❐

The graph of *Square*, *graph*(*Square*), is given by the algebraic expression (x, x^2). In other cases, no such algebraic expression exists. For example, *Voice* is specified through its graph in figure 1.1, not through an algebraic expression. Thus, graphs can be used to define functions that cannot be conveniently given by declarative assignments.

PROBING FURTHER

Relations

The graph of a function $f: X \to Y$ is a subset of $X \times Y$, as defined in (2.4). An arbitrary subset of $X \times Y$ is called a **relation**. A relation is a set of tuples (x, y) that pair an element $x \in X$ with an element $y \in Y$, as suggested in figure 2.4. For relations, it is common to call X the **domain** and Y the **codomain**. A function is a special kind of relation in which, for every $x \in X$, there is exactly one $y \in Y$ such that (x, y) is an element of the relation. Thus, a particular relation $R \subset X \times Y$ is a function if, for every $x \in X$, there is a $y_1 \in Y$ such that $(x, y_1) \in R$, and if, in addition, $(x, y_2) \in R$, then $y_1 = y_2$.

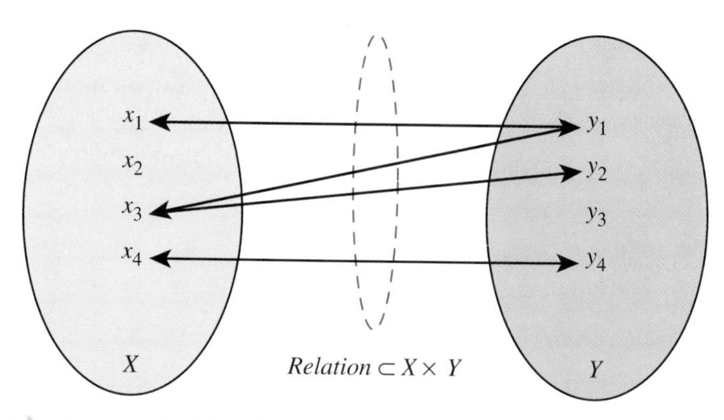

FIGURE 2.4: Any subset of $X \times Y$ is a relation.

Consider again the prototype in (2.2):

$$\forall\, x \in X, \quad f(x) = \text{ expression in } x.$$

The graph of f is

$$graph(f) = \{(x, y) \in X \times Y \mid y = \text{ expression in } x\}.$$

The expression "$y =$ expression in x" is a predicate in the variable (x, y), and so this prototype definition conforms to the prototype new set constructor given in (A.4) of appendix A:

$$\boxed{NewSet = \{z \in Set \mid Pred(z)\}.}$$

Because the graph of f is a set, we can define the function f through its graph by using the same techniques we use to define sets.

2.1.3 *Tables*

If $f: X \to Y$ has finite domain, then $graph(f) \subset X \times Y$ is a finite set, and so it can be specified simply by a list of all its elements. This list can be put in the form of a table. This table defines the function.

TABLE 2.1
Tabular representation of *Score*

Name	Mark
John Brown	90.0
Jane Doe	91.2
...	...

Example 2.7: Suppose the function

$$Score: Students \to [0, 100]$$

gives the outcome of the first midterm exam for each student in the class. Obviously, this function cannot be given by an algebraic declarative assignment. But it can certainly be given as a table, as shown in table 2.1.

❒

Example 2.8: The command `nslookup` on a networked computer is a function that maps host names into their IP (Internet) address. For example, if you type

```
nslookup cory.eecs.berkeley.edu
```

you get the IP address 128.32.134.240. The **domain name server** attached to your machine stores the nslookup function as a table. ❒

2.1.4 *Procedures*

Sometimes the value $f(x)$ that a function f assigns to an element $x \in domain(f)$ is obtained by executing a procedure.

Example 2.9: Here is a MATLAB®* procedure for computing the factorial function

$$fact: \{1, \ldots, 10\} \to Naturals,$$

where *Naturals* is the set of natural numbers:

```
fact(1) = 1;
for n = 2:10
   fact(n) = n * fact(n-1);
end
```

❒

* MATLAB is a registered trademark of The MathWorks, Inc.

Unlike previous mechanisms for defining a function, this one gives a constructive method for determining an element in the range, given an element in the domain. This style is called **imperative**, in distinction from declarative. The relationship between these two styles is interesting and quite subtle. It is explored further in section 2.1.6.

2.1.5 *Composition*

Functions can be combined to define new functions. The simplest mechanism is to connect the output of one function to the input of another. We have been doing this informally to define systems by connecting components in block diagrams, such as that in figure 1.18.

If the first function is f_1 and the second is f_2, then we write the composed function as $f_2 \circ f_1$; that is, for every x in the domain of f_1,

$$\boxed{(f_2 \circ f_1)(x) = f_2(f_1(x)).}$$

This is called **function composition**. A fundamental requirement for such a composition to be valid is that the range of f_1 must be a subset of the domain of f_2. In other words, any output from the first function must be in the set of possible inputs for the second. Thus, for example, the output of *modulator* in figure 1.18 is a voice-like signal, which is precisely what the system *telephone network* is able to accept as an input. Thus, we can compose the modulator with the telephone network. Without this input–output **connection restriction**, the interconnection would be meaningless.*

It is worth pausing to study the notation $f_2 \circ f_1$. Assume $f_1 : X \to Y$ and $f_2 : X' \to Y'$. If $Y \subset X'$, we can define

$$f_3 = f_2 \circ f_1,$$

where $f_3 : X \to Y'$ such that

$$\forall\, x \in X, \quad f_3(x) = f_2(f_1(x)). \tag{2.5}$$

Notice that f_1 is applied first and then f_2. Why is f_1 listed second in $f_2 \circ f_1$? This convention simply mirrors the ordering of $f_2(f_1(x))$ in (2.5). We can visualize f_3 as in figure 2.5.

* We just called an element in the domain of a function its input and the corresponding value of the function its output. This interpretation of domain as inputs and range as outputs is natural, and it is the reason that systems are described by functions.

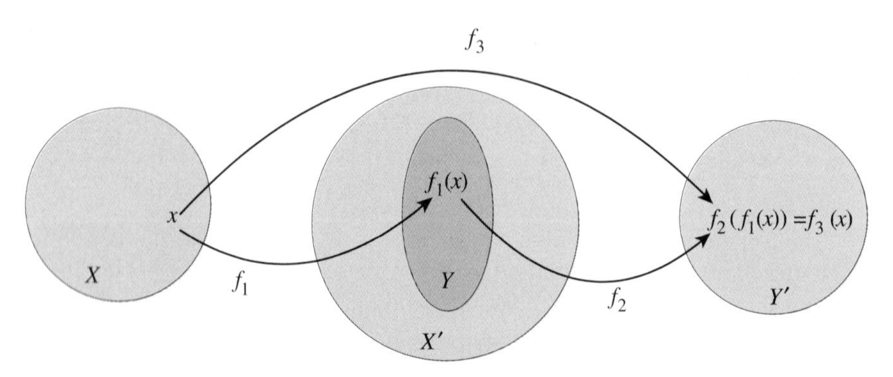

FIGURE 2.5: Function composition: $f_3 = f_2 \circ f_1$.

Example 2.10: Consider the representation of a color image with a colormap. The decoding of the image is depicted in figure 1.7. The image itself might be given by the function

$$ColorMapImage: DiscVerticalSpace \times$$
$$DiscHorizontalSpace \to ColorMapIndexes.$$

The function

$$Display: ColorMapIndexes \to Intensity^3$$

decodes the colormap indexes. If *ColorMapIndexes* has 256 values, it could be identified with the set *Integers8* of all eight-bit words, as we have seen. If we compose these functions

$$ColorComputerImage = Display \circ ColorMapImage,$$

then we get the decoded representation of the image

$$ColorComputerImage: DiscVerticalSpace \times$$
$$DiscHorizontalSpace \to Intensity^3.$$

ColorComputerImage describes how an image looks when it is displayed, whereas *ColorMapImage* describes how it is stored in the computer. ❒

If $f: X \to X$—that is, the domain and range of f are the same—we can form the function

$$\boxed{f^2 = f \circ f.}$$

We can compose f^2 with f to form f^3, and so on.

Example 2.11: Consider the function $S\colon Reals^2 \to Reals^2$, where the assignment $(y_1, y_2) = S(x_1, x_2)$ is defined by matrix multiplication,

$$\begin{bmatrix} y_1 \\ y_2 \end{bmatrix} = \begin{bmatrix} 1 & 2 \\ 3 & 4 \end{bmatrix} \begin{bmatrix} x_1 \\ x_2 \end{bmatrix}. \tag{2.6}$$

The function $S^2 = S \circ S\colon Reals^2 \to Reals^2$ is also defined by matrix multiplication, and the corresponding matrix is the square of the matrix in (2.6).

To see this, let $(y_1, y_2) = S(x_1, x_2)$ and $(z_1, z_2) = S(y_1, y_2) = (S \circ S)(x_1, x_2)$. Then we see that

$$\begin{bmatrix} z_1 \\ z_2 \end{bmatrix} = \underbrace{\begin{bmatrix} 1 & 2 \\ 3 & 4 \end{bmatrix}}_{A} \begin{bmatrix} y_1 \\ y_2 \end{bmatrix} = \begin{bmatrix} 1 & 2 \\ 3 & 4 \end{bmatrix} \left(\begin{bmatrix} 1 & 2 \\ 3 & 4 \end{bmatrix} \begin{bmatrix} x_1 \\ x_2 \end{bmatrix} \right) = \begin{bmatrix} 1 & 2 \\ 3 & 4 \end{bmatrix}^2 \begin{bmatrix} x_1 \\ x_2 \end{bmatrix}$$

$$= \underbrace{\begin{bmatrix} 7 & 10 \\ 15 & 22 \end{bmatrix}}_{A^2} \begin{bmatrix} x_1 \\ x_2 \end{bmatrix} = A^2 \begin{bmatrix} x_1 \\ x_2 \end{bmatrix}. \quad \square$$

Example 2.12: Consider another example in the context of the telephone system. Let *Voices* be the set of possible voice input signals of the form

$$Voice\colon Time \to Pressure.$$

Voices is a function space,

$$Voices = [Time \to Pressure].$$

A telephone converts a *Voice* signal into a signal in the set

$$LineSignals = [Time \to Voltages].$$

Thus, we could define

$$Mouthpiece\colon Voices \to LineSignals.$$

The twisted wire pair may distort this signal, and so we define a function

$$LocalLoop\colon LineSignals \to LineSignals.$$

The input to the line card is therefore

$$(LocalLoop \circ Mouthpiece)(Voice).$$

Similarly, let *BitStreams* be the set of possible bitstreams of the form

$$BitStream: DiscreteTime \to Binary,$$

where $DiscreteTime = \{0, 1/64{,}000, 2/64{,}000, \ldots\}$, inasmuch as there are 64,000 bits/sec. Therefore,

$$BitStreams = [DiscreteTime \to Binary].$$

The encoder in a line card can be mathematically described as a function

$$Encoder: LineSignals \to BitStreams$$

or, with more detail, as a function

$$Encoder: [Time \to Voltages] \to [DiscreteTime \to Binary].$$

The digital telephone network itself may be modeled as a function,

$$Network: BitStreams \to BitStreams.$$

We can continue in this manner until we model the entire path of a voice signal through the telephone network as the function composition

$$Earpiece \circ LocalLoop_2 \circ Decoder \circ Network \circ Encoder \circ LocalLoop_1 \circ Mouthpiece. \tag{2.7}$$

Given a complete definition of each of these functions, we would be well equipped to understand the degradations experienced by a voice signal in the telephone network. ❐

2.1.6 *Declarative versus imperative*

Declarative definitions of functions assert a relationship between elements in the domain and elements in the range. Imperative definitions provide a procedure for finding an element in the range, given one in the domain. Often, both types of specifications can be given for the same function. However, sometimes the specifications are subtly different.

Consider the function

$$SquareRoot: Reals_+ \to Reals_+,$$

defined by the statement "$SquareRoot(x)$ is the unique value of $y \in Reals_+$ such that $y^2 = x$." This declarative definition of *SquareRoot* does not tell us how to

PROBING FURTHER

Declarative interpretation of imperative definitions

The declarative approach establishes a relation between the domain and the range of a function. For example, the equation

$$y = \sin(x)/x$$

can be viewed as defining a subset of $Reals \times Reals$. This subset is the graph of the function $Sinc \colon Reals \to Reals$.

The imperative approach also establishes a function, but it is a function that maps the program state before the statement is executed into a program state after the statement is executed. Consider, for example, the Java statement

```
y = Math.sin(x)/x.
```

When only this statement (rather than a larger program) is considered, the program state is the value of the two variables, `x` and `y`. Suppose that these have been declared to be of type `double`, which in Java represents double-precision floating-point numbers encoding according to an Institute of Electrical and Electronics Engineers (IEEE) standard. Let the set *Doubles* be the set of all numbers so encoded, and note that `NaN` (not a number) $\in Doubles$ is the result of division by zero. The set of possible program states is thus $Doubles \times Doubles$. The Java statement therefore defines a function

$$Statement \colon (Doubles \times Doubles) \to (Doubles \times Doubles).$$

calculate its value at any point in its domain. Nevertheless, it defines *SquareRoot* perfectly well. In contrast, an imperative definition of *SquareRoot* would provide a procedure, or algorithm, for calculating $SquareRoot(x)$ for a given x. Call the result of such an algorithm $\hat{y}$. Because the algorithm would yield an approximation in most cases, $\hat{y}^2$ would not be exactly equal to x. Thus, the declarative and imperative definitions are not always the same.

Any definition of a function that follows the prototype (2.2) is a declarative definition. It does not give a procedure for evaluating the function.

Example 2.13: As another example in which declarative and imperative definitions differ in subtle ways, consider the following mathematical equation:

$$y = \frac{\sin(x)}{x}. \tag{2.8}$$

Consider the Java statement

```
y = Math.sin(x)/x
```

or an equivalent MATLAB statement,

```
y = sin(x)/x.
```

Superficially, these look very similar to (2.8). There are minor differences in syntax in the Java statement; otherwise, it is hard to tell the difference. But there are differences. For one, the mathematical equation (2.8) has meaning if y is known and x is not. It declares a relationship between x and y. The Java and MATLAB statements define a procedure for computing `y` given `x`. Those statements have no meaning if y is known and x is not.

The mathematical equation (2.8) can be interpreted as a predicate that defines a function: For example, the function *Sinc*: *Reals* → *Reals*, where

$$graph(Sinc) = \{(x, y) \mid x \in Reals, y = \sin(x)/x\}. \tag{2.9}$$

The Java and MATLAB statements can be interpreted as imperative definitions of a function;* that is, given an element in the domain, they specify how to compute an element in the range. However, these two statements do not define the same function as in (2.9). To see this, consider the value of y when $x = 0$. Given the mathematical equation, it is not entirely trivial to determine the value of y. You can verify that $y = 1$ when $x = 0$ by using l'Hôpital's rule.† In contrast, the meaning of the Java and MATLAB statements is that $y = 0/0$ when $x = 0$, which Java and MATLAB (and most modern languages) define to be `NaN`.‡ Thus, given $x = 0$, the procedures yield different values for y than does the mathematical expression. ❐

We can see from this example some of the strengths and weaknesses of imperative and declarative approaches. Given only a declarative definition, it

*A source of confusion is that many programming languages, including MATLAB, use the term "function" to mean something a bit different from a mathematical function. They use it to mean a **procedure** that can compute an element in the range of a function, given an element in its domain. Under certain restrictions (avoiding global variables, for example), MATLAB functions do in fact compute mathematical functions. In general, however, they do not.

†**l'Hôpital's rule** states that if $f(a) = g(a) = 0$, then

$$\lim_{x \to a} \frac{f(x)}{g(x)} = \lim_{x \to a} \frac{\dot{f}(x)}{\dot{g}(x)},$$

if the limit exists, where $\dot{f}(x)$ is the derivative of f with regard to x, and $\dot{g}(x)$ is the derivative of g with regard to x.

‡An exception is symbolic algebra programs, such as Mathematica or Maple, which will evaluate $\sin(x)/x$ to 1 when $x = 0$. These programs use sophisticated, rule-based solution techniques and, in effect, recognize the need for and apply l'Hôpital's rule.

is difficult for a computer to determine the value of y. In general, the use of declarative definitions in computers requires quite a bit more sophistication than does the use of imperative definitions.

Imperative definitions are easier for computers to work with. But the Java and MATLAB statements illustrate one weakness of the imperative approach: It is arguable that $y =$ `NaN` is the wrong answer, and so the Java and MATLAB statements have a bug. This bug is unlikely to be detected unless, in testing, these statements happen to be executed with the value $x = 0$. A correct Java program might look like this:

```
if (x == 0.0) y = 1.0;
else y = Math.sin(x)/x.
```

Thus, a weakness of the imperative approach is that ensuring correctness is more difficult. Humans have developed a huge arsenal of techniques and skills for thoroughly understanding declarative definitions (thus lending confidence in their correctness), but we are only beginning to learn how to ensure correctness in imperative definitions.

2.2 Defining signals

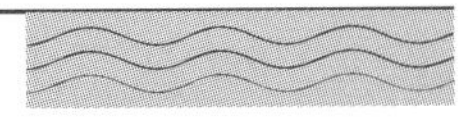

Signals are functions. Thus, both declarative and imperative approaches can be used to define them.

2.2.1 Declarative definitions

Consider, for example, an audio signal s, a pure tone at 440 Hz (middle A on the piano keyboard). Recall that audio signals are functions *Sound*: *Time* $\to$ *Pressure*, where the set *Time* $\subset$ *Reals* represents a range of time and the set *Pressure* represents air pressure.* To define this function, we might give the declarative description

$$\forall\, t \in \mathit{Time}, \quad s(t) = \sin(440 \times 2\pi t). \tag{2.10}$$

In many texts, you will see the shorthand

$$s(t) = \sin(440 \times 2\pi t)$$

used as the definition of the function s. Using the shorthand is acceptable only when the domain of the function is well understood from the context. This shorthand can be particularly misleading in the consideration of systems, and

* Recall further that we normalize *Pressure* so that zero represents the ambient air pressure. We also use arbitrary units, rather than a physical unit such as millibars.

so we use it only sparingly. A portion of the graph of the function (2.10) is shown in figure 1.3.

2.2.2 *Imperative definitions*

We can also give an imperative description of such a signal. When we think of signals rather than more abstractly of functions, a subtle question arises when we attempt to construct an imperative definition: Do we give the value of $s(t)$ for a particular t? Or for all t in the domain? Suppose we want the latter, which seems to be a more complete definition of the function. Then we have a problem. The domain of this function may be any time interval, or all time! Suppose we want just one second of sound. Define $t = 0$ to be the start of that one second. Then the domain is $[0, 1]$. But there is an (uncountably) infinite number of values for t in this range! No Java or MATLAB program could provide the value of $s(t)$ for all these values of t.

Because a signal is a function, we give an imperative description of the signal exactly as we did for functions. We provide a procedure that has the potential of providing values for $s(t)$, given any t.

Example 2.14: We could define a Java **method** as follows:*

```
double s(double t) {
    return (Math.sin(440*2*Math.PI*t));
}
```

Calling this method with a value for `t` as an argument yields a value for `s(t)`. Java (and most object-oriented languages) use the term "method" for most procedures. ❐

Another alternative is to provide a set of samples of the signal.

Example 2.15: In MATLAB, we could define a vector `t` that gives the values of time that we are interested in:

```
t = [0:1/8000:1].
```

In the vector `t` there are 8,001 values evenly spaced between 0 and 1, and so our sampling rate is 8,000 samples per second. Then we can compute values of `s` for these values of `t` and listen to the resulting sound:

```
s = cos(2*pi*440*t);
sound(s,8000)
```

* It is uncommon, and not recommended, to use single-character names for methods and variables in programs. We do it here only to emphasize the correspondence with the declarative definition. In mathematics, it is common to use single-character names.

The vector s also has 8,001 elements, representing evenly spaced samples of one second of A-440. ❒

2.2.3 *Physical modeling*

An alternative way to define a signal is to construct a model for a physical system that produces that signal.

Example 2.16: A pure tone might be defined as a solution to a differential equation that describes the physics of a tuning fork.

A tuning fork consists of a metal finger (called a **tine**) that is displaced by striking it with a hammer. After being displaced, it vibrates. If the tine encounters no friction, it will vibrate forever. We can denote the displacement of the tine after being struck at time zero as a function $y\colon Reals_+ \to Reals$. If we assume that the initial displacement introduced by the hammer is one unit, then, using our knowledge of physics, we can determine that for all $t \in Reals_+$, the displacement satisfies the differential equation

$$\ddot{y}(t) = -\omega_0^2 y(t), \tag{2.11}$$

where ω_0 is the constant that depends on the mass and stiffness of the tine and $\ddot{y}(t)$ denotes the second derivative with respect to time of y (see Probing further: Physics of a tuning fork box).

It is easy to verify that y given by

$$\forall\, y \in Reals_+, \quad y(t) = \cos(\omega_0 t) \tag{2.12}$$

is a solution to this differential equation (just take its second derivative). Thus, the displacement of the tuning fork is sinusoidal. This displacement couples directly with air around the tuning fork, creating vibrations in the air (sound). If we choose materials for the tuning fork so that $\omega_0 = 2\pi \times 440$, then the tuning fork will produce the tone of A-440 on the musical scale. ❒

PROBING FURTHER

Physics of a tuning fork

A tuning fork consists of two fingers (tines), as shown in figure 2.6. If you displace one of these tines by hitting it with a hammer, it vibrates with a nearly perfect sinusoidal characteristic. As it vibrates, it pushes the air, creating a nearly perfect sinusoidal variation in air pressure that propogates as sound. Why does it vibrate this way?

continued on next page

PROBING FURTHER

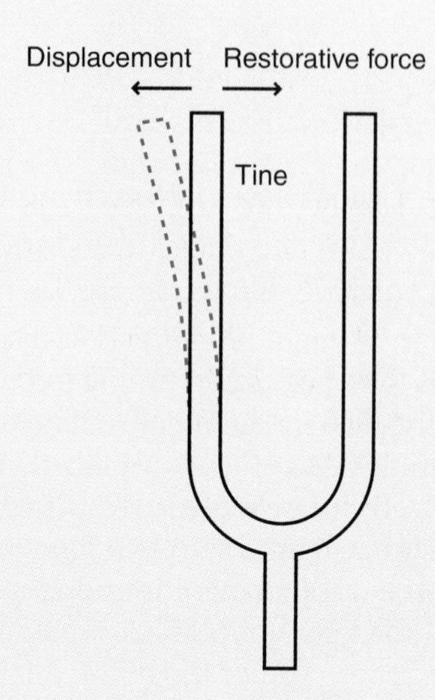

FIGURE 2.6: A tuning fork.

Suppose the displacement of the tine (in relation to its position at rest) at time t is given by $x(t)$, where x: *Reals* → *Reals*. There is a force on the tine pushing it toward its at-rest position. This is the restorative force of the elastic material used to make the tine. The force is proportional to the displacement (the greater the displacement, the greater the force), and so

$$f(t) = -kx(t),$$

where k is the proportionality constant that depends on the material and geometry of the tine. In addition, Newton's second law of motion tells us the relationship between force and acceleration,

$$f(t) = ma(t),$$

where m is the mass and $a(t)$ is the acceleration at time t. Of course,

$$a(t) = \frac{d^2}{dt}x(t) = \ddot{x}(t);$$

therefore,

$$m\ddot{x}(t) = -kx(t)$$

or

$$\ddot{x}(t) = -(k/m)x(t).$$

Comparing with (2.11), we see that $\omega_0^2 = k/m$.

A solution to this equation needs to be some signal that is proportional to its own second derivative. A sinusoid as in (2.12) has exactly this property. The sinusoidal behavior of the tine is called **simple harmonic motion**.

2.3 Defining systems

All of the methods that we have discussed for defining functions can be used, in principle, to define systems. However, in practice, the situation is much more complicated for systems than for signals. Recall from section 1.2.1 that a system is a function in which the domain and range are sets of signals called signal spaces. Elements of these domains and ranges are considerably more difficult to specify than, say, an element of *Reals* or *Integers*. For this reason, it is almost never reasonable to use a graph or a table to define a system. Much of the rest of this book is devoted to giving precise ways to define systems for which some analysis is possible. Here we consider some simple techniques that can be immediately understood. Then we show how more complicated systems can be constructed from simpler ones by using block diagrams. We give a rigorous meaning to these block diagrams so that we can use them without resorting to perilous intuition to interpret them.

Consider a system S where

$$S\colon [D \to R] \to [D' \to R']. \tag{2.13}$$

Suppose that $x \in [D \to R]$ and $y = S(x)$. Then we call the pair (x, y) a **behavior** of the system. A behavior is an input–output pair. The set of all behaviors is

$$\boxed{\mathit{Behaviors}(S) = \{(x, y) \mid x \in [D \to R] \text{ and } y = S(x)\}.}$$

Giving the set of behaviors is one way to define a system. Explicitly giving the set *Behaviors*, however, is usually impractical, because it is a huge set, typically infinite (see sidebars on pages 69 and 71–72). Thus, we seek other ways of talking about the relationship between a signal x and a signal y when $y = S(x)$.

To describe a system, we must specify its domain (the space of input signals), its range (the space of output signals), and the rule by which the system assigns an output signal to each input signal. This assignment rule is more difficult to describe and analyze than the input and output signals themselves. A table is almost never adequate, for example. Indeed, for most systems we do not have effective mathematical tools for describing or understanding their behavior. Thus, it is useful to restrict our system designs to those we can understand. We first consider some simple examples.

2.3.1 *Memoryless systems and systems with memory*

Memoryless systems are characterized by the property that previous input values are not remembered when determining the current output value. More precisely, a system $F\colon [\mathit{Reals} \to Y] \to [\mathit{Reals} \to Y]$ is **memoryless** if there is a function $f\colon Y \to Y$ such that

$$\forall\, t \in \mathit{Reals} \text{ and } \forall\, x \in [\mathit{Reals} \to Y], \quad (F(x))(t) = f(x(t)).$$

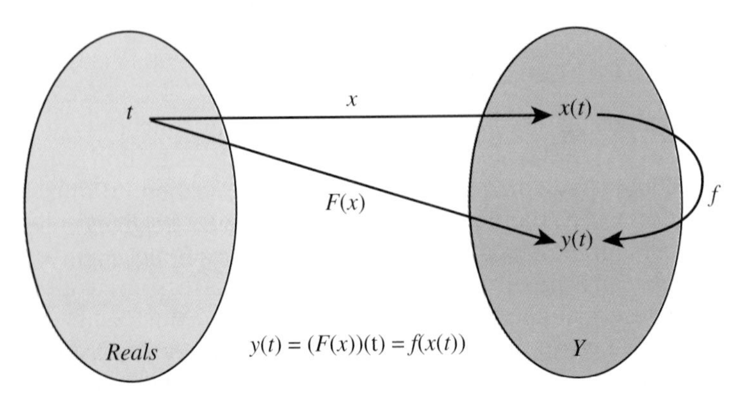

FIGURE 2.7: A memoryless system F has an associated function f that can be used to determine its output $y(t)$, given only the current input $x(t)$ at time t. In particular, it depends neither on values of the function x for other values of time nor on t.

This is illustrated in figure 2.7. In other words, at any time t, the output $(F(x))(t)$ depends only on the input $x(t)$ at that same time t; in particular, it does not depend on t or on previous or future values of x.

Specification of a memoryless system reduces to specification of the function f. If Y is finite, then a table may be adequate.

Example 2.17: Consider a continuous-time system with input x and output y, where

$$\forall\, t \in Reals, \quad y(t) = x^2(t).$$

This example defines a simple system, in which the value of the output signal at each time depends only on the value of the input signal at that time. Such systems are said to be memoryless because you do not have to remember previous values of the input in order to determine the current value of the output. ❐

In contrast, here is an example of a system with memory.

Example 2.18: Consider a continuous-time system with input x and output $y = F(x)$, where

$$\forall\, t \in Reals, \quad y(t) = \frac{1}{M}\int_{t-M}^{t} x(\tau)d\tau.$$

With a change of variables, this can also be written

$$y(t) = \frac{1}{M}\int_{0}^{M} x(t-\tau)d\tau.$$

This system is clearly not memoryless. It has the effect of smoothing the input signal. We study it and many related systems in detail in later chapters. ❐

2.3.2 Differential equations

Consider a class of systems given by functions $S\colon ContSignals \to ContSignals$, where $ContSignals$ is a set of **continuous-time signals**. Depending on the scenario, we could have $ContSignals = [Time \to Reals]$ or $ContSignals = [Time \to Complex]$, where $Time = Reals$ or $Time = Reals_+$. These are often called **continuous-time systems** because they operate on continuous-time signals. Frequently, such systems can be defined by **differential equations** that relate the input signal to the output signal.

Example 2.19: Consider a particle constrained to move forward or backward along a straight line with an externally imposed force. We consider this particle to be a system in which the output is its position and the externally imposed force is the input.

Denote the position of the particle by $x\colon Time \to Reals$, where $Time = Reals_+$. By considering only the nonnegative real numbers, we are assuming that the model has a starting time. We denote the acceleration by $a\colon Time \to Reals$. By Newton's law, which relates force, mass, and acceleration,

$$f(t) = ma(t),$$

where $f(t)$ is the force at time t, and m is the mass. By the definition of acceleration,

$$\forall\, t \in Reals_+, \quad \ddot{x}(t) = a(t) = f(t)/m,$$

where $\ddot{x}(t)$ denotes the second derivative with respect to time of x. If we know the initial position $x(0)$ and initial speed $\dot{x}(0)$ of the particle at time 0, and if we are given the input force f, we can evaluate the position at any t by integrating this differential equation:

$$x(t) = x(0) + \dot{x}(0)t + \int_0^t \Big[\int_0^s (f(\tau)/m)d\tau\Big]ds. \qquad (2.14)$$

We can regard the initial position and velocity as inputs, together with force, in which case the system is a function

$$Particle\colon Reals \times Reals \times [Reals_+ \to Reals] \to [Reals_+ \to Reals],$$

where for any inputs $(x(0), \dot{x}(0), f)$, $x = Particle(x(0), \dot{x}(0), f)$ must satisfy (2.14).

Suppose, for example, that the input is $(1, -1, f)$, where $m = 1$ and $\forall\, t \in Reals_+,\ f(t) = 1$. We can calculate the position by carrying out the integration in (2.14) to find that

$$\forall\, t \in Reals_+, \quad x(t) = 1 - t + 0.5t^2.$$

Suppose instead that $x(0) = \dot{x}(0) = 0$ and $\forall\, t \in Reals_+, f(t) = \cos(\omega_0 t)$, where ω_0 is some fixed number. Again, we can carry out the integration to get

$$\int_0^t \int_0^s \cos(\omega_0 u) du\, ds = -\frac{\cos(\omega_0 t) - 1}{\omega_0^2}.$$

Notice that the position of the particle is sinusoidal. Notice further that the amplitude of this sinusoid decreases as ω_0 increases. Intuitively, this has to be the case. If the externally imposed force is varying more rapidly back and forth, the particle has less time to respond to each direction of force, and hence its excursion is less. In subsequent chapters, we study how the response of certain kinds of systems varies with the frequency of the input. ❒

2.3.3 *Difference equations*

Consider a class of systems given by functions S: *DiscSignals* → *DiscSignals*, where *DiscSignals* is a set of **discrete-time signals**. Depending on the scenario, we could have *DiscSignals* = [*Integers* → *Reals*]; *DiscSignals* = [*Integers* → *Complex*]; *DiscSignals* = [$Naturals_0$ → *Reals*]; or even *DiscSignals* = [$Naturals_0$ → *Complex*]. These are often called **discrete-time systems** because they operate on discrete-time signals. Frequently, such systems can be defined by **difference equations** that relate the input signal to the output signal.

Example 2.20: Consider a system

$$S: [Naturals_0 \to Reals] \to [Naturals_0 \to Reals],$$

where for all $x \in [Naturals_0 \to Reals]$, $S(x) = y$ is given by

$$\forall\, n \in Integers, \quad y(n) = (x(n) + x(n-1))/2.$$

The output at each index is the average of two of the inputs. This is a simple example of a **moving average** system, in which typically more than two input values get averaged to produce an output value.

Suppose that $x = u$, the **unit step** function, defined by

$$\forall\, n \in Integers, \quad u(n) = \begin{cases} 1 & \text{if } n \geq 0 \\ 0 & \text{otherwise.} \end{cases} \tag{2.15}$$

We can easily calculate the output y,

$$\forall\, n \in Integers, \quad y(n) = \begin{cases} 1 & \text{if } n \geq 1 \\ 1/2 & \text{if } n = 0 \\ 0 & \text{otherwise.} \end{cases}$$

The system smoothes the transition of the unit step a bit.

A slightly more interesting input is a sinusoidal signal given by

$$\forall\, n \in Integers, \quad x(n) = \cos(2\pi f n).$$

The output is given by

$$\forall\, n \in Integers, \quad y(n) = (\cos(2\pi f n) + \cos(2\pi f(n-1)))/2.$$

By trigonometric identities,* this can be written as

$$y(n) = R\cos(2\pi f n + \theta),$$

where

$$\theta = \arctan(\frac{\sin(-2\pi f)}{1 + \cos(-2\pi f)})/2$$

and

$$R = \sqrt{2 + 2\cos(2\pi f)}.$$

As in the previous example, a sinusoidal input stimulates a sinusoidal output with the same frequency. In this case, the amplitude of the output varies (in a fairly complicated way) as a function of the input frequency. We examine this phenomenon in more detail in subsequent chapters by studying the *frequency response* of such systems. ❐

* $$A\cos(\theta + \alpha) + B\cos(\theta + \beta) = C\cos\theta - S\sin\theta = R\cos(\theta + \phi)$$

where

$$C = A\cos\alpha + B\cos\beta,$$
$$S = A\sin\alpha + B\sin\beta,$$
$$R = \sqrt{C^2 + S^2}, \text{ and}$$
$$\phi = \arctan(S/C).$$

Example 2.21: The general form for a moving average is given by*

$$\forall\, n \in Integers, \quad y(n) = \frac{1}{M} \sum_{k=0}^{M-1} x(n-k),$$

where x is the input and y is the output. This system is called an M-point **moving average**, because at any n it gives the average of the M most recent values of the input. It computes an average, just as does example 2.18, but the integral has been replaced by its discrete counterpart, the sum. ❒

Moving averages are widely used on Wall Street to smooth out momentary fluctuations in stock prices to try to determine general trends. We will study the smoothing properties of this system. We will also study more general forms of difference equations, of which the moving average is a special case.

These examples give declarative definitions of systems. Imperative definitions require a procedure for computing the output signal given the input signal. It is clear how to do that with the memoryless system, assuming that an imperative definition of the function f is available, and with the moving average. The integral equation, however, is harder to define imperatively. An imperative description of such systems that is suitable for computation on a computer requires approximation through solvers for differential equations. Simulink®, for example, which is part of the MATLAB package, provides such solvers. Alternatively, an imperative description can be given in terms of analog circuits or other physical systems that operate directly on the pertinent continuous domain. Discrete-time systems often have reasonable imperative definitions as **state machines**, considered in detail in the next chapter.

2.3.4 *Composing systems by using block diagrams*

We have been using **block diagrams** informally to describe systems. But it turns out that block diagrams can have as rigorous and formal a meaning as mathematical notations. We begin the exploration of this concept here and pursue it much further in chapter 4.

A block diagram is a **visual syntax** for describing a system as an interconnection of other (component) systems, each of which emphasizes one particular input-to-output transformation of a signal. A block diagram is a collection of

* If this notation is unfamiliar, see Basics: Summations box.

Summations

BASICS

In example 2.21, a discrete-time moving average system is defined by

$$\forall n \in Integers, \quad y(n) = \frac{1}{M} \sum_{k=0}^{M-1} x(n-k),$$

where x is the input and y is the output. The notation $\sum_{k=0}^{M-1}$ indicates a sum of M terms. The terms are all given by $x(n-k)$, where k takes on values from 0 to $M-1$. Thus,

$$\sum_{k=0}^{M-1} x(n-k) = x(n) + x(n-1) + \cdots + x(n-M+1).$$

Such summations are related to integrals. Example 2.18 describes a continuous-time system with input x and output y, where

$$\forall t \in Reals, \quad y(t) = \frac{1}{M} \int_{t-M}^{t} x(\tau) d\tau.$$

This is similar to the discrete-time moving average in that it sums values of x, but here it sums over a continuum of values of the dummy variable τ. In the discrete-time version, the sum is over discrete values of the dummy variable k, which has only integer values.

The summation notation has an ambiguity that it does not share with the integral notation. In particular, it is not clear how to interpret an expression such as

$$\sum_{k=0}^{M-1} 1 + 2.$$

There are two possibilities, and they are not equal:

$$\left(\sum_{k=0}^{M-1} 1 + 2\right) \neq \left(\sum_{k=0}^{M-1} 1\right) + 2.$$

The left sum is equal to $3M$, whereas the right sum is $M + 2$. In integration, this ambiguity does not occur because of the explicit reference to the dummy variable as $d\tau$. In particular, it is clear that

$$\int_0^T 1 + 2 d\tau \neq \int_0^T 1 d\tau + 2.$$

The left integral is equal to $3T$, whereas the right integral is $T + 2$.

blocks interconnected by arrows. Arrows are labeled by signals. Each block represents an individual system that transforms an incoming signal, or **input signal**, into an outgoing signal, or **output signal**. For example, *modulator* in figure 1.18 is a system that transforms a bit sequence into a voice-like signal.

A block diagram, which is a composition of systems, is itself a system. We can use function composition, as discussed in section 2.1.5, to give a precise meaning to this larger system. A block represents a function, and the connection of an output from one block to the input of another represents the composition of their two functions. The only requirement for interconnecting two blocks is that the output of the first block must be an acceptable input for the second. Thus, for example, the output of *modulator* in figure 1.18 is a voice-like signal, which is precisely what the system *telephone network* is able to accept as an input.

Block diagrams can be much more readable than symbolic function composition, particularly for complicated interconnections. They also offer a natural hierarchy, in which we can combine blocks to hide certain signals and component systems and to emphasize others. For instance, the *telephone network* in figure 1.18 hides the details shown in figure 1.14. This emphasizes the POTS capabilities of the telephone network, while hiding its other features.

For certain sorts of blocks, composing them in a block diagram results in a new system whose properties are easy to determine. In chapter 4 we show how to combine state machine blocks to define a new state machine. In chapter 8 we show how to combine filter blocks to define new filter blocks. Here, we consider the composition of blocks when all we know about the blocks is that they represent functions with a given domain and range and when no further structure is available.

The simplest block diagram has a single block, as in figure 2.8. The block represents a system with input signal x and output signal y. Here, x denotes a variable over the set X, and y denotes a variable over the set Y. The system is described by the function $S: X \to Y$. Both X and Y are sets of functions or signals. Therefore, the variables x and y themselves denote functions.

In general, a system obtained by a **cascade composition** of two blocks is given by the composition of the functions describing those blocks. In figure 2.9,

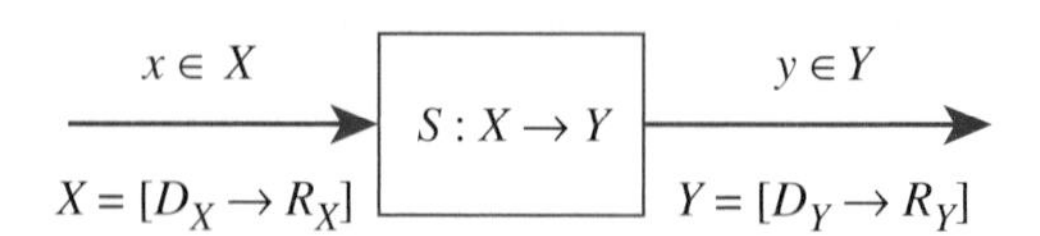

FIGURE 2.8: The simplest block diagram represents a function S that maps an input signal $x \in X$ to an output signal $y \in Y$. The domain and range of the input are D_X and R_X, repectively, and those of the output are D_Y and R_Y.

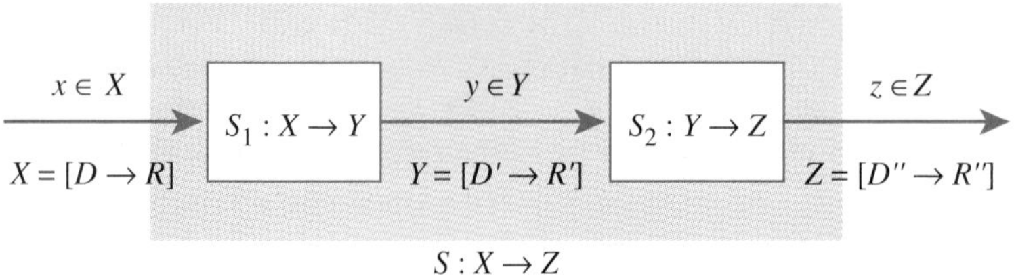

$x \in X$ $S_1 : X \to Y$ $y \in Y$ $S_2 : Y \to Z$ $z \in Z$

$X = [D \to R]$ $Y = [D' \to R']$ $Z = [D'' \to R'']$

$S : X \to Z$

FIGURE 2.9: The cascade composition of the two systems is described by $S = S_2 \circ S_1$.

the function S describes the system obtained by connecting the systems S_1 and S_2, with $S = S_2 \circ S_1$; that is,

$$\forall x \in X, \quad S(x) = S_2(S_1(x)).$$

The combined system has input signal x, output signal z, and internal signal y. (The internal signal is not visible in the input or output of the combined system.) For this composition to make sense, the range of the first system must, of course, be contained by the domain of the second. Figure 2.9 shows the typical case in which this range and domain are the same. The voice path in (2.7) is an example of cascade composition.

PROBING FURTHER

Composition of graphs

We suggest a general method for writing down a declarative specification of the interconnected system S in figure 2.9 in terms of the subsystems S_1 and S_2 and the **connection restriction** that the output of S_1 be acceptable as an input of S_2.

We describe S_1 and S_2 by their graphs,

$$graph(S_1) = \{(x, y_1) \in X \times Y \mid y_1 = S_1(x)\}$$

and

$$graph(S_2) = \{(y_2, z) \in Y \times Z \mid z = S_2(y_2)\},$$

and we specify the connection restriction as the predicate

$$y_1 = y_2.$$

We use the different dummy variables y_1 and y_2 to distinguish between the two systems and the connection restriction.

continued on next page

PROBING FURTHER

The graph of the combined system S is then given by

$$graph(S) = \{(x, z) \in X \times Z \mid \exists y_1, \exists y_2$$
$$(x, y_1) \in graph(S_1) \wedge (y_2, z) \in graph(S_2) \wedge y_1 = y_2\}.$$

Here, the existential quantifier symbol $\exists$ means "there exists," and $\wedge$ denotes logical conjunction, "and." It is now straightforward to show that $graph(S) = graph(S_2 \circ S_1)$ so that $S = S_2 \circ S_1$.

In the case of the cascade composition of figure 2.9, this elaborate method is unnecessary, because we can write down $S = S_2 \circ S_1$ simply by inspecting the figure. But for feedback connections, we may not be able to write down the combined system directly.

There are three other reasons for understanding this method. First, we use it later to obtain a description of interconnected state machines from their component machines. Second, this method is used to describe electronic circuits. Third, if we want a computer to figure out the description of the interconnected system from a description of the subsystems and the connection restrictions, we have to design an algorithm that the computer must follow. Such an algorithm can be based on this general method.

Consider two more block diagrams with slightly more complicated structures. Figure 2.10 is similar to figure 2.9. The system described by S_1 is the same as before, but the system described by S_2 has a pair of input signals $(w, y) \in W \times Y$. The combined system has the pair $(x, w) \in X \times W$ as input signal, z as output signal, and y as internal signal, and it is described by the function $S: X \times W \to Z$, where

$$\forall\, (x, w) \in X \times W, \quad S(x, w) = S_2(w, S_1(x)). \tag{2.16}$$

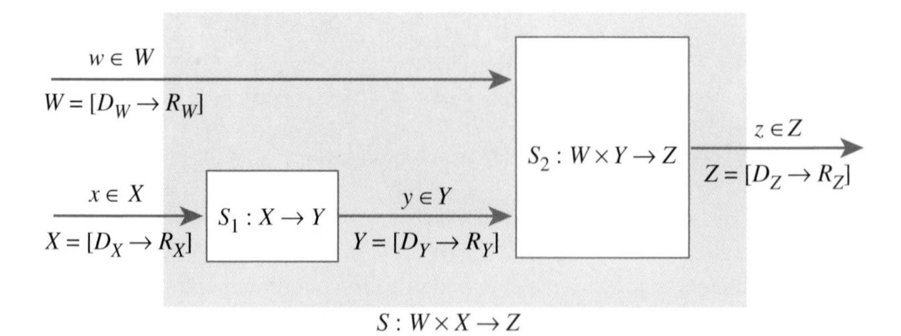

FIGURE 2.10: The combined system with input signals x, w and output signal z is described by the function S, where $\forall\, (x, w)$, $S(x, w) = S_2(w, S_1(x))$.

Notice that it is now much harder to define this system by using the function composition notation, $\circ$, but the block diagram makes its definition evident. In fact, the block diagram notation is much more flexible.

The system of figure 2.11 is obtained from that of figure 2.10 by connecting the output signal z to the input signal w. As a result, the new system has input signal x, output signal z, and internal signals y and w, and it is described by the function $S'\colon X \to Z$, where

$$\forall x \in X, \quad S'(x) = S_2(S'(x), S_1(x)). \tag{2.17}$$

The connection of z to w is called a **feedback** connection because the output z is fed back as input w. Of course, such a connection has meaning only if Z, the range of S_2, is a subset of W. The system in figure 2.11 is also difficult to define by using function composition notation, $\circ$, and yet again the block diagram definition is clear.

There is one enormous difference between (2.16) and (2.17). Expression (2.16) serves as a definition of the function S: To every (x, w) in its domain, S assigns the value given by the right side, which is uniquely determined by the given functions S_1 and S_2. But in expression (2.17), the value $S'(x)$ assigned to x may not be determined by the right side of (2.17), inasmuch as the right side itself depends on $S'(x)$. In other words, (2.17) is an *equation* that must be solved to determine the value of $S'(x)$ for a given x; that is, $S'(x) = y$, where y is a solution of

$$y = S_2(y, S_1(x)). \tag{2.18}$$

Such a solution, if it exists, is called a **fixed point**. We now face the difficulty that this equation may have no solution, exactly one solution, or several solutions.

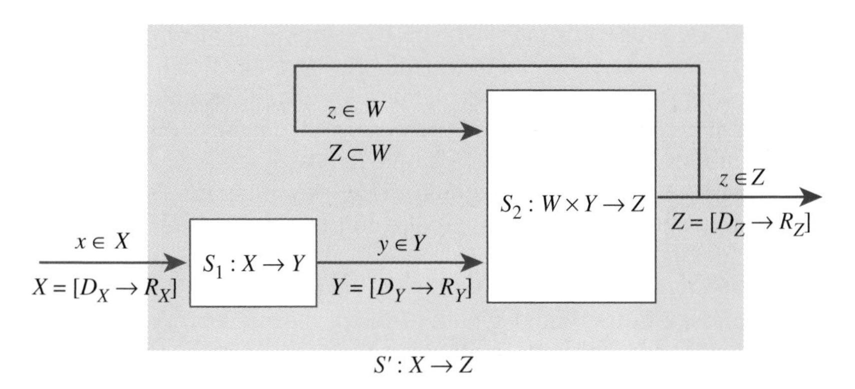

FIGURE 2.11: The combined system is described by the function S', where $S'(x) = S_2(S'(x), S_1(x))$.

Another difficulty is that the value y that solves (2.18) is not a number but a function. So it will not be easy to solve such an equation. Because feedback connections always arise in control systems, we will study how to solve them. We first solve them in the context of state machines, which are introduced in the next chapter.

All these block diagrams follow the same principles. Component systems are used to define composite systems. For them to be useful, of course, it is necessary to be able to infer properties of the composite systems. Fortunately, this is often the case, although feedback connections prove subtle. This idea is explored further in chapters 4 and 8.

2.4 Summary

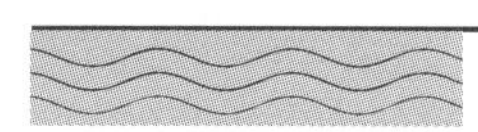

Both signals and systems are modeled as functions. It is often straightforward to figure out the domain and range of a particular signal and system. It is more difficult to specify the function's assignment rule. Because the domain and range of a system are themselves signal spaces, the assignment rule for a system is more complex than that for a signal. The domain and range signal spaces of a system can be quite different: A modem converts bit sequences into sounds, and an encoder converts sounds into bit sequences.

The assignment rule of a function takes a declarative or imperative form. The declarative form is usually mathematical; for example, define *Chirp*: $[-1, 1] \to$ *Reals* by

$$\forall\, t \in [-1, 1], \quad Chirp(t) = \cos(20\pi t^2).$$

The imperative form is a procedure to evaluate the function at an arbitrary point in its domain. The procedure may involve a table lookup (if the domain is finite) or a computer program. If the domain is infinite, the evaluation procedure may yield only an approximation of the declarative form of the "same" function.

A physical system is often described by using differential or difference equations that embody its law of motion. A mechanical system's law of motion is derived from Newton's laws. An electrical circuit's law of motion is derived from Kirchhoff's laws and the laws of the circuit's constitutive elements: resistors, capacitors, inductors, and transistors.

Most systems are built by composing smaller subsystems. The composition may be expressed in the visual syntax of block diagrams or, mathematically, by using function composition. Feedback is the most complex form of system composition: A feedback specification requires the solution of a fixed-point equation.

KEY: E = mechanical T = requires plan of attack C = more than 1 answer

EXERCISES

E 1. The broadcast signal of an AM radio station located at 110 on the radio dial has a carrier frequency of 110 kHz. An AM signal that includes the carrier has the form

$$\forall t \in Time, \quad AMSignal(t) = (1 + m(t)) \sin(2\pi \times 110{,}000t),$$

where m is an audio signal like *Voice* in figure 1.1, except that $\forall t \in Time$, $|m(t)| < 1$. Because you cannot easily plot such a high-frequency signal, give an expression for and plot *AMSignal* (using MATLAB) for the case in which $Time = [0, 1]$, $m(t) = \cos(\pi t)$, and the carrier frequency is 20 Hz.

T 2. This problem is a study of the relationship between the notion of **delay** and the graph of a function.

(a) Consider two functions f and g from *Reals* into *Reals*, where $\forall t \in Reals$, $f(t) = t$ and $g(t) = f(t - t_0)$, where t_0 is a fixed number. Sketch a plot of f and g for $t_0 = 1$ and $t_0 = -1$. Observe that if $t_0 > 0$, then $graph(g)$ is obtained by moving $graph(f)$ to the right, and if $t_0 < 0$, by moving it to the left.

(b) Show that if $f\colon Reals \to Reals$ is any function whatsoever, and $\forall t$, $g(t) = f(t - t_0)$, then if $(t, y) \in graph(f)$, then $(t + t_0, y) \in graph(g)$. This is another way of saying that if $t_0 > 0$, then the graph is moved to the right, and if $t_0 < 0$, then the graph is moved to the left.

(c) If t represents time, and if $t_0 > 0$, we say that g is obtained by *delaying* f. Why is it reasonable to say this?

E 3. Indicate whether the following statements are true or false.

(a) $[\{1, 2, 3\} \to \{a, b\}] \subset [Naturals \to \{a, b\}]$.

(b) $\{g \mid g = graph(f) \wedge f\colon X \to Y\} \subset X \times Y$.

(c) $F\colon [Reals \to Reals] \to [Reals \to Reals]$, such that $\forall t \in Reals$ and $\forall x \in [Reals \to Reals]$,

$$(F(x))(t) = \sin(2\pi \cdot 440t),$$

is a memoryless system.

(d) Let $f{:}\mathit{Reals} \to \mathit{Reals}$ and $g{:}\mathit{Reals} \to \mathit{Reals}$, where g is obtained by delaying f by $\tau \in \mathit{Reals}$; that is,

$$\forall\, t \in \mathit{Reals}, \quad g(t) = f(t - \tau).$$

Then $\mathit{graph}(g) \subset \mathit{graph}(f)$.

E 4. Figure 2.12 shows graphs of two functions: $f{:}[-1, 1] \to [-1, 1]$ and $g{:}[-1, 1] \to [-1, 1]$. For each case, define the function by giving an algebraic expression for its value at each point in its domain. This expression will have several parts, similar to the definition of the *signum* function in (2.3). Note that $g(0) = 0$ for the graph on the right. Plot $\mathit{graph}(f \circ g)$ and $\mathit{graph}(g \circ f)$.

T 5. Let $X = \{a, b, c\}$ and $Y = \{1, 2\}$. For each of the following subsets $G \subset X \times Y$, determine whether G is the graph of a function from X to Y and, if it is, describe the function as a table.

(a) $G = \{(a, 1), (b, 1), (c, 2)\}$.

(b) $G = \{(a, 1), (a, 2), (b, 1), (c, 2)\}$.

(c) $G = \{(a, 1), (b, 2)\}$.

C 6. A router in the Internet is a switch with several input ports and several output ports. A packet containing data arrives at an input port at an arbitrary time, and the switch forwards the packet to one of the outgoing ports. The ports of different routers are connected by transmission links. When a packet arrives at an input port, the switch examines the packet, extracting from it a destination address d. The switch then looks up the output port in its routing table, which contains entries of the form

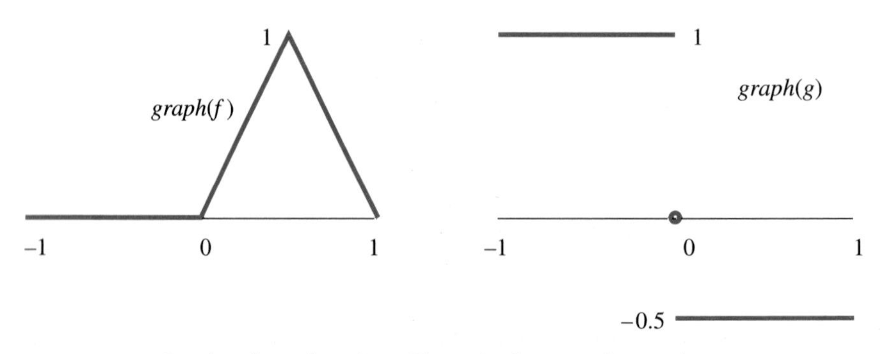

FIGURE 2.12: Graphs of two functions. The color lines are the graphs.

$(d, outputPort)$. It then forwards the packet to the specified output port. The Internet works by setting up the routing tables in the routers.

Consider a simplified router with one input port and two output ports, named O_1 and O_2. Let D be the set of destination addresses.

(a) Explain why the routing table can be described as a subset $T \subset D \times \{O_1, O_2\}$.

(b) Is it reasonable to constrain T to be the graph of a function from $D \to \{O_1, O_2\}$? Why?

(c) Assume the signal at the input port is a sequence of packets. How would you describe the space of input signals to the router and output signals from the router?

(d) How would you describe the switch as a function from the space of input signals to the space of output signals?

C 7. For each of the following expressions, state whether it can be interpreted as an assignment, an assertion, or a predicate. More than one choice may be valid, because the full context is not supplied.

(a) $x = 5$.

(b) $A = \{5\}$.

(c) $x > 5$.

(d) $3 > 5$.

(e) $x > 5 \wedge x < 3$.

T 8. A logic circuit with m binary inputs and n binary outputs is shown in figure 2.13. It is described by a function $F: X \to Y$, where $X = Binary^m$ and $Y = Binary^n$. (In a circuit, the signal values 1 and 0 in *Binary* correspond to voltages *High* and *Low*, respectively.) How many such distinct logic functions F are there?

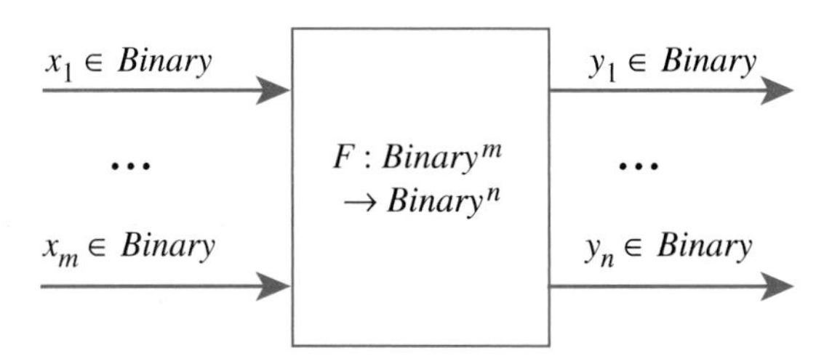

FIGURE 2.13: The logic circuit has m binary inputs $(x_1, \ldots, x_m)$ and n binary outputs $(y_1, \ldots, y_n)$.

E 9. The function $H\colon [Reals_+ \to Reals] \to [Naturals_0 \to Reals]$ given by $\forall\, x \in [Reals_+ \to Reals]$,

$$\forall\, n \in Naturals_0, \quad H(x)(n) = x(10n),$$

is a mathematical example of a system with input signal space $[Reals_+ \to Reals]$ and output signal space $[Naturals_0 \to Reals]$. Give a mathematical example of a system H in which

(a) both its input and output signal spaces are $[Naturals_0 \to Binary]$.

(b) its input signal space is $[Naturals_0 \to Reals]$ and its output signal space is $[Naturals_0 \to \{0, 1\}]$.

(c) the input signal space is $[Integers \to Reals]$ and its output signal space is $[Reals \to Reals]$.

E 10. Consider the functions

$$g\colon Y \to Reals \quad \text{and} \quad f\colon Naturals \to Y,$$

where Y is a set.

(a) Draw a block diagram for $(g \circ f)$, with one block for each of g and f, and label the inputs and outputs of the blocks with the domain and range of g and f.

(b) Suppose Y is given by

$$Y = [\{1, \ldots, 100\} \to Reals].$$

(Thus, the function f takes a natural number and returns a sequence of length 100, while the function g takes a sequence of length 100 and returns a real number.)

Suppose further that g is given by

$$\forall\, y \in Y, \quad g(y) = \sum_{i=1}^{100} y(i) = y(1) + y(2) + \cdots + y(100),$$

and f by $\forall\, x \in Naturals$, $\forall\, z \in \{1, \ldots, 100\}$,

$$(f(x))(z) = \cos(2\pi z/x).$$

(Thus, x gives the period of a cosine waveform, and f gives 100 samples of that waveform.) Give a one-line MATLAB expression that evaluates $(g \circ f)(x)$ for any $x \in Naturals$. Assume the value of x is already in a MATLAB variable called `x`.

(c) Find $(g \circ f)(1)$.

T 11. The following system S takes a discrete-time signal $x \in X$ and transforms it into a discrete-time signal $y \in Y$ whose value at index n is the average of the previous four values of x. Such a system is called a **moving average**. Suppose that $X = Y = [Naturals \to Reals]$, where *Naturals* is the set of natural numbers. More precisely, the system is described by the function S such that for any $x \in X$, $y = S(x)$ is given by

$$y(n) = \begin{cases} [x(1) + \cdots + x(n)]/4 & \text{for } 1 \leq n < 4 \\ [x(n-3) + x(n-2) + x(n-1) + x(n)]/4 & \text{for } n \geq 4 \end{cases}$$

Notice that the first three samples are averages only if we assume that samples before those that are available have a value of zero. Thus, there is an initial **transient** while the system collects enough data to begin computing meaningful averages.

Write a MATLAB program to calculate and plot the output signal y for time $1 \leq n \leq 20$ for the following input signals:

(a) x is a unit step delayed by 10; that is, $x(n) = 0$ for $n \leq 9$ and $x(n) = 1$ for $n \geq 10$.

(b) x is a unit step delayed by 15.

(c) x alternates between 1 and -1; that is, $x(n) = 1$ for n odd, and $x(n) = -1$ for n even. Hint: Try computing $\cos(\pi n)$ for $n \in Naturals$.

(d) Comment on what this system does. Qualitatively, how is the output signal different from the input signal?

T 12. The following system is similar to problem 11, but time is continuous. Now $X = Y = [Reals \to Reals]$ and the system $F: X \to Y$ is defined as follows. For all $x \in X$ and $t \in Reals$,

$$(F(x))(t) = \frac{1}{10} \int_{t-10}^{t} x(s)ds.$$

Show that if x is the sinusoidal signal

$$\forall\, t \in Reals, \quad x(t) = \sin(\omega t),$$

then y is also sinusoidal:

$$\forall\, t \in Reals, \quad y(t) = A \sin(\omega t + \phi).$$

You do not need to give precise expressions for A and ϕ; just show that the result has this form. Also, show that as ω gets large, the amplitude of the output gets small. Higher frequencies, which represent more abrupt changes in the input, are attenuated to a greater degree by the system than are lower frequencies.

Hint: The following fact from calculus may be useful:

$$\int_a^b \sin(\omega s)ds = \frac{1}{\omega}(\cos(\omega a) - \cos(\omega b)).$$

Also, the identity in the footnote on page 67 might be useful for showing that the output is sinusoidal with the appropriate frequency.

E 13. Suppose that $f\colon Reals \to Reals$ and $g\colon Reals \to Integers$ such that for all $x \in Reals$,

$$g(x) = \begin{cases} 1 & \text{if } x > 0 \\ 0 & \text{if } x = 0 \\ -1 & \text{if } x < 0 \end{cases}$$

and

$$f(x) = 1 + x.$$

(a) Define $h = g \circ f$.

(b) Suppose that

$$F\colon [Reals \to Reals] \to [Reals \to Reals]$$

and

$$G\colon [Reals \to Reals] \to [Reals \to Integers]$$

such that for all $s \in [Reals \to Reals]$ and $x \in Reals$,

$$(F(s))(x) = f(s(x)),$$

and

$$(G(s))(x) = g(s(x)),$$

where f and g are as just given. Sketch a block diagram for $H = G \circ F$, where you have one block for each of G and F. Label the inputs and outputs of your blocks with the domain and range of the functions in the blocks.

(c) Let $s \in [Reals \to Reals]$ be such that for all $x \in Reals$,

$$s(x) = \cos(\pi x).$$

Define u by

$$u = (G \circ F)(s).$$

T 14. Let $D = DiscSignals = [Integers \to Reals]$ and let

$$G: D \times D \to D$$

such that for all $x, y \in D$ and for all $n \in Integers$,

$$(G(x, y))(n) = x(n) - y(n-1).$$

Now suppose we construct a new system H as follows:

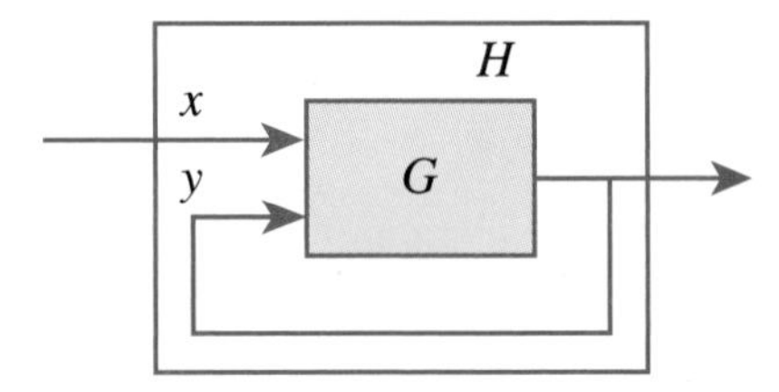

Define H (as much as you can).

T 15. Consider a system H similar to that in the previous problem,

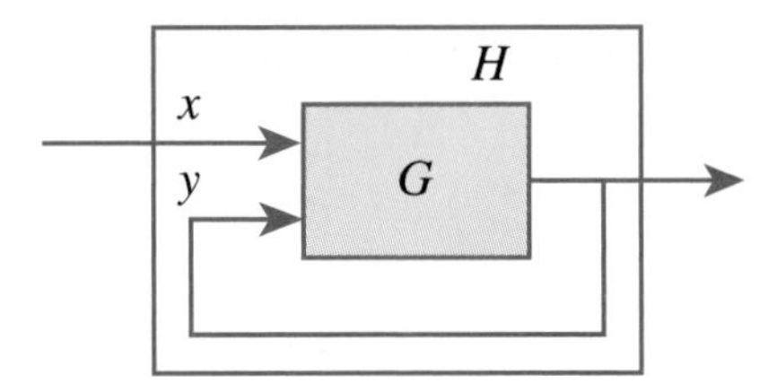

but where now $x \in Reals_+$ and $y \in Reals_+$. The inputs and outputs are no longer signals but rather just nonnegative real numbers. Therefore, $G: Reals_+ \times Reals_+ \to Reals_+$. Block diagrams, of course, work just as well for such simpler functions.

In this problem, we explore fixed points by considering a classic algorithm for calculating the square root of a nonnegative real number. Let the aforementioned function G be given by

$$\forall\, x, y \in Reals_+, \quad G(x, y) = 0.5(y + x/y). \tag{2.19}$$

(a) Show that for a given $x \in Reals_+$, if the fixed point exists, then $H: Reals_+ \to Reals_+$ is given by

$$\forall x \in Reals_+, \quad H(x) = \sqrt{x}.$$

(b) To use this system to calculate a square root, we simply start with a guess for y—say, 1—and calculate $G(x, y)$ repeatedly until it converges to a stable value for y. That stable value is the fixed point. Do this calculation for $x = 4$ and for $x = 12$, repeating the evaluation of G until you obtain a close approximation to the true square root. You may want to use MATLAB for this.

INTERVIEW
Panos Antsaklis

Panos Antsaklis is the Brosey Professor of Electrical Engineering and the Director of the Center for Applied Mathematics at the University of Notre Dame. His research addresses problems of control and automation and examines ways to design engineering systems of high autonomy, focusing on the behavior of networked embedded systems and on hybrid and discrete event dynamical systems. He is a recipient of the Kaneb Award for Excellence in Teaching at the University of Notre Dame, and he has served as President of the IEEE Control Systems Society as an IEEE Fellow. He also enjoys history and learning from its lessons.

How did you decide to study electrical engineering?

I have been interested in how things work since an early age! I disassembled and assembled everything I was allowed to get my hands on—bicycles, clocks, electric appliances—to the amusement of my father and to the worry of my mother, who was afraid that I may not be able to put them back together on time or at all. I guess the reason behind all this was not only to learn how and why things worked, but also to try to make them work better or even use ideas to design new devices!

What was your first job in the industry?

My first job was as an undergraduate spending a summer in Bern, Switzerland working at a small company that made small electric appliances. The next summer I worked at a research center in Greece programming control algorithms. I found the research work in controls much more exciting and challenging. In fact I did my graduate work in the systems and control area.

Which person in the field has inspired you most? In what ways?

I certainly admire the contributions, the dedication, and the insight of several engineers and scientists. Here I will mention just two names from long ago. The first one is Ktesibios of Alexandria (the third century BC), an engineer and the inventor of the ancient water clock, the first feedback control mechanism on record. The second one is James C. Maxwell. In the 1860s he captured the basic laws of electricity in mathematical form in his famous Maxwell equations, and also predicted the existence of radio waves. His contributions signaled the beginning of a new era where mathematics and experimental sciences work side by side and depend on one another for truly remarkable advances.

Do you have any advice for students studying electrical and computer engineering?

As an engineer you have to have your feet firmly on the ground, but at the same time your eyes should be on the horizon. Use your scientific and mathematical expertise to make sure that you are correct, but let your imagination ride on your intuition to make the next great leap forward. Great innovations do not come along very often, so make sure you are well versed in the fundamentals so that you are prepared to recognize them.

What is your vision for the future of electrical and computer engineering?

The area of electrical and computer engineering has been the driving force of a great part of our technological civilization today, and I expect this to continue for many years. Methodologies that have emerged in our discipline have played a leading role in many of the advances in a great variety of areas—from the medical field and instrumentation, to finance and analysis tools, to lasers, to MEMS and nanotechnology, from communication networks to car electronics, to aircraft avionics, and to automatic controls. It is truly remarkable how successful our discipline has been. I can only see more amazing contributions in the future. It cannot be any other way!

Are there any special projects you are currently working on that you'd like to tell students about?

Imagine a cluster of mini-satellites that communicate with each other to coordinate their actions; or a cluster of MEMS; or a million individual segments in the lens of a space telescope that periodically need to be adjusted individually to compensate for temperature changes, aging, or failures.

These are examples of systems that consist of many subsystems distributed over space, which interact with each other via communication channels that may be wireless. Such a system of systems can change dynamically with time as units come in and out of the network and typically interact with the real world and so they must meet hard and soft time constraints, as they are real time systems. Each individual unit is an information processor, a computer, and a node in a network. We have been interested in designing such systems using concepts and methodologies from the areas of hybrid systems and controls, communication networks, and computer science.

CHAPTER 3
State machines

Systems are functions that transform signals. Both the domain and the range of these functions are signal spaces; this significantly complicates specification of the functions. A broad class of systems can be characterized according to the concept of **state** and the idea that a system evolves through a sequence of changes in state, or **state transitions**. Such characterizations are called **state-space models**.

A state-space model describes a system procedurally, giving a sequence of step-by-step operations for the evolution of a system. It shows how the input signal drives changes in state and how the output signal is produced. It is thus an imperative description. Implementing a system described by a state-space model in software or hardware is straightforward. The hardware or software simply needs to sequentially perform the steps given by the model. Conversely, it is often useful to describe a piece of software or hardware as a system using a state-space model, which yields better to analysis than do more informal descriptions.

In this chapter, we introduce state-space models by discussing systems with a finite (and relatively small) number of states. Such systems typically operate on event streams, often implementing control logic. For example, the decision logic of modem negotiation described in chapter 1 can be described by a finite-state model. Such a model is much more precise than the English-language descriptions that are commonly used for such systems.

3.1 Structure of state machines

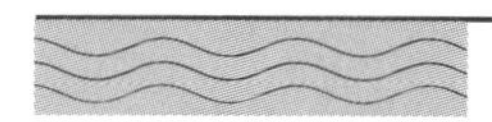

A description of a system as a function involves three entities: the set of input signals, the set of output signals, and the function itself, $F\colon InputSignals \to OutputSignals$. For a **state machine**, the input and output signals have the form

$$EventStream\colon Naturals_0 \to Symbols,$$

where $Naturals_0 = \{0, 1, 2, \ldots\}$, and *Symbols* is an arbitrary set. The domain of these signals represents *ordering* but not necessarily time (neither discrete nor continuous time). The ordering of the domain means that we can say that one event occurs *before* or *after* another event. But we cannot say how much time elapses between these events. In chapter 5 we study how state-space models can be used with functions of time.

A state machine constructs the output signal one symbol at a time by observing the input signal one symbol at a time. Specifically, a state machine *StateMachine* is a five-tuple,

$$StateMachine = (States, Inputs, Outputs, update, initialState), \qquad (3.1)$$

where *States*, *Inputs*, and *Outputs* are sets; *update* is a function; and $initialState \in States$. The meanings of these names are as follows:

States is the **state space**.
Inputs is the **input alphabet**.
Outputs is the **output alphabet**.
$initialState \in States$ is the **initial state**.
$update\colon States \times Inputs \to States \times Outputs$ is the **update function**.

This five-tuple is called the **sets and functions model** of a state machine.

Inputs and *Outputs* are the sets of possible input and output symbols. The set of **input signals** consists of all infinite sequences of input symbols,

$$InputSignals = [Naturals_0 \to Inputs].$$

The set of **output signals** consists of all infinite sequences of output symbols,

$$OutputSignals = [Naturals_0 \to Outputs].$$

Let $x \in InputSignals$ be an input signal. A particular symbol in the signal can be written $x(n)$ for any $n \in Naturals_0$. We write the entire input signal as a sequence

$$(x(0), x(1), \ldots, x(n), \ldots).$$

This sequence defines the function x in terms of symbols $x(n) \in Inputs$, which represent particular input symbols.

We reiterate that the index n in $x(n)$ refers not to time but rather to the **step number**. This is only an **ordering constraint**: Step n occurs after step $n-1$ and before step $n+1$. The state machine evolves (i.e., moves from one state to the next) in **steps**.*

3.1.1 *Updates*

The interpretation of *update* is this: If $s(n) \in States$ is the current state at step n, and $x(n) \in Inputs$ is the current input symbol, then the current output symbol $y(n)$ and the next state $s(n+1)$ are given by

$$(s(n+1), y(n)) = update(s(n), x(n)).$$

Thus, the *update* function makes it possible for the state machine to construct the output signal step by step by observing the input signal step by step.

The state machine *StateMachine* of (3.1) defines a function

$$F\colon InputSignals \to OutputSignals \tag{3.2}$$

such that for any input signal $x \in InputSignals$, the corresponding output signal is $y = F(x)$. However, it does much more than just define this function. It also gives us a procedure for evaluating this function on a particular input signal. The **state response** $(s(0), s(1), \ldots)$ and output signal y are constructed as follows:

$$s(0) = initialState \tag{3.3}$$

and

$$\forall\, n \geq 0,\ \ (s(n+1), y(n)) = update(s(n), x(n)). \tag{3.4}$$

Observe that if the initial state is changed, the function F changes, and so the initial state is an essential part of the definition of a state machine.

Each evaluation of (3.4) is called a **reaction** because it defines how the state machine reacts to a particular input symbol. Note that exactly one output symbol is produced for each input symbol. Thus, it is not necessary to have access to the entire input sequence in order to start producing output symbols. This feature proves extremely useful in practice, because it is usually impractical to have access to the entire input sequence (it is infinite in size!). The procedure

*Of course, the steps could last a fixed duration of time, in which case there would be a simple relationship between step number and time. The relationship may be a mixed one, in which some input symbols are separated by a fixed amount of time and some are not.

summarized by (3.3) and (3.4) is **causal**, in that the next state $s(n+1)$ and current output symbol $y(n)$ depend only on the initial state $s(0)$ and on current and past input symbols $x(0), x(1), \ldots, x(n)$.

It is sometimes convenient to decompose *update* into two functions:

nextState: *States* × *Inputs* → *States* is the **next state function**

and

output: *States* × *Inputs* → *Outputs* is the **output function**.

The interpretation is this: If $s(n)$ is the current state, and $x(n)$ is the current input symbol at step n, the next state is

$$s(n+1) = nextState(s(n), x(n)),$$

and the current output symbol is

$$y(n) = output(s(n), x(n)).$$

Evidently, for all $s(n) \in States$ and $x(n) \in Inputs$,

$$\begin{aligned}(s(n+1), y(n)) &= update(s(n), x(n)) \\ &= (nextState(s(n), x(n)), output(s(n), x(n))).\end{aligned}$$

3.1.2 *Stuttering*

A state machine produces exactly one output symbol for each input symbol. For each input symbol, it may also change state (of course, it could also remain in the same state by changing back to the same state). This means that with no input symbol, there is neither an output symbol nor a change of state.

Later, when we compose simpler state machines to construct more complicated ones, it will prove convenient to be explicit in the model about the fact that a lack of input triggers no output and no state change. We do that by insisting that the input and output symbol sets include a **stuttering symbol**, typically denoted *absent*; that is,

$$absent \in Inputs, \text{ and } absent \in Outputs.$$

Moreover, we require that for any $s \in States$,

$$update(s, absent) = (s, absent). \tag{3.5}$$

This is called a **stuttering reaction** because no progress is made. An absent input symbol triggers an absent output symbol and no state change. Now any number of *absent* symbols may be inserted anywhere into the input sequence without changing the nonabsent output symbols. Stuttering reactions prove essential for hybrid systems models, considered in chapter 6.

Example 3.1: Consider a 60-minute parking meter. There are three (nonstuttering) input symbols: *in5* and *in25*, which represent feeding the meter 5 and 25 cents, respectively; and *tick*, which represents the passage of one minute. The meter displays the time in minutes remaining before the meter expires. When *in5* occurs, this time is increased by 5, and when *in25* occurs, increased by 25, up to a maximum of 60 minutes. When *tick* occurs, the time is decreased by 1, down to a minimum of 0. When the remaining time is 0, the display reads *expired*.

We can construct a finite-state machine model for this parking meter. The set of states is

$$States = \{0, 1, 2, \ldots, 60\}.$$

The input and output alphabets are

$$Inputs = \{in5, in25, tick, absent\}$$

and

$$Outputs = \{expired, 1, 2, \ldots, 60, absent\}.$$

The initial state is

$$initialState = 0.$$

The update function

$$update: States \times Inputs \to States \times Outputs$$

is given by $\forall\, s(n) \in States,\ x(n) \in Inputs$,

$$update(s(n), x(n)) =$$

$$\begin{cases} (0, expired) & \text{if } x(n) = tick \wedge (s(n) = 0 \vee s(n) = 1) \\ (s(n) - 1, s(n) - 1) & x(n) = tick \wedge s(n) > 1 \\ (\min(s(n) + 5, 60), \min(s(n) + 5, 60)) & x(n) = in5 \\ (\min(s(n) + 25, 60), \min(s(n) + 25, 60)) & x(n) = in25 \\ (s(n), absent) & x(n) = absent, \end{cases}$$

where min is a function that returns the minimum of its arguments.

If the input sequence is $(in25, tick^{20}, in5, tick^{10}, \ldots)$,* for example, then the output sequence is

$$(expired, 25, 24, \ldots, 6, 5, 10, 9, 8, \ldots, 2, 1, expired, \ldots).\ \square$$

3.2 Finite-state machines

Often, *States* is a finite set. In this case, the state machine is called a **finite-state machine** (**FSM**). FSMs yield to powerful analytical techniques because, in principle, it is possible to explore all possible sequences of states. The parking meter just described is a finite-state machine. The remainder of this chapter and all of the next chapter focus on finite-state machines. The discussion returns to infinite-state systems in chapter 5. In lab 3 (see Lab Manual) software implementation of finite-state machines is considered.

When the number of states is small, and when the input and output alphabets are finite (and small), we can describe the state machine by using a very readable and intuitive diagram called a **state transition diagram**.

Example 3.2: A verbal description of an automatic telephone answering machine might go like this:

> When a call arrives, the phone rings. If the phone is not picked up by the third ring, the machine answers. It plays a prerecorded greeting requesting that the caller leave a message ("Hello; sorry I can't answer your call right now. . . . Please leave a message after the beep"), records the caller's message, and then automatically hangs up. If the phone is answered before the third ring, the machine does nothing.

Figure 3.1 shows a state transition diagram for the state machine model of this answering machine. You can probably read the diagram in figure 3.1 without any further explanation. It is sufficiently intuitive. However, we will explain it precisely. ❐

3.2.1 State transition diagrams

Figure 3.1 consists of bubbles linked by arcs. (The arcs are also called arrows.) In this bubbles-and-arcs syntax, each bubble represents one state of the answering machine, and each arc represents a **transition** from one state to another. The bubbles and arcs are annotated; that is, they are labeled with some text. The execution of the state machine consists of a sequence reactions, in which each reaction involves a transition from one state to another (or back to the same state)

* We are using the common notation $tick^{10}$ to mean a sequence of 10 consecutive *ticks*.

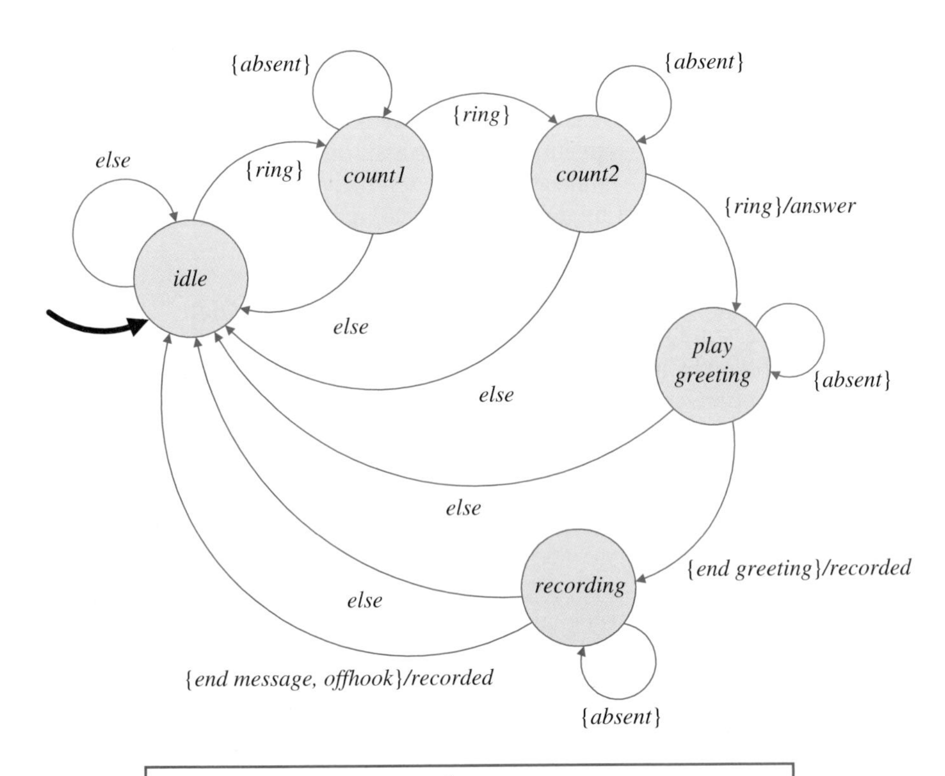

States
idle — nothing is happening *count1* — one ring has arrived *count2* — two rings have arrived *play greeting* — playing the greeting message *recording* — recording the message

Inputs
ring — incoming ringing signal *offhook* — a telephone extension is picked up *end greeting* — greeting message is finished playing *end message* — end of message detected (e.g., dialtone) *absent* — no input of interest

Outputs
answer — answer the phone and start the greeting message *record* — start recording the incoming message *recorded* — recorded an incoming message *absent* — default output when there is nothing interesting to say

FIGURE 3.1: State transition diagram for the telephone answering machine.

along one of the arcs. The tables at the bottom of the figure are not part of the state transition diagram, but they improve our understanding of the diagram by giving the meanings of the names of the states, input symbols, and output symbols.

The notation for state transition diagrams is summarized in figure 3.2. Each bubble is labeled with the name of the **state** it represents. The state names can be anything, but they must be distinct. The state machine of figure 3.1 has five states. The state names define the state space:

$$States = \{idle, count1, count2, play\ greeting, recording\}.$$

Each arc is labeled by a **guard** and (optionally) an **output**. If an output symbol is given, it is separated from the guard by a forward slash, as in the example $\{ring\}/answer$ going from state *count2* to *play greeting*. A guard specifies which

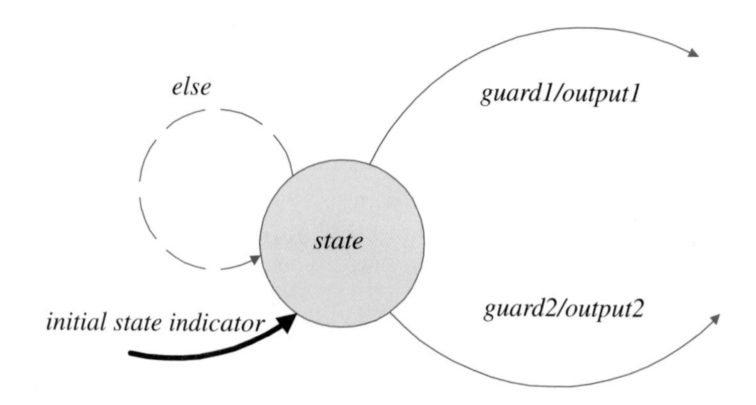

State machine
$(States, Inputs, Outputs, update, initialState)$
$update: States \times Inputs \to State \times Outputs$
$initialState \in States$

Elements
$State \in States$
$output1, output2 \in Outputs$
$guard1, guard2 \subset Inputs$
$else = \{i \in Inputs \mid i \notin (guard1 \cup guard2)\}$

Determinacy
(There is at most one possible reaction to an input symbol)
$guard1 \cap guard2 = \emptyset$

FIGURE 3.2: Summary of notation in state transition diagrams, shown for a single state with two outgoing arcs and one self loop.

input symbols may trigger the associated transition. It is a subset of the *Inputs*, the input alphabet, which, for the answering machine, is

$$Inputs = \{ring, offhook, end\ greeting, end\ message, absent\}.$$

In figure 3.1, some guards are labeled "*else*." This special notation designates an arc that is taken when there is no match on any other guard emerging from a given state. The arc with the guard *else* is called the **else arc**. Thus, *else* is the set of all input symbols not included in any other guard emerging from the state. More precisely, for a given state, *else* is the complement with respect to *Inputs* of the union of the guards on emerging arcs. For example, in figure 3.1, for the state *recording*,

$$else = \{ring, offhook, end\ greeting\}.$$

In the example in figure 3.2, *else* is defined as

$$else = \{i \in Inputs \mid i \notin (guard1 \cup guard2)\}.$$

If no *else* arc is specified and the set *else* is not empty, then the *else* arc is implicitly a **self-loop**, as shown by the dashed arc in figure 3.2. A self-loop is an arc that transitions back to the same state. When the else arc is a self-loop, then the stuttering symbol may be a member of the set *else*.

Initially, the system of figure 3.1 is in the *idle* state. The **initial state** is indicated by the bold arc on the left that leads into the state *idle*. Each time an input symbol arrives, the state machine reacts. It checks the guards on arcs going out of the current state and determines which of them contains the input symbol. It then chooses that transition.

Two problems may occur:

1. The input symbol may not be contained in the guard of any outgoing arc. In our state machine models, for every state, there is at least one outgoing transition that matches the input symbol (because of the *else* arc). This property is called **receptiveness**; it means that the machine can always react to an input symbol. In other words, there is always a transition out of the current state that is enabled by the current input symbol. (The transition may lead back to the current state if it is a self-loop.) Our state machines are said to be **receptive**.
2. More than one guard going out from the current state may contain the input symbol. A state machine that has such a structure is said to be **nondeterministic**. The machine is free to choose any arc whose guard contains the input symbol, and so more than one behavior is possible for the machine. Nondeterministic state machines are discussed later in this chapter. Until then, we assume that the guards are always defined to give deterministic state machines. Specifically, the guards on outgoing arcs from any state are mutually

exclusive. In other words, the intersection of any two guards on outgoing arcs of a state is empty, as indicated in figure 3.2. Of course, by the definition of the *else* set, for any *guard* that is not *else*, it is true that $guard \cap else = \emptyset$.

A sequence of input symbols thus triggers a sequence of state transitions. The resulting sequence of states is called the **state response**.

Example 3.3: In figure 3.1, if the input sequence is

$$(ring, ring, offhook, \ldots),$$

then the state response is

$$(idle, count1, count2, idle, \ldots).$$

The ellipses are there because the answering machine generally responds to an infinite input sequence, and we are showing only the beginning of that response. This behavior can be compactly represented by a **trace**,

$$idle \xrightarrow{ring} count1 \xrightarrow{ring} count2 \xrightarrow{offhook} idle \ldots.$$

A trace represents the state response together with the input sequence that triggers it. This trace describes the behavior of the answering machine when someone picks up a telephone extension after two rings.*

A more elaborate trace illustrates the behavior of the answering machine when it takes a message:

$$idle \xrightarrow{ring} count1 \xrightarrow{ring} count2 \xrightarrow{ring} play\ greeting \xrightarrow{end\ greeting} recording \xrightarrow{end\ message} idle \ldots. \tag{3.6}$$

❒

A state machine also produces outputs. In figure 3.1, the output alphabet is

$$Outputs = \{answer, record, recorded, absent\}.$$

* When you lift the handset of a telephone to answer, your phone sends a signal called "offhook" to the telephone switch. The reason for the name "offhook" is that in the earliest telephone designs, the handset hung from a hook on the side of the phone. In order to answer, you had to pick the handset off the hook. When you finished your conversation, you replaced the handset on the hook, generating an "onhook" signal. The onhook signal is irrelevant to the answering machine, so it is not included in the model.

An output symbol is produced as part of a reaction. The output symbol that is produced is indicated after a slash on an arc. If the arc annotation shows no output symbol, then the output symbol is *absent*.

Example 3.4: The output sequence for the trace (3.6) is

$$(absent, absent, answer, record, recorded, \ldots).$$

There is an output symbol for every input symbol, and some of the output symbols are *absent*. ❒

It should be clear how to obtain the state response and output sequence for any input sequence. We begin in the initial state and then follow the state transition diagram to determine the successive state transitions for successive input symbols. Knowing the sequence of transitions, we also know the sequence of output symbols.

Shorthand

State transition diagrams can get verbose, containing many arcs with complicated labels. A number of shorthand options can make a diagram clearer by reducing the clutter.

- If no guard is specified on an arc, then that transition is always taken when the state machine reacts and is in the state from which arc emerges, as long as the input is not the stuttering symbol. In other words, giving no guard is equivalent to giving the entire set *Inputs* as a guard, minus the stuttering symbol. Recall that the stuttering symbol always triggers a transition back to the same state and always produces a stuttering symbol on the output.
- Any clear notation for specifying subsets can be used to specify guards. For example, if $Inputs = \{a, b, c\}$, then the guard $\{b, c\}$ can be given by $\neg a$ (read "not a").
- An *else* transition for a state need not be given explicitly. It is an implied self-loop if it is not given. This is why it is shown with a dashed line in figure 3.2.
- The output symbol is the stuttering symbol of *Outputs* if it is not given.

These shorthand notations are not always a good idea. For example, the *else* transitions often correspond to exceptional (unexpected) input sequences, and staying in the same state might not be the correct behavior. For instance, in figure 3.1, all *else* transitions are shown explicitly, and all exceptional input sequences result in the machines ending up in the state *idle*. This is probably reasonable behavior, allowing the machine to recover. Had we left the *else* transitions implicit, we would probably have ended up with less reasonable behavior. Use your judgment in deciding whether to explicitly include *else* transitions.

3.2.2 Update table

An alternative way to describe a finite state machine is by an **update table**. This is simply a tabular representation of the state transition diagram.

For the diagram of figure 3.1, the table is shown in figure 3.3. The first column lists the current state. The remaining columns list the next state and the output symbol for each of the possible input symbols.

The first row, for example, corresponds to the current state *idle*. If the input symbol is *ring*, the next state is *count1* and the output symbol is *absent*. Under any of the other input symbols, the state remains *idle* and the output symbol remains *absent*.

Current state	(*next state, output symbol*) Under specified input symbol				
	ring	*offhook*	*end greeting*	*end message*	*absent*
idle	(*count1*, *absent*)	(*idle*, *absent*)	(*idle*, *absent*)	(*idle*, *absent*)	(*idle*, *absent*)
count1	(*count2*, *absent*)	(*idle*, *absent*)	(*idle*, *absent*)	(*idle*, *absent*)	(*count1*, *absent*)
count2	(*play greeting*, *answer*)	(*idle*, *absent*)	(*idle*, *absent*)	(*idle*, *absent*)	(*count2*, *absent*)
play greeting	(*idle*, *absent*)	(*idle*, *absent*)	(*recording*, *record*)	(*idle*, *absent*)	(*play greeting*, *absent*)
recording	(*idle*, *absent*)	(*idle*, *recorded*)	(*idle*, *absent*)	(*idle*, *recorded*)	(*recording*, *absent*)

FIGURE 3.3: Update table for the telephone answering machine specifies next state and current output symbol as a function of current state and current input symbol.

Types of state machines

The type of state machines introduced in this section are known as **Mealy machines**, after G. H. Mealy, who studied them in 1955. Their distinguishing feature is that output symbols are associated with state transitions; that is, when a transition is taken, an output symbol is produced. Alternatively, we could have associated output symbols with states, which would result in a model known as **Moore machines**, after F. Moore, who studied them in 1956. In a Moore machine, an output symbol is produced while the machine is in a particular state. Mealy machines turn out to be more useful when they are composed synchronously, as is done in the next chapter. This is why we choose this variant of the model.

It is important to realize that state machine models, like most models, are not unique. A great deal of engineering judgment goes into a picture like figure 3.1, and two engineers might devise very different pictures for what they believe to be the same system. Often, the differences are in the amount of detail shown. One picture may show the operation of a system in more detail than another. The less detailed picture is called an **abstraction** of the more detailed picture. Also likely are differences in the names chosen for states, input symbols, and output symbols and differences even in the meaning of the input and output symbols. There may be differences in how the machine responds to exceptional circumstances (input sequences that are not expected). For example, what should the answering machine do if it gets the input sequence (*ring*, *end greeting*, *end message*)? This probably reflects a malfunction in the system. In figure 3.1, the reaction to this sequence is easy to see: The machine ends up in the *idle* state.

In view of these likely differences, it becomes important to be able to talk about **abstraction relations** and **equivalence relations** between state machine models. This turns out to be a fairly sophisticated topic, one that we touch upon later in section 3.3.

The meaning of state

We have three equivalent ways of describing a state machine: the sets and functions description, the state transition diagram, and the update table. These descriptions have complementary uses. The table makes obvious the sparsity of output symbols in the answering machine example. Both the table and the diagrams are useful for a human studying the system to follow its behavior in different circumstances. The sets and functions description and the table are useful for building the state machine in hardware or software. The sets and functions description is also useful for mathematical analysis.

Of course, the tables and the transition diagram can be used only if there are finitely many states and finitely many input and output symbols—that is, if the sets *States*, *Inputs*, and *Outputs* are finite. The sets and functions description is often equally comfortable with finite- and infinite-state spaces. We discuss infinite-state systems in chapter 5.

Like any state machine, the telephone answering machine is a **state-determined** system. Once we know its current state, we can tell what its future behavior is for any future input symbols. We do not need to know what input symbols in the past led to the current state in order to predict how the system will behave in the future. In this sense we can say the current state of the system summarizes the history of the system. This is, in fact, the key intuitive notion of state.

The number of states equals the number of patterns that we need for summarizing the history. If this is intrinsically finite, then a finite-state model exists for the system. If it is intrinsically infinite, then no finite-state model exists. We can often determine which of these two situations applies by using simple intuition. We can also show that a system has a finite-state model by finding one. Showing that a system does not have a finite-state model is a bit more challenging.

Example 3.5: Consider the example of a system called *CodeRecognizer* whose input and output signals are sequences of 0 and 1 (with arbitrarily inserted stuttering symbols, which have no effect). The system outputs *recognize* at the end of every subsequence 1100 in the input; otherwise, it outputs *absent*. If the input x is given by a sequence

$$(x(0), x(1), \ldots)$$

and the output y is given by the sequence

$$(y(0), y(1), \ldots),$$

then, if none of the input symbols is *absent*,

$$y(n) = \begin{cases} \textit{recognize} & \text{if } (x(n-3), x(n-2), x(n-1), x(n)) = (1, 1, 0, 0) \\ \textit{absent} & \text{otherwise.} \end{cases} \tag{3.7}$$

Intuitively, in order to determine $y(n)$, it is enough to know whether the previous pattern of (nonabsent) inputs is 0, 1, 11, or 110. If this intuition is correct, we can implement *CodeRecognizer* by a state machine with four states that remember the patterns 0, 1, 11, 110. The machine of figure 3.4 does the job. The fact that we have a finite-state machine model of this system shows that this is a finite-state system. ❒

The relationship in this example between the number of states and the number of input patterns that need to be stored suggests how to construct functions mapping input sequences to output sequences that cannot be realized by finite-state machines. Here is a particularly simple example of such a function called *Equal*.

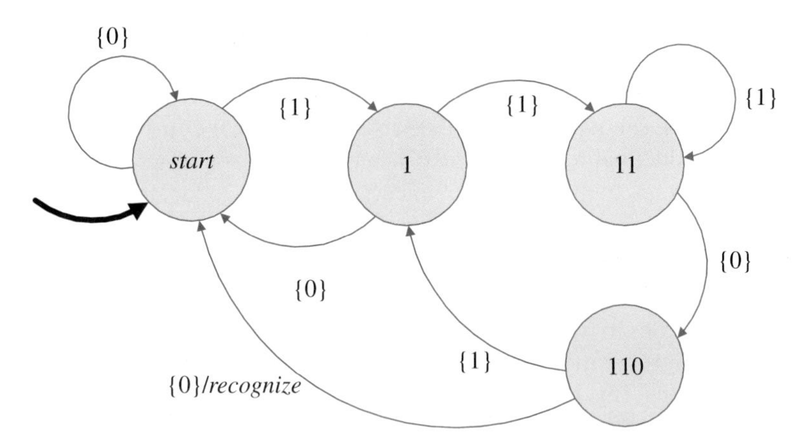

FIGURE 3.4: A machine that implements *CodeRecognizer*. It outputs *recognize* at the end of every input subsequence 1100; otherwise it outputs *absent*.

Example 3.6: An input signal of *Equal* is a sequence of 0 and 1 (again with stuttering symbols arbitrarily inserted). At each step, *Equal* outputs *equal* if the previous inputs contain an equal number of 0s and 1s; otherwise, *Equal* outputs *notEqual*. In other words, if the input sequence x is the sequence $(x(0), x(1), \dots)$, with no stuttering symbols, then the output sequence $y = F(x)$ is given by

$$\forall\, n \in Naturals_0, \quad y(n) = \begin{cases} equal & \text{if number of 1s is the same as 0s} \\ & \quad \text{in } x(0), \dots, x(n) \\ notEqual & \text{otherwise.} \end{cases} \tag{3.8}$$

Intuitively, in order to realize *Equal*, the machine must remember the difference between the number of 1s and 0s that have occurred in the past. Because these numbers can be arbitrarily large, the machine must have infinite memory, and so *Equal* cannot be realized by a finite-state machine.

We give a mathematical argument to show that *Equal* cannot be realized by any finite-state machine. The argument uses contradiction.

Suppose that a machine with N states realizes *Equal*. Consider an input sequence that begins with N 1s: $(1, \dots, 1, x(N), \dots)$. Let the state response be

$$(s(0), s(1), \dots, s(N), \dots).$$

Because there are only N distinct states, and because the state response has a length of at least $N + 1$, the state response must visit at least one state twice. Call that state α. Suppose $s(m) = s(n) = \alpha$, with $m < n \le N$. Then the two sequences $1^m 0^m$ and $1^n 0^m$ must lead to the same state and, hence, yield the same last output symbol α.* But the last output symbol for 1^m0^m should be *equal*, and the last output symbol for 1^n0^m should be *notEqual*, which is a contradiction. Therefore, our hypothesis that a finite-state machine realizes *Equal* must be wrong! Exercise 6 asks you to construct an infinite-state machine that realizes *Equal*. ❒

3.3 Nondeterministic state machines

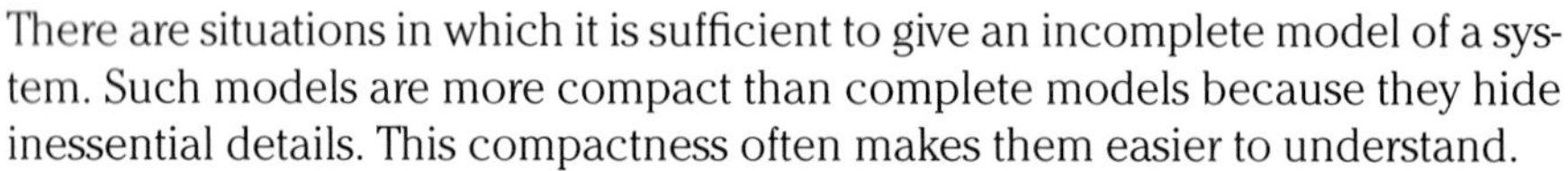

There are situations in which it is sufficient to give an incomplete model of a system. Such models are more compact than complete models because they hide inessential details. This compactness often makes them easier to understand.

A useful form of incomplete model is a **nondeterministic state machine**. Many nondeterministic state machines have fewer states and transitions than would be required by a complete model. The state machines we have studied so far are **deterministic**.

3.3.1 *State transition diagram*

The state transition diagram for a state machine contains one bubble for each state and one arc for each state transition. Nondeterministic machines are no different. Each arc is labeled "*guard/output*," where

$$guard \subset Inputs, \text{ and } output \in Outputs.$$

In a deterministic machine, the guards on arcs emerging from any given state are mutually exclusive; that is, they have no common symbols. This is precisely what makes the machine deterministic. For nondeterministic machines, this constraint is relaxed: Guards can overlap. Thus, a given input symbol may appear in the guard of more than one transition, which means that one of several transitions can be taken when that input symbol arrives. This is precisely what makes the machine nondeterministic.

Example 3.7: Consider the state machine shown in figure 3.5. It begins in state a and transitions to state b the first time it encounters a 1 on the input. It then stays in state b arbitrarily long. If it receives a 1 at the input, it must stay in state b. If it receives a 0, then it can either stay in b or transition to a. With the input sequence

$$(0, 1, 0, 1, 0, 1, \ldots),$$

* Recall that 1^m means a sequence of m consecutive 1s, similarly for 0^m.

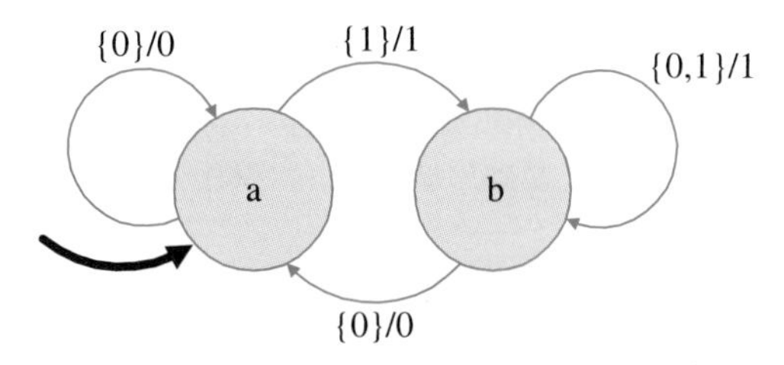

FIGURE 3.5: A simple nondeterministic state machine.

the following are all possible state responses and output sequences:

$$(a, a, b, a, b, a, b, \ldots)$$
$$(0, 1, 0, 1, 0, 1, \ldots)$$

$$(a, a, b, b, b, a, b, \ldots)$$
$$(0, 1, 1, 1, 0, 1, \ldots)$$

$$(a, a, b, b, b, b, b, \ldots)$$
$$(0, 1, 1, 1, 1, 1, \ldots)$$

$$(a, a, b, a, b, b, b, \ldots)$$
$$(0, 1, 0, 1, 1, 1, \ldots). \square$$

Nondeterminism can be used to construct an **abstraction** of a complicated machine; an abstraction is a simpler machine that has all the behaviors of the more complicated machine.

Example 3.8: Consider again the 60-minute parking meter. Its input alphabet is

$$Inputs = \{coin5, coin25, tick, absent\}.$$

Upon arrival of *coin5*, the parking meter increases its count by five, up to a maximum of 60 minutes. Upon arrival of *coin25*, it increases its count by 25, again up to a maximum of 60. Upon arrival of *tick*, it decreases its count by one, down to a minimum of zero.

A deterministic state machine model is illustrated schematically in figure 3.6(a). The state space is

$$States = \{0, 1, \ldots, 60\},$$

which contains too many states to draw conveniently. Patterns in the state space are therefore suggested with the ellipsis.

Suppose that we are interested in modeling the interaction between this parking meter and a police officer. The police officer does not care what state the parking meter is in, except to determine whether the meter has expired. Thus, we need only two nonstuttering output symbols:

$$Outputs = \{safe, expired, absent\}.$$

The symbol *expired* is produced whenever the machine enters state 0.

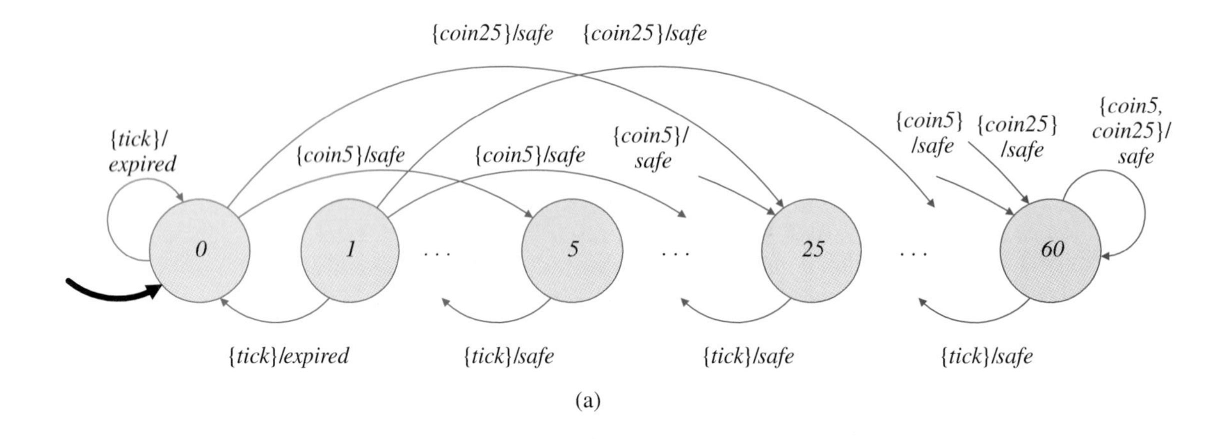

(a)

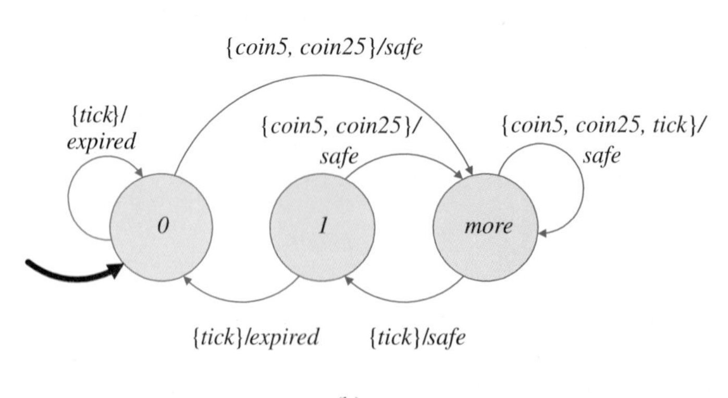

(b)

FIGURE 3.6: Deterministic and nondeterministic models for a 60-minute parking meter.

If the model has enough states that a full state transition diagram is tedious and complex, it might not be useful for generating insight about the design. Moreover, the detail that is modeled may not add insight about the interaction with a police officer.

Figure 3.6(b) is a nondeterministic model of the same parking meter. It has three states:

$$States = \{0, 1, more\}.$$

The input symbols *coin5* and *coin25* in state 0 or 1 cause a transition to state *more*. The input symbol *tick* in state *more* nondeterministically moves the state to 1 or leaves it in *more*.

The top state machine (a) has more detail than the bottom machine (b). Later, we give a precise meaning to the phrase "has more detail" by using the concept of *simulation*. For the moment, note that the bottom machine can generate any output sequence that the top machine generates, for the same input sequence. But the bottom machine can also generate output sequences that the top machine cannot. For example, the sequence

$$(expired, safe, safe, expired, \ldots),$$

in which there are two *safe* output symbols between two *expired* output symbols is not a possible output sequence of the top machine, but it is a possible output sequence of the bottom machine. In the top machine, successive *expired* output symbols must be separated by 0 or at least five *safe* output symbols. This detail is not captured by the bottom machine. But in modeling the interaction with a police officer, this detail may not be important, and so omitting it may be entirely appropriate. ❐

The machines that are designed and built, including parking meters, are usually deterministic. However, the state spaces of many of these machines are very large, much larger than in this example, and it can be difficult to understand their behavior. We use simpler nondeterministic machine models that hide inessential details of the deterministic machine. The analysis of the simpler model reveals some but not all properties of the more complex machine. The art, of course, is in choosing the model that reveals the properties of interest.

3.3.2 *Sets and functions model*

The state machines we have been studying, with definitions of the form (3.1), are deterministic. If we know the initial state and the input sequence, then the entire state trajectory and output sequence can be determined. This is because any current state $s(n)$ and current input symbol $x(n)$ uniquely determine the next state and output symbol: $(s(n+1), y(n)) = update(s(n), x(n))$.

In a nondeterministic state machine, the next state is not completely determined by the current state and input symbol. For a given current state $s(n)$ and input symbol $x(n)$, there may be more than one next state. So we cannot characterize the machine by the function $update(s(n), x(n))$ because there is no single next state. Instead, we define a function *possibleUpdates* so that $possibleUpdates(s(n), x(n))$ is the *set* of possible next states and output symbols. Whereas a deterministic machine has update function

$$update\colon States \times Inputs \to States \times Outputs,$$

a nondeterministic machine has a (nondeterministic) state transition function,

$$possibleUpdates\colon States \times Inputs \to P(States \times Outputs), \tag{3.9}$$

where $P(State \times Outputs)$ is the **power set** of $States \times Outputs$. Recall that the power set is the set of all subsets; that is, any subset of $States \times Outputs$ is an element of $P(States \times Outputs)$.

In order for a nondeterministic machine to be **receptive**, it is necessary that

$$\forall\, s(n) \in States, \text{ and } \forall\, x(n) \in Inputs, \quad possibleUpdates(s(n), x(n)) \neq \emptyset.$$

Recall that a receptive machine accepts any input symbol in any state, makes a state transition (possibly back to the same state), and produces an output symbol. In other words, there is no situation in which the reaction to an input symbol is not defined.

Operationally, a nondeterministic machine arbitrarily selects the next state and current output symbol from *possibleUpdates*, given the current state and current input symbol. The model says nothing about how the selection is made.

As with deterministic machines, we can collect the specification of a nondeterministic state machine into a five-tuple:

$$StateMachine = (States, Inputs, Outputs, possibleUpdates, initialState). \tag{3.10}$$

The *possibleUpdates* function is different from the *update* function of a deterministic machine.

Deterministic state machines of the form (3.1) are a special case of nondeterministic machines in which $possibleUpdates(s(n), x(n))$ consists of a single element: namely, $update(s(n), x(n))$. In other words,

$$possibleUpdates(s(n), x(n)) = \{update(s(n), x(n))\}.$$

Thus, any deterministic machine, as well as any nondeterministic machine, can be described by (3.10).

In the nondeterministic machine of (3.10), a single input sequence may give rise to many state responses and output sequences. If $(x(0), x(1), x(2), \ldots)$ is an input sequence, then $(s(0), s(1), s(2), \ldots)$ is a (possible) state trajectory and $(y(0), y(1), y(2), \ldots)$ is a (possible) output sequence, provided that

$$s(0) = initialState$$
$$\forall\, n \geq 0, \quad (s(n+1), y(n)) \in possibleUpdates(s(n), x(n)).$$

A deterministic machine defines a function from an input sequence to an output sequence,

$$F\colon InputSignals \to OutputSignals,$$

where

$$InputSignals = [Naturals_0 \to Inputs],$$

and

$$OutputSignals = [Naturals_0 \to Outputs].$$

We define a **behavior** of the machine to be a pair (x, y) such that $y = F(x)$; that is, a behavior is a possible input–output pair. In a deterministic machine, for each $x \in InputSignals$, there is exactly one $y \in OutputSignals$ such that (x, y) is a behavior.

We define the set

$$Behaviors \subset InputSignals \times OutputSignals, \tag{3.11}$$

where

$$Behaviors = \{(x, y) \in InputSignals \times OutputSignals \\ \mid y \text{ is a possible output sequence for input sequence } x\}.$$

For a deterministic state machine, the set *Behaviors* is the graph of the function F.

For a nondeterministic machine, for a given $x \in InputSignals$, there may be more than one $y \in OutputSignals$ such that (x, y) is a behavior. The set *Behaviors* is therefore no longer the graph of a function. Instead, it defines a **relation**—a generalization of a function in which there can be two or more distinct elements in the range corresponding to the same element in the domain. The interpretation is still straightforward, however: If $(x, y) \in Behaviors$, then input sequence x may produce output sequence y.

3.4 Simulation and bisimulation

Two different state machines with the same input and output alphabets may be equivalent in the sense that for the same input sequence, they produce the same output sequence. We explore this concept of equivalence in this section.

Example 3.9: The three state machines in figure 3.7 have the same input and output alphabets:

$$Inputs = \{1, absent\} \text{ and } Outputs = \{0, 1, absent\}.$$

Machine (a) has the most states. However, its behavior is identical to that of (b). Both machines produce an alternating sequence of two 1s and one 0 as they receive a sequence of 1s at the input. The machine in (c) is nondeterministic. It has more behaviors than does (a) or (b): It can produce any alternating sequence of at least two 1s and one 0. Thus (c) is more general, or more abstract, than the machine (a) or (b). ❐

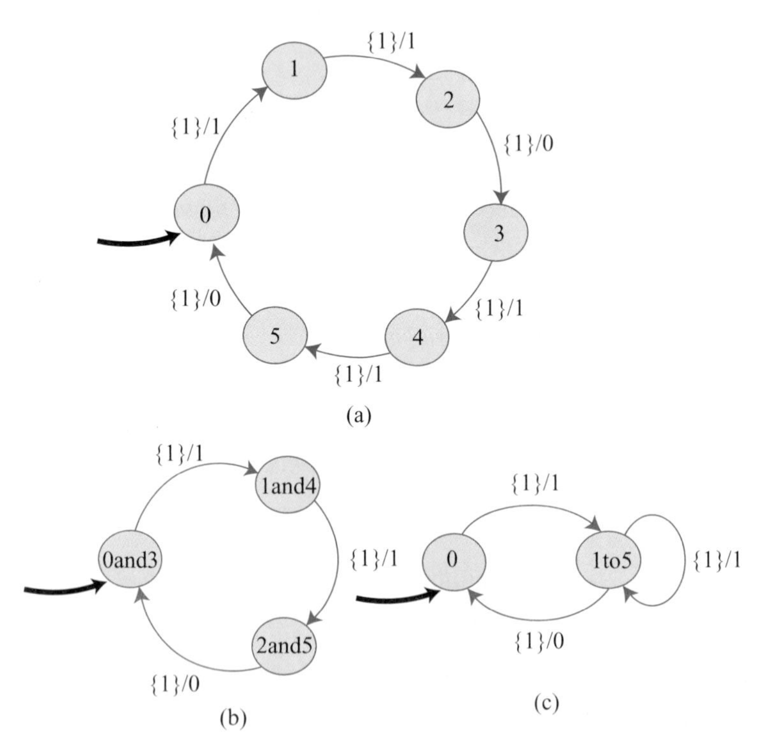

FIGURE 3.7: Three state machines, of which (a) and (b) simulate one another and (c) simulates (a) and (b).

To study the relationships between the machines in figure 3.7, we introduce the concepts of **simulation** and **bisimulation**. The machine in (c) is said to **simulate** (b) and (a). The machine in (b) is said to **bisimulate** (a), or to **be bisimilar to** (a). Bisimulation can be viewed as a form of equivalence between state machines. Simulation can be viewed as a form of abstraction of state machines.

Example 3.10: In figure 3.6, the bottom machine can generate any output sequence that the top machine generates, for the same input sequence. The reverse is not true; there are output sequences that can be generated by the bottom machine but not by the top machine. The bottom machine is an abstraction of the top one.

We will see that the bottom machine simulates the top machine (but not vice versa). ❒

To understand simulation, it is easiest to consider a **matching game** between one machine and the other, more abstract machine that simulates the first. The game starts with both machines in their initial states. The first machine is allowed to react to an input symbol. If this machine is nondeterministic, it may have more than one possible reaction; it is permitted to choose any one of these reactions. The second, more abstract machine must react to the same input symbol in such a way that it produces the same output symbol. If it is nondeterministic, it is free to pick from among the possible reactions any one that matches the output symbol of the first machine and that permits it to continue to match the output symbols of the first machine in future reactions. The second machine "wins" this matching game if it can always match the output symbol of the first machine. We then say that the second machine simulates the first one. If the first machine can produce an output symbol that the second one cannot match, then the second machine does not simulate the first machine.

Example 3.11: We wish to determine whether (c) simulates (b) in figure 3.7. The game starts with the two machines in their initial states, which we jointly denote by the pair

$$s_0 = (0and3, 0) \in States_b \times States_c.$$

Machine (b) (the one being simulated) moves first. Given an input symbol, it reacts. If it is nondeterministic, then it is free to react in any way possible; in this case, however, (b) is deterministic, and so it will have only one possible reaction. Machine (c) then has to **match** the move taken by (b); given the same input symbol, it must react in such a way that it produces the same output symbol.

There are two possible input symbols to machine (b): 1 and *absent*. If the input symbol is *absent*, the machine reacts by stuttering. Machine (c) can match this by stuttering as well. For this example, it will always do this to match stuttering moves, and so we will not consider them further.

Excluding the stuttering input symbol, there is only one possible input symbol to machine (b): 1. The machine reacts by producing the output symbol 1 and changing to state *1and4*. Machine (c) can match this by taking the only possible transition out of its current state and thereby also produce output symbol 1. The resulting states of the two machines are denoted

$$s_1 = (1and4, 1to5) \in States_b \times States_c.$$

From here, there is again only one nonstuttering input symbol possible, and so (b) reacts by moving to *2and5* and producing the output symbol 1. Now (c) has two choices, but in order to match (b), it chooses the self-loop transition, which produces 1. The resulting states are

$$s_2 = (2and5, 1to5) \in States_b \times States_c.$$

From here, (b) reacts to the nonstuttering input symbol by moving to *0and3* and producing output symbol 0. To match this move, (c) selects the transition that moves the state to 0, producing 0. The resulting states are s_0, back to where we started. So we know that (c) can always match (b).

The "winning" strategy of the second machine can be summarized by the set

$$S_{b,c} = \{s_0, s_1, s_2\} \subset States_b \times States_c. \square$$

The set $S_{b,c}$ in this example is called a **simulation relation**; it shows how (c) simulates (b). A simulation relation associates states of the two machines. Suppose we have two state machines, X and Y, which may be deterministic or nondeterministic. Let

$$X = (States_X, Inputs, Outputs, possibleUpdates_X, initialState_X)$$

and

$$Y = (States_Y, Inputs, Outputs, possibleUpdates_Y, initialState_Y).$$

The two machines have the same input and output alphabets. If either machine is deterministic, then its *possibleUpdates* function always returns a set with only one element in it.

If Y simulates X, the simulation relation is given as a subset of $States_X \times States_Y$. Note the ordering here; the machine that moves first in the game, X, the one being simulated, is first in the expression $States_X \times States_Y$.

To consider the reverse scenario, if X simulates Y, then the relation is given as a subset of $States_Y \times States_X$. In this version of the game, Y must move first.

If simulation relations occur in both directions, then the machines are bisimilar.

We can state the "winning" strategy mathematically. We say that Y **simulates** X if there is a subset $S \subset States_X \times States_Y$ such that

1. $(initialState_X, initialState_Y) \in S$, and
2. If $(s_X(n), s_Y(n)) \in S$, then $\forall\, x(n) \in Inputs$, and $\forall\, (s_X(n+1), y_X(n)) \in possibleUpdates_X(s_X(n), x(n))$, there is a $(s_Y(n+1), y_Y(n)) \in possibleUpdates_Y(s_Y(n), x(n))$ such that

   (a) $(s_X(n+1), s_Y(n+1)) \in S$ and
   (b) $y_X(n) = y_Y(n)$.

This set S, if it exists, is called the **simulation relation**. It establishes a correspondence between states in the two machines.

Example 3.12: Consider again the state machines in figure 3.7. The machine in (b) simulates the one in (a). The simulation relation is a subset

$$S_{a,b} \subset \{0, 1, 2, 3, 4, 5\} \times \{0and3, 1and4, 2and5\}.$$

The names of the states in (b) (which are arbitrary) are suggestive of the appropriate simulation relation. Specifically,

$$S_{a,b} = \{(0, 0and3), (1, 1and4), (2, 2and5), (3, 0and3), (4, 1and4), (5, 2and5)\}.$$

The first condition of a simulation relation, that the initial states match, is satisfied because $(0, 0and3) \in S_{a,b}$. The second condition can be tested by playing the game, starting in each pair of states in $S_{a,b}$.

Start with the two machines in one pair of states in $S_{a,b}$, such as the initial states 0 and *0and3*. Then consider the moves that machine (a) can make in a reaction. Ignoring stuttering, if we start with $(0, 0and3)$, (a) must move to state 1 (given input 1). If (b) is given the same input symbol, can it match the move? To match the move, it must react to the same input symbol, produce the same output symbol, and move to a state so that the new state of (a) paired with the new state of (b) is in $S_{a,b}$. Indeed, given input symbol 1, (b) produces output symbol 1 and moves to state *1and4*, which is matched to state 1 of (a).

It is easy (albeit somewhat tedious) to check that this matching can be done from any starting point in $S_{a,b}$. ❐

This example shows how to use the game to check that a particular subset of $States_X \times States_Y$ is a simulation relation. Thus, the game can be used either to construct a simulation relation or to check whether a particular set is a simulation relation.

For the machines in figure 3.7, we have shown that (c) simulates (b) and that (b) simulates (a). Simulation is **transitive**, which means that we can immediately conclude that (c) simulates (a). In particular, if we are given simulation relations $S_{a,b} \subset States_a \times States_b$ ((*b*) simulates (*a*)) and $S_{b,c} \subset States_b \times States_c$ ((*c*) simulates (*b*)), then

$$\begin{aligned} S_{a,c} = \{(s_a, s_c) \in States_a \times States_c \\ \mid \exists\, s_b \in States_b \text{ where } (s_a, s_b) \in S_{a,b} \text{ and } (s_b, s_c) \in S_{b,c}\} \end{aligned} \tag{3.12}$$

is the simulation relation showing that (*c*) simulates (*a*).

Example 3.13: For the examples in figure 3.7, we have already determined

$$S_{a,b} = \{(0, 0and3), (1, 1and4), (2, 2and5), (3, 0and3), (4, 1and4), (5, 2and5)\}$$

and

$$S_{b,c} = \{(0and3, 0), (1and4, 1to5), (2and5, 1to5)\}.$$

From (3.12) we can conclude that

$$S_{a,c} = \{(0, 0), (1, 1to5), (2, 1to5), (3, 0), (4, 1to5), (5, 1to5)\},$$

which further supports the suggestive choices of state names. ❐

Simulation relations are not necessarily symmetric.

Example 3.14: For the examples in figure 3.7, (b) does not simulate (c). To see this, we can attempt to construct a simulation relation by playing the game. Starting in the initial states,

$$s_0 = (0and3, 0),$$

we allow (c) to move first. Presented with a nonstuttering input symbol, 1, it produces 1 and moves to *1to5*. Machine (b) can match this by producing 1 and moving to *1and4*. But from state *1to5*, (c) can now produce 0 with input symbol 1, which (b) cannot match. Thus, the game gets stuck, and we fail to construct a simulation relation. ❐

Consider another example, one that illustrates that there may be more than one simulation relation between two machines.

Example 3.15: In figure 3.8, it is easy to check that (c) simulates (a) and (b). We now verify that (b) simulates (a) and also (a) simulates (b) by

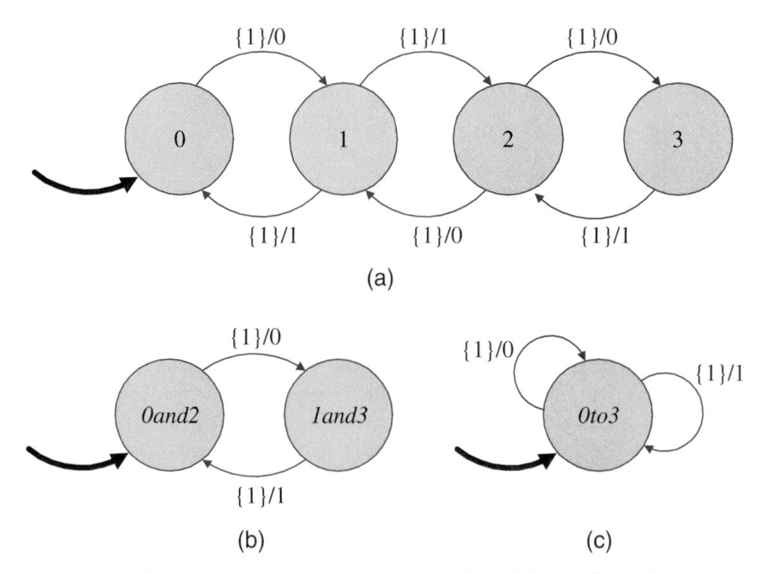

FIGURE 3.8: Three state machines, of which (a) and (b) are bisimilar and (c) simulates (a) and (b).

determining that not only can (b) match any move (a) makes but also (a) can match any move (b) makes. In fact, because (a) is nondeterministic, in two of its states it has two distinct ways of matching the moves of (b). Because it moves second, it can arbitrarily choose from among these possibilities.

If from state 1 it always chooses to return to state 0, then the simulation relation is

$$S_{b,a} = \{(0and2, 0), (1and3, 1)\}.$$

Otherwise, if from state 2 it always chooses to return to state 1, then the simulation relation is

$$S_{b,a} = \{(0and2, 0), (1and3, 1), (0and2, 2)\}.$$

Otherwise, the simulation relation is

$$S_{b,a} = \{(0and2, 0), (1and3, 1), (0and2, 2), (1and3, 3)\}.$$

Thus, the simulation relation is not unique. ❐

A common use of simulation is to establish a relationship between a more abstract model and a more detailed model. In the example just given, (c) is a more abstract model of either (b) or (a). It is more abstract in the sense that it loses detail. For example, it has lost the property that 0s and 1s alternate in the output sequence. We now give a more compelling example of such abstraction, in which the abstraction dramatically reduces the number of states while still preserving some properties of interest.

Example 3.16: In the case of the parking meter, the bottom machine in figure 3.6 simulates the top machine. Let A denote the top machine, and let B denote the bottom machine. We now identify the simulation relation.

The simulation relation is a subset $S \subset \{0, 1, \ldots, 60\} \times \{0, 1, \mathit{more}\}$. It is intuitively clear that 0 and 1 of the bottom machine correspond to 0 and 1, respectively, of the top machine. Thus, $(0, 0) \in S$ and $(1, 1) \in S$. It is also intuitive that *more* corresponds to all of the remaining states $2, 3, \ldots, 60$ of the top machine. So we propose to define the simulation relation as

$$S = \{(0,0), (1,1)\} \cup \{(s_A, \mathit{more}) \mid 2 \leq s_A \leq 60\}. \tag{3.13}$$

We now check that S is indeed a simulation relation.

The first condition of a simulation relation, that the initial states match, is satisfied because $(0, 0) \in S$. The second condition is more tedious to verify. It says that for each pair of states in S and for each input symbol, the two machines can transition to a pair of new states that is also in S and that these two transitions produce the same output symbol. Because machine A is deterministic, there is no choice about which transition it takes and which output symbol it produces. In machine B, there are choices, but all we require is that one of the choices match.

The only state of machine B that actually offers choices is *more*. Upon receiving *tick*, the machine can transition back to *more* or down to 1. In either case, the output symbol is *safe*. It is easy to see that these two choices are sufficient for state *more* to match states $2, 3, \ldots, 60$ of machine A.

Thus the bottom machine does indeed simulate the top machine with the simulation relation (3.13). ❐

3.4.1 *Relating behaviors*

A simulation relation establishes a correspondence between two state machines, one of which is typically much simpler than the other. The relation lends confidence that analyzing the simpler machine does indeed reveal properties of the more complicated machine.

This confidence rests on a theorem and corollary that we develop in this section. These results relate the input–output behaviors of state machines that are related by simulation.

Given an input sequence $x = (x(0), x(1), x(2), \ldots)$, if a state machine can produce the output sequence $y = (y(0), y(1), y(2), \ldots)$, then (x, y) is said to be a behavior of the state machine. The set of all behaviors of a state machine obviously satisfies

$$Behaviors \subset InputSignals \times OutputSignals.$$

Theorem. Let B simulate A. Then

$$Behaviors_A \subset Behaviors_B.$$

This theorem is easy to prove. Consider a behavior $(x, y) \in Behaviors_A$. We need to show that $(x, y) \in Behaviors_B$.

Let the simulation relation be S. Find all possible state responses for A,

$$s_A = (s_A(0), s_A(1), \ldots),$$

that result in behavior (x, y). (If A is deterministic, then there is only one response.) The simulation relation assures us that we can find a state response for B,

$$s_B = (s_B(0), s_B(1), \ldots),$$

where $(s_A(i), s_B(i)) \in S$, such that given input symbol x, B produces y. Thus, $(x, y) \in Behaviors_B$.

Intuitively, the theorem simply states that B can match every move of A and produce the same output sequence. It also implies that if B cannot produce a particular output sequence, then neither can A. This is stated formally in the following corollary.

Corollary. Let B simulate A. If

$$(x, y) \notin Behaviors_B,$$

then

$$(x, y) \notin Behaviors_A.$$

The theorem and corollary are useful for analysis. The general approach is as follows: we have a state machine A. We wish to show that its input–output function satisfies some property; that is, every behavior satisfies some condition. We construct a simpler machine B that simulates A and whose input–output relation satisfies the same property. Then the theorem guarantees that A will satisfy this property, too; that is, because all behaviors of B satisfy the property, all behaviors of A must also. This technique is useful because it is often easier to understand a simple state machine than a complex state machine with many states.

Conversely, if there is a particular property that no behavior of A may have, it is sufficient to find a simpler machine B that simulates A and does not have this property. This scenario is typical of a **safety** problem, for which we must show that dangerous outputs from our system are not possible.

Example 3.17: For the parking meter of figure 3.6, for example, we can use the nondeterministic machine to show that if a coin is inserted at step n, the output symbol at steps n and $n+1$ is *safe*. By the corollary, it is sufficient to show that the nondeterministic machine cannot behave any differently. ❐

It is important to understand what the theorem says and what it does not say. It does not say, for example, that if $Behaviors_A \subset Behaviors_B$, then B simulates A. In fact, this statement is not true. Consider the two machines in figure 3.9, in which

$$Inputs = \{1, absent\},$$

$$Outputs = \{0, 1, absent\}.$$

These two machines have the same behaviors. The nonstuttering output symbols are $(1, 0)$ or $(1, 1)$, selected nondeterministically, with the assumption that

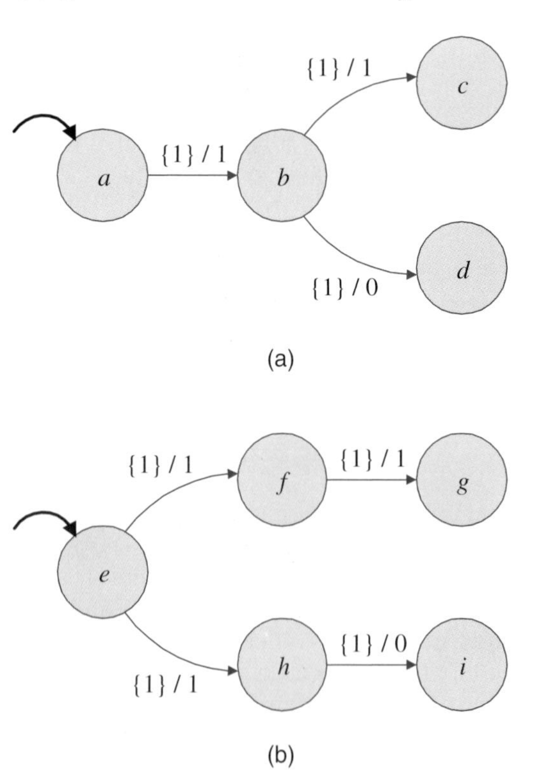

FIGURE 3.9: Two state machines with the same behaviors that are not bisimilar.

the input sequence has at least two nonstuttering symbols. However, they are not bisimilar. In particular, (b) does not simulate (a). To see this, we play the matching game. Machine (a) is allowed to move first. Ignoring stuttering, it has no choice but to move from a to b and produce output symbol 1. Machine (b) can match this two ways; it has no basis on which to prefer one way to match it over another, so it picks one: say, moving to state f. Now it is the turn of machine (a), which has two choices. If it choses to move to d, then machine (b) cannot match its move. (A similar argument works if (b) picks state h.) Thus, machine (b) does not simulate machine (a), despite the fact that $Behaviors_A \subset Behaviors_B$.*

3.5 Summary

State machines are models of systems whose input and output signal spaces consist of sequences of symbols. There are three ways of defining state machines: a sets and functions description, the state transition diagram, and the update table. The state machine model gives a step-by-step procedure for evaluating the output signal. It describes a state-determined system: Once we know the current state, we can tell the future behavior for any future input symbols.

A state machine can be nondeterministic: Given the current state and current input symbol, it may have more than one possible next state and current output symbol. Nondetermistic machines typically arise through abstraction of deterministic machines. Two state machines, with the same input and output alphabets, may be related through simulation and bisimulation. Simulation is used to understand properties of the behavior of one machine in terms of the behaviors of another (presumably simpler) machine. Bisimulation is used to establish a form of equivalence between state machines.

EXERCISES

KEY: E = mechanical T = requires plan of attack C = more than 1 answer

E 1. A state machine with

$$Inputs = \{a, b, c, d, absent\}$$

has a state s with two emerging arcs with guards

$$guard1 = \{a\}$$

and

$$guard2 = \{a, b, d\}.$$

* Recall that in our notation, $\subset$ allows the two sets to be equal.

(a) Is this state machine deterministic?

(b) Define the *else* transition for state s, and specify the source and destination state for the *else* arc.

E 2. For the answering machine example of figure 3.1, assume that the input sequence is

$$(offhook, offhook, ring, offhook, ring, ring, ring, offhook, \ldots).$$

This corresponds to a user of the answering machine making two phone calls, answering a third call after the first ring, and answering a second call after the third ring.

(a) Give the state response of the answering machine.

(b) Give the trace of the answering machine.

(c) Give the output sequence.

E 3. Consider the alphabets

$$Inputs = Outputs = Binary = \{0, 1\}.$$

Note that there are no stuttering input or output symbols here. This simplifies the notation in the problem somewhat.

(a) Construct a state machine that uses these alphabets such that if $(x(0), x(1), \ldots)$ is any input sequence without stuttering symbols, the output sequence is given by

$$\forall\, n \in Naturals_0,$$
$$y(n) = \begin{cases} 1 & \text{if } n \geq 2 \wedge (x(n-2), x(n-1), x(n)) = (1, 1, 1) \\ 0 & \text{otherwise.} \end{cases}$$

In words, the machine outputs 1 if the current input symbol and the two previous input symbols are all 1s, otherwise it outputs 0. (Had we included a stuttering symbol, this equation would be a bit more complicated.)

(b) For the same input and output alphabet, construct a state machine that outputs 1 if the current input symbol and two previous input symbols are either $(1, 1, 1)$ or $(1, 0, 1)$ and that otherwise outputs 0.

E 4. A modulo N counter is a device that can output any integer between 0 and $N - 1$. The device has three input symbols—*increment*, *decrement*, and *reset*—plus, as always, a stuttering symbol, *absent*; *increment* increases the output integer by 1; *decrement* decreases this integer by 1; and *reset* sets the output symbol to 0. Here, *increment* and *decrement* are modulo N operations.

Note: Modulo N numbers work as follows: For any integer m, m *mod* $N = k$, where $0 \leq k \leq N - 1$ is the unique integer such that N divides $(m - k)$. Thus there are only N distinct modulo N numbers: namely, $0, \ldots, N - 1$.

(a) Give the state transition diagram of this counter for $N = 4$.

(b) Give the update table of this counter for $N = 4$.

(c) Give a description of the state machine by specifying the five entities that appear in (3.1); again assume $N = 4$.

(d) Take $N = 3$. Calculate the state response for the input sequence

$$(increment^4, decrement^3, \ldots)$$

starting with initial state 1, where s^n means s repeated n times.

T 5. The state machine *UnitDelay* is defined to behave as follows. On the first nonstuttering reaction (when the first nonstuttering input symbol arrives), the output symbol a is produced. On subsequent reactions (when subsequent input symbols arrive), the input symbol that arrived at the previous nonstuttering reaction is produced as an output symbol.

(a) Assume the input and output alphabets are

$$Inputs = Outputs = \{a, b, c, absent\}.$$

Construct a finite-state machine that implements *UnitDelay* for this input set. Devise both a state transition diagram and a definition of each of the components in (3.1).

(b) Assume the input and output sets are

$$Inputs = Outputs = Naturals_0 \cup \{absent\},$$

and that on the first nonstuttering reaction, the machine produces 0 instead of a. Give an informal argument that no finite-state machine can implement *UnitDelay* for this input set. Construct an infinite-state machine by defining each of the components in (3.1).

T 6. Construct an infinite-state machine that realizes *Equal*.

C 7. An elevator connects two floors, 1 and 2. It can go up (if it is on floor 1), go down (if it is on floor 2), and stop on either floor. Passengers at any floor may press a button requesting service. Design a controller (a state machine) for the elevator so that (1) every request is served, and (2) if there is no pending request, the elevator is stopped. For simplicity, do not be concerned about responding to requests from passengers inside the elevator.

T 8. The state machine in figure 3.10 has the property that it outputs at least one 1 between any two 0s. Construct a nondeterministic two-state machine that simulates this one and preserves that property.

T 9. For the nondeterministic state machine in figure 3.11, the input and output alphabets are

$$Inputs = Outputs = \{0, 1, absent\}.$$

(a) Define the *possibleUpdates* function (3.9) for this state machine.

(b) Define the relation *Behaviors* in (3.11) for this state machine. Part of the challenge here is to find a way to describe this relation compactly. For simplicity, ignore stuttering; that is, assume the input symbol is never *absent*.

E 10. The state machine in figure 3.12 implements *CodeRecognizer* but has more states than does the one in figure 3.4. Show that it is equivalent by giving a bisimulation relation with the machine in figure 3.4.

E 11. The state machine in figure 3.13 has input and output alphabets

$$Inputs = \{1, a\}$$

and

$$Outputs = \{0, 1, a\},$$

where a (short for *absent*) is the stuttering symbol. State whether each of the following is in the set *Behaviors* for this machine. In each of the

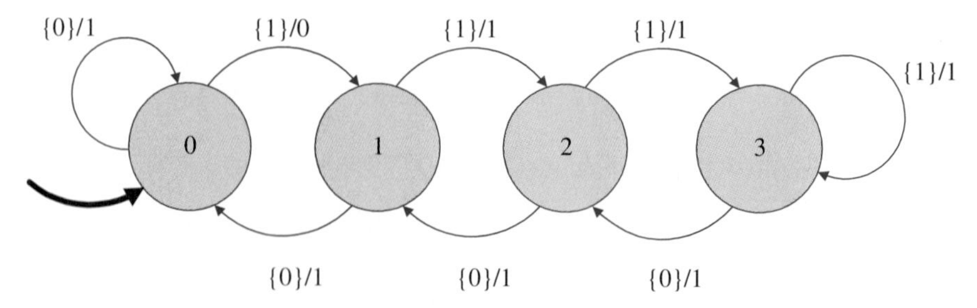

FIGURE 3.10: Machine that outputs at least one 1 between any two 0s.

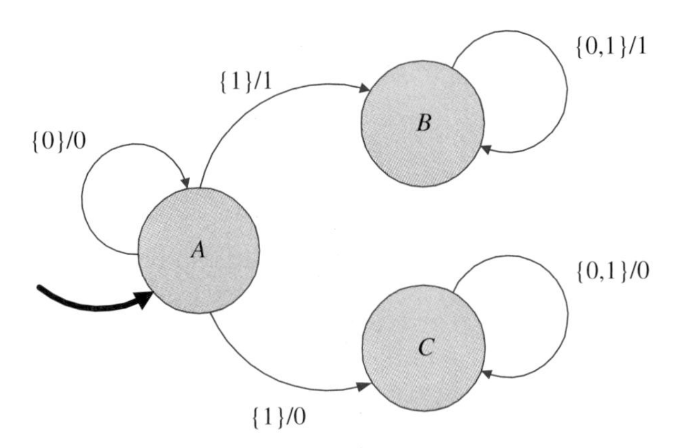

FIGURE 3.11: Nondeterministic state machine.

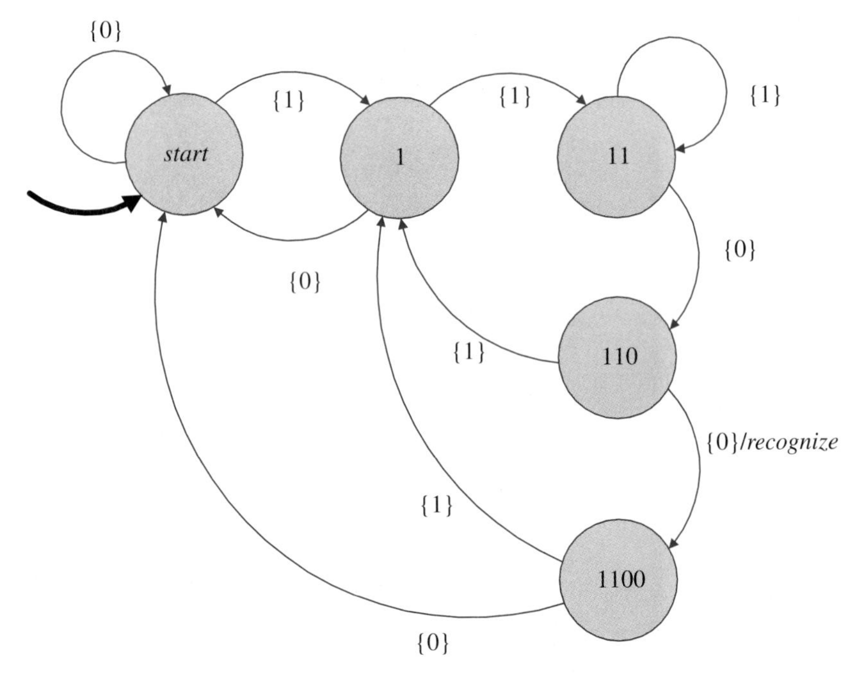

FIGURE 3.12: A machine that implements *CodeRecognizer* but has more states than does the one in figure 3.4.

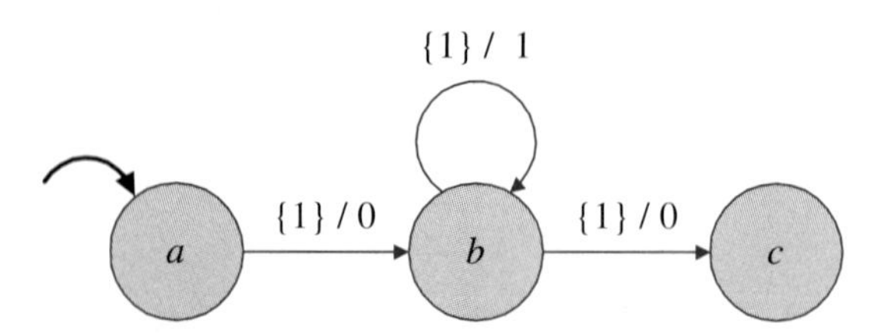

FIGURE 3.13: State machine.

following, the ellipsis means that the last symbol is repeated forever. Also, in each case, the input and output signals are given as sequences.

(a) $((1, 1, 1, 1, 1, \ldots), (0, 1, 1, 0, 0, \ldots))$

(b) $((1, 1, 1, 1, 1, \ldots), (0, 1, 1, 0, a, \ldots))$

(c) $((a, 1, a, 1, a, \ldots), (a, 1, a, 0, a, \ldots))$

(d) $((1, 1, 1, 1, 1, \ldots), (0, 0, a, a, a, \ldots))$

(e) $((1, 1, 1, 1, 1, \ldots), (0, a, 0, a, a, \ldots))$

E 12. The state machine in figure 3.14 has alphabets

$$Inputs = \{1, absent\}$$

and

$$Outputs = \{0, 1, absent\}.$$

Find a bisimilar state machine with only two states, and give the bisimulation relation.

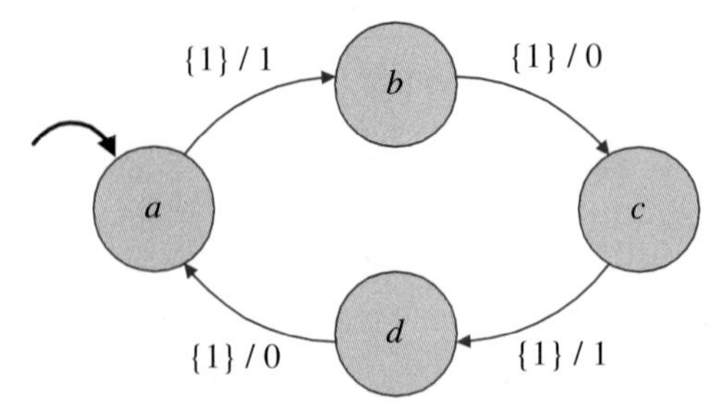

FIGURE 3.14: A machine that has more states than it needs.

E 13. You are told that state machine A has the alphabets

$$Inputs = \{1, 2, absent\},$$

$$Outputs = \{1, 2, absent\}, \text{ and}$$

$$States = \{a, b, c, d\},$$

but you are told nothing further. Do you have enough information to construct a state machine B that simulates A? If so, construct such a state machine, and devise the simulation relation.

E 14. Construct a state machine with $Inputs = \{0, 1, absent\}$ and $Outputs = \{r, absent\}$ that outputs r whenever the input signal (without stuttering symbols) contains the sequence $(0, 0, 0)$ and otherwise outputs *absent*. More precisely, if $x = (x(0), x(1), \ldots)$ is the input sequence, then $y = (y(0), y(1), \ldots)$ is the output sequence, where

$$y(n) = \begin{cases} r & \text{if } (x(n-2), x(n-1), x(n)) = (0, 0, 0) \\ absent & \text{otherwise.} \end{cases}$$

T 15. Consider a state machine in which

$$Inputs = \{1, absent\},$$

$$Outputs = \{0, 1, absent\},$$

$$States = \{a, b, c, d, e, f\}, \text{and}$$

$$initialState = a$$

and the *update* function is given by the following table (ignoring stuttering):

(currentState, inputSymbol)	(nextState, outputSymbol)
$(a, 1)$	$(b, 1)$
$(b, 1)$	$(c, 0)$
$(c, 1)$	$(d, 0)$
$(d, 1)$	$(e, 1)$
$(e, 1)$	$(f, 0)$
$(f, 1)$	$(a, 0)$

(a) Draw the state transition diagram for this machine.

(b) Ignoring stuttering, give the *Behaviors* relation for this machine.

(c) Find a three-state machine that is bisimilar to this one. Draw that state machine, and give the bisimulation relation.

CHAPTER 4

Composing state machines

We design interesting systems by composing simpler components. Because systems are functions, their composition is function composition, as discussed in section 2.1.5. State machines, however, are not given directly as functions that map input sequences into output sequences. Instead, they are given procedurally, where the *update* function defines how to progress from one state to the next. This chapter explains how to define a new state machine that describes a composition of multiple state machines.

In section 2.3.4 we used a block diagram syntax to define compositions of systems. We use the same syntax here, and we similarly build up an understanding of composition by first considering easy cases. The hardest cases are those in which there is feedback, because the input of one state machine may depend on its own output. It is challenging in this case to come up with a procedure for updating the state of the composite machine. For some compositions, in fact, it is not even possible. Such compositions are said to be ill-formed.

4.1 Synchrony

We consider a set of interconnected components, in which each component is a state machine. By "interconnected" we mean that the outputs of one component may be inputs of another. We wish to construct a state machine model for the composition of components. Composition has two aspects. The first aspect is straightforward: It specifies which outputs of one component are the inputs of another component. These input–output connections are specified with the use of block diagrams.

The second aspect of composition concerns the timing relationships between inputs and outputs. We choose a particular style of composition called **synchrony**. This style dictates that each state machine in the composition reacts *simultaneously* and *instantaneously*. Thus, a reaction of the composite machine consists of a set of simultaneous reactions of each of the component machines.

A reaction of the composite machine is triggered by inputs from the environment. Thus, *when* a reaction occurs is externally determined. This is also true for a single machine. Like a single state machine, a composite machine may stutter. This simply means that each component machine stutters.

A system that reacts only in response to external stimulus is said to be **reactive**. Because our compositions are synchronous, they are often called **synchronous/reactive** systems.

The reactions of the component machines and of the composite machine are viewed as being instantaneous; that is, a reaction does not take time. In particular, the output symbol from a state machine is viewed as being *simultaneous* with the input symbol, without delay. This creates some interesting subtleties, especially for feedback composition when the input of a state machine is connected to its own output. We discuss the ramifications of the synchronous/reactive interpretation later.

Synchrony is a very useful model of the behavior of physical systems. Digital circuits, for example, are almost always designed with the help of this model. Circuit elements are viewed as taking no time to calculate their outputs when given their inputs, and time overall is viewed as progressing in a sequence of discrete time steps according to ticks of a clock. Of course, the time that it takes for a circuit element to produce an output cannot ultimately be ignored, but the model is useful because for most circuit elements in a complex design, this time *can* be ignored. Only the time delay of the circuit elements along a **critical path** affects the overall performance of the circuit.

More recently than for circuits, synchrony has come to be used in software as well. Concurrent software modules interact according to the synchronous model. Languages built on this principle are called **synchronous languages**. They are used primarily in real-time embedded system* design.

* An **embedded system** is a computing system (a computer and its software) that is embedded in a larger system that is not first and foremost a computer. A digital cellular telephone, for example, contains computers that realize the radio modem and the speech codec. New-model cars contain computers for ignition control, antilock brakes, and traction control. Aircraft contain computers for navigation and flight control. In fact, most modern electronic controllers of physical systems are realized as embedded systems.

4.2 Side-by-side composition

A simple form of composition of two state machines is shown in figure 4.1. We call this **side-by-side composition**. Side-by-side composition by itself is not useful, but it is useful in combination with other types of composition. The two state machines in figure 4.1 do not interact with one another. Nonetheless, we wish to define a single state machine representing the synchronous operation of the two-component state machines.

The state space of the composite state machine is simply

$$States = States_A \times States_B.$$

Taking the cross product in the opposite order results in a *different* but *bisimilar* composite state machine. The initial state is

$$initialState = (initialState_A, initialState_B).$$

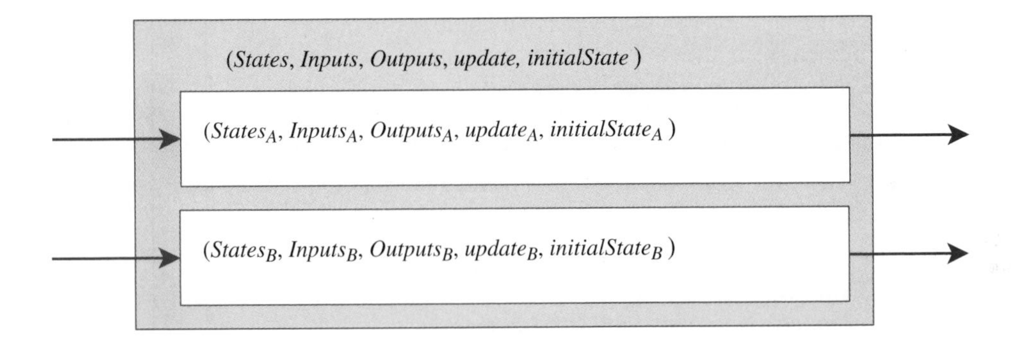

Definition of the side-by-side composite machine

$States = States_A \times States_B$

$Inputs = Inputs_A \times Inputs_B$

$Outputs = Outputs_A \times Outputs_B$

$initialState = (initialState_A, initialState_B)$

$((s_A(n+1), s_B(n+1)), (y_A(n), y_B(n))) = update((s_A(n), s_B(n)), (x_A(n), x_B(n))),$

where

$(s_A(n+1), y_A(n)) = update_A(s_A(n), x_A(n))$ and

$(s_B(n+1), y_B(n)) = update_B(s_B(n), x_B(n))$

FIGURE 4.1: Summary of side-by-side composition of state machines.

The input and output alphabets are

$$Inputs = Inputs_A \times Inputs_B \tag{4.1}$$

and

$$Outputs = Outputs_A \times Outputs_B. \tag{4.2}$$

The update function of the composite machine, *update*, consists of the update functions of the component machines, side by side:

$$((s_A(n+1), s_B(n+1)), (y_A(n), y_B(n))) = update((s_A(n), s_B(n)), (x_A(n), x_B(n))),$$

where

$$(s_A(n+1), y_A(n)) = update_A(s_A(n), x_A(n)),$$

and

$$(s_B(n+1), y_B(n)) = update_B(s_B(n), x_B(n)).$$

Recall that $Inputs_A$ and $Inputs_B$ include a stuttering element. This is convenient because it allows a reaction of the composite when we really want only one of the machines to react. Suppose the stuttering elements are $absent_A$ and $absent_B$. If the second component of the input symbol is $absent_B$, the reaction of the composite consists only of the reaction of the first machine. The stuttering element of the composite is the pair of stuttering elements of the component machines $(absent_A, absent_B)$.

Example 4.1: The side-by-side composition in the top of figure 4.2 has the composite machine with state space

$$States = States_A \times States_B = \{(1, 1), (2, 1)\}$$

and alphabets

$$Inputs = \{(0, 0), (1, 0), (absent_A, 0), (0, absent_B), (1, absent_B), (absent_A, absent_B)\}$$

and

$$Outputs = \{(a, c), (b, c), (absent_A, c), (a, absent_B), (b, absent_B), (absent_A, absent_B)\}.$$

The initial state is

$$initialState = (1, 1).$$

The *update* function can be given as a table, only a part of which is in the following display. The state transition diagram in the lower part of figure 4.2 describes the composition.

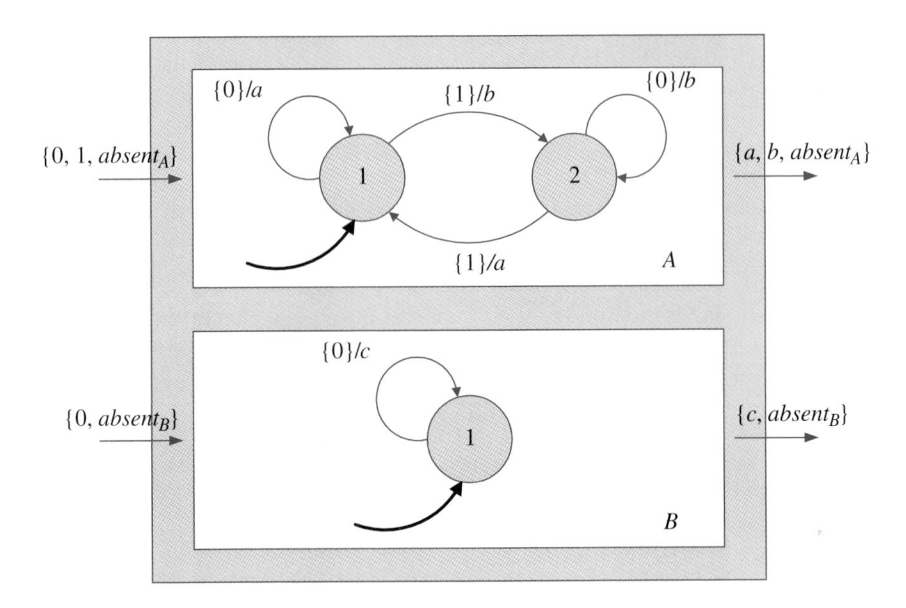

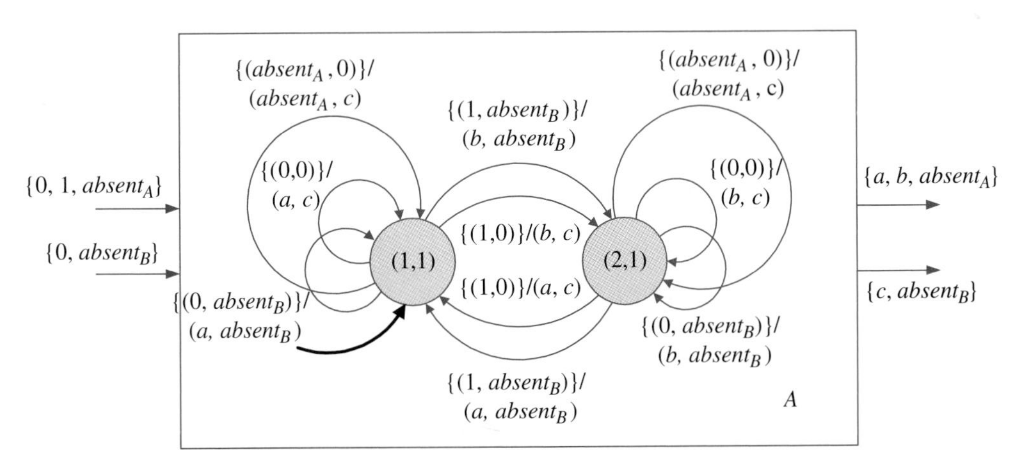

FIGURE 4.2: Example of a side-by-side composition.

	(*next state, output*) for input			
current state	**(0,0)**	**(1,0)**	**($absent_A$,0)**	**. . .**
(1,1)	((1, 1), (*a*, c))	((2, 1), (*b*, c))	((1,1)(($absent_A$, c))	. . .
(2,1)	((2, 1), (*b*, c))	((1, 1), (*a*, c))	((2,1), (($absent_A$, c))	. . .

Notice that if the second component of the input sequence is always $absent_B$, then the side-by-side composition behaves essentially like machine *A*, and if the first component is always $absent_A$, then it behaves like machine *B*. The stuttering element of the composite is of course the pair ($absent_A$, $absent_B$). ❐

4.3 *Cascade composition*

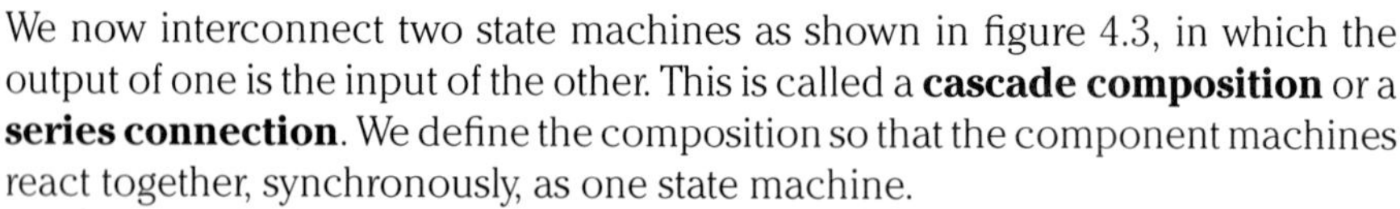

We now interconnect two state machines as shown in figure 4.3, in which the output of one is the input of the other. This is called a **cascade composition** or a **series connection**. We define the composition so that the component machines react together, synchronously, as one state machine.

Suppose the two state machines are given by

$$StateMachine_A = (States_A, Inputs_A, Outputs_A, update_A, initialState_A)$$

and

$$StateMachine_B = (States_B, Inputs_B, Outputs_B, update_B, initialState_B).$$

Let the composition be given by

$$StateMachine = (States, Inputs, Outputs, update, initialState).$$

Clearly, for a composition like that in figure 4.3 to be possible,

$$Outputs_A \subset Inputs_B.$$

Then any output sequence produced by machine *A* can be an input sequence for machine *B*. As a result,

$$OutputSignals_A \subset InputSignals_B.$$

This is analogous to a **type constraint** in programming languages, in which in order for two pieces of code to interact, they must use compatible data

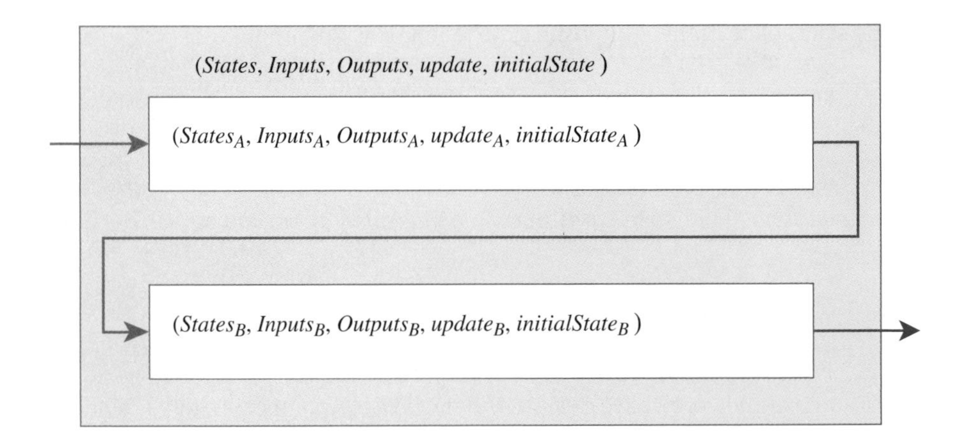

Assumptions about the component machines

$Outputs_A \subset Inputs_B$

Definition of the cascade composite machine

$States = States_A \times States_B$

$Inputs = Inputs_A$

$Outputs = Outputs_B$

$initialState = (initialState_A, initialState_B)$

$((s_A(n+1), s_B(n+1)), y_B(n)) = update((s_A(n), s_B(n)), x(n)),$

where

$(s_A(n+1), y_A(n)) = update_A(s_A(n), x(n))$ and

$(s_B(n+1), y_B(n)) = update_B(s_B(n), y_A(n)).$

FIGURE 4.3: Summary of cascade composition of state machines.

types. We encountered a similar constraint in discussing function composition in section 2.1.5.

We are ready to construct a state machine model for this series connection. As noted in figure 4.3, the input alphabet of the composite is

$$Inputs = Inputs_A.$$

The stuttering element of *Inputs* is, of course, just the stuttering element of $Inputs_A$. The output alphabet of the composite is

$$Outputs = Outputs_B.$$

The state space of the composite state machine is the product set

$$States = States_A \times States_B. \tag{4.3}$$

This asserts that the composite state machine is in state $(s_A(n), s_B(n))$ when $StateMachine_A$ is in state $s_A(n)$ and $StateMachine_B$ is in state $s_B(n)$. The initial state is

$$initialState = (initialState_A, initialState_B).$$

We could equally well have defined the states of the composite state machine in the opposite order,

$$States = States_B \times States_A.$$

This would result in a different but bisimilar state machine description (either one simulates the other). Intuitively, it does not matter which of these two choices we make, and we choose (4.3).

To complete the description of the composite machine, we need to define its *update* function in terms of the component machines. Here, a slight subtlety arises. Because we are using synchronous composition, the output symbol of machine A is *simultaneous* with its input symbol. Thus, in a reaction, the output symbol of machine A in that reaction must be available to machine B in the same reaction. This seems intuitive, but it has some counterintuitive consequences. Although the reactions of machine A and B are simultaneous, we must determine the reaction of A before we can determine the reaction of B. This apparent paradox is an intrinsic feature of synchronous composition. We will have to deal with it carefully in feedback composition, where it is not immediately evident which reactions need to be determined first.

In the cascade composition, it is intuitively clear what we need to do to define the update function. We first determine the reaction of machine A. Suppose that at the nth reaction the input symbol is $x(n)$ and the state is $s(n) = (s_A(n), s_B(n))$, where $s_A(n)$ is the state of machine A and $s_B(n)$ is the state of machine B. Machine A reacts by updating its state to $s_A(n+1)$ and producing output symbol $y_A(n)$:

$$(s_A(n+1), y_A(n)) = update_A(s_A(n), x(n)). \tag{4.4}$$

Its output symbol $y_A(n)$ becomes the input symbol to machine B. Machine B reacts by updating its state to $s_B(n+1)$ and producing output symbol $y_B(n)$:

$$(s_B(n+1), y_B(n)) = update_B(s_B(n), y_A(n)). \tag{4.5}$$

The output of the composite machine, of course, is just the output of machine B, and the next state of the composite machine is just $(s_A(n+1), s_B(n+1))$, so the composite machine's *update* is

$$((s_A(n+1), s_B(n+1)), y_B(n)) = update((s_A(n), s_B(n)), x(n)),$$

where $s_A(n+1)$, $s_B(n+1)$, and $y_B(n)$ are given by (4.4) and (4.5). The definition of the composite machine is summarized in figure 4.3.

Example 4.2: The cascade composition in figure 4.4 has the composite machine with state space

$$States = States_A \times States_B = \{(0,0), (0,1), (1,0), (1,1))\}$$

and alphabets

$$Inputs = Outputs = \{0, 1, absent\}.$$

The initial state is

$$initialState = (0,0).$$

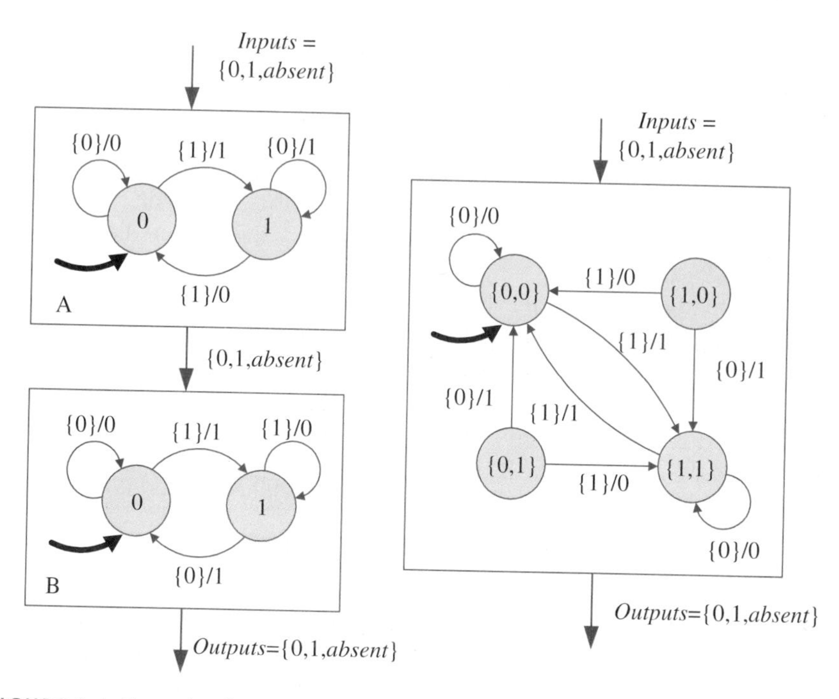

FIGURE 4.4: Example of a cascade composition. The composed state machine is on the right.

The *update* function is given by the following table:

	(*next state, output*) for input		
current state	**(0,0)**	**(1,0)**	**($absent_A$,0)**
(0,0)	((0,0),0)	((1,1),1)	((0,0), *absent*)
(0,1)	((0,0),1)	((1,1),0)	((0,1), *absent*)
(1,0)	((1,1),1)	((0,0),0)	((1,0), *absent*)
(1,1)	((1,1),0)	((0,0),1)	((1,1), *absent*)

The update function is also presented in the state transition diagram of figure 4.4. The self-loops corresponding to the stuttering input symbol *absent* are not shown in the diagram.

Observe from the table or the diagram that states (0, 1) and (1, 0) are not reachable from the initial state. A state s is said to be **reachable** if some sequence of input symbols can take the state machine from the initial state to s. This suggests that a simpler machine with fewer states would exhibit the same input–output behaviors. In fact, notice from the table that the input is always equal to the output. A trivial one-state machine can exhibit the same input–output behaviors. (Exercise 8 at the end of this chapter gives a procedure for calculating the reachable states of an arbitrary state machine.)

The simple behavior of the composite machine is not immediately apparent from figure 4.4. We have to systematically construct the composite machine to derive this simple behavior. In fact, this composite machine can be viewed as an encoder and decoder, because the input bit sequence is encoded by a distinctly different bit sequence (the intermediate signal in figure 4.4), and then the second machine, given the intermediate signal, reconstructs the original.

This particular encoder is known as a **differential precoder**. It is "differential" in that when the input symbol is 0, the intermediate signal sample is unchanged from the previous sample (whether it was 0 or 1), and when the input symbol is 1, the sample is changed. Thus, the input symbol indicates *change* in the input with a 1, and *no change* with a 0. Differential precoders are used when it is important that the average number of 1s and 0s is the same, regardless of the input sequence that is encoded. ❐

4.4 Product-form inputs and outputs

In the state machine model of (3.1), at each step the environment selects one input symbol to which the machine reacts and produces one output symbol. Sometimes we wish to model the fact that some input values are selected by one part of the environment and other input values are simultaneously selected by another part. Also, some output values are sent to one part of the environment and other output values are simultaneously sent to another part. The product-form composition permits these models.

The machine in figure 4.5 is shown as a block with two distinct input arrows and two distinct output arrows. The figure suggests that the machine receives inputs from two sources and sends outputs to two destinations. In the answering machine example of chapter 3, for instance, the *end greeting* input value may originate in a physically different piece of hardware in the machine than may the *offhook* value.

The distinct arrows into and out of a block are called **ports**. Each port is associated with a set of values called the **port alphabet**, as shown in figure 4.5. Each port alphabet must include a stuttering element. The set *Inputs* of input values to the state machine is the product of the input sets associated with the ports. Of course, the product can be constructed in any order; each ordering results in a distinct (but bisimilar) state machine model.

In figure 4.5 there are two input ports and two output ports. The upper input port can present to the state machine any value in the alphabet $Inputs_A$, which includes *absent*, its stuttering element. The lower port can present any value in the set $Inputs_B$, which also includes *absent*. The input value actually presented to the state machine in a reaction is taken from the set

$$Inputs = Inputs_A \times Inputs_B.$$

The stuttering element for this alphabet is the pair $(absent, absent)$. The output value produced by a reaction is taken from the set

$$Outputs = Outputs_A \times Outputs_B.$$

If the output of the nth reaction is $(y_A(n), y_B(n))$, then the upper port shows $y_A(n)$ and the lower port shows $y_B(n)$. These can be separately presented as inputs to downstream state machines. Again, the stuttering element is $(absent, absent)$.

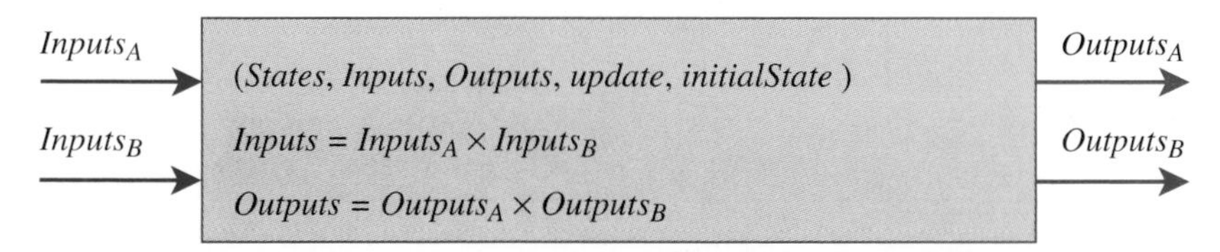

FIGURE 4.5: State machine with product-form inputs and outputs.

Example 4.3: The answering machine of figure 3.1 has input alphabet

$$Inputs = \{ring, offhook, end\ greeting, end\ message\}.$$

In a typical realization of an answering machine, *ring* and *offhook* come from a subsystem (often an **application-specific integrated circuit** [ASIC]) that interfaces to the telephone line. The value *end greeting* comes from another subsystem, such as a magnetic tape machine or digital audio storage device, that plays the answer message. The value *end message* comes from a similar but distinct subsystem that records incoming messages. Thus, a more convenient model will show three separate factors for the inputs, as in figure 4.6. That figure also shows the outputs in product form, anticipating that the distinct output values will need to be sent to distinct subsystems.

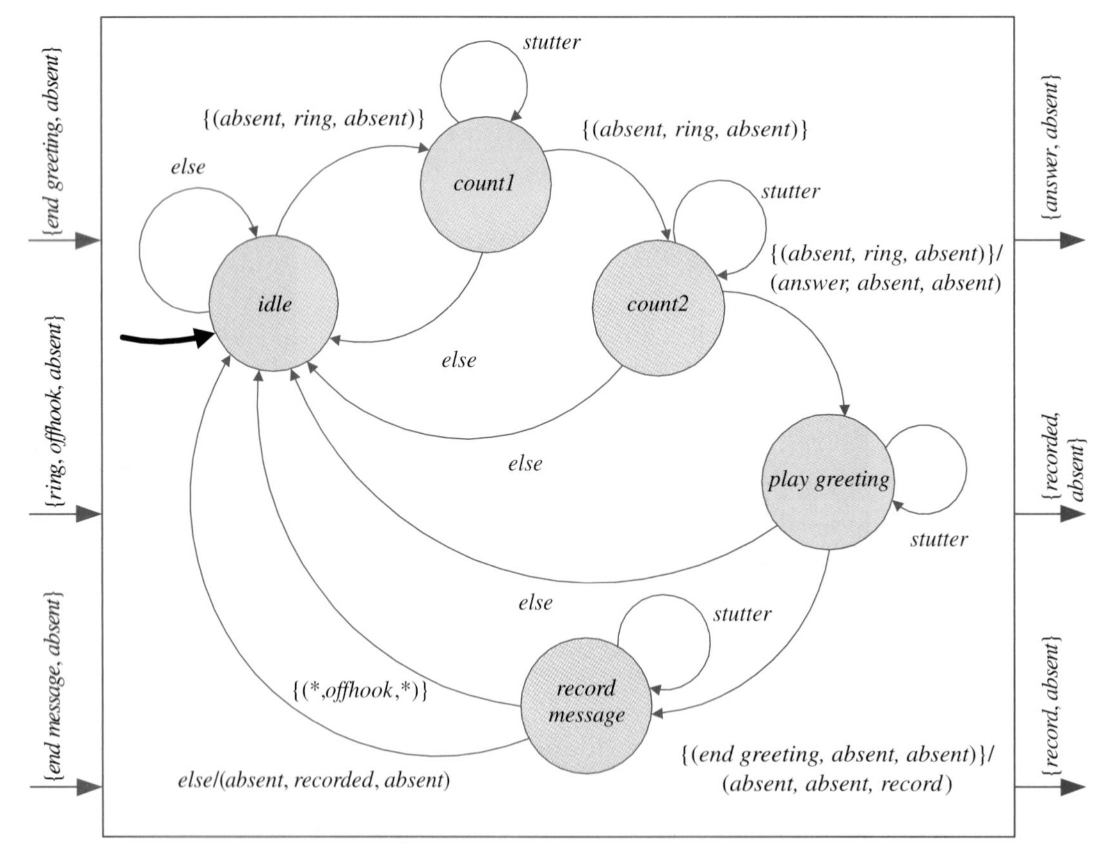

FIGURE 4.6: Answering machine with product-form inputs and outputs has three input ports and three output ports. *stutter* = {(*absent*, *absent*, *absent*)}

Several features distinguish the diagram in figure 4.6 from that of figure 3.1. Each state except the *idle* state has acquired a self-loop labeled *stutter*, which is a name for the guard

$$stutter = \{(absent, absent, absent)\}.$$

This self-loop prevents the state machine from returning to the idle state (via the *else* transition) when nothing interesting is happening on the inputs. Usually, there is no reaction if nothing interesting is happening on the inputs, but because of synchrony, this machine may be composed with others, and all machines have to react at the same time. Therefore, if anything interesting is happening elsewhere in the system, this machine has to react even though nothing interesting is happening here. Recall that such a reaction is called a stutter. The state does not change, and the output symbol produced is the stuttering element of the output alphabet.

Each guard now consists of a set of triples, inasmuch as the product-form input has three components. The shorthand "(*, *offhook*, *)" on the arc from the *record message* state to the *idle* state represents a set,

$$(*, offhook, *) = \{(absent, offhook, absent), (end\ greeting, offhook, absent),$$
$$(absent, offhook, end\ message), (end\ greeting, offhook, end\ message)\}.$$

The "*" is a **don't care** or **wildcard** notation. Anything in its position will trigger the guard.

Because there are three output ports, the output symbols are also triples, but most of them are implicitly $(absent, absent, absent)$. ❒

4.5 General feed-forward composition

Because state machines can have product-form inputs and outputs, it is easy to construct a composition of state machines that combines features of both the cascade composition of figure 4.3 and the side-by-side composition of figure 4.1. An example is shown in figure 4.7. In that figure,

$$Outputs_A = Outputs_{A1} \times Outputs_{A2}$$

and

$$Inputs_B = Inputs_{B1} \times Inputs_{B2}.$$

Notice that the path from the bottom port of machine A goes both to the output of the composite machine and to the top port of machine B. Sending a value to multiple destinations like this is called **forking**. In exercise 1 at the end of this chapter, you are asked to define the composite machine for this example.

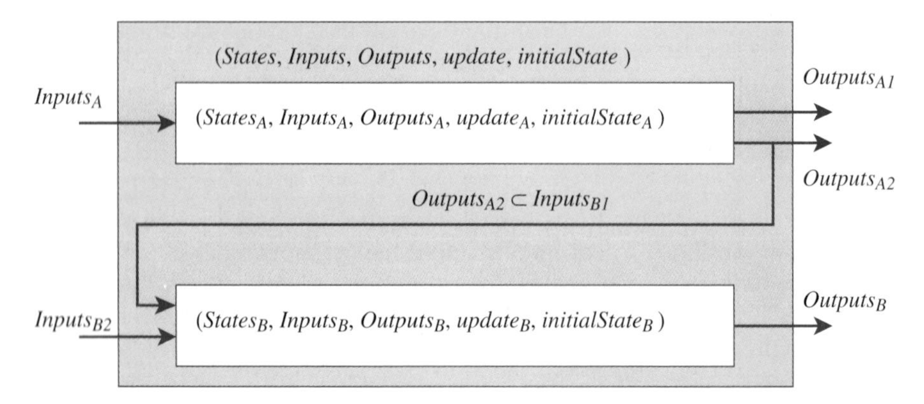

FIGURE 4.7: More complex composition.

Example 4.4: We compose the answering machine of figure 4.6 with a playback system, shown in figure 4.8, that plays messages that have been recorded by the answering machine. The playback system receives the *recorded* input symbol from the answering machine whenever the answering machine is done recording a message. Its task is to light an indicator that a message is pending and to wait for a user to press a play button on the answering machine to request that the pending messages be played back. When that button is pressed, all pending messages are played back. When they are done being played back, the indicator light is turned off.

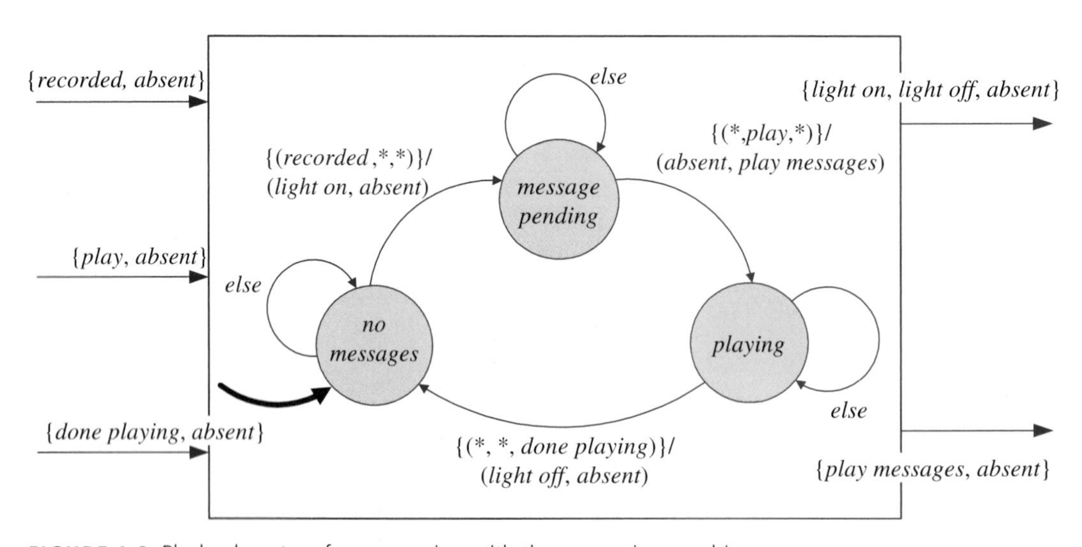

FIGURE 4.8: Playback system for composing with the answering machine.

The composition is shown in figure 4.9. The figure shows a number of other components, not modeled as state machines, to help understand how everything works in practice. These other components are shown as three-dimensional objects, to emphasize their physicality. We have simplified the figure by omitting the *absent* elements of all the sets. They are implicit.

A telephone line interface provides *ring* and *offhook* when these are detected. Detection of one of these can trigger a reaction of the composite machine. In fact, any output symbol from any of the physical components can trigger a reaction of the composite machine. When *AnsweringMachine* generates the *answer* output symbol, the "greeting playback device" plays back the greeting. From the perspective of the state machine model, all that happens is that time passes (during which some reactions may occur) and then an *end greeting* input symbol is received. The recording device works similarly. When *AnsweringMachine* generates a *recorded* output symbol, the

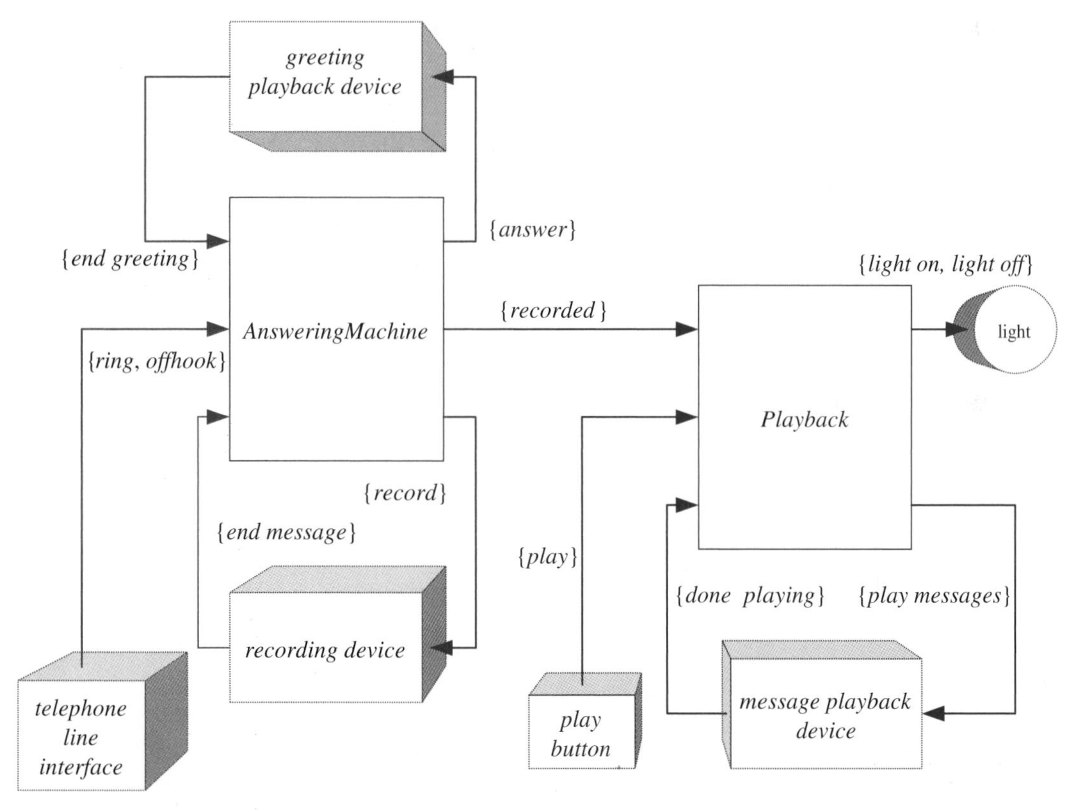

FIGURE 4.9: Composition of an answering machine with a message playback machine. The three-dimensional boxes are physical components that are not modeled as state machines. They are the sources of some inputs and the destinations of some outputs.

Playback machine responds by lighting the indicator light. When a user presses the play button, the input symbol *play* is generated, the composite machine reacts, and the *Playback* machine issues a *play messages* output symbol to the "message playback device." This device also allows time to pass and then generates a *done playing* input symbol to the composite state machine.

If we wish to model the playback or recording subsystem in more detail by using finite-state machines, then we need to be able to handle feedback compositions. These are considered later. ❐

4.6 Hierarchical composition

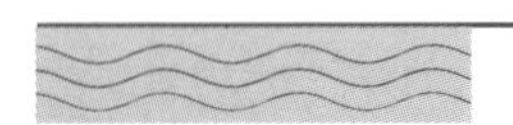

By using the compositions just discussed, we can now handle any interconnection of state machines that does not have feedback. Consider, for example, the cascade of the three state machines shown in figure 4.10. The composition techniques we have discussed so far involved only two state machines. It is easy to generalize the composition in figure 4.3 to handle three state machines (see exercise 2), but a more systematic method might be to apply the composition of figure 4.3 to compose two of the state machines and then apply it again to compose the third state machine with the result of the first composition. This is called **hierarchical composition**.

In general, given a collection of interconnected state machines, there are several ways to hierarchically compose them. For example, in figure 4.10, we

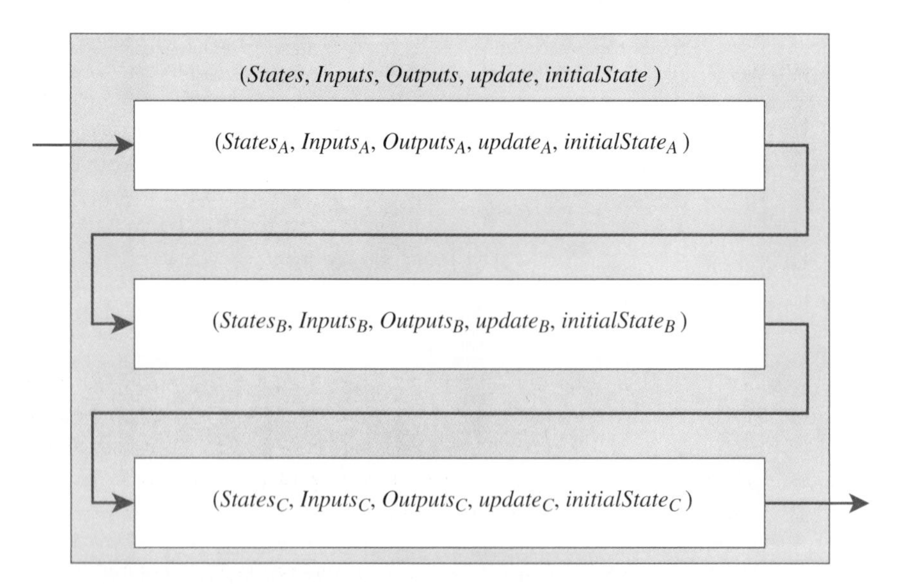

FIGURE 4.10: Cascade composition of three state machines. They can be composed in different ways into different, but bisimilar, state machines.

could first compose machines A and B to get, say, machine D, and then compose D with C. Alternatively, we could first compose B and C to get, say, machine E, and then compose E and A. These two procedures result in different but bisimilar state machine models (each simulates the other).

4.7 Feedback

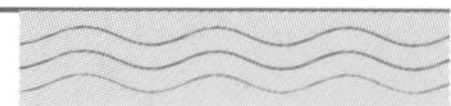

In simple feedback systems, an output from a state machine is fed back as an input to the same state machine. In more complicated feedback systems, several state machines might be connected in a loop; the output of one eventually affects its own input through some intervening state machines.

Feedback is a subtle form of composition in the synchronous model. In synchronous composition, in a reaction, the output symbol of a state machine is simultaneous with the input symbol. So the output symbol of a machine in feedback composition depends on an input symbol that depends on its own output symbol!

We frequently encounter such situations in mathematics. A common problem is to find x such that

$$x = f(x) \tag{4.6}$$

for a given function f. A solution to this equation, if it exists, is called a **fixed point** in mathematics. It is analogous to feedback because the "output" $f(x)$ of f is equal to its "input" x, and vice versa. The top diagram in figure 4.11 illustrates a similar relationship: The state machine's output symbol is the same as its (simultaneous) input symbol.

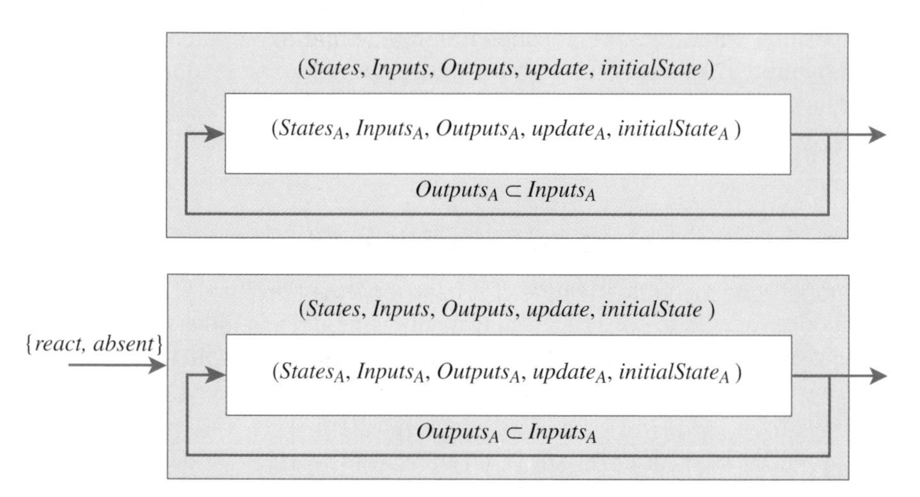

FIGURE 4.11: Feedback composition with no inputs.

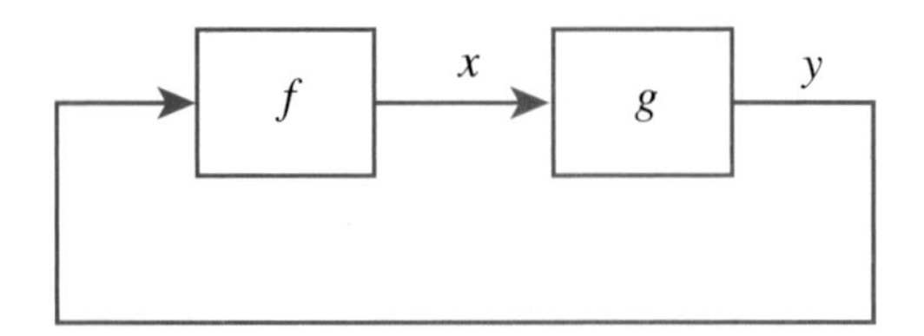

FIGURE 4.12: Illustration of a fixed-point problem.

A more complicated problem, involving two equations, is to find x and y so that

$$x = f(y) \text{ and } y = g(x). \tag{4.7}$$

The analogous feedback composition has two state machines in feedback, with the structure of figure 4.12.*

A fixed-point equation like (4.6) or (4.7) may have no fixed point, a unique fixed point, or multiple fixed points. Take, for example, the function $f\colon Reals \to Reals$, where $\forall\, x \in Reals,\ f(x) = 1 + x^2$. In this case, (4.6) becomes $x = 1 + x^2$, which has no fixed point in the real numbers. If $f(x) = 1 - x$, (4.6) becomes $x = 1 - x$, which has a unique fixed point, $x = 0.5$. Last, if $f(x) = x^2$, (4.6) becomes $x = x^2$, which has two fixed points, $x = 0$ and $x = 1$.

In the context of state machines, a feedback composition with no fixed point in some reachable state is a defective design; we call such a composition **ill-formed**. We cannot evaluate an ill-formed composition. Usually, we also wish to exclude feedback compositions that have more than one nonstuttering fixed point in some reachable state. Thus, these too are ill-formed. A feedback composition with a unique nonstuttering fixed point in all reachable states is **well-formed**. Fortunately, it is easy to construct well-formed feedback compositions, and they prove surprisingly useful. We explore this further, beginning with a somewhat artificial case of feedback composition with no inputs.

4.7.1 *Feedback composition with no inputs*

The upper state machine in figure 4.11 has an output port that feeds back to its input port. We wish to construct a state machine model that hides the feedback, as suggested by the figure. The result will be a state machine with no input. This

* Figure 4.9 would be a feedback composition if any of the three recording or playback devices were modeled as state machines. In that figure, however, these devices are part of the environment.

does not fit our model, which requires the environment to provide inputs to which the machine reacts. So we provide an artificial input alphabet,

$$Inputs = \{react, absent\},$$

as suggested in the lower machine in figure 4.11. We interpret the input symbol *react* as a command for the internal machine to react and the input symbol *absent* as a command for the internal machine to stutter. The output alphabet is

$$Outputs = Outputs_A.$$

This is an odd example of a synchronous/reactive system because of the need for this artificial input alphabet. Typically, however, such a system is composed with others, as suggested in figure 4.13. That composition does have an external input. So the overall composition, including the component with no external input, reacts whenever an external input symbol is presented, and there is no need for the artificial inputs. Of course, when a stuttering element is provided to the composite, all components stutter.

Although it is not typical, we first consider the example in figure 4.11 because the formulation of the composition is simplest. We augment the model to allow inputs after this.

In figure 4.11, for the feedback connection to be possible, of course, we must have

$$Outputs_A \subset Inputs_A.$$

Suppose the current state at the nth reaction is $s(n) \in States_A$. The problem is to find the output symbol $y(n) \in Outputs_A$. Because $y(n)$ is also the input symbol, it must satisfy

$$(s(n+1), y(n)) = update_A(s(n), y(n)),$$

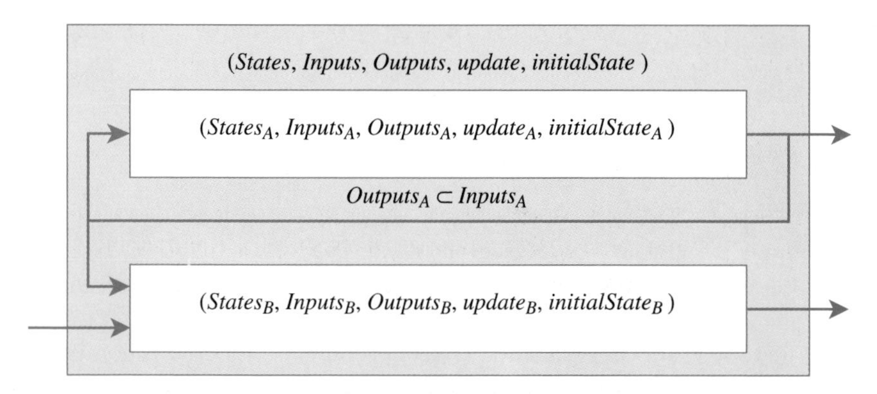

FIGURE 4.13: Feedback composition composed with another state machine.

where $s(n+1)$ is the next state. The difficulty here is that the "unknown" $y(n)$ appears on both sides. Once we find $y(n)$, $s(n+1)$ is immediately determined by the $update_A$ function. To simplify the discussion, we get rid of $s(n+1)$ by working with the function

$$output_A : States_A \times Inputs_A \to Outputs_A.$$

This function gives the output symbol as a function of the current state and the current input symbol, as we saw in section 3.1.1. Therefore, our problem is this: Given the current state $s(n)$ and the known function $output_A$, find $y(n)$ such that

$$y(n) = output_A(s(n), y(n)). \tag{4.8}$$

Here $s(n)$ is a known constant; therefore, the equation is of the form (4.6), and its solution, if it exists, is a fixed point.

One solution that is always available is to stutter; that is,

$$y(n) = absent, \text{ (and then } s(n+1) = s(n)),$$

because $absent = output_A(s(n), absent)$, assuming that *absent* is the stuttering input symbol for machine *A*. But this is not an interesting solution, because the state does not change. We want to find a nonstuttering solution for $y(n)$.

We say that the composition of figure 4.11 is **well-formed** if for every reachable $s(n) \in States_A$, there is a unique nonstuttering output symbol $y(n)$ that solves (4.8); otherwise, the composition is **ill-formed**. If the composition is well-formed, the composite machine definition is as follows:

$States = States_A$;

$Inputs = \{react, absent\}$;

$Outputs = Outputs_A$;

$initialState = initialState_A$; *and*

$$update(s(n), x(n)) = \begin{cases} update_A(s(n), y(n)), & \\ \text{where } y(n) \neq absent \text{ uniquely satisfies (4.8)} & \text{if } x(n) = react \\ (s(n), x(n)) & \text{if } x(n) = absent. \end{cases}$$

Notice that the composite machine is defined only if the composition is well-formed—that is, if there is a unique $y(n)$ that satisfies (4.8). If there is no such $y(n)$, the composition is ill-formed, and the composite machine is not defined.

The next example illustrates the difference between well-formed and ill-formed compositions. It suggests a procedure for solving (4.8) in the important special case of systems with state-determined output.

Example 4.5: Consider the three feedback compositions in figure 4.14. In all cases, the input and output alphabets of the component machines are

$$Inputs_A = Outputs_A = \{true, false, absent\}.$$

The input alphabet to the *composite* machine is $\{react, absent\}$, as in figure 4.11, but we do not show this (to reduce clutter in the figure). We want to find a nonstuttering solution $y(n)$ of (4.8). Because the output symbol is also the input symbol, we are looking for a nonstuttering input symbol.

Consider the top machine first. Suppose the current state is the initial state, $s(n) = 1$. There are two outgoing arcs, and for a nonstuttering input symbol, both produce $y(n) = false$, so we can conclude that the output symbol of the machine is *false*. Because the output symbol is *false*, the input symbol is also *false*, and the nonstuttering fixed point of (4.8) is unique:

$$output_A(1, false) = false.$$

The state transition taken by the reaction goes from state 1 to state 2.

Suppose next that the current state is $s(n) = 2$. Again, there are two outgoing arcs. Both arcs produce output symbol *true* for a nonstuttering input symbol, and so we can conclude that the output symbol is *true*. Because the output symbol is *true*, the input symbol is also *true*; there is again a unique nonstuttering fixed point,

$$output_A(2, true) = true;$$

and the state transition taken goes from 2 to 1. Because there is a unique nonstuttering fixed point in every reachable state, the feedback composition is well-formed.

The composite machine alternates states on each reaction and produces the output sequence

$$(false, true, false, true, false, true, \ldots)$$

for the input sequence

$$(react, react, react, \ldots).$$

The composite machine is shown in figure 4.15.

Now consider the second machine in figure 4.14. If the initial state is 1, the analysis is the same as that just given. There is a unique nonstuttering

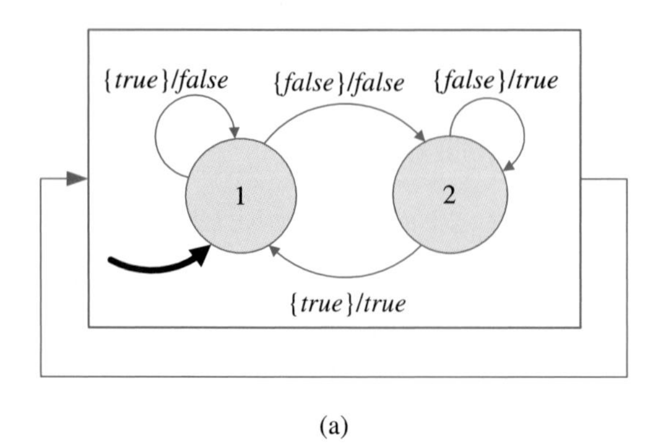

(a)

{true}/false
{false}/false
{false}/true
1
2
{true}/false

(b)

{true}/true
{false}/false
{false}/false
1
2
{true}/true

(c)

FIGURE 4.14: Three examples of feedback composition. Examples (b) and (c) are ill-formed. Composition (b) has no nonstuttering fixed point in state 2, whereas composition (c) has two nonstuttering fixed points in either state.

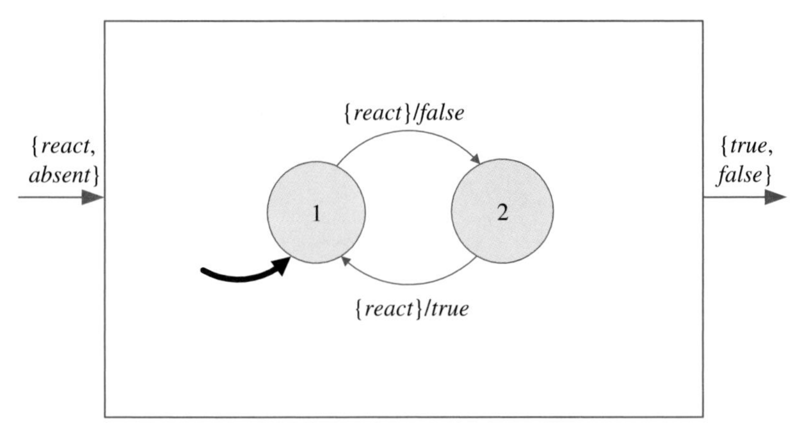

FIGURE 4.15: Composite machine for figure 4.14(a).

fixed point, the output and input symbols are both *false*, and the state transition goes from 1 to 2. But if the initial state is 2 and the unknown input symbol is *true*, the output symbol is *false*; and if the unknown input symbol is *false*, the output symbol is *true*. Thus, there is no nonstuttering fixed point $y(n)$ that solves

$$output_A(2, y(n)) = y(n),$$

and hence the feedback composition is not well-formed.

Consider the third machine in figure 4.14. This feedback composition is also ill-formed but for a different reason. If the initial state is 1 and the unknown input symbol is *true*, the output symbol is also *true*, and so *true* is a fixed point, and the output symbol can be *true*. However, the output symbol can also be *false*, because if it is, then a transition will be taken that produces the input symbol *false*. Therefore, *false* is also a fixed point. Thus, the problem here is that there are more than one nonstuttering solution, not that there is none!

Our conclusion is that machines like the second and third cannot be connected in a feedback composition as shown. The second is rejected because it has no solution; the third, because it has more than one. We accept feedback compositions only when there is exactly one nonstuttering solution in each reachable state. ❒

4.7.2 *State-determined output*

In the first machine of figure 4.14, in each state, all outgoing arcs produce the *same* output symbol, independently of the input symbol. In other words, the output symbol $y(n)$ depends only on the state; in the example, $y(n) = false$ if

$s(n) = 1$, and $y(n) = true$ if $s(n) = 2$. The unique fixed point of (4.8) is this output symbol, and we can immediately conclude that the feedback composition is well-formed.

We say that a machine A has **state-determined output** if in every reachable state $s(n) \in States_A$, there is a unique output symbol $y(n) = b$ (which depends on $s(n)$) that is independent of the nonstuttering input symbol; that is, for every $x(n) \neq absent$,

$$output_A(s(n), x(n)) = b.$$

In this special case of state-determined output, the composite machine is defined as follows:

$States = States_A$;

$Inputs = \{react, absent\}$;

$Outputs = Outputs_A$;

$initialState = initialState_A$; and

$update(s(n), x(n))$

$$= \begin{cases} update_A(s(n), b), & \\ \text{where } b \text{ is the unique output symbol in state } s(n) & \text{if } x(n) = react \\ (s(n), x(n)) & \text{if } x(n) = absent. \end{cases}$$

When a machine with state-determined output is combined with any other state machines in a feedback composition, the resulting composition is also well-formed, as illustrated in the next example.

Example 4.6: In figure 4.16, the machine A is combined with machine B in a feedback composition. A is the same as the first machine, and B is the same as the second machine in figure 4.14. (The output port of B is drawn on the left and the input port on the right so that the block diagram looks neater.) A has state-determined output, but B does not. The composition is well-formed.

To see this, suppose both machines are in their initial states 1. A produces output symbol *false*, which is independent of its input symbol. This output is the input of B, which then produces output symbol *false* and makes the transition to state 2. The output symbol *false* of B is the input to A, which makes the transition to its state 2. (A and B make their state transitions together in our synchronous/reactive model.) We can determine the output symbol and transition in the same way for all other states. The state diagram of the composite machine is shown in the bottom of the figure. Note that

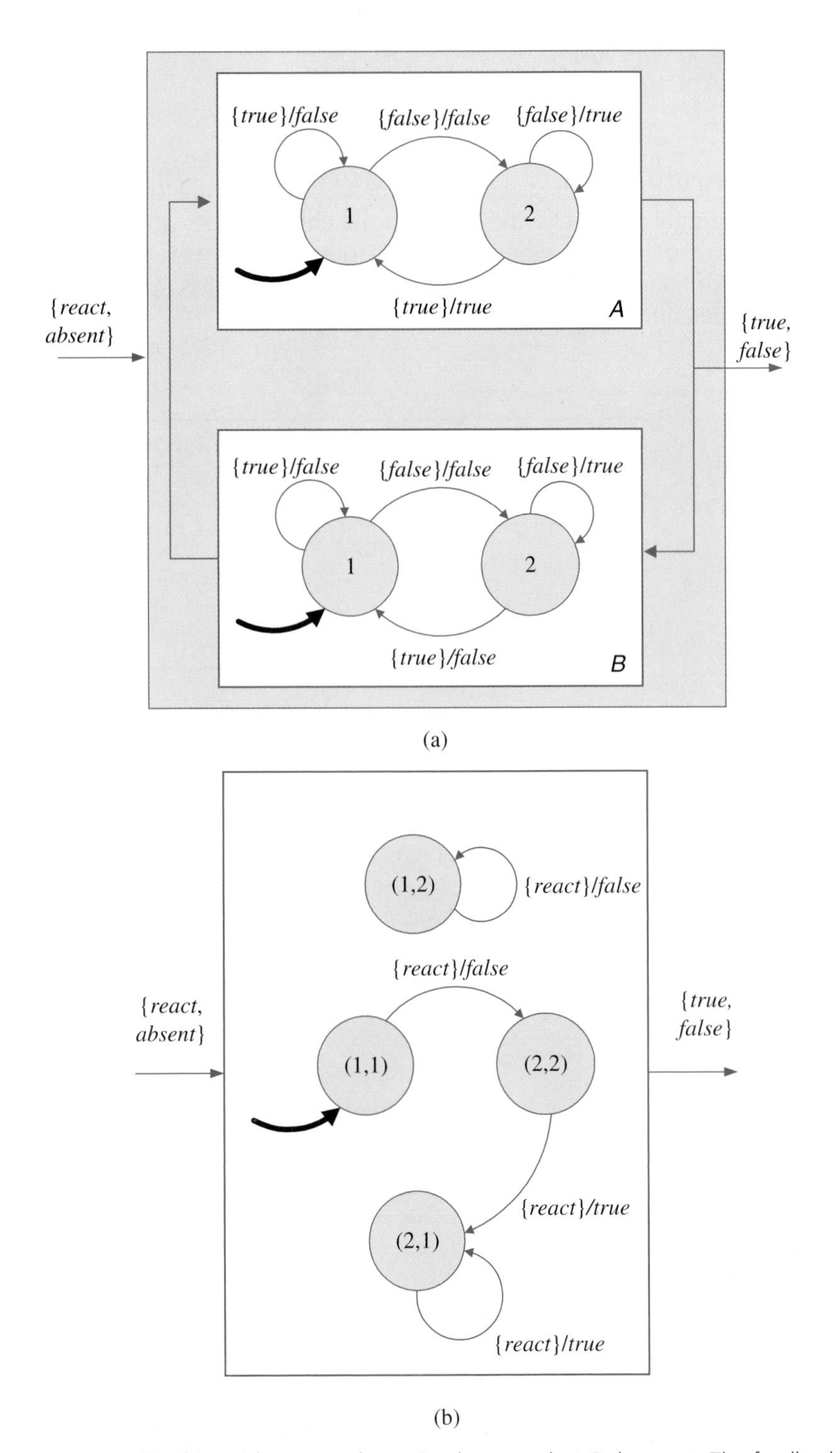

FIGURE 4.16: Machine *A* has state-determined output, but *B* does not. The feedback composition is well-formed, and the composite machine is shown on the bottom. Note that state `(1,2)` is not reachable.

state $(1,2)$ is not reachable from the initial state $(1,1)$, and so we could have ignored it in determining whether the composition is well-formed.

The input alphabet of the composite machine is $\{react, absent\}$, in which *absent* is the stuttering input symbol. The output alphabet is the same as the output alphabet of A, $\{true, false, absent\}$. The state space is $States_A \times States_B$. The *update* function is given by the following table:

	(next state, output) for input	
current state	***react***	***absent***
(1,1)	((2,2),*false*)	((1,1),*absent*)
(2,2)	((2,1),*true*)	((2,2), *absent*)
(1,2)	((1,2),*false*)	((1,2), *absent*)
(2,1)	((2,1),*true*)	((2,1), *absent*)

❒

It is possible for a machine without state-determined outputs to be placed in a well-formed feedback composition, as illustrated in the next example.

Example 4.7: Consider the example in figure 4.17. For the component machine, the output alphabet is $Outputs_A = \{true, false, maybe, absent\}$, and the input alphabet is $Inputs_A = \{true, false, absent\}$. The stuttering element is *absent*. The machine does not have state-determined output because, for instance, the outgoing arcs from state 1 can produce both *maybe* and *false*. Nevertheless, equation (4.8) has a unique nonstuttering fixed point in each state:

$$output_A(1, false) = false \text{ and } output_A(2, true) = true.$$

Thus, the feedback composition is well-formed. The composite machine is shown on the bottom. ❒

It can be considerably harder to find the behavior of a feedback composition without state-determined outputs, even if the composition is well-formed. In section 4.7.4, we give a constructive procedure that often works to find a fixed point quickly and to determine whether it is unique. However, even that procedure does not always work (and, in fact, fails in example 4.7). If the input alphabet is finite, the only strategy that always works is to try all possible output values $y(n)$ in (4.8) for each reachable state $s(n)$. Before discussing this procedure, we generalize to more interesting feedback compositions.

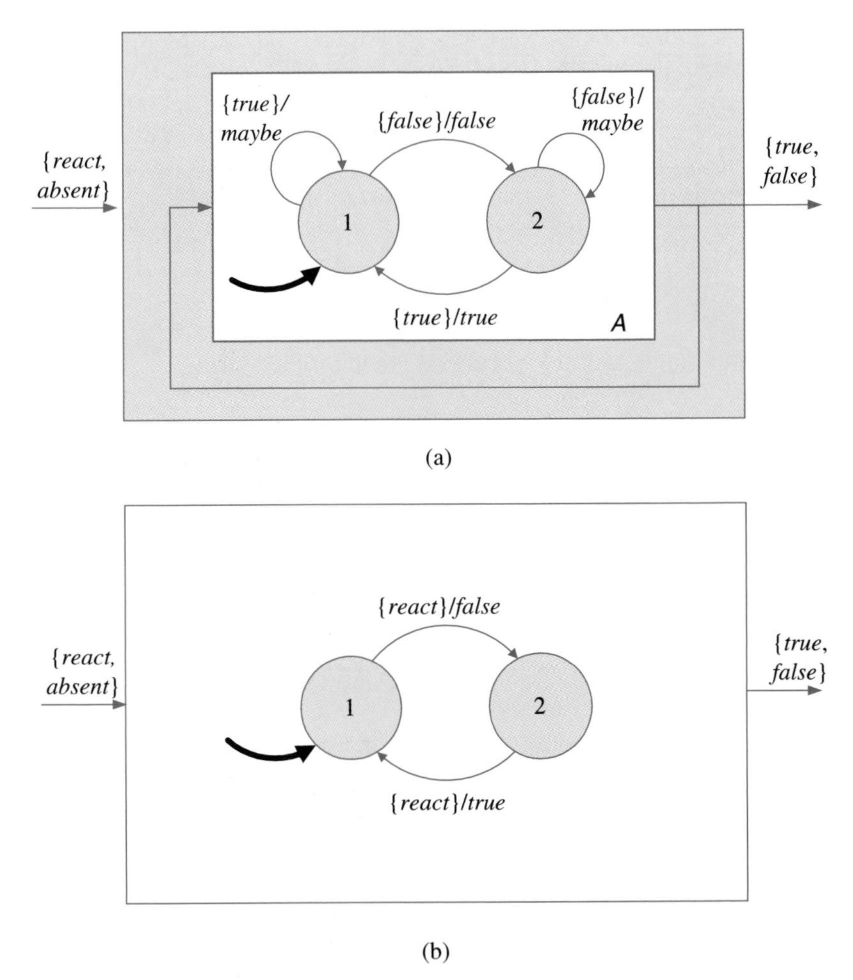

FIGURE 4.17: Machine *A* does not have state-determined outputs, but the feedback composition is well-formed. The machine on the bottom is the composite machine.

4.7.3 *Feedback composition with inputs*

Now consider the state machine in figure 4.18. It has two input and output ports. The second output port feeds back to the second input port. We wish to construct a state machine model that hides the feedback, as suggested by the figure, and becomes a simple input–output state machine. This is similar to the example in figure 4.11, but now there is an additional input and an additional output. The procedure for finding the composite machine is similar, but the notation is more cumbersome. Given the current state and the current external input symbol, we must determine the "unknown" output symbol.

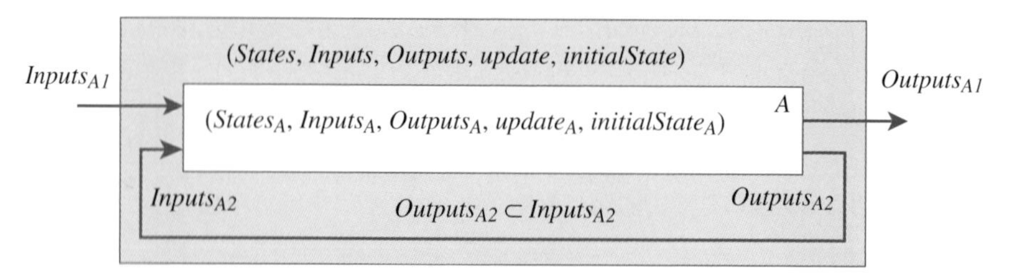

FIGURE 4.18: Feedback composition of a state machine.

The inputs and outputs of machine A are in product form:

$$Inputs_A = Inputs_{A1} \times Inputs_{A2}$$

and

$$Outputs_A = Outputs_{A1} \times Outputs_{A2}.$$

For the feedback composition to be possible, we must have

$$Outputs_{A2} \subset Inputs_{A1}.$$

The output function of A is

$$output_A : States_A \times Inputs_A \to Outputs_A.$$

It is convenient to write it in product form as

$$output_A = (output_{A1}, output_{A2}),$$

where

$$output_{A1} : States_A \times Inputs_A \to Outputs_{A1}$$

gives the output symbol at the first output port and

$$output_{A2} : States_A \times Inputs_A \to Outputs_{A2}$$

gives the output symbol at the second output port.

Suppose that at the nth reaction, the current state of A is $s(n)$ and the current external input symbol is $x_1(n) \in Inputs_{A1}$. Then the problem is to find the "unknown" output symbol $(y_1(n), y_2(n)) \in Outputs_A$ such that

$$output_A(s(n), (x_1(n), y_2(n))) = (y_1(n), y_2(n)). \tag{4.9}$$

The symbol $y_2(n)$ appears on both sides because the second input $x_2(n)$ to machine A is equal to $y_2(n)$. In terms of the product form, (4.9) is equivalent to two equations:

$$output_{A1}(s(n), (x_1(n), y_2(n))) = y_1(n) \tag{4.10}$$

and

$$output_{A2}(s(n), (x_1(n), y_2(n))) = y_2(n). \tag{4.11}$$

In these equations, $s(n)$ and $x_1(n)$ are known, whereas $y_1(n)$ and $y_2(n)$ are unknown. Observe that if (4.11) has a unique solution $y_2(n)$, then the input symbol to A is $(x_1(n), y_2(n))$ and the next state $s(n+1)$ and output symbol $y_1(n)$ are determined. Thus, the fixed point equation (4.11) plays the same role as (4.8).

We say that the composition of figure 4.18 is well-formed if for every reachable state $s(n) \in States_A$ and for every external input symbol $x_1(n) \in Inputs_{A1}$, there is a unique nonstuttering output symbol $y_2(n) \in Outputs_{A2}$ that solves (4.11). If the composition is well-formed, the composite machine definition is:

$$States = States_A;$$

$$Inputs = Inputs_{A1};$$

$$Outputs = Outputs_{A1};$$

$$initialState = initialState_A;$$

$$update(s(n), x(n)) = (nextState(s(n), x(n)), output(s(n), x(n)));$$

$$nextState(s(n), x(n)) = nextState_A(s(n), (x(n), y_2(n))); \quad \text{and}$$

$$output(s(n), x(n)) = output_A(s(n), (x(n), y_2(n))), \text{ where } y_2(n)$$

is the unique solution of (4.11).

(The *nextState* function is defined in section 3.1.1.)

In the following example, we illustrate the procedure for defining the composition machine, given a sets and functions description (3.1) for the component machine A.

Example 4.8: Figure 4.19 shows a feedback composition in which component machine A has two input ports and one output port,

$$Inputs_A = Reals \times Reals, \qquad Outputs_A = Reals,$$

and states $States_A = Reals$. Thus, A has infinite input and output alphabets and infinitely many states. At the nth reaction, the pair of input values is

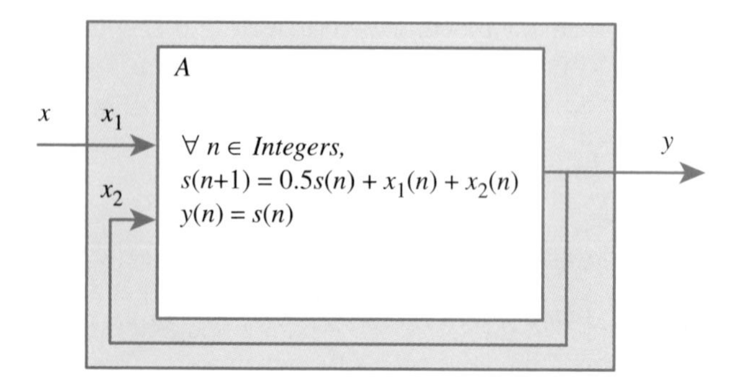

FIGURE 4.19: Machine *A* has two input ports and one output port. The output port is connected to the second input port. The composition is well-formed. The composite machine is shown at the bottom.

denoted by $(x_1(n), x_2(n))$, the current state by $s(n)$, the next state by $s(n+1)$, and the output symbol by $y(n)$. In terms of these, the update function is given by

$$\begin{aligned}(s(n+1), y(n)) &= update_A(s(n), (x_1(n), x_2(n))) \\ &= (0.5s(n) + x_1(n) + x_2(n), s(n)).\end{aligned}$$

Equivalently,

$$s(n+1) = nextState_A(s(n), (x_1(n), x_2(n))) = 0.5s(n) + x_1(n) + x_2(n)$$

and

$$y(n) = output_A(s(n), (x_1(n), x_2(n))) = s(n).$$

Thus, the component machine *A* has state-determined output. The feedback connects the output port to the second input port, so $x_2(n) = y(n)$. Given the current state $s(n)$ and the external input symbol $x(n)$ at the first input port, (4.11) becomes

$$output_A(s(n), (x_1(n), x_2(n))) = x_2(n),$$

which yields

$$x_2(n) = s(n).$$

Thus, the composite machine is defined by

$$Inputs = Reals, \; Outputs = Reals, \; States = Reals$$

and

$$update(s(n), x(n)) = (0.5s(n) + x(n) + s(n), s(n)) = (1.5s(n) + x(n), s(n)).$$

Note that the input to the composite machine is a scalar. The composite machine is shown in the lower part of the figure. ❐

4.7.4 *Constructive procedure for feedback composition*

Our examples so far involve one or two state machines and a feedback loop. If any machine in the loop has state-determined output, then finding the fixed point is easy. Most interesting designs are more complicated, involving several state machines and several feedback loops, and the loops do not necessarily include machines with state-determined output.

In this section, we describe a constructive procedure for finding the fixed point that often (but not always) works. It is "constructive" in the sense that it can it be applied mechanically and will, in a finite number of steps, either identify a fixed point or give up. The approach is simple. At each reaction, begin with all unspecified signals having value *unknown*. Then with what is known about the input symbols, try each state machine to determine as much as possible about the output symbols. You can try the state machines in any order. Taking what you learn about the output symbols, update what you know about the feedback input symbols, and repeat the process, trying each state machine again. Repeat this process until all signal values are specified or until you learn nothing more about the output symbols. We illustrate the procedure in an example involving only one machine, but keep in mind that the procedure works for any number of machines.

Example 4.9: Figure 4.20 shows a feedback composition without state-determined output. Nonetheless, our constructive procedure can be used to find a unique fixed point for each reaction. Suppose that the current state is a and that the input to the composition is *react*. Begin by assuming that the symbol on the feedback connection is *unknown*. Try component machine A (this is the only component machine in this example, but if there were more, we could try them in any order). Examining machine A, we see that in its current state, a, the output symbol cannot be fully determined. Thus, this machine does not have state-determined output. However, more careful examination reveals that in state a, the second element of the output tuple *is* determined. That second element has value 1. Fortunately, this changes the value on the feedback connection from *unknown* to 1.

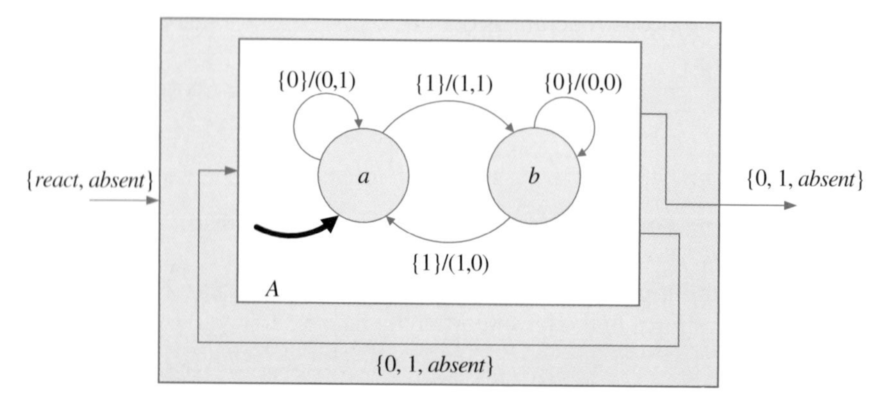

FIGURE 4.20: Feedback composition without state-determined output.

Now we repeat the procedure. We choose a state machine to try. Again, there is only one state machine in this example, so we try *A*. This time, we know that the input symbol is 1, so we know that the machine must take the transition from *a* to *b* and produce the output tuple (1, 1). The result is that all symbols become known for the reaction, and so we are done evaluating the reaction.

Now assume that the current state is *b*. Again, the feedback symbol is initially *unknown*, but once again, trying *A*, we see that the second element of the output tuple must be 0. Thus, we change the feedback symbol from *unknown* to 0 and try the machine again. This time, its input is 0, so it must take the self-loop back to *b* and produce the output tuple (0, 0).

Recall that the set *Behaviors* is the set of all (x, y) such that x is an input sequence and y is an output sequence. For this machine, ignoring stuttering, the only possible input sequence is $(react, react, react, \ldots)$. We have just determined that the resulting output sequence is $(1, 0, 0, 0, \ldots)$. Thus, ignoring stuttering,

$$Behaviors = \{((react, react, react, \ldots), (1, 0, 0, 0, \ldots))\}.$$

Of course, we should take into account stuttering, so this set needs to be augmented with all (x, y) pairs that look like the one above but have stuttering symbols inserted. ❐

This procedure can be applied in general to any composition of state machines. If the procedure can be applied successfully (nothing remains *unknown*) for all reachable states of the composition, then the composition is well-formed. The following example applies the procedure to a more complicated example.

Example 4.10: We add more detail to the message recorder in figure 4.9. In particular, as shown in figure 4.21, we wish to model the fact that the message

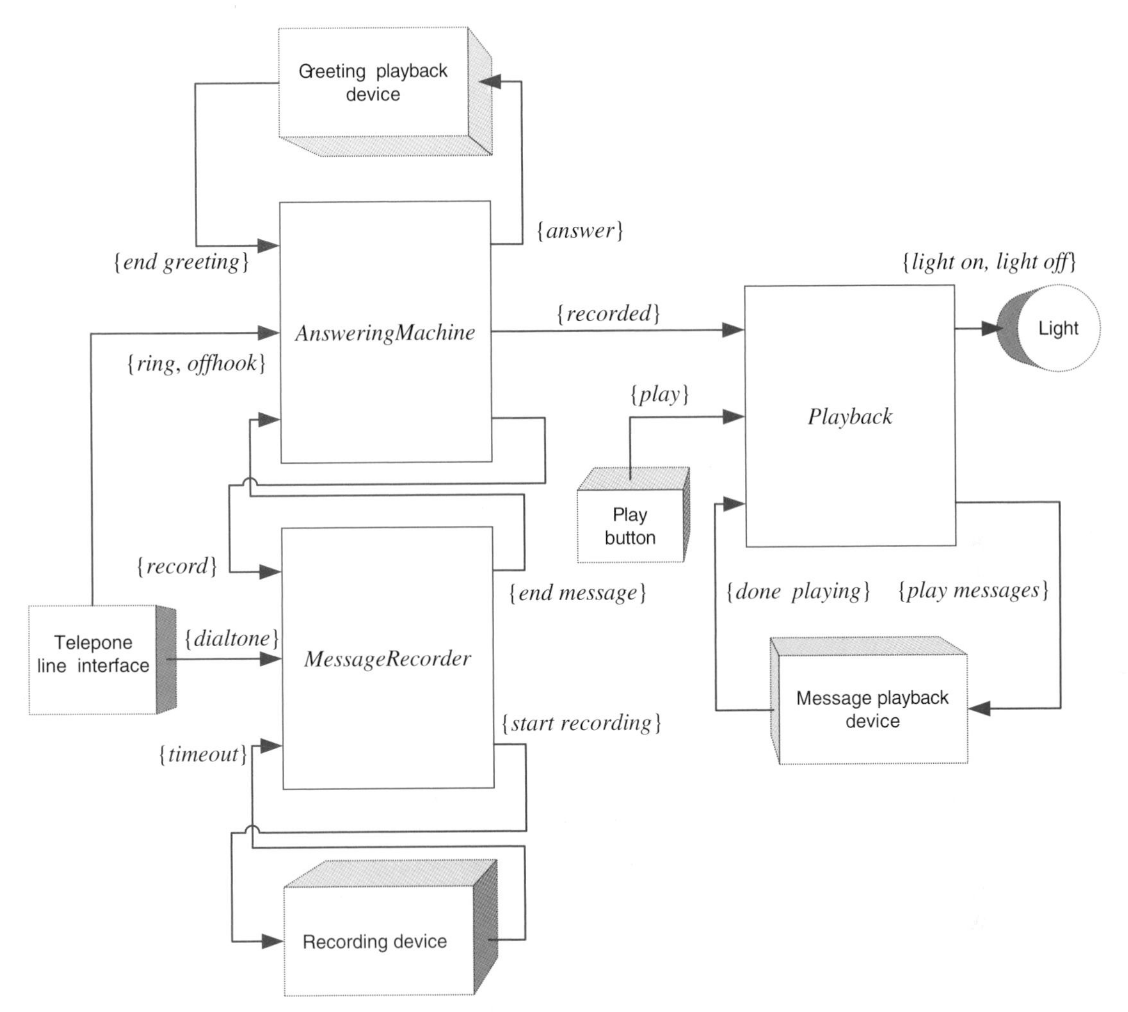

FIGURE 4.21: Answering machine composition with feedback. The *absent* elements are not shown (to reduce clutter).

recorder stops recording when either it detects a dial tone or a timeout period is reached. This is modeled by a two-state finite-state machine, shown in figure 4.22. Note that this machine does not have state-determined output. For example, in state *idle*, the output could be (*absent*, *start recording*) or it could be (*absent*, *absent*) when the input is not the stuttering input.

The *MessageRecorder* and *AnsweringMachine* state machines form a feedback loop. Let us verify that composition is well-formed. First, note that in the *idle* state of the *MessageRecorder*, the upper output symbol is known to be *absent* (see figure 4.22). Thus, only in the *recording* state is there any possibility of a problem that would lead to the composition's

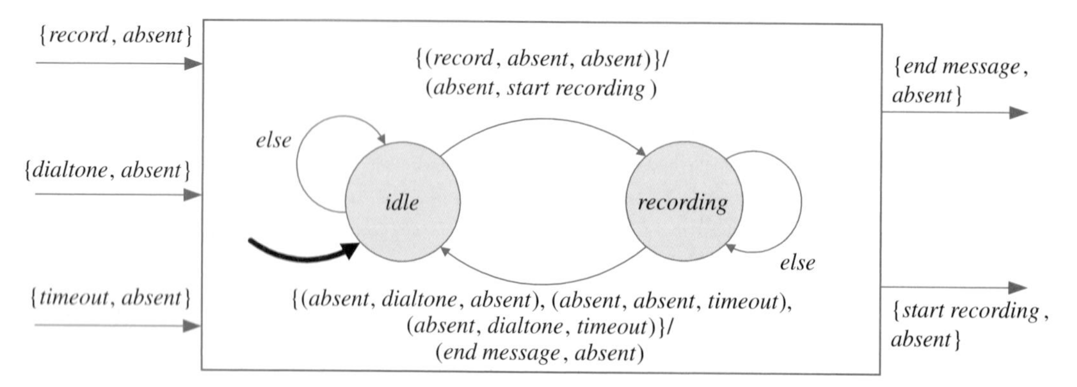

FIGURE 4.22: Message recorder subsystem of the answering system.

being ill-formed. In that state, the output symbol is not known unless the input symbols are known. However, notice that the *recording* state is entered only when a *record* input symbol is received. Figure 4.6 shows that the *record* value is generated only when the machine enters the state *record message*. But in all arcs emerging from that state, the lower output symbol of *AnsweringMachine* is always absent; the input symbol does not need to be known for that to be apparent. Continuing this reasoning by considering all possible state transitions from this point, we can convince ourselves that the feedback loop is well-formed. ❐

The sort of reasoning in this more complicated example is difficult and error-prone for even moderate compositions of state machines. It is best automated. Compilers for synchronous languages do exactly this. Successfully compiling a program involves proving that feedback loops are well-formed.

4.7.5 *Exhaustive search*

If a feedback composition has one or more machines with state-determined output, then finding a unique fixed point is easy. Without such state-determined output, we can apply the procedure in the previous section. Unfortunately, if the procedure fails, we cannot conclude that the composition is ill-formed. The procedure fails for example 4.7, shown in figure 4.17, despite the fact that this example is well-formed. For that example, we can determine the unique fixed point by exhaustive search; that is, for each reachable state of the composition, and for each possible input to the composition, we try all possible transitions out of the current states of the component machines. We reject those that lead to a contradiction. For example, in figure 4.17, assuming the current state is 1, the output of the component machine cannot be *maybe* because then the input would have to be *maybe*, which would result in the output's being *absent*. If after all contradictions are rejected there remains exactly one possibility in each reachable state, then the composition is well-formed.

Exhaustive search works in figure 4.17 only because the number of reachable states is finite and the number of transitions out of each state is finite. If either of these conditions is violated, then exhaustive search will not work. Thus, there are state machines that when put in a feedback loop are well-formed but for which there is no constructive procedure for evaluating a reaction (see Probing further: Constructive semantics). Even when exhaustive search is theoretically possible, in practice the number of possibilities that must be tried grows extremely fast.

PROBING FURTHER

Constructive semantics

The term **semantics** means meaning. We have defined the meaning of compositions of state machines by using the notion of synchrony, which makes feedback compositions particularly interesting. When we define "well-formed," we are, in effect, limiting the types of compositions that are valid. Compositions that are not well-formed fall outside our synchronous semantics. They have no meaning.

One way to define the semantics of a composition is to provide a procedure for evaluating the composition (the resulting procedure is called an **operational semantics**). We have given three successively more difficult procedures for evaluating a reaction of a composition of state machines with feedback. If at least one machine in each directed loop has state-determined output, then it is easy to evaluate a reaction. If not, we can apply the constructive procedure of section 4.7.4. However, that procedure may result in some feedback connection's remaining *unknown* even though the composition is well-formed. The ultimate procedure is exhaustive search, as described in section 4.7.5. However, exhaustive search is not always possible, and even when it is theoretically possible, the number of possibilities to explore may be so huge that it is not practical. There are state machines that when put in a feedback loop are well-formed but for which there is no constructive procedure for evaluating a reaction and no constructive way to demonstrate that they are well-formed. Thus, there is no operational semantics for our feedback compositions.

This situation is not uncommon in computing and in mathematics. Kurt Gödel's famous incompleteness theorem of 1931, for example, states (loosely) that in any formal logical system, there are statements that are true but not provable. This is analogous in that we can have feedback compositions that are well-formed, but we have no procedure that will always work to demonstrate that they are well-formed. Around the same time, Alan Turing and Alonzo Church demonstrated that there are functions that cannot be computed by any procedure.

To deal with this issue, Gerard Berry proposed that synchronous compositions have a **constructive semantics**, which means precisely that well-formed compositions are defined to be those for which the constructive procedure of section 4.7.4 works. When that procedure fails, we simply declare the composition to be unacceptable. This is a pragmatic solution, and in many situations, it is adequate. However, it proves too restrictive; particularly in the field of feedback control. Most feedback control systems in practical use would be rejected by this semantics.

See G. Berry, *The Constructive Semantics of Esterel*, Book Draft, ftp://ftp.esterel.org/esterel/pub/papers/constructiveness3.ps.

4.7.6 Nondeterministic machines

Nondeterministic state machines can be composed in the same way that deterministic state machines are composed. In fact, because deterministic state machines are a special case, the two types of state machines can be mixed in a composition. Compositions without feedback are straightforward and operate almost exactly as described earlier (see exercises 14 and 15). Compositions with feedback require a small modification to our evaluation process.

Recall that to evaluate the reaction of a feedback composition, we begin by setting to *unknown* any input symbols that are not initially known. We then proceed through a series of rounds in which, in each round, we attempt to determine the output symbols of the state machines in the composition, given what we know about the input symbols. After some number of rounds, no more information is gained. At this point, if all the input and output symbols are known, the composition is well-formed. This procedure works for most (but not all) well-formed compositions.

This process needs to be modified slightly for nondeterministic machines because in each reaction, a machine may have several possible output symbols and several possible next states. For each machine, at each reaction, we define the sets $PossibleInputs \subset Inputs$, $PossibleNextStates \subset States$, and $PossibleNextOutputs \subset Outputs$. If the inputs to a particular machine in the composition are known completely, then *PossibleInputs* has exactly one element. If they are completely unknown, then *PossibleInputs* is empty.

The rounds proceed in a manner similar to that described earlier. For each state machine in the composition, given what is known about the input symbols (i.e., given *PossibleInputs*), determine what you can about the next state and output symbols. This may result in the addition of elements to *PossibleNextStates* and *PossibleNextOutputs*. When a round results in no such added elements, the process has converged. If none of the *PossibleInputs* or *PossibleOutputs* sets is empty, then the composition is well-formed.

4.8 Summary

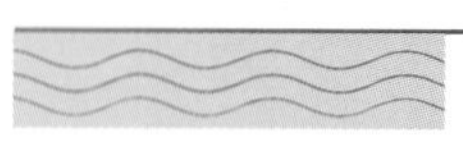

Many systems are designed as state machines. Usually the design is structured by composing component state machines. In this chapter, we considered synchronous composition. Feedback composition proves particularly subtle because the input symbol of a state machine in a reaction may depend on its own output symbol in the same reaction. We call a feedback composition well-formed if every signal has a unique nonstuttering symbol in each reaction. In lab 4 (see Lab Manual), you contruct a well-formed feedback composition of state machines.

Describing systems as compositions of state machines helps in many ways. It promotes understanding. The block diagram syntax that describes the structure often shows the individual components that are responsible for distinct functions of the overall system. Some components may already be available, and so we can reuse their designs. The design of the answering machine in figure 4.9 takes into account the availability of the telephone line interface, recording device, and so forth. Composition also simplifies description; once we specify the component state machines and the composition, the overall state machine is automatically defined by the rules of composition. Compilers for synchronous programming languages and tools for verification do this defining automatically.

We have three successively more difficult procedures for evaluating a reaction of a composition of state machines with feedback. If at least one machine in each directed loop has state-determined output, then it is easy to evaluate a reaction. If it does not, we can apply the constructive procedure of section 4.7.4. However, that procedure may be inconclusive. The ultimate procedure is exhaustive search, as described in section 4.7.5. However, exhaustive search is not always possible, and even when it is theoretically possible, the number of possibilities to explore may be so huge that it is not practical.

EXERCISES

KEY: E = mechanical T = requires plan of attack C = more than 1 answer

E 1. Define the composite state machine in figure 4.7 in terms of the component machines, as done for the simpler compositions in figures 4.3 and 4.1. Be sure to state any required assumptions.

E 2. Define the composite state machine in figure 4.10 in terms of the component machines, as done for the simpler compositions in figures 4.3 and 4.1. Be sure to state any required assumptions. Give the definition in two different ways:

(a) Directly form a product of the three state spaces.

(b) First compose the A and B state machines to get a new D state machine, and then compose D with C.

(c) Comment on the relationship between the models in part (a) and (b).

T 3. Consider the state machine *UnitDelay* studied in part (a) of exercise 5 at the end of chapter 3.

(a) Construct a state machine model for a cascade composition of two such machines. Give the sets and functions model (it is easier than the state transition diagram or table).

(b) Are all of the states in the state space of your model in part (a) reachable? If not, give an example of an unreachable state.

(c) Give the state space (only) for cascade compositions of three- and four-unit delays. How many elements are there in each of these state spaces?

(d) Give an expression for the size of the state space as function of the number N of cascaded delays in the cascade composition.

C 4. Consider the parking meter example of chapter 3, example 3.1, and the modulo N counter of exercise 4 at the end of chapter 3. Use these two machines to model (1) a citizen who parks at the meter when the machines start and inserts 25 cents every 30 minutes and (2) a police officer who checks the meter every 45 minutes and issues a ticket if the time on the meter has expired. For simplicity, assume that the police officer issues a new ticket each time he finds the meter expired and that the citizen remains parked forever.

You may construct the model at the block diagram level, as in figure 4.9, but describe in words any changes you need to make to the designs of chapter 3. Construct state transition diagrams for any additional state machines you need. How long does it take for the citizen to get the first parking ticket?

Assume you have an eternal clock that emits an event *tick* every minute.

Note that the output alphabet of the modulo N counter does not match the input alphabet of the parking meter. Neither does its input alphabet match the output alphabet of the parking meter. Thus, one or more intermediate state machines are needed to translate these alphabets. You should fully specify these state machines (i.e., don't just give them at the block diagram level). Hint: These state machines, which perform an operation called **renaming**, need only one state.

C 5. Consider a machine in which

$$\begin{aligned} \mathit{States} &= \{0, 1, 2, 3\}, \\ \mathit{Inputs} &= \{\mathit{increment}, \mathit{decrement}, \mathit{reset}, \mathit{absent}\}, \\ \mathit{Outputs} &= \{\mathit{zero}, \mathit{absent}\}, \text{ and} \\ \mathit{initialState} &= 0, \end{aligned}$$

such that *increment* increases the state by 1 (modulo 4), *decrement* decreases the state by 1 (modulo 4), *reset* resets the state to 0, and the output symbol is *absent* unless the next state is 0, in which case the output symbol is *zero*. Thus, for example, if the current state is 3 and the input is *increment*, then the new state will be 0 and the output will be *zero*. If the current state is 0 and the input is *decrement*, then the new state will be 3 and the output will be *absent*.

(a) Give the *update* function for this machine, and sketch the state transition diagram.

(b) Design a cascade composition of two state machines, each with two states, such that the composition has the same behaviors as the one just described. Construct a diagram of the state machines and their composition, and carefully define all the input and output alphabets.

(c) Devise a bisumulation relation between the single machine and the cascade composition.

C 6. A road has a pedestrian crossing with a traffic light. The light is normally green for vehicles, and the pedestrian is told to wait. However, if a pedestrian presses a button, the light turns yellow for 30 seconds and then red for 30 seconds. When it is red, the pedestrian is told "cross now." After the 30 seconds of red, the light turns back to green. If a pedestrian presses the button again while the light is red, then the red is extended to a full minute.

Construct a composite model for this system that has at least two state machines, *TrafficLight* for the traffic light seen by the cars, and *WalkLight* for the walk light seen by the pedestrians. The state of machine should represent the state of the lights. For example, *TrafficLight* should have at least three states: one for green, one for yellow, and one for red. Each color may, however, be associated with more than one state. For example, there may be more than one state in which the light is red. It is typical in modeling systems for the states of the model to represent states of the physical system.

Assume you have a timer such that if you emit an output *start timer*, then 30 seconds later an input symbol *timeout* will appear. It is sufficient to give the state transition graphs for the machines. State any assumptions you need to make.

E 7. Suppose you are given two state machines A and B. Suppose the sizes of the input alphabets are i_A and i_B, respectively; the sizes of the output alphabets are o_A and o_B, respectively; and the numbers of states are s_A and s_B, respectively. Give the sizes of the input and output alphabets and the number of states for the following compositions:

(a) side-by-side,

(b) cascade, and

(c) feedback, where the structure of the feedback follows the pattern in figure 4.16(a).

T 8. Example 4.2 shows a state machine in which a state is not reachable from the initial state. Here is a recursive algorithm for calculating the reachable states for any nondeterministic machine:

$$StateMachine = (States, Inputs, Outputs, possibleUpdates, initialState).$$

Recursively define subsets $ReachableStates(n)$, $n = 0, 1, \ldots$ of *States* by the following:

$$ReachableStates(0) = \{initialState\},$$

and for $n \geq 0$,

$$\begin{aligned} ReachableStates(n+1) = \{&s(n+1) \mid \exists\, s(n) \in ReachableStates(n), \\ &\exists\, x(n) \in Inputs, \exists\, y(n) \in Outputs, \text{ where} \\ &(s(n+1), y(n)) \in possibleUpdates(s(n), x(n))\} \\ &\cup\, ReachableStates(n). \end{aligned}$$

In words: $ReachableStates(n+1)$ is the set of states that can be reached from $ReachableStates(n)$ in one step by using any input symbol, together with $ReachableStates(n)$.

(a) Show that for all n, $ReachableStates(n) \subset ReachableStates(n+1)$.

(b) Show that $ReachableStates(n)$ is the set of states that can be reached in n or fewer steps, starting in *initialState*. Now show that if for some n

$$ReachableStates(n) = ReachableStates(n+1), \tag{4.12}$$

then $ReachableStates(n) = ReachableStates(n+k)$ for all $k \geq 0$.

(c) Suppose (4.12) holds for $n = N$. Show that $ReachableStates(N)$ is the set of all reachable states—that is, that this set comprises all the states that can be reached by using any input sequence starting in *initialState*.

(d) Suppose there are N states. Show that (4.12) holds for $n = N$.

(e) Compute $ReachableStates(n)$ for all n for the machine in figure 4.4.

(f) Suppose *States* is infinite. Show that the set of reachable states is given by

$$\cup_{n=0}^{\infty} ReachableStates(n).$$

T 9. The algorithm in Exercise 8 has a fixed-point interpretation. For a nondeterministic state machine

$$StateMachine = (States, Inputs, Outputs, possibleUpdates, initialState),$$

define the function $nextStep: P(States) \to P(States)$ that maps subsets of *States* into subsets of *States* (recall that $P(A)$ is the power set of A) as follows: for any $S(n) \subset States$,

$$nextStep(S(n)) = \{s(n+1) \mid \exists\, s(n) \in S(n), \exists\, x(n) \in Inputs, \exists\, y(n) \in Outputs,$$

$$(s(n+1), y(n)) \in possibleUpdates(s(n), x(n))\} \cup S(n).$$

By definition, a fixed point of *nextStep* is any subset $S \subset States$ such that $nextStep(S) = S$.

(a) Show that both $\emptyset$ and *States* are fixed points of *nextStep*.

(b) Let *ReachableStates* be the set of all states that can be reached starting in *initialState*. Show that *ReachableStates* is also a fixed point.

(c) Show that *ReachableStates* is the least fixed point of *nextStep* containing *initialState*.

C 10. Recall the playback machine of figure 4.8 and the *CodeRecognizer* machine of figure 3.4. Enclose *CodeRecognizer* in a block and compose it with the playback machine so that someone can play back the recorded messages only if she correctly enters the code 1100. You will need to modify the playback machine appropriately.

E 11. Consider the following state machine in a feedback composition, in which the input and output alphabets for the state machine are {1, 2, 3, *absent*}:

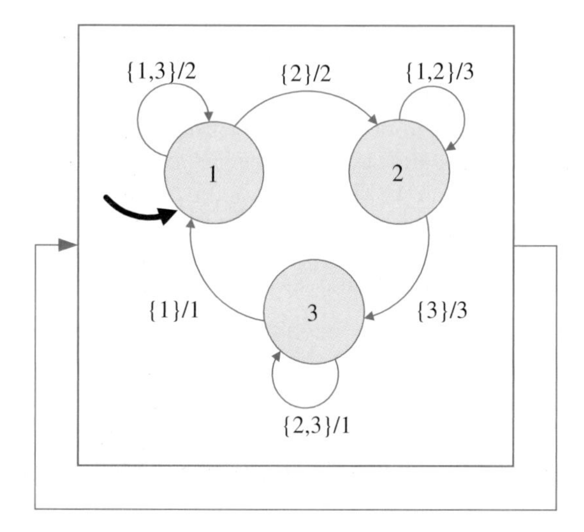

Is it well-formed? If so, then find the output symbols for the first 10 reactions.

E 12. In this problem, we explore the fact that a carefully defined delay in a feedback composition always makes the composition well-formed.

(a) For an input and output alphabet

$$\mathit{Inputs} = \mathit{Outputs} = \{\mathit{true}, \mathit{false}, \mathit{absent}\},$$

design a state machine that outputs *false* on the first reaction and then, in subsequent reactions, outputs the value observed at the input in the previous reaction. This is similar to *UnitDelay* of exercise 5 at the end of chapter 3, the only difference being that it outputs an initial *false* instead of *absent*.

(b) Compose the machine in figure 4.14(b) with the delay from part (a) of this problem in a feedback loop (as in figure 4.16). Give an argument that the composition is well-formed. Then do the same for figure 4.14(c) instead of (b).

C 13. Construct a feedback state machine with the structure of figure 4.11 that outputs the periodic sequence $a, b, c, a, b, c, \ldots$ (with, as usual, any number of intervening stuttering outputs between the nonstuttering outputs).

E 14. Modify figure 4.1 as necessary so that the machines in the side-by-side composition are both nondeterministic.

E 15. Modify figure 4.3 as necessary so that both the machines in the cascade composition are nondeterministic.

C,T 16. Data packets are to be reliably exchanged between two computers over communication links that may lose packets. The following protocol has been suggested. Suppose computer *A* is sending and *B* is receiving. Then *A* sends a packet and starts a timer. If *B* receives the packet, it sends back an acknowledgment. (The packet or the acknowledgment or both may be lost.) If *A* does not receive the acknowledgment before the timer expires, it retransmits the packet. If the acknowledgment arrives before the timer expires, *A* sends the next packet.

(a) Construct two state machines, one for *A* and one for *B*, that implement the protocol.

(b) Construct a two-state nondeterministic state machine to model the link from *A* to *B*, and another copy to model the link from *B* to *A*. Remember that the link may correctly deliver a packet or may lose it.

(c) Compose the four machines to model the entire system.

(d) Suppose the link delivers a packet correctly but after a delay that exceeds the timer setting. What will happen?

T 17. Consider the following three state machines:

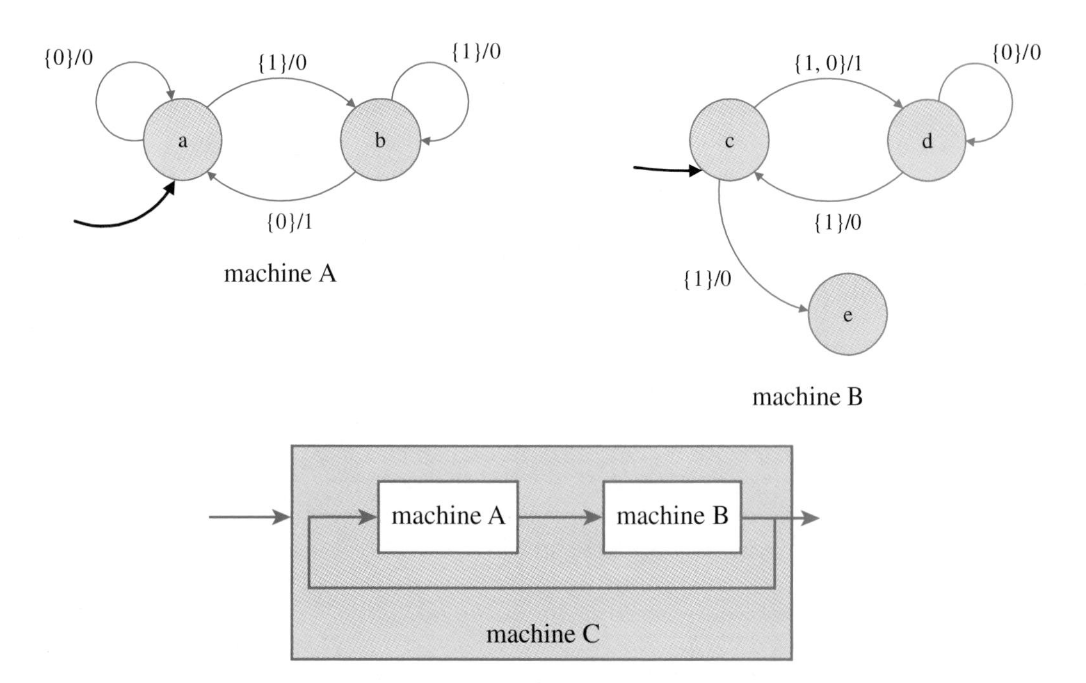

Machines A and B have input and output alphabets

$$Inputs = Outputs = \{0, 1, absent\}.$$

Machine C has the same output alphabet but an input alphabet $Inputs_C = \{react, absent\}$.

(a) Which of these machines are deterministic?

(b) Draw the state transition diagram for the composition (machine C), showing only states that are reachable from the initial state.

(c) Give the $Behaviors_C$ relation for the composition of machine C, ignoring stuttering.

T 18. The feedback composition in figure 4.14(c) is ill-formed because it has two nonstuttering fixed points in each of the two states of the component machine. Instead of declaring it to be ill-formed, we could have interpreted the composition as representing a nondeterministic state machine; that is, in each state, we accept either of the two possible fixed points as possible reactions of the machine. Using this interpretation, construct the nondeterministic machine for the feedback composition by giving its sets and functions description and a state transition diagram.

INTERVIEW
Gerard Berry

Gerard Berry is the Chief Scientist Officer at Esterel Technologies. His research activities include programming language design, semantics and implementation, reactive and real-time programming, automation verification of finite-state systems, and mathematical logic. Gerard Berry got the Bronze Medal of CNRS in 1979, the Monpetit Price of the French Academy of Sciences in 1989, and the Science and Defense French award in 1999. Most recently, he was a Finalist for the 2001 World Technology Awards in the IT Software Award Category.

How did you decide to study computer science?
I was a student at Ecole Polytechnique, in France, and then at Corps des Mines in Ecole des Mines de Paris. In 1968, when at Polytechnique, I met my first computer, a very special machine with mechanical "magneto-striction" memory. I got caught forever in computer programming.

What was your first job in the industry?
Around 1970, I decided to become a researcher in computer science, part-time at Ecole des Mines and part-time at the newly created IRIA Institute in France, which later became INRIA. At that time, computer science was a fascinating nascent subject. Unfortunately, France had no computers, which meant that we had to become very good in theory. Our world was initially quite small, and we could have regular seminars with the best people in the world (they could all fit in a small building).

What is the most challenging part of your job?
I am a programming language designer and implementer, and my language is dedicated to safety-critical applications. Therefore, the most challenging part of my work is to *get things right for the long term:* Any construct in a language is bound to last forever because users never want to remove something. Therefore, all constructs must be semantically right from the very beginning. This is a real challenge.

Why is it important to know a spectrum of languages, as opposed to focusing on one or a select few?
It is very hard to make several ways of thinking and several conceptual frameworks fit in a single language. A programming language is a fundamental vehicle to translate what we think into action. It also implies a fundamental and frustrating limitation: It makes it almost impossible to express the concepts that are not built-in. The art is to find the right balance between expressivity and consistency. Notice that the language problem is perceived quite differently in Europe than in the United States. In Europe, there are a dozen spoken languages and everybody is used to the idea that knowing more than one may be necessary. This is not needed at all in the United States!

What would you say have been the greatest advances of the last decade in electrical and computer engineering?

Historians name human periods by tools—the age of stone, of copper, of bronze, of iron, of the steam engine, of electricity, of mass media, and so forth. One should never underestimate the importance of the effect of tooling on individual and collective behavior.

How does the Esterel Language fit into the world of electrical and computer engineering?

We are trying to promote Esterel as a very clean and efficient language for the design, synthesis, and verification of hardware and embedded software systems. The operational model of cycle-based computation can fit a very wide variety of situations, and the Esterel cycle is in no way bound to machine cycles. The control-dominated part of Esterel has unique power to help program complex controllers, and we are now unifying it (really) with the more synchronous functional data-flow like constructs of Lustre, to reduce the gap between control and data handling. The deterministic and well-defined semantics makes synthesis easier and ensures that the synthesized netlist or code fully respects the semantics. The newly available verification techniques (e.g., SAT) make it possible to attack larger verification problems. The fact that the same program can be compiled into hardware or software is important for codesign.

CHAPTER 5
Linear systems

Recall that the **state** of a system is a summary of its past. It is what the system needs to remember about the past in order to react at the present and move into the future. The systems examined in previous chapters typically had a finite number of possible states. Many useful and interesting systems are not like that, however: They have an infinite number of states. The analytical approaches used to analyze finite-state systems, such as simulation and bisimulation, are more difficult when the number of states is not finite.

In this chapter, we begin considering infinite-state systems. We impose two key constraints. First, we require that the state space, input alphabet, and output alphabet be numeric sets; that is, we must be able to do arithmetic on members of these sets. (Contrast this with the answering machine example, in which the states are symbolic names; arithmetic does not make sense.) Second, we require that the *update* function be linear. We will define what this means precisely. In exchange for these two constraints, we gain a very rich set of analytical methods for designing and understanding systems. In fact, most of the remaining chapters are devoted to developing these methods.

In particular, we study state machines in which

$$
\begin{aligned}
States &= Reals^N, \\
Inputs &= Reals^M, \text{ and} \\
Outputs &= Reals^K.
\end{aligned}
\tag{5.1}
$$

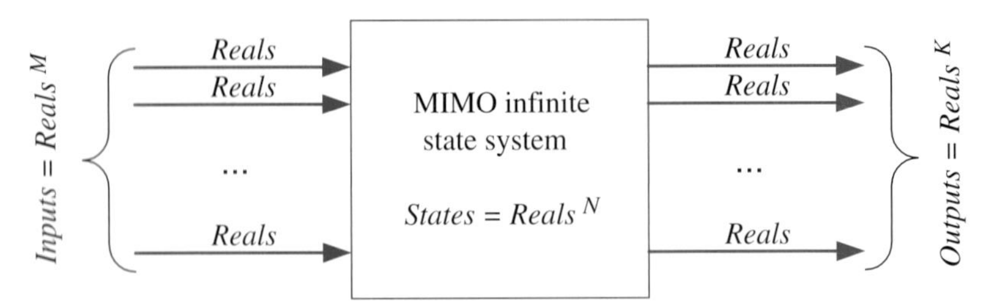

FIGURE 5.1: Block representing a multiple-input, multiple-output (MIMO) system.

Such state machines are shown schematically in figure 5.1. The inputs and outputs are in product form, as discussed for general state machines in section 4.4. The system can therefore be viewed as having M distinct inputs and K distinct outputs. Thus, the input is a tuple with M real numbers, and the output is a tuple with K real numbers. Such a system is called a **multiple-input, multiple-output (MIMO) system**. When $M = K = 1$, it is called a **single-input, single-output (SISO) system**. The state is a tuple with N real numbers. N (rather than M or K) is called the **dimension** of the system.

Example 5.1: A stereo audio system processes two channels of audio; therefore, in this system, $M = K = 2$. "Surround sound," used in movie theaters and some home audio systems, has five channels, and so in that system, $M = K = 5$. ❐

Example 5.2: Some modern cars have **traction control** systems, in which the torque applied to each wheel is controlled individually to avoid skidding. The key input comes from the accelerator pedal, which specifies a desired acceleration. The output is the torque applied to each of four wheels. Hence, $M = 1$ and $K = 4$. A more sophisticated traction control system may also use as input the steering angle, in which case $M = 2$. ❐

5.1 Operation of an infinite-state machine

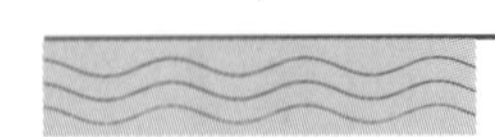

Recall that a deterministic state machine is a five-tuple,

$$M = (States, Inputs, Outputs, update, initialState), \qquad (5.2)$$

where *States* is the state space, *Inputs* is the input space, *Outputs* is the output space, *update*: *States* × *Inputs* → *States* × *Outputs* is the update function, and *initialState* is the initial state.

In this chapter, the *update* function has the form

$$update\colon Reals^N \times Reals^M \to Reals^N \times Reals^K.$$

The result of evaluating this function is an N-tuple (the next state) and a K-tuple (the current output). It is useful in this chapter to break this function into two parts, as done in section 3.1.1: one part consisting of the new state and one part consisting of the output,

$$update = (nextState, output),$$

where

$$nextState\colon Reals^N \times Reals^M \to Reals^N$$

and

$$output\colon Reals^N \times Reals^M \to Reals^K,$$

such that

$$\forall\, s \in Reals^N, \forall\, x \in Reals^M, \quad update(s, x) = (nextState(s, x), output(s, x)).$$

These two functions separately show the next state and the current output as a function of the current state and input. Given an input sequence $x(0), x(1), \ldots$ of M-tuples in $Reals^M$, the system recursively generates a **state response** $s(0), s(1), \ldots$ of N-tuples in $Reals^N$ and an **output response** $y(0), y(1), \ldots$ of K-tuples in $Reals^K$ as follows:

$$s(0) = initialState$$

and

$$(s(n+1), y(n)) = update(s(n), x(n)), \quad n \geq 0. \tag{5.3}$$

The second equation can be rewritten as a separate **state update equation**,

$$\boxed{\forall\, n \in Integers,\ n \geq 0, \quad s(n+1) = nextState(s(n), x(n)),} \tag{5.4}$$

and an **output equation**,

$$\boxed{\forall\, n \in Integers,\ n \geq 0, \quad y(n) = output(s(n), x(n)).} \tag{5.5}$$

Equations (5.4) and (5.5) together are called a **state-space model** of the system, because instead of their showing the output directly as a function of the input, the state is explicitly described. The equations suggest a detailed procedure for calculating the response of a system. We start with a given initial state, $s(0) =$ *initialState*, and an input sequence, $x(0), x(1), \ldots$. At step $n = 0$, we evaluate the right side of (5.4) at the known values of $s(0)$ and $x(0)$, and we assign the result to $s(1)$. At step $n = 1$, we evaluate the right side at the known values of $s(1)$ and $x(1)$, and we assign the result to $s(2)$. To proceed from step n to step $n + 1$, we need to remember only $s(n)$ and know only the new input $x(n)$. At each step n, we evaluate the output $y(n)$, using (5.5). This procedure is no different from that used for state machines in previous chapters. However, starting with section 5.2, we specialize the *nextState* and *output* functions so that they are linear, which then leads to a powerful set of analytical tools.

BASICS

Functions yielding tuples

The ranges of the *nextState* and *output* functions are tuples. Their role is more readily understood if we break them down further into an N-tuple and K-tuple of functions, one for each element of the result tuple; that is, we define the functions

$$nextState_i\colon Reals^N \times Reals^M \to Reals,\ i = 1, \ldots, N,$$

such that $\forall\, s \in Reals^N$ and $\forall\, x \in Reals^M$,

$$nextState(s, x) = (nextState_1(s, x), \ldots, nextState_N(s, x)).$$

We write simply

$$nextState = (nextState_1, \ldots, nextState_N).$$

The output function can be written similarly as

$$output = (output_1, \ldots, output_K),$$

where

$$output_i\colon Reals^N \times Reals^M \to Reals,\ i = 1, \ldots, K.$$

With these expressions, the state update equation and output equation can be written as follows. For all $n \in Integers$, $n \geq 0$,

$$\begin{aligned}
s_1(n+1) &= nextState_1((s_1(n), \ldots, s_N(n)), (x_1(n), \ldots, x_M(n))),\\
s_2(n+1) &= nextState_2((s_1(n), \ldots, s_N(n)), (x_1(n), \ldots, x_M(n))),\\
&\cdots \\
s_N(n+1) &= nextState_N((s_1(n), \ldots, s_N(n)), (x_1(n), \ldots, x_M(n))),
\end{aligned} \tag{5.6}$$

BASICS

and

$$\begin{aligned}
y_1(n) &= output_1((s_1(n), \ldots, s_N(n)), (x_1(n), \ldots, x_M(n))), \\
y_2(n) &= output_2((s_1(n), \ldots, s_N(n)), (x_1(n), \ldots, x_M(n))), \\
&\ldots \\
y_K(n) &= output_K((s_1(n), \ldots, s_N(n)), (x_1(n), \ldots, x_M(n))).
\end{aligned} \tag{5.7}$$

This system of equations shows the detailed structure of the operation of such a state machine.

5.1.1 *Time*

The index n in the equations just shown denotes the **step** number, the count of reactions, as with any state machine. For general state machines, it is rare to associate a fixed time interval with a step. Thus, there is normally no simple relation between the step number and the real time at which the corresponding reaction occurs. For example, in the answering machine, if the initial state is *idle*, there may be an arbitrary amount of time before *ring* occurs and the state moves to *count1*.

The systems we study in this and the next several chapters, however, usually evolve with a fixed time interval between updates. Suppose this interval is δ seconds. Then step n occurs at time $n\delta$ seconds, relative to time 0. Such systems are **discrete-time systems**, and the index n is called the **time index**.

We require that for each time index n, the input $x(n)$ be in $Reals^M$ and the output $y(n)$ be in $Reals^K$. We *disallow* stuttering input or output values such as *absent*. This is consistent with the interpretation of n as a real (physical) time index: The system's input and output must have some physical value at each n, and *absent* is not such a value.

Example 5.3: Compact discs (CDs) store digital audio signals as discrete-time signals. For each channel of audio, there are 44,100 numbers (samples) representing each second of sound. The nth number, therefore, represents the sound value at time $n\delta$ from the beginning of the CD, where $\delta = 1/44{,}100$, about 23 microseconds. In contrast, the telephone network transmits speech signals by sending only 8,000 samples per second, and so the sampling interval is $\delta = 1/8{,}000$, or 125 microseconds. We will see in chapter 11 that this difference partly accounts for the poorer audio quality over the telephone network in comparison with the quality of CDs. ❐

The systems in this chapter are **time-invariant** systems, meaning that the *nextState* and *output* functions do not change with the time index n. It is a common convention to have time start at 0 rather than at $-\infty$, or, say, 50. Under this convention the set of input signals is

$$InputSignals = [Integers_+ \to Reals^M],$$

where $Integers_+ = Naturals_0 = \{0, 1, 2, \ldots\}$. Correspondingly,

$$OutputSignals = [Integers_+ \to Reals^K].$$

The state response is therefore a function

$$s\colon Integers_+ \to Reals^N,$$

where $s(0) = initialState$.

BASICS

Matrices and vectors

An $M \times N$ **matrix** A is written as

$$A = \begin{bmatrix} a_{1,1} & a_{1,2} & \cdots & a_{1,N} \\ a_{2,1} & a_{2,2} & \cdots & a_{2,N} \\ \cdots & \cdots & \cdots & \cdots \\ a_{M,1} & a_{M,2} & \cdots & a_{M,N} \end{bmatrix}.$$

The **dimension** of the matrix is said to be $M \times N$, where M is always the number of rows and N is always the number of columns. In general, the coefficients of the matrix are real or complex numbers, and so they support all the standard arithmetic operations. We write the matrix more compactly as

$$A = [a_{i,j}, 1 \le i \le M, 1 \le j \le N],$$

or, even more simply, as $A = [a_{i,j}]$ when the dimension of A is understood. The matrix entries $a_{i,j}$ are called the coefficients of the matrix.

A **vector** is a matrix with only one row or only one column. An N-dimensional **column vector** s is written as an $N \times 1$ matrix:

$$s = \begin{bmatrix} s_1 \\ s_2 \\ \cdots \\ s_N \end{bmatrix}.$$

BASICS

An N-dimensional **row vector** z^T is written as a $1 \times N$ matrix:

$$z^T = [z_1, z_2, \ldots, z_N].$$

The **transpose** of an $M \times N$ matrix $A = [a_{i,j}]$ is the $N \times M$ matrix $A^T = [a_{j,i}]$. Therefore, the transpose of an N-dimensional column vector s is the N-dimensional row vector s^T, and the transpose of an N-dimensional row vector z is the N-dimensional column vector z^T.

From now on, unless explicitly stated otherwise, all vectors denoted s, x, y, b, c, and so forth (i.e., *without* the transpose notation) are column vectors, and vectors denoted s^T, x^T, y^T, b^T, c^T, and so forth (i.e., *with* the transpose notation) are row vectors. We follow convention and use lowercase letters to denote vectors and uppercase letters to denote matrices.

A tuple of numeric values is often represented as a vector. A tuple, however, is neither a "row" nor a "column." Thus, the representation as a vector carries the additional information that it is either a row or a column vector.

BASICS

Matrix arithmetic

Two matrices (or vectors, inasmuch as they are matrices) can be added or subtracted, provided that they have the same dimension. Just as with adding or subtracting tuples, the elements are added or subtracted. Thus, if $A = [a_{i,j}]$ and $B = [b_{i,j}]$ and both have dimension $M \times N$, then

$$A + B = [a_{i,j} + b_{i,j}].$$

Under certain circumstances, matrices can also be multiplied. If A has dimension $M \times N$ and B has dimension $N \times P$, then the product AB is defined. The number of columns of A must equal the number of rows of B. Suppose the matrices are given by

$$A = \begin{bmatrix} a_{1,1} & a_{1,2} & \ldots & a_{1,N} \\ a_{2,1} & a_{2,2} & \ldots & a_{2,N} \\ \ldots & \ldots & \ldots & \ldots \\ a_{M,1} & a_{M,2} & \ldots & a_{M,N} \end{bmatrix} \text{ and } B = \begin{bmatrix} b_{1,1} & b_{1,2} & \ldots & b_{1,P} \\ b_{2,1} & b_{2,2} & \ldots & b_{2,P} \\ \ldots & \ldots & \ldots & \ldots \\ b_{N,1} & b_{N,2} & \ldots & b_{N,P} \end{bmatrix}.$$

Then the i, j element of the product $C = AB$ is*

$$c_{i,j} = \sum_{m=1}^{N} a_{i,m} b_{m,j}. \tag{5.8}$$

The product C has dimension $M \times P$.

* If this notation is unfamiliar, see Basics: Summations box on page 69.

continued on next page

BASICS

Of course, matrix multiplication also works if one of the matrices is a vector. If b is a column vector of dimension N, then $c = Ab$, as defined by (5.8), is a column vector of dimension M. If, on the other hand, b^T is a row vector of dimension M, then $c^T = b^T A$, as defined by (5.8), is a row vector of dimension N.

Multiplying a matrix by a vector can be interpreted as applying a function to a tuple. The vector is the tuple, and the matrix (together with the definition of matrix multiplication) defines the function. Thus, in introducing matrix multiplication into our systems, we are doing nothing new except introducing a more compact notation for defining a particular class of functions.

A matrix A is a **square matrix** if it has the same number of rows and columns. A square matrix may be multiplied by itself. Thus, A^n for some integer $n > 0$ is defined to be A multiplied by itself n times. A^0 is defined to be the **identity matrix**, also written I, which has ones along the diagonal and zeros everywhere else. If A has an inverse, then that inverse is denoted A^{-1}, and $AA^{-1} = A^{-1}A = I$.

5.2 Linear functions

A function $f\colon Reals \to Reals$ is a **linear function** if $\forall\, u \in Reals$ and $a \in Reals$,

$$f(au) = af(u),$$

and $\forall\, u, v \in Reals$,

$$f(u + v) = f(u) + f(v).$$

The first property is called **homogeneity**, and the second property is called **additivity**. When both its domain and range are *Reals*, a linear function can be represented as

$$\forall\, x \in Reals, \quad f(x) = ax, \tag{5.9}$$

for some constant a. The term "linear" comes from the fact that the graph of this function is a straight line through the origin, with slope a. (If the line does not pass through the origin, the function is said to be **affine**. It would satisfy neither homogeniety nor additivity.) We need a more general notion of linear function, one that operates on tuples.

A function $f\colon Reals^N \to Reals^M$ is a **linear function** if $\forall\, u, v \in Reals^N$ and $\forall\, a \in Reals$,

$$f(au) = af(u), \tag{5.10}$$

and

$$f(u + v) = f(u) + f(v). \tag{5.11}$$

As before, (5.10) is the **homogeneity** property and (5.11) is the **additivity** property. The two properties can be combined into the **superposition** property:

$$f \text{ is } \textbf{linear} \text{ if } \forall\, u, v \in Reals^N \text{ and } \forall\, a, b \in Reals, \quad f(au + bv) = af(u) + bf(v). \tag{5.12}$$

In (5.12) superposition is defined as a linear combination $au + bv$ of two elements u and v in the domain of f. In fact, for linear functions, superposition holds for a linear combination of any number of elements; that is, $\forall\, u_1, \ldots, u_n \in Reals^N$ and $\forall\, a_1, \ldots, a_n \in Reals$,

$$f(a_1 u_1 + \cdots + a_n u_n) = a_1 f(u_1) + \cdots + a_n f(u_n). \tag{5.13}$$

In (5.13), $u_1, \ldots, u_n$ are vectors in $Reals^N$, $f(u_1), \ldots, f(u_n)$ are vectors in $Reals^M$, and $a_1, \ldots, a_n$ are scalars (numbers).

Every matrix defines a linear function in the following way. Let A be an $M \times N$ matrix. Then the function $f: Reals^N \to Reals^M$ defined by

$$\forall\, x \in Reals^N, \quad f(x) = Ax \tag{5.14}$$

is a linear function, as can be checked by using the basics of matrix arithmetic.

Of more interest is that every linear function can be represented by such a matrix multiplication, as in the scalar case (5.9). To show this, we show how to find the appropriate matrix, given any linear function. Define the vectors

$$e_1 = \begin{bmatrix} 1 \\ 0 \\ \cdots \\ 0 \end{bmatrix}, \; e_2 = \begin{bmatrix} 0 \\ 1 \\ \cdots \\ 0 \end{bmatrix}, \ldots, e_N = \begin{bmatrix} 0 \\ 0 \\ \cdots \\ 1 \end{bmatrix}.$$

Then note that we can express any vector $x \in Reals^N$ as a sum

$$x = x_1 e_1 + \cdots + x_N e_N,$$

where x_i (a scalar) is the ith element of the vector x. Now consider a linear function $f: Reals^N \to Reals^M$. Then, according to (5.13),

$$y = f(x) = x_1 f(e_1) + \cdots + x_N f(e_N). \tag{5.15}$$

Write the column vector $f(e_j) \in Reals^M$ as

$$f(e_j) = \begin{bmatrix} a_{1,j} \\ a_{2,j} \\ \cdots \\ a_{M,j} \end{bmatrix}.$$

Then we can rewrite (5.15) as

$$y = \begin{bmatrix} y_1 \\ y_2 \\ \cdots \\ y_M \end{bmatrix} = x_1 \begin{bmatrix} a_{1,1} \\ a_{2,1} \\ \cdots \\ a_{M,1} \end{bmatrix} + x_2 \begin{bmatrix} a_{1,2} \\ a_{2,2} \\ \cdots \\ a_{M,2} \end{bmatrix} + \cdots + x_N \begin{bmatrix} a_{1,N} \\ a_{2,2} \\ \cdots \\ a_{M,N} \end{bmatrix}$$

$$= \begin{bmatrix} a_{1,1} & a_{1,2} & \cdots & a_{1,N} \\ a_{2,1} & a_{2,2} & \cdots & a_{2,N} \\ \cdots & \cdots & \cdots & \cdots \\ a_{M,1} & a_{M,2} & \cdots & a_{M,N} \end{bmatrix} \begin{bmatrix} x_1 \\ x_2 \\ \cdots \\ x_N \end{bmatrix}.$$

More compactly,

$$y = Ax,$$

where A is the $M \times N$ matrix

$$A = [a_{i,j}, 1 \leq i \leq M, 1 \leq j \leq N].$$

Thus, there is a straightforward association between linear functions with domain $Reals^N$ and range $Reals^M$ and $M \times N$ matrices. This association is very important for us.

Example 5.4: The function $g\colon Reals^3 \to Reals$ given by

$$\forall\, x \in Reals^3, \quad g(x_1, x_2, x_3) = 0.5x_1 - 0.4x_3$$

is linear, as can be checked by verifying (5.12). Here, x_i refers to the ith element of the vector x. The matrix representation is

$$g(x) = [0.5,\ 0,\ -0.4] \begin{bmatrix} x_1 \\ x_2 \\ x_3 \end{bmatrix}.$$

The function $f\colon Reals^N \to Reals^N$ given by

$$\forall\, x \in Reals^N, \quad f(x)_n = \begin{cases} x_{n+1} & \text{if } n < N \\ x_1, & \text{if } n = N \end{cases}$$

is linear. Here $f(x)_n$ refers to the nth element of the vector $f(x)$. The matrix representation is

$$f(x) = \begin{bmatrix} 0 & 1 & 0 & \dots & 0 \\ 0 & 0 & 1 & \dots & 0 \\ \dots & \dots & \dots & \dots & \dots \\ 1 & 0 & 0 & \dots & 0 \end{bmatrix} x. \square$$

5.3 The [A,B,C,D] representation of a discrete linear system

We consider an arbitrary system in which

$$States = Reals^N, \ Inputs = Reals^M, \text{ and } \ Outputs = Reals^K.$$

Its state-space model is given by the state update and output equations: $\forall\, n \in Integers_+$,

$$s(n+1) = nextState(s(n), x(n))$$

and

$$y(n) = output(s(n), x(n)).$$

This system is said to be a **linear system** if the initial state is an N-tuple of zeros and if the *nextState* and *output* functions are linear. If it is also time invariant (the *nextState* and *output* functions do not change with time), then it is a **linear time-invariant (LTI) system**. We can then represent the *nextState* function by an $N \times (N+M)$ matrix and the *output* function by a $K \times (N+M)$ matrix.

Consider the $N \times (N+M)$ matrix representing the *nextState* function. This matrix has $N+M$ columns. We denote the $N \times N$ matrix that comprises the first N columns by A and the $N \times M$ matrix that comprises the last M columns as B, so that

$$nextState(s(n), x(n)) = As(n) + Bx(n).$$

We similarly partition the $K \times (N+M)$ matrix representing the *output* function into the $K \times N$ matrix C that comprises the first N columns and the $K \times M$ matrix D that comprises the last M columns. Then

$$output(s(n), x(n)) = Cs(n) + Dx(n).$$

With this notation, the state-space model is represented as

$$\boxed{s(n+1) = As(n) + Bx(n) \text{ and } y(n) = Cs(n) + Dx(n).} \tag{5.16}$$

This is the **[*A, B, C, D*] representation** of the LTI system. This compact representation is very powerful. All the results in this chapter are in terms of these four matrices. The A matrix is the most important, because it characterizes the system dynamics, as is shown in the following sections.

Example 5.5: Assume that $[A, B, C, D]$ is

$$A = \begin{bmatrix} 1 & 1 & 0 \\ 0 & 1 & 1 \\ 0 & 0 & 1 \end{bmatrix}, \; B = \begin{bmatrix} 0 \\ 0 \\ 1 \end{bmatrix}, \; C = [\,1 \quad 0 \quad 0\,], \text{ and } D = [\,1\,].$$

Thus, $N = 3$, $M = 1$, and $K = 1$. (This is a three-dimensional SISO system.) Using these in (5.16) yields the state-space model in tuple form:

$$\begin{aligned} s_1(n+1) &= s_1(n) + s_2(n), \\ s_2(n+1) &= s_2(n) + s_3(n), \\ s_3(n+1) &= s_3(n) + x(n), \text{ and} \\ y(n) &= s_1(n) + x(n). \end{aligned}$$

The first three equations together yield the *nextState* function; the fourth equation yields the *output* function. ❐

Observe in (5.16) that if *initialState* is zero, $s(0) = 0$, and the input $x(n) = 0$, for all $n \geq 0$, then the state is unchanged, $s(n) = 0$. That is why we say that 0 is an **equilibrium**, or **rest**, state of the $[A, B, C, D]$ system.

Example 5.6: An echo effect can be obtained for audio signals from the following difference equation,

$$y(n) = x(n) + \alpha y(n - N),$$

where the (discrete-time) audio input is x, the output is y, $0 \leq \alpha < 1$ is a real constant, and $N \in$ *Naturals* is an integer constant. Together, α and N determine how long the echo lasts and what it sounds like. This is another SISO system, with $M = 1$ and $K = 1$. It works simply because the output sample at n is a combination of the current input $x(n)$ and a scaled old output $\alpha y(n - N)$, called the **echo term**. Typically, N needs to be a large number in order for this to be heard as an echo. For example, if the sample rate is 8,000 samples per second, then if $N = 8{,}000$, the echo term $\alpha y(n-N)$ is the previous output one second earlier. This is heard as a distinct echo.

To construct a state-space model, we need to figure out what will work as the state. Because the output depends on $y(n-N)$, which is in the past, the state must yield $y(n-N)$. The following state definition works:

$$s(n) = \begin{bmatrix} y(n-1) \\ y(n-2) \\ \cdots \\ y(n-N) \end{bmatrix}.$$

It is easy to check that with this state definition, $[A, B, C, D]$ is given by

$$A = \begin{bmatrix} 0 & 0 & \ldots & 0 & 0 & \alpha \\ 1 & 0 & \ldots & 0 & 0 & 0 \\ 0 & 1 & \ldots & 0 & 0 & 0 \\ \vdots & \vdots & \vdots & \vdots & \vdots & \vdots \\ 0 & 0 & \ldots & 1 & 0 & 0 \\ 0 & 0 & \ldots & 0 & 1 & 0 \end{bmatrix}, \quad B = \begin{bmatrix} 1 \\ 0 \\ \vdots \\ 0 \end{bmatrix},$$

$$C = [\,0 \quad 0 \quad \ldots \quad 0 \quad \alpha\,], \text{ and } D = [\,1\,]. \square$$

5.3.1 *Impulse response*

The $[A, B, C, D]$ representation provides a complete description of an LTI system. It is complete in the sense that, given any input sequence, we can calculate the output sequence by using the state and output equations of (5.16) or, for the scalar case, (5.19) and (5.20). There are several other descriptions of LTI systems that are also complete. The first of these that we consider is the **impulse response**. For systems that are initially at rest, the impulse response gives enough information to calculate the output sequence, given any input sequence. The calculation is performed by using what is known as a **convolution sum**.

Suppose that $M = 1$ and the input sequence x is given by $x = \delta$, where

$$\forall\, n \in Integers, \quad \delta(n) = \begin{cases} 1 & \text{if } n = 0 \\ 0 & \text{if } n \neq 0. \end{cases} \tag{5.17}$$

This function $\delta\colon Integers \to Reals$ is called an **impulse**, or a **Kronecker delta function** (this function figures prominently in chapters 8 and 9). It is remarkable that the response of an LTI system to this particular input is a complete description of the LTI system.

Assuming the system is initially at rest, we can use (5.16) to write the state response to input δ as

$$
\begin{aligned}
s(0) &= 0 \\
s(1) &= B \\
s(2) &= AB \\
s(3) &= A^2B \\
&\cdots \\
s(n) &= A^{n-1}B \\
&\cdots
\end{aligned}
$$

This is because if $x = \delta$, then $x(0) = 1$ and $x(n) = 0$ for all $n \neq 0$.

To avoid confusion, we use the name h instead of y for the output of a system when the input is the impulse δ. In other words, whereas y is the output for any input x, h is the output for the specific input δ. This output is called the **impulse response**. Using (5.16) we can write it as

$$
\begin{aligned}
h(0) &= D \\
h(1) &= CB \\
h(2) &= CAB \\
h(3) &= CA^2B \\
&\cdots \\
h(n) &= CA^{n-1}B \\
&\cdots.
\end{aligned}
$$

Of course, because the system is initially at rest, $h(n) = 0$ for all $n < 0$. We can recognize the pattern and write the impulse response as

$$
\boxed{\forall n \in Integers, \quad h(n) = \begin{cases} 0 & \text{if } n < 0 \\ D & \text{if } n = 0 \\ CA^{n-1}B & \text{if } n \geq 1\,. \end{cases}} \tag{5.18}
$$

This formula is rarely the best way to compute an impulse response (usually it is easier to directly determine the output when the input is an impulse), but it does relate the impulse response to the $[A, B, C, D]$ representation. The remarkable fact, developed later, is that knowing the impulse response is sufficient for calculating the output given *any* input (assuming the system is initially at rest).

The impulse response makes sense only if the dimension M of the input is one (otherwise the input could not be an impulse as defined previously). We can gain insight by considering even more special systems in which $N = K = 1$, as in the next section.

5.3.2 One-dimensional SISO systems

The simplest LTI system is the one-dimensional, single-input, single-output (SISO) system. Because $N = M = K = 1$, the $[A, B, C, D]$ representation is simply $[a, b, c, d]$, where a, b, c, d are scalar constants. The state-space model (5.16) is $\forall n \in$ *Integers*$_+$,

$$s(n+1) = as(n) + bx(n) \tag{5.19}$$

and

$$y(n) = cs(n) + dx(n). \tag{5.20}$$

The initial state $s(0) = initialState$. For this system, the state at a given time index is a real number, as are the input and the output.

Consider an example in which we construct a state-space model from an input–output description of a system.

Example 5.7: In section 2.3.3 we considered a simple **moving average** example, in which the output y is given in terms of the input x by

$$\forall n \in Integers_+, \quad y(n) = (x(n) + x(n-1))/2. \tag{5.21}$$

This is not a state-space model, because it gives the output directly in terms of the current and *past* input. To construct a state-space model for it, we first need to decide what the state is. Usually, there are multiple answers, and so we face a choice. The state is a summary of the past. An examination of (5.21) shows that we need to remember the previous input, $x(n-1)$, in order to produce an output $y(n)$ (of course, we also need the current input, $x(n)$, but that is not part of the past; that is the present). Therefore, we can define the state to be

$$\forall n \in Integers_+, \quad s(n) = x(n-1).$$

We assume that the system is initially at rest; that is, $s(0) = 0$. (If we knew $x(-1)$, we would take that to be the initial state.) With this choice of state, we need to choose a, b, c, and d so that (5.19) and (5.20) are equivalent to (5.21). Look first at (5.20), which reads

$$y(n) = cs(n) + dx(n).$$

Observing that $s(n) = x(n-1)$, can you determine c and d? From (5.21), it is obvious that $c = d = 1/2$.

Next, we determine a and b in

$$s(n+1) = as(n) + bx(n).$$

Because $s(n) = x(n-1)$, it follows that $s(n+1) = x(n)$, and this becomes

$$x(n) = ax(n-1) + bx(n),$$

from which we can see that $a = 0$ and $b = 1$.

Note that we could have chosen the state differently. For example,

$$\forall\, n \in Integers_+, \quad s(n) = x(n-1)/2$$

would work fine. How would that change a, b, c, and d? (The fact that different choices of state can yield the same input–output relation is connected with the notion of bisimulation in section 3.4.) ❐

In the preceding example, we started with an input–output description and obtained a state-space model. The next example shows that we can also reverse that procedure.

Example 5.8: Consider a one-dimensional SISO system in which

$$s(n+1) = as(n) + bx(n)$$

and

$$y(n) = cs(n) + dx(n).$$

From these equations, we get

$$y(n) = c[as(n-1) + bx(n-1)] + dx(n)$$

and

$$y(n-1) = cs(n-1) + dx(n-1).$$

Multiplying the second equation by a and subtracting from the first eliminates the state to yield a difference equation description of the system,

$$y(n) - ay(n-1) = dx(n) + (cb - ad)x(n-1).$$

There is a generalization of this example that works for state-space models of any dimension. ❐

In the following example, we use a state-space model to calculate the output of a system, given an input sequence.

Example 5.9: Suppose the state $s(n)$ is your bank balance at the beginning of day n, and $x(n)$ is the amount you deposit or withdraw during day n. If $x(n) > 0$, it means that you are making a deposit of $x(n)$ dollars, and if $x(n) < 0$, it means that you are withdrawing $x(n)$ dollars. The output of the system at time index n is the bank balance on day n. Thus,*

$$States = Inputs = Outputs = Reals.$$

Suppose that the daily interest rate is r. Then your balance at the beginning of day $n + 1$ is given by

$$\forall\, n \in Integers, \quad s(n+1) = (1+r)s(n) + x(n). \tag{5.22}$$

The output of the system is your current balance,

$$\forall\, n \in Integers, \quad y(n) = s(n).$$

Comparing to (5.19) and (5.20), we have $a = 1 + r$, $b = 1$, $c = 1$, and $d = 0$. The initial condition is *initialState*, your bank balance at the beginning of day 0. Suppose the daily interest rate is 0.01, or one percent.† Suppose that *initialState* = 100, and you deposit \$1,000 on day 0 and withdraw \$30 every subsequent day for the next 30 days. What is your balance $s(31)$ on day 31?

You can compute $s(31)$ recursively from

$$s(0) = 100,$$

$$s(1) = 1.01s(0) + 1{,}000,$$

$$\dots$$

$$s(n+1) = 1.01s(n) - 30, \quad n = 1, \dots, 30,$$

but this would be tedious. We can instead develop a formula that is a bit easier to use. We do this for a general one-dimensional $[a, b, c, d]$ system. ❒

* These sets are probably not, strictly speaking, equal to *Reals*, because deposits and withdrawals can be only a whole number of cents. Also, the bank will most probably round your balance to the nearest cent. Thus, our model here is an approximation. Using *Reals* is a considerable simplification.

† This would be reasonable only in an economy with hyperinflation.

Suppose we are given an input sequence $x(0), x(1), \ldots$ for an $[a, b, c, d]$ system. As in example 5.9, if we repeatedly use (5.19), we obtain the first few terms of a sequence:

$$s(0) = initialState, \tag{5.23}$$

$$s(1) = as(0) + bx(0), \tag{5.24}$$

$$\begin{aligned} s(2) &= as(1) + bx(1) \\ &= a\{as(0) + bx(0)\} + bx(1) \\ &= a^2 s(0) + abx(0) + bx(1), \end{aligned} \tag{5.25}$$

$$\begin{aligned} s(3) &= as(2) + bx(2) \\ &= a\{a^2 s(0) + abx(0) + bx(1)\} + bx(2) \\ &= a^3 s(0) + a^2 bx(0) + abx(1) + bx(2), \end{aligned} \tag{5.26}$$

$$\ldots.$$

From this, it is not difficult to guess the general pattern for the state response and the output response. The state response of (5.19) is given by

$$\boxed{s(n) = a^n initialState + \sum_{m=0}^{n-1} a^{n-1-m} bx(m)} \tag{5.27}$$

for all $n \geq 0$, and the output response of (5.20) is given by

$$\boxed{y(n) = ca^n initialState + \left\{ \sum_{m=0}^{n-1} ca^{n-1-m} bx(m) \right\} + dx(n)} \tag{5.28}$$

for all $n \geq 0$.

We use induction to show that these are correct. In induction, we show that these are correct for some fixed n and then show that if they are correct for any n, then they are correct for $n+1$. For $n = 0$, (5.27) yields $s(0) = a^0 initialState = initialState$, which matches (5.23) and hence is correct.* Now suppose that the right side of (5.27) gives the correct value of the response for some $n \geq 0$. We

* For any real number a, $a^0 = 1$ by definition of exponentiation.

must show that it gives the correct value of the response for $n+1$. From (5.19), and using the hypothesis that (5.27) is the correct expression for $s(n)$, we get

$$\begin{aligned} s(n+1) &= as(n) + bx(n) \\ &= a\{a^n initialState + \sum_{m=0}^{n-1} a^{n-1-m} bx(m)\} + bx(n) \\ &= a^{n+1} initialState + \sum_{m=0}^{n-1} a^{n-m} bx(m) + bx(n) \\ &= a^{n+1} initialState + \sum_{m=0}^{n} a^{n-m} bx(m), \end{aligned}$$

which is the expression on the right side of (5.27) for $n+1$. It follows by induction that the response is indeed given by (5.27) for all $n \geq 0$. The fact that the output response is given by (5.28) follows immediately from (5.20) and (5.27).

Example 5.10: We use formula (5.27) in example 5.9 to figure out the monthly payment of $\$w$ on a \$10,000, 32-month loan with a monthly interest charge of 0.01. Therefore, in (5.27) we substitute $n = 32$, $a = 1.01$, $b = 1$, $s(0) = -10,000$, $s(32) = 0$, and $x(0) = \ldots = x(31) = w$, to get

$$0 = -1.01^{32} \times 10{,}000 + \sum_{m=0}^{31} 1.01^{31-m} w.$$

Using the identity (valid for $\rho \neq 1$)

$$\sum_{m=0}^{M} \rho^m = \frac{\rho^{1+M} - 1}{\rho - 1},$$

we get

$$\sum_{m=0}^{31} 1.01^{31-m} w = \frac{1.01^{32} - 1}{0.01} w.$$

Therefore, the monthly payment is

$$w = 0.01 \times \frac{10000 \times 1.01^{32}}{1.01^{32} - 1} = \$366.70. \;\square$$

In the next example, we identify the unknown parameter a of an $[a, b, c, d]$ representation, given input–output data.

Example 5.11: Financial analysts compare alternative investment opportunities by using a measure called the **internal rate of return**. Suppose an investment of \$10,000 at the beginning of year 0 yields an income stream of \$2,000 for each of 10 successive years and nothing thereafter. By definition, the investment's internal rate of return is the annual interest rate r that a bank should pay so that you can get the same income stream. We therefore need to find r such that

$$s(n+1) = (1+r)s(n) + x(n),$$

with $s(0) = 10{,}000$, $s(10) = 0$, and $x(0) = \cdots = x(9) = -2{,}000$. We recognize this as a state update with $a = 1 + r$ and $b = 1$. Substituting in (5.27) yields

$$\begin{aligned} 0 &= (1+r)^{10}10{,}000 - \sum_{m=0}^{9}(1+r)^{9-m} \times 2{,}000 \\ &= (1+r)^{10}10{,}000 - \frac{(1+r)^{10} - 1}{r} \times 2{,}000, \end{aligned}$$

and so

$$1 - (1+r)^{-10} = 5r,$$

which we can solve by trial and error to obtain $r \approx 0.15$, or 15 percent. Suppose there is another investment opportunity for the same \$10,000 investment that yields an internal rate of return smaller than 15 percent. Then, all else being equal, the first opportunity would be the better choice. ❐

5.3.3 *Zero-state and zero-input response*

Both the expressions for the state response (5.27) and for the output (5.28) are the sums of two terms. The role of these two terms is better understood if they are considered in isolation.

If the system is initially at rest, *initialState* $= 0$, the first term vanishes, and only the second term is left. This second term is called the **zero-state response**. It gives the response of the system to an input sequence when the initial state is zero. For many applications, particularly when a physical system is modeled, the zero-state response is what we are interested in.

If the input sequence is zero, that is, $0 = x(0) = x(1) = \ldots$, the second term vanishes, and only the first term is left. The first term is called the **zero-input response**. It gives the response of the system to some initial condition, with zero-input stimulus applied. Of course, if the system is initially at rest and the input is zero, then the state remains at zero. Thus, the zero-input response is interesting only if the system is not initially at rest.

Therefore, the right sides of both equations (5.27) and (5.28) are a sum of a zero-state response and a zero-input response. To clarify which equation we are talking about, we use the following terminology:

> zero-state state response: The state sequence $s(n)$ when the initial state is zero.
>
> zero-input state response: The state sequence $s(n)$ when the input is zero.
>
> zero-state output response: The output sequence $y(n)$ when the initial state is zero.
>
> zero-input output response: The output sequence $y(n)$ when the input is zero.

Note that "zero-state" really means that the *initial* state is zero, whereas "zero-input" means that the input is *always* zero.

Focus on the zero-state output response. First, note from (5.18) that in this scalar case, the impulse response can be written

$$\forall n \in Integers, \quad h(n) = \begin{cases} 0 & \text{if } n < 0 \\ d & \text{if } n = 0 \\ ca^{n-1}b & \text{if } n \geq 1. \end{cases} \tag{5.29}$$

We can use this sequence in (5.28) to simplify its form. This simplified form shows us how to calculate the response to an arbitrary input when we are given only the response to an impulse.

Combining (5.28) and (5.29), we can write the zero-state output response as

$$\forall n \geq 0, \quad y(n) = \sum_{m=0}^{n} h(n-m)x(m). \tag{5.30}$$

Thus, we see that h is a complete description of the system, in the sense that it is all we need to know in order to find the output when given any input x (assuming an initial state of zero).

Let $x(n) = 0$ for all $n < 0$; noting that $h(n) = 0$ for all $n < 0$, we can write this as

$$\boxed{\forall n \in Integers, \quad y(n) = \sum_{m=-\infty}^{\infty} h(n-m)x(m).} \tag{5.31}$$

The additional terms in the summation are harmless because they all have a value of zero. A summation of this form is called a **convolution sum**. We say that y is the convolution of h and x and write it in the shorthand expression

$$\boxed{y = h * x.}$$

The "*" symbol represents convolution. By changing variables, defining $k = n - m$, we can see that the convolution sum can also be written in the equivalent form

$$\boxed{\forall\, n \in Integers, \quad y(n) = \sum_{k=-\infty}^{\infty} h(k)x(n-k);} \tag{5.32}$$

that is, $h * x = x * h$. Convolution sums are studied in much more detail in chapter 9.

> The impulse response of a system is simply the output when the input is an impulse. If the system is initially at rest, then its output is given by the convolution of the input and the impulse response.

Example 5.12: For our bank example, $a = 1 + r$, $b = 1$, $c = 1$, and $d = 0$ in (5.20). The impulse response of the bank system is given by (5.18):

$$h(n) = \begin{cases} 0 & \text{if } n \le 0 \\ (1+r)^{n-1} & \text{if } n \ge 1. \end{cases}$$

This represents the balance of a bank account with daily interest rate r if an initial deposit of one dollar is made on day 0 and no further deposits or withdrawals are made. Notice that because $1 + r > 1$, the balance continues to increase forever; see figure 5.2. This system is said to be **unstable** because even though the input is always bounded, the output grows without bound. The system is also an **infinite impulse response (IIR) system**.

Writing the output as a convolution, using (5.30), we see that

$$\forall\, n \ge 0, \quad y(n) = \sum_{m=0}^{n-1} (1+r)^{n-m-1} x(m).$$

This gives a simple formula that we can use to calculate the bank balance on any given day (although it is tedious for large n, for which a computer would be useful). ❐

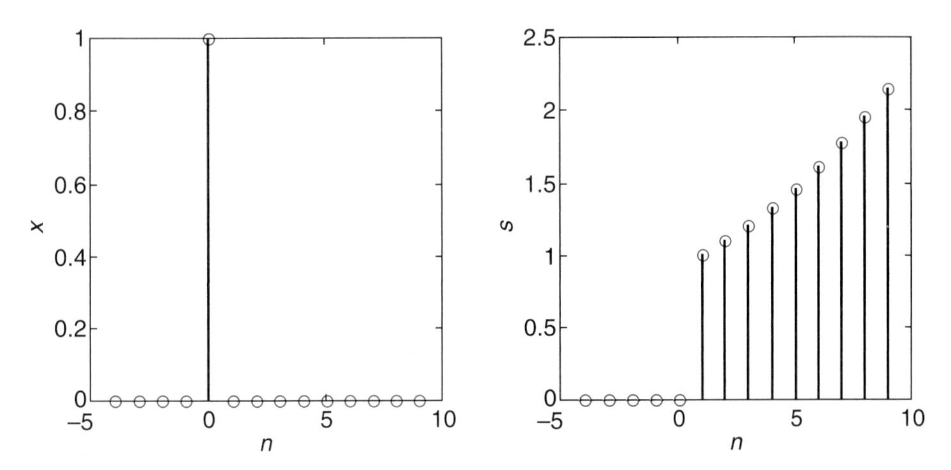

FIGURE 5.2: Plots of the impulse (left) and the impulse response of example 5.12 (right) for $r = 0.1$.

The zero-input state response of the $[a, b, c, d]$ system is

$$s(n) = a^n s(0), n \geq 0.$$

This is a geometric or **exponential sequence**. If $s(0) \neq 0$, the response will eventually die out; that is, $s(n) \to 0$ as $n \to \infty$, if and only if $|a| < 1$, in which case we say the system is **stable**.

5.3.4 *Multidimensional SISO systems*

In the previous section we considered the simplest systems of the form of figure 5.1 with $M = K = N = 1$. Systems with larger dimension, $N > 1$, occur more frequently in practice. In this section, we allow the dimension N to be arbitrarily large but keep the simplification that $M = K = 1$, so that the system is still SISO. In the $[A, B, C, D]$ representation, A is $N \times N$, B is $N \times 1$, C is $1 \times N$, and D is 1×1. Hence, for an SISO system, we may write $B = b$, $C = c^T$, and $D = d$, where b and c are N-dimensional column vectors and d is a scalar. Thus, SISO systems have an $[A, b, c, d]$ representation, and their state-space model is

$$s(n+1) = As(n) + bx(n) \tag{5.33}$$

and

$$y(n) = c^T s(n) + dx(n). \tag{5.34}$$

The result of evaluating the *nextState* function (5.33) is an N-dimensional vector, $s(n+1)$. The $N \times N$ matrix A defines the linear combination of the N elements of $s(n)$ that are used to calculate the N elements $s(n+1)$. The N-dimensional column vector b defines the weights used to include $x(n)$ in the linear combination for each element of $s(n+1)$.

The result of evaluating the *output* function (5.34) is a scalar, $y(n)$. The N-dimensional row vector c^T defines the linear combination of elements of $s(n)$, and d defines the weight used to include $x(n)$ in the output $y(n)$.

Example 5.13: We previously constructed a state-space model for a two-point moving average. The general form of this is the M-point moving average, given by

$$\forall n \in Integers, \quad y(n) = \frac{1}{M} \sum_{k=0}^{M-1} x(n-k). \tag{5.35}$$

To be specific, take $M = 3$. Equation (5.35) becomes

$$\forall n \in Integers, \quad y(n) = \frac{1}{3}(x(n) + x(n-1) + x(n-2)). \tag{5.36}$$

We can construct a state-space model for this in a manner similar to what we did for the two-point moving average. First, we need to decide what the state is. Recall that the state is the summary of the past. Equation (5.36) tells us that we need to remember $x(n-1)$ and $x(n-2)$, the two past inputs. We could define these to be the state, collected as a column vector,

$$s(n) = \begin{bmatrix} x(n-1) \\ x(n-2) \end{bmatrix}.$$

(Of course, we could have equally well put the elements in the other order; see exercise 9 at the end of this chapter.)

Consider the output equation (5.34). We need to determine c^T and d. The vector c^T is a row vector with dimension $N = 2$, and so we must fill in the blanks in the following output equation:

$$y(n) = [\text{--}, \text{--}] \begin{bmatrix} x(n-1) \\ x(n-2) \end{bmatrix} + [\text{--}]x(n).$$

It is easy to see that each of the three blanks must be filled with $1/M = 1/3$ in order to get (5.36). Thus,

$$c = \begin{bmatrix} 1/3 \\ 1/3 \end{bmatrix}, \quad d = 1/3.$$

Consider next the state equation (5.33). We need to determine A and b. The matrix A is 2×2. The vector b is the column vector of dimension 2. Therefore, we can fill in the blanks in the following state equation:

$$s(n+1) = \begin{bmatrix} x(n) \\ x(n-1) \end{bmatrix} = \begin{bmatrix} \text{--} & \text{--} \\ \text{--} & \text{--} \end{bmatrix} \begin{bmatrix} x(n-1) \\ x(n-2) \end{bmatrix} + \begin{bmatrix} \text{--} \\ \text{--} \end{bmatrix} x(n).$$

From this, we can fill in the blanks, getting

$$A = \begin{bmatrix} 0 & 0 \\ 1 & 0 \end{bmatrix} \text{ and } b = \begin{bmatrix} 1 \\ 0 \end{bmatrix}.$$

Note that once the state is specified, there is only one way to fill in the blanks and obtain the $[A, b, c, d]$ representation. ❐

The state response of the SISO system (5.33) and (5.34) is given by an expression that is similar to (5.27) but involves matrices and vectors rather than just scalars:

$$\boxed{s(n) = A^n initialState + \sum_{m=0}^{n-1} A^{n-1-m} bx(m),} \tag{5.37}$$

for all $n \geq 0$. The state response is also sometimes called the **state trajectory**. The output response of (5.34) is given by

$$\boxed{y(n) = c^T A^n initialState + \left\{ \sum_{m=0}^{n-1} c^T A^{n-1-m} bx(m) + dx(n) \right\}} \tag{5.38}$$

for all $n \geq 0$. (Exercise 13 at the end of this chapter asks you to derive these equations.)

Notice again that the state response (5.37) and the output response (5.38) are each the sums of the **zero-input response** and the **zero-state response**.

The impulse response, in terms of the state-space model, is the sequence of real numbers

$$\boxed{h(n) = \begin{cases} 0 & \text{if } n < 0 \\ d & \text{if } n = 0 \\ c^T A^{n-1} b & \text{if } n \geq 1. \end{cases}} \tag{5.39}$$

This is just like (5.18) except that b and c are vectors and d is a scalar. Applying this formula can be quite tedious; it is usually easier to simply let $x = \delta$, the

Kronecker delta function, and observe or calculate the output. The zero-state output response is given by convolution of this impulse response with the input, (5.32).

Example 5.14: We can find the impulse response h of the moving average system of (5.35) by letting $x = \delta$, where δ is given by (5.17); that is,

$$\forall\, n \in Integers, \quad h(n) = \frac{1}{M}\sum_{k=0}^{M-1} \delta(n-k).$$

Now, $\delta(n-k) = 0$ except when $n = k$, at which point it equals one. Thus,

$$h(n) = \begin{cases} 0 & \text{if } n < 0 \\ 1/M & \text{if } 0 \le n < M \\ 0 & \text{if } n \ge M. \end{cases}$$

This function is therefore the impulse response of an M-point moving average system. This result could also have been obtained by comparing (5.35) with (5.32), the output as a convolution, or it could have been obtained by constructing a state-space model for the general M-point moving average and applying (5.39). In this case, however, the latter method would have proved the most tedious. ❐

Notice that in the previous example, the impulse response is finite in extent (it starts at 0 and stops at $M - 1$). For this reason, such a system is called a **finite impulse response (FIR) system**.

Example 5.15: The M-point moving average can be viewed as a special case of the more general FIR system, given by

$$\forall\, n \in Integers, \quad y(n) = \sum_{k=0}^{M-1} h(k)x(n-k).$$

Letting $h(k) = 1/M$ for $0 \le k < M$, we get the M-point moving average. Choosing other values for $h(k)$, however, we can get other responses (this is explored in chapter 8).

A state-space model for the FIR system is constructed by again deciding on the state. A reasonable choice is the $M - 1$ past samples of the input,

$$s(n) = [x(n-1), x(n-2), \ldots, x(n-M+1)]^T,$$

a column vector. The state-space model is then given by (5.33) and (5.34) with

$$A = \begin{bmatrix} 0 & 0 & 0 & 0 & \cdots & 0 & 0 \\ 1 & 0 & 0 & 0 & \cdots & 0 & 0 \\ 0 & 1 & 0 & 0 & \cdots & 0 & 0 \\ \cdots & \cdots & \cdots & \cdots & \cdots & \cdots & \cdots \\ 0 & 0 & 0 & 0 & \cdots & 1 & 0 \end{bmatrix}, \quad b = \begin{bmatrix} 1 \\ 0 \\ \cdots \\ 0 \\ 0 \end{bmatrix},$$

$$c = \begin{bmatrix} h(1) \\ h(2) \\ \cdots \\ h(M-2) \\ h(M-1) \end{bmatrix}, \quad \text{and } d = h(0).$$

Notice that the model that we found in example 5.13 has this form. The $(M-1) \times (M-1)$ matrix A has coefficients $a_{i+1,i} = 1$, whereas all other coefficients are zero. This is a rather special form of the A matrix, limited to FIR systems. In the vector b, the first coefficient is equal to one, whereas all others are zero. The vector c contains the coefficients of the impulse response. ❐

Many interesting systems, unlike this example, have an infinite impulse response and are referred to as **IIR systems**.

Example 5.16: Recall from example 5.6 that an echo effect can be obtained for audio signals from the following difference equation:

$$\forall n \in Integers, \quad y(n) = x(n) + \alpha y(n - N).$$

The impulse response h can be obtained by simply letting the input be an impulse, $x = \delta$, and finding the output $y = h$; that is,

$$\forall n \in Integers, \quad h(n) = \delta(n) + \alpha h(n - N). \tag{5.40}$$

However, this of course means that

$$h(n - N) = \delta(n - N) + \alpha h(n - 2N).$$

Substituting this back into (5.40), we get

$$h(n) = \delta(n) + \alpha\delta(n - N) + \alpha^2 h(n - 2N).$$

However, this of course means that

$$h(n-2N) = \delta(n-2N) + \alpha h(n-3N),$$

and so

$$h(n) = \delta(n) + \alpha\delta(n-N) + \alpha^2\delta(n-2N) + \alpha^3 h(n-3N).$$

Continuing in this manner, we see that the impulse response of the echo system is the original impulse and an infinite set of echos (delayed and scaled impulses). This can be written compactly as follows:

$$\forall\, n \in Integers, \quad h(n) = \sum_{k=0}^{\infty} \alpha^k \delta(n-kN).$$

This impulse response is plotted in figure 5.3 for $N = 4$ and $\alpha = 0.7$. In that figure, you can see that the original impulse gets through the system at $n = 0$, whereas the first echo is scaled by 0.7 and delayed to $n = 4$, and the second echo is scaled by 0.7^2 and delayed to $n = 8$.

We can check this impulse response by using the state-space representation of example 5.6 in (5.39). From example 5.6, $d = 1$, and so (5.39) is obviously correct for $n < 0$ and $n = 0$. Checking it for $n > 0$ is somewhat more involved. Assuming $N = 4$, we have from example 5.6

$$A = \begin{bmatrix} 0 & 0 & 0 & \alpha \\ 1 & 0 & 0 & 0 \\ 0 & 1 & 0 & 0 \\ 0 & 0 & 1 & 0 \end{bmatrix}, \quad b = \begin{bmatrix} 1 \\ 0 \\ 0 \\ 0 \end{bmatrix}, \quad c = \begin{bmatrix} 0 \\ 0 \\ 0 \\ \alpha \end{bmatrix}, \quad \text{and } d = [\,1\,].$$

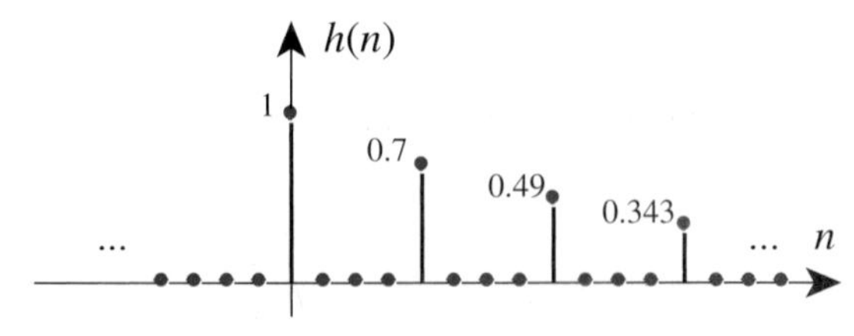

FIGURE 5.3: Impulse response of the echo example 5.16 for $\alpha = 0.7$ and $N = 4$.

To use (5.39), we need to know A^{n-1}. It is easy to check that

$$A^2 = \begin{bmatrix} 0 & 0 & \alpha & 0 \\ 0 & 0 & 0 & \alpha \\ 1 & 0 & 0 & 0 \\ 0 & 1 & 0 & 0 \end{bmatrix}, \; A^3 = \begin{bmatrix} 0 & \alpha & 0 & 0 \\ 0 & 0 & \alpha & 0 \\ 0 & 0 & 0 & \alpha \\ 1 & 0 & 0 & 0 \end{bmatrix},$$

$$\text{and } A^4 = \begin{bmatrix} \alpha & 0 & 0 & 0 \\ 0 & \alpha & 0 & 0 \\ 0 & 0 & \alpha & 0 \\ 0 & 0 & 0 & \alpha \end{bmatrix} = \alpha I,$$

where I is the 4×4 identity matrix. Thus,

$$A^5 = \alpha A, \; A^6 = \alpha A^2, \; A^7 = \alpha A^3, \; \text{and } A^8 = \alpha A^4 = \alpha^2 I.$$

The pattern continues. Note that because of the particular structure of b, $A^{n-1}b$ is simply the first column of A^{n-1}. Thus,

$$h(1) = c^T A^0 b = [0\,0\,0\,\alpha] \begin{bmatrix} 1 \\ 0 \\ 0 \\ 0 \end{bmatrix} = 0.$$

Similarly,

$$h(2) = c^T A b = [0\,0\,0\,\alpha] \begin{bmatrix} 0 \\ 1 \\ 0 \\ 0 \end{bmatrix} = 0,$$

and

$$h(3) = c^T A^2 b = [0\,0\,0\,\alpha] \begin{bmatrix} 0 \\ 0 \\ 1 \\ 0 \end{bmatrix} = 0.$$

Only for $h(4)$ do we get a nonzero result:

$$h(4) = c^T A^3 b = [0\,0\,0\,\alpha] \begin{bmatrix} 0 \\ 0 \\ 0 \\ 1 \end{bmatrix} = \alpha.$$

Continuing in this manner, we can determine that three of every four samples of the impulse response are zero and that the nonzero ones have the form $\alpha^{n/4}$, where n is a multiple of four, in perfect agreement with figure 5.3. ❒

The echo system of the previous example is called an IIR because the response to an impulse never completely dies out. Here is another example of an IIR system.

Example 5.17: An audio **oscillator** is a system that produces a sinusoidal signal of a given frequency. We can construct one with the two-dimensional system given by

$$A = \begin{bmatrix} \cos(\omega) & -\sin(\omega) \\ \sin(\omega) & \cos(\omega) \end{bmatrix}, \; b = \begin{bmatrix} 0 \\ 1 \end{bmatrix}, \; c = \begin{bmatrix} 1 \\ 0 \end{bmatrix}, \text{ and } d = 0,$$

where ω is a constant that will turn out to be the oscillation frequency. It can be shown (see exercise 14 at the end of this chapter) that for all $n = 0, 1, 2, \ldots,$

$$A^n = \begin{bmatrix} \cos(n\omega) & -\sin(n\omega) \\ \sin(n\omega) & \cos(n\omega) \end{bmatrix}.$$

Suppose the initial state is

$$initialState = [0, 1]^T.$$

Then the zero-input state response is

$$\begin{aligned} s_{zero-input}(n) &= A^n \; initialState \\ &= \begin{bmatrix} \cos(n\omega) & -\sin(n\omega) \\ \sin(n\omega) & \cos(n\omega) \end{bmatrix} \begin{bmatrix} 0 \\ 1 \end{bmatrix} \\ &= \begin{bmatrix} -\sin(n\omega) \\ \cos(n\omega) \end{bmatrix}, \end{aligned}$$

and the zero-input output response is

$$\begin{aligned} y_{zero-input}(n) &= c^T A^n \; initialState \\ &= [1 \;\; 0] \begin{bmatrix} -\sin(n\omega) \\ \cos(n\omega) \end{bmatrix} \\ &= -\sin(n\omega). \end{aligned}$$

Notice that without any input, as long as the initial state is nonzero, the system will produce a sinusoidal output in perpetuity.

If we instead consider the situation in which the system is initially at rest, then we find that the impulse response is

$$h(n) = \begin{cases} d & \text{if } n = 0 \\ c^T A^{n-1} b & \text{if } n \geq 1 . \end{cases}$$

For the values of A, b, and c in this example, we obtain

$$h(n) = \begin{cases} 0 & \text{if } n = 0 \\ -\sin((n-1)\omega) & \text{if } n \geq 1 . \end{cases}$$

If the oscillator is initially at rest, it can be started with an impulse at the input, and its output will henceforth be sinusoidal, like the output of the zero-input response. ❐

5.3.5 *Multidimensional MIMO systems*

In the preceding sections, both the input and output were scalars. A MIMO system is only slightly more complicated. A state-space model for such a system is

$$\forall n \in Integers_+, \quad s(n+1) = As(n) + Bx(n) \tag{5.41}$$

and

$$y(n) = Cs(n) + Dx(n), \tag{5.42}$$

where $s(n) \in Reals^N$, $x(n) \in Reals^M$, and $y(n) \in Reals^K$, for any integer n. In the $[A, B, C, D]$ representation, A is an $N \times N$ (square) matrix, B is an $N \times M$ matrix, C is a $K \times N$ matrix, and D is a $K \times M$ matrix. Now that all these are matrices, it is conventional to write them with capital letters.

Let $initialState \in Reals^N$ be a given initial state and let $x(0), x(1), x(2), \ldots,$ be a sequence of inputs in $Reals^M$ (each input is a vector of dimension M). The state response of (5.41) is given by

$$s(n) = A^n initialState + \sum_{m=0}^{n-1} A^{n-1-m} Bx(m) \tag{5.43}$$

for all $n \geq 0$, and the output sequence of (5.42) is given by

$$y(n) = CA^n initialState + \left\{ \sum_{m=0}^{n-1} CA^{n-1-m} Bx(m) + Dx(n) \right\} \tag{5.44}$$

for all $n \geq 0$ (each output is a vector of dimension K).

As before, the right sides of these equations are the sum of a zero-input response and a zero-state response. Consider the zero-state output response,

$$y(n) = \sum_{m=0}^{n-1} CA^{n-1-m}Bx(m) + Dx(n), \tag{5.45}$$

for all $n \geq 0$. Define the sequence $h(0), h(1), h(2), \ldots$ of $K \times M$ matrices by

$$h(n) = \begin{cases} D & \text{if } n = 0 \\ CA^{n-1}B & \text{if } n \geq 1. \end{cases} \tag{5.46}$$

This can no longer be called an impulse response, because the input cannot be an impulse; it has the wrong dimension (see the following box).

From (5.45) it follows that the zero-state output response is given by

$$y(n) = \sum_{m=0}^{n} h(n-m)x(m), \; n \geq 0. \tag{5.47}$$

This is once again a **convolution sum**.

The i, jth element of the matrix sequence $h(0), h(1), \ldots$, namely, $h_{i,j}(0), h_{i,j}(1), \ldots$, is indeed the impulse response of the SISO system whose input is x_j and whose output is y_i. In terms of figure 5.1, this is the SISO system obtained by setting all inputs except the jth input to zero and considering only the ith output. If B^j denotes the jth column of B and C_i^T denotes the ith row of C, then $h_{i,j}$ is the impulse response of the SISO system $[A, B^j, C_i, D_{i,j}]$. All these SISO systems have the same A matrix.

PROBING FURTHER

Impulse responses of MIMO systems

The h function in (5.46) is not the impulse response of a MIMO system with representation $[A, B, C, D]$. However, with some care, it is possible to relate it to a set of impulse responses that characterize the system.

The system has M inputs and K outputs, and for each integer n, $h(n)$ in (5.46) is a $K \times M$ matrix. Each input symbol $x(n)$ is an M-tuple of reals. Let $x_m(n)$ represent the mth element in this tuple, and let x_m represent the sequence of such elements. Similarly, let $y_k(n)$ represent the kth element of the output tuple, and let y_k represent the sequence of such outputs. Finally, let $h_{k,m}(n)$ represent the k, mth element of the matrix $h(n)$, and let $h_{k,m}$ represent the sequence of such elements.

The SISO system comprising the kth output and mth input has impulse response $h_{k,m}$. Specifically, if $x_m = \delta$ and $x_p(n) = 0$ for all $p \neq m$ and for all $n \in Integers$, then $y_k = h_{k,m}$. Thus, h can be viewed as a matrix of impulse responses, one for each possible pairing of input and output signals.

5.3.6 *Linear input–output function*

In the systems that this chapter considers, the *nextState* and *output* functions are linear. Recall that a function $f: X \to Y$ is linear if (and only if) it satisfies the superposition property: For all $x_1, x_2 \in X$, and for all $w, u \in Reals$,

$$\boxed{f(wx_1 + ux_2) = wf(x_1) + uf(x_2).}$$

What does it mean for a *system* to be linear? Recall that a system S is a function $S: X \to Y$, where X and Y are signal spaces. For a MIMO system, $X = [Integers \to Reals^M]$, and $Y = [Integers \to Reals^K]$. The function S is defined by (5.44), which yields $y = S(x)$, given x. Therefore, the answer is obvious: S is a **linear system** if S is a linear function. (It is an LTI system if it is also time invariant).

An examination of (5.44)—or its simpler SISO versions, (5.28) or (5.38)—shows that superposition is satisfied if *initialState* is zero. This is because with zero initial state, the output $y(n)$ is a linear combination of the input samples $x(n)$. Hence, a system given by a state-space model that is initially at rest is a linear system. The superposition property turns out to be an extremely useful property, as we discover in the next chapters.

5.4 Continuous-time state-space models

A continuous-time state-space model for an LTI SISO system has the form

$$\forall\, t \in Reals_+, \quad \dot{z}(t) = Az(t) + bv(t) \tag{5.48}$$

and

$$w(t) = c^T z(t) + dv(t), \tag{5.49}$$

where

- $z: Reals_+ \to Reals^N$ gives the state response;
- $\dot{z}(t)$ is the derivative with respect to time of z evaluated at $t \in Reals_+$;
- $v: Reals_+ \to Reals$ is the input signal; and
- $w: Reals_+ \to Reals$ is the output signal.

As with the discrete-time SISO model, A is an $N \times N$ matrix, b and c are $N \times 1$ column vectors, and d is a scalar. As before, this is the $[A, b, c, d]$ representation of the system, but the equations are different.

Continuous-time systems are no longer state machines, inasmuch as inputs, outputs, and state transitions do not occur at discrete instances. Nevertheless, they share many properties of discrete systems. The major difference between

PROBING FURTHER

Approximating continuous-time systems

Discrete LTI systems often arise as approximations of continuous-time systems that are described by **differential equations**. Those differential equations may describe the physics. A differential equation is of the following form (see, for example, lab 6 in Lab Manual):

$$\forall\, t \in Reals, \quad \dot{z}(t) = g(z(t), v(t)). \tag{5.50}$$

Here, $t \in Reals$ stands for continuous time, $z : Reals \to Reals^N$ is the state response, and $v : Reals \to Reals^M$ is the input signal; that is, at any time t, $z(t)$ is the state and $v(t)$ is the input. The notation $\dot{z}$ stands for derivative of the state response z with respect to t, and so $g\colon Reals^N \times Reals^M \to Reals^N$ is a given function specifying the derivative of the state. Specifying the derivative of the state is similar to specifying a state update. Recall that a derivative is a normalized difference over an infinitesimally small interval. A continuous-time system can be thought of as one in which the state updates occur in intervals that are so small that the state appears to evolve continuously rather than discretely.

In general, z is an N-tuple, $z = (z_1, \ldots, z_N)$, where $z_i\colon Reals_+ \to Reals$. The derivative of an N-tuple is simply the N-tuple of derivatives, $\dot{z} = (\dot{z}_1, \ldots, \dot{z}_N)$. We know from calculus that

$$\dot{z}(t) = \frac{dz}{dt} = \lim_{\delta \to 0} \frac{z(t+\delta) - z(t)}{\delta},$$

and so, if $\delta > 0$ is a small number, we can approximate this derivative by

$$\dot{z}(t) \approx \frac{z(t+\delta) - z(t)}{\delta}.$$

Using this for the derivative in the left side of (5.50) yields

$$z(t+\delta) - z(t) = \delta g(z(t), v(t)). \tag{5.51}$$

Look at this equation at sample times $t = 0, \delta, 2\delta, \ldots$. We denote the value of the state response at the nth sample time by $s(n) = z(n\delta)$ and the value of the input at this same sample time by $x(n) = v(n\delta)$. In terms of these variables, (5.51) becomes

$$s(n+1) - s(n) = \delta g(s(n), x(n)),$$

which we can write as a state update equation:

$$s(n+1) = s(n) + \delta g(s(n), x(n)).$$

the model of (5.48) and that of (5.33) and (5.34) is that instead of specifying the new state as a function of the input and the old state, (5.48) specifies the derivative of the state. The derivative of a vector z is simply the vector consisting of the derivative of each element of the vector. A derivative, of course, characterizes the trend of the state at any particular time. Characterizing a trend makes more sense than characterizing a new state for a continuous-time system, because the state evolves continuously.

All the methods that we have developed for discrete-time systems can also be developed for continuous-time systems. However, they are somewhat more challenging mathematically because the summations are integrals. We study continuous-time systems in chapter 13.

A continuous-time state-space model may be approximated by a discrete-time state-space model (see the following box). In fact, this approximation forms the basis for most computer simulations of continuous-time systems. Simulation of continuous-time systems is explored in lab 6 (see Lab Manual).

5.5 Summary

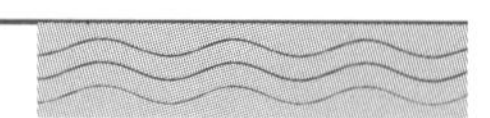

This chapter has begun the exploration of state machines whose state update and output functions are linear. The number of possible states for such systems is typically infinite, and so brute-force methods that enumerate the states are ineffective. Instead, powerful mathematical tools leverage the linearity of the key functions. This chapter barely begins an exploration of the very rich set of tools that engineers have developed for such systems. Subsequent chapters continue that exploration.

EXERCISES

KEY: **E** = mechanical **T** = requires plan of attack **C** = more than 1 answer

E 1. Use induction to obtain (5.13) as a consequence of (5.12).

E 2. Use induction to show that if

$$A = \begin{bmatrix} a & 1 \\ 0 & a \end{bmatrix},$$

then for all $n \geq 0$,

$$A^n = \begin{bmatrix} a^n & na^{n-1} \\ 0 & a^n \end{bmatrix}.$$

E 3. Let $f\colon Reals^N \to Reals^N$ be a linear function given by $\forall\, x \in Reals^N,\ f(x) = Ax$ for some $N \times N$ matrix A. Give a similar definition for the composition $f \circ f$. Is it also linear?

E 4. What would be the monthly payment in example 5.10 if the monthly interest were 0.015? Also, what would be the total payment over the 32 months?

E 5. Construct a SISO state-space model for a system whose input and output are related by

$$\forall\, n \in Integers, \quad y(n) = x(n-1) + x(n-2).$$

You may assume the system is initially at rest. It is sufficient to give the A matrix, vectors b and c, and scalar d of (5.33) and (5.34).

E 6. The A matrix in (5.33) for a SISO system is

$$A = \begin{bmatrix} 1 & 1 \\ 0 & 1 \end{bmatrix}.$$

Calculate the zero-input state response if

(a) the initial state is $[1, 0]^T$,

(b) the initial state is $[0, 1]^T$, and

(c) the initial state is $[1, 1]^T$.

E 7. Consider the one-dimensional state-space model, $\forall\, n \in Integers_+$,

$$s(n+1) = s(n) + x(n)$$

and

$$y(n) = s(n).$$

Suppose the initial state is $s(0) = a$ for some given constant a. Find another constant b such that if the first three inputs are $x(0) = x(1) = x(2) = b$, then $y(3) = 0$. Note: In general, problems of this type are concerned with **controllability**. The question is whether you can find an input (in this case, constrained to be constant) such that some particular condition on the output is met. The input becomes a **control signal**.

E 8. A SISO LTI system has the A matrix given by

$$A = \begin{bmatrix} 0 & 1 \\ 0 & 0 \end{bmatrix}$$

and the b vector by $[0, 1]^T$. Suppose that $s(0) = [0, 0]^T$. Find the input values $x(0)$, $x(1)$ such that the state at step 2 is $s(2) = [1, 2]^T$.

E 9. In example 5.13, the state was chosen to be $s(n) = [x(n-1), x(n-2)]^T$. How would the $[A, b, c, d]$ representation change if the state were chosen to be

(a) $[x(n-2), x(n-1)]^T$?

(b) $[x(n-1) + x(n-2), x(n-1) - x(n-2)]^T$?

E 10. Suppose the A matrix of a two-dimensional SISO system is

$$A = \sigma \begin{bmatrix} \cos(\pi/6) & \sin(\pi/6) \\ -\sin(\pi/6) & \cos(\pi/6) \end{bmatrix}.$$

Suppose the initial state is $s(0) = [1, 0]^T$ and the input is zero. Sketch the zero-input state response for $n = 0, 1, \ldots, 12$ for the cases

(a) $\sigma = 0$,

(b) $\sigma = 0.9$, and

(c) $\sigma = 1.1$.

E 11. In this problem, we further consider example 5.9. As in the example, suppose *initialState* $= 100$, and you deposit \$1,000 on day 0 and withdraw \$30 every subsequent day for the next 30 days.

(a) Write a MATLAB®* program to compute your bank balance $s(n)$, $0 \leq n \leq 31$, and plot the result.

(b) Use formula (5.27) to calculate your bank balance at the beginning of day 31. The following identity may prove useful:

$$\sum_{m=0}^{N} a^m = \frac{1 - a^{N+1}}{1 - a},$$

where $a \neq 1$.

* MATLAB is a registered trademark of The MathWorks, Inc.

E 12. Use MATLAB to calculate and plot the impulse response of the system

$$s(n+1) = as(n) + bx(n)$$

and

$$y(n) = cs(n)$$

for the following cases:

(a) $a = 1.1$,

(b) $a = 1.0$,

(c) $a = 0.9$, and

(d) $a = -0.5$,

where in all cases, $b = c = 1$ and $d = 0$.

E 13. Use induction to derive the SISO response expressions (5.37) and (5.38).

E 14. Consider the two-dimensional system given in example 5.17, in which

$$A = \begin{bmatrix} \cos(\omega) & -\sin(\omega) \\ \sin(\omega) & \cos(\omega) \end{bmatrix}.$$

Show that for all $n = 0, 1, 2, \ldots,$

$$A^n = \begin{bmatrix} \cos(n\omega) & -\sin(n\omega) \\ \sin(n\omega) & \cos(n\omega) \end{bmatrix}.$$

Hint: Use induction and the identities

$$\cos(\alpha + \beta) = \cos(\alpha)\cos(\beta) - \sin(\alpha)\sin(\beta)$$

and

$$\sin(\alpha + \beta) = \sin(\alpha)\cos(\beta) + \cos(\alpha)\sin(\beta).$$

E 15. A **damped oscillator** is a variant of the oscillator in example 5.17, in which the sinusoidal signal decays with time. The damped oscillator is identical except that the A matrix is given by

$$A = \alpha \begin{bmatrix} \cos(\omega) & -\sin(\omega) \\ \sin(\omega) & \cos(\omega) \end{bmatrix},$$

where $0 < \alpha < 1$ is a constant damping factor.

(a) Find the zero-input state response and the zero-input response for the initial state

$$initialState = [0, 1]^T.$$

(b) Find the zero-state impulse response.

T 16. Consider the audio echo system in example 5.6. A more interesting effect (sometimes called **reverberation**) can be obtained by simultaneously combining multiple echos. In this problem, consider the following difference equation, which combines two echos:

$$y(n) = x(n) + \alpha y(n - M) + \beta y(n - N),$$

where the (discrete-time) audio input is x, the output is y, and $0 \leq \alpha < 1$ and $0 \leq \beta < 1$ are constants that specify two distinct echo terms. Suppose for simplicity that $M = 4$ and $N = 5$. These numbers are not large enough to yield an audible echo, but the mathematical model for larger numbers is similar, and so these are adequate for study. Note that the zero-state impulse response of this system is much more complicated than that of the simple echo system (try to find it), which accounts for a considerably more realistic echo effect.

(a) Give a state-space representation $[A, B, C, D]$ of this system.

(b) Modify the system so that the audio input has two channels (stereo) but the output is still one channel. Combine the two input channels by adding them with equal weight. Give $[A, B, C, D]$.

CHAPTER 6
Hybrid systems

This text models signals and systems as functions. To develop understanding, we study the structure of the domain and range of these functions, as well as the structure of the mapping from the domain to the range. Despite the uniformity of this approach, we have begun to evolve two distinct families of models. In chapters 3 and 4, this mapping is structured through the use of state machines. In chapter 5, these state machines are generalized so that the number of possible states is infinite, and they are specialized so that the systems are linear and time-invariant (LTI). LTI systems prove to yield to powerful analytical techniques, which are only hinted at in chapter 5. In chapters 7 to 14 these analytical techniques are further developed by structuring the system mapping with the use of frequency-domain concepts.

The analytical methods available for LTI systems prove so compelling that we wish to apply them even to systems that are not LTI. In fact, no real-world system is truly LTI. At a minimum, the system's properties were certainly different during the initial stages of the big bang, so it cannot be time invariant. More practically, systems change over time; they are turned on and off, they deteriorate, and so forth. Moreover, systems that behave as linear systems typically do so only over some operating range. For example, if the magnitude of the inputs exceeds some threshold, a real-world system will overload and will no longer behave linearly. A similar effect may result when the state wanders beyond some modest range. This chapter shows how models that are applicable only some of the time can be used effectively.

In chapters 3 and 4, signals are sequences of events. Their domain is typically $Naturals_0$, and their range is typically a finite and arbitrary set of symbols. The domain is interpreted not as time but rather as indexes of a sequence. In chapters 5 and 7 to 14, the domain of signals is interpreted as time. For continuous-time signals the domain is either *Reals* or $Reals_+$, whereas for discrete-time signals it is either *Integers* or $Naturals_0$; in either case, however, the domain is interpreted as representing advancing time. This interpretation of the domain as time is essential to the notion of frequency that is used throughout the forthcoming chapters.

Chapter 5 and this one provide a bridge between **state machine models** and **time-based models** by developing state machine models for time-based systems. In this chapter, we build another bridge between these two families of models by showing that they can often be combined usefully and used *simultaneously* in the *same* model, rather than as alternative views of a system. The resulting models are called **hybrid systems**. They are a powerful tool for understanding real-world systems.

To understand the value of hybrid systems, it is useful to reflect on the relative strengths and weaknesses of time-based models and state machine models. Chapter 5 demonstrates that state machine models are more general by showing how they can be used to describe time-based models. Because they are more general, why not just always use state machine models? The methods of state machine models, such as composition by forming a product of the state spaces, simulation, and bisimulation, do not yield the depth of understanding that we achieve in the subsequent chapters from examining frequency response. Why not always use frequency response? Frequency response is a rather specialized analytical tool. It applies only to LTI systems. Most real-world systems are not LTI, and so such powerful analytical tools must be used with careful caveats about the range of operation over which they do apply.

Consider, for example, a home audio system. It takes data from a compact disc and converts it into auditory stimulus. Is it LTI? Obviously not, because its system function changes rather drastically when it is turned on and off. The acts of turning it on and off, however, seem to match well the state transitions of a state machine. Can we devise a model in which there is a state machine with two states, "on" and "off," and, associated with each state, there is an LTI system that describes the behavior of the system in the corresponding mode of operation? Indeed we can. Such a model is called a hybrid system.

In order to get state machine models to coexist with time-based models, we need to interpret state transitions on the time line used for the time-based portion of the system, be it continuous time or discrete time. In the audio system, for example, we need to associate a time with the acts of turning it on or off. The models used in chapters 3 and 4 do not naturally do this, because their signals are sequences of events; that is, they are functions whose domain is $Naturals_0$, in which there is no temporal association with an $n \in Naturals_0$.

Recall from chapter 3 that the input and output alphabets of a state machine are required to include a stuttering element, typically denoted *absent*. Whenever the state machine reacts, if its input is the stuttering element, then it does not change state, and its output is the stuttering element. This is key to hybrid system models because it allows us to embed the state machine into a time-based model. At any time when there is no interesting input event, the machine stutters.

A hybrid system combines time-based signals with sequences of events. The time-based signals are of the form $x: T \to R$, where R is some range (such as *Reals* or *Complex*), and T is *Reals*, $Reals_+$, *Integers*, or $Naturals_0$, depending on whether the time domain is discrete or continuous and whether the model includes a time origin. In chapters 3 and 4, the event signals had the form $u: Naturals_0 \to Symbols$, where the set *Symbols* has a stuttering element. For a hybrid system, however, these have to share a common time base with the time-based signals, so they have the form $u: T \to Symbols$. Thus, events occur in time. Typically, for most $t \in T$, $u(t) = absent$, the stuttering element. The nonstuttering element is used only at the discrete values of time when an event occurs.

6.1 Mixed models

A state machine model becomes a time-based model if it reacts at all times in the time base T. This means that state machines and time-based models can interact as peers, sending time-based signals to one another.

> **Example 6.1:** Moving averages are popular on Wall Street for detecting trends in stock prices. In their use, however, a key question arises: How long should the moving average be? A short-term moving average might detect short-term trends, whereas a long-term moving average might detect long-term trends. A classical method combines the two and compares them to generate signals to buy and sell. If the short-term trend is more sharply upward than the long-term trend, a signal to buy is generated. If the short-term trend is more sharply downward than the long-term trend, a signal to sell is generated.
>
> A system implementing this **moving average cross-over method** is shown in figure 6.1. The input is the discrete-time signal *price*: *Integers* → *Reals* that represents the closing price of a stock each day. Both LTI systems *shortTerm* and *longTerm* are moving average systems, but *shortTerm* averages fewer successive inputs than does *longTerm*. The outputs of these systems are the discrete-time signals x and y. The finite-state machine reacts on each sample from these signals. It begins in the state *short over long*. The transition out of this state has the guard
>
> $$\{(x(n), y(n)) \mid x(n) > y(n)\}.$$
>
> When this transition is taken, a *buy* signal is generated. The sell signal is generated similarly. The plots below show the buy and sell signals generated by a (synthetic) sequence of stock prices.

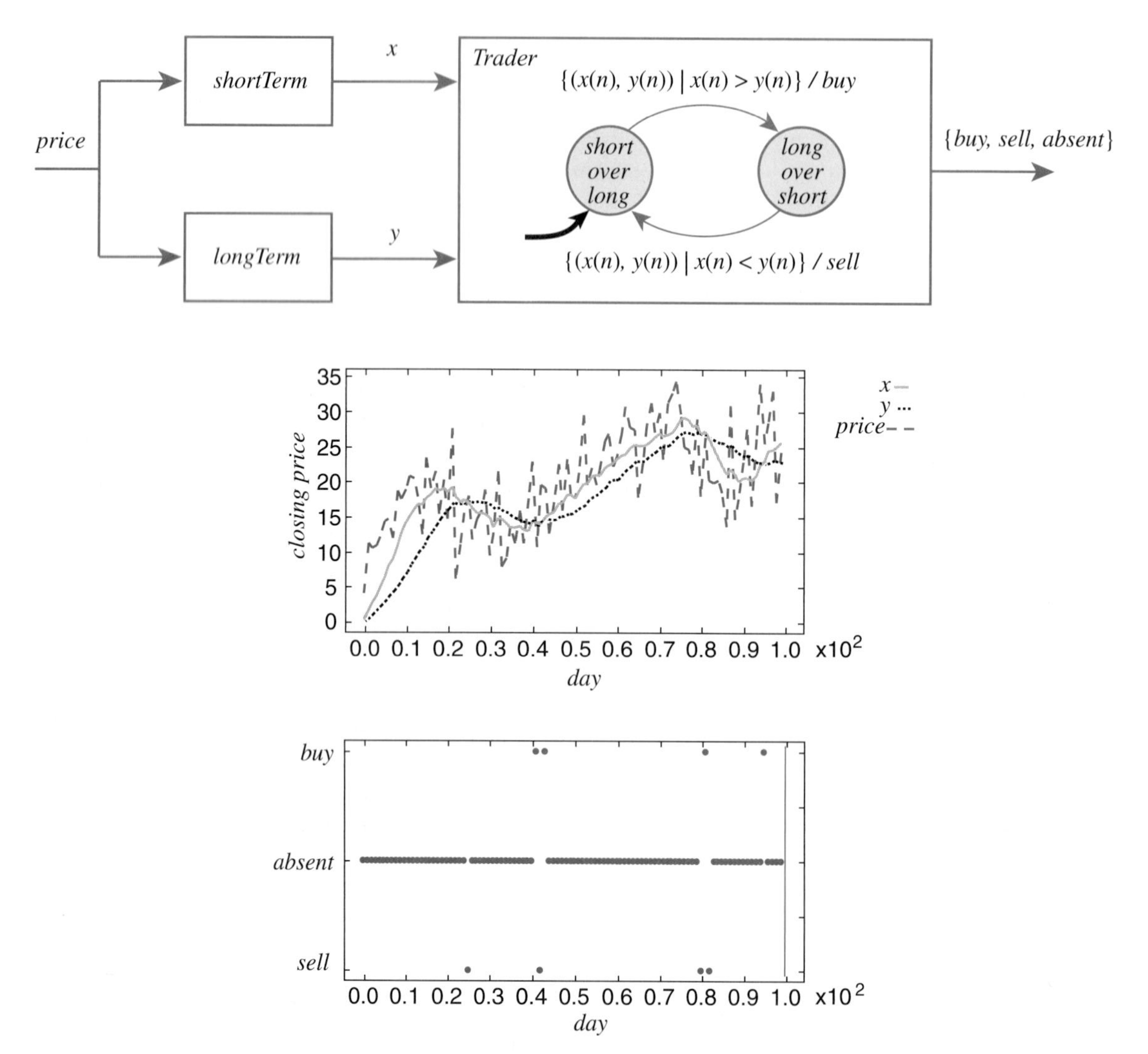

FIGURE 6.1: An implementation of the classical moving average cross-over method for trading stocks.

This example illustrates a simple form of technical stock trading. In this extreme form, it has a controversial feature: It ignores the fundamentals of the company whose stock is being traded. The stock price alone is used as the indicator of worth. In fact, much more sophisticated signal processing methods are used by technical stock traders, and they often do take as inputs other quantifiers of company worth, such as reported revenues and profits. ❐

6.2 Modal models

In the previous section, time-based systems are combined with state machines as peers. A richer interaction is possible with a hierarchical combination. The general structure of a hierarchical hybrid system model is shown in figure 6.2. In that figure, there is a two-state finite-state machine. There are some changes to the notation, however, from what was used in chapters 3 and 4.

First, notice that the inputs and outputs include both event signals and time-based signals. Second, notice that each state of the state machine is associated with a time-based system, called the **refinement** of the state. The refinement of a state yields the time-based behavior of *HybridSystem* while the machine is in that state. Thus, the states of the state machine define **modes** of operation of the system, in which the behavior in a given mode is determined by the refinement. A hybrid system is sometimes called a **modal model** for this reason. The refinement has access to all the inputs of *HybridSystem* and produces the time-based output signals of *HybridSystem* while the machine is in its mode.

Note that the term "state" for such a hybrid system can become confusing. The state machine has states, but so do the refinement systems (unless they are memoryless). When there is any possibility of confusion, we explicitly refer to the states of the machine as modes and to the states of the refinement as **refinement states**. The complete state of the hybrid system is a pair (m, s), where m is the mode and s is the state of the time-based refinement system associated with mode m.

Another difference from the notation used in chapters 3 and 4 is that state transitions in the machine have, in addition to the usual guard and output

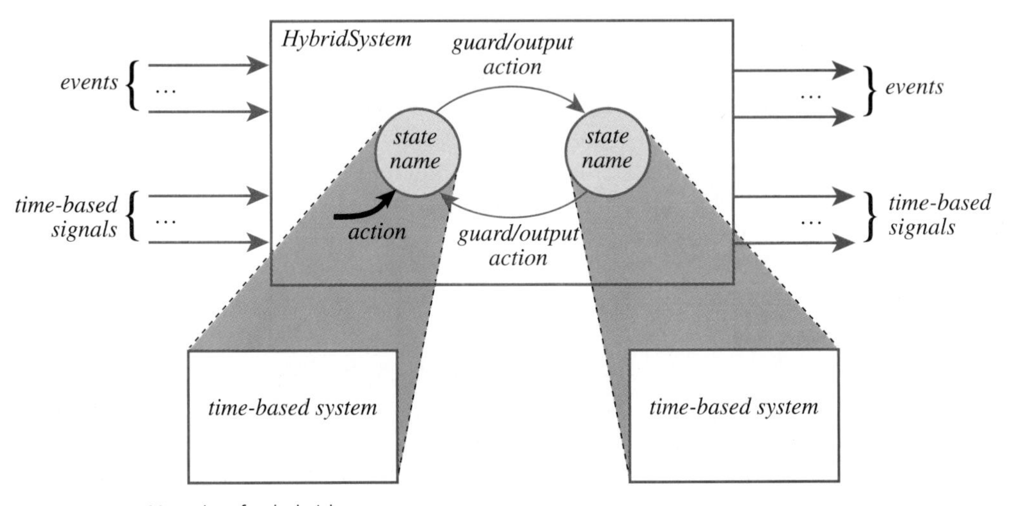

FIGURE 6.2: Notation for hybrid systems.

notations, an **action**. The action typically sets the initial refinement state of the time-based system in the destination mode.

The guards are, as usual, sets. However, the guards must be expressive enough that a transition can be triggered by a particular value of a refinement state or by a value of a time-based input. Thus, the elements of the guards are tuples containing values of input events, time-based signals, and the refinement states. In the state machines in chapters 3 and 4, the elements of the guards contained only values of input events. For hybrid systems, we add time-based signals and refinement states.

Example 6.2: Overload of an electronic system may be modeled by a state transition that is triggered when the magnitude of the current refinement state exceeds some threshold. ❐

On the other hand, when the system is in some mode, the refinement state is affected only by the time-based inputs. It is not affected by the event inputs. This keeps the time-based models simple, so that they do not have to deal with stuttering inputs.

Correspondingly, the time-based outputs are generated by the refinement and hence need not be mentioned after the slash on the transitions.

The state machine may react at any time in the time base T. The mode in which it is before this reaction is called the **current mode**. It takes a discrete state transition and switches to the **destination mode** if the input values and the refinement state at that time match a guard. If it does not make a discrete state transition, then the state machine stutters. In either case, the refinement of the current mode also reacts to the time-based inputs, changes its state, and produces outputs.

Example 6.3: Many high-end audio systems offer "digital signal processing." Such a system typically has an embedded computer (a digital signal processor [DSP]; see Probing further box on page 357). This computer is used to process the audio signal in various ways: for example, to add reverberation or to perform frequency selective filtering. A particularly simple function that might be performed is **loudness** compensation, something offered by all but the cheapest audio systems.

At low volumes, the human ear is less sensitive to low frequencies (base notes) than to high frequencies. Loudness compensation boosts the low frequencies. This is done simply by implementing a filter, which is an LTI system that can be described by a state-space model, as in chapter 5. Thus, there are two modes: one in which the low frequencies are boosted (by means of the filter) and one in which they are not.

In a simple mechanism for loudness compensation, a switch on a control panel turns on and off the compensation. Figure 6.3 shows a hybrid system that reacts to input events from this switch to select from between

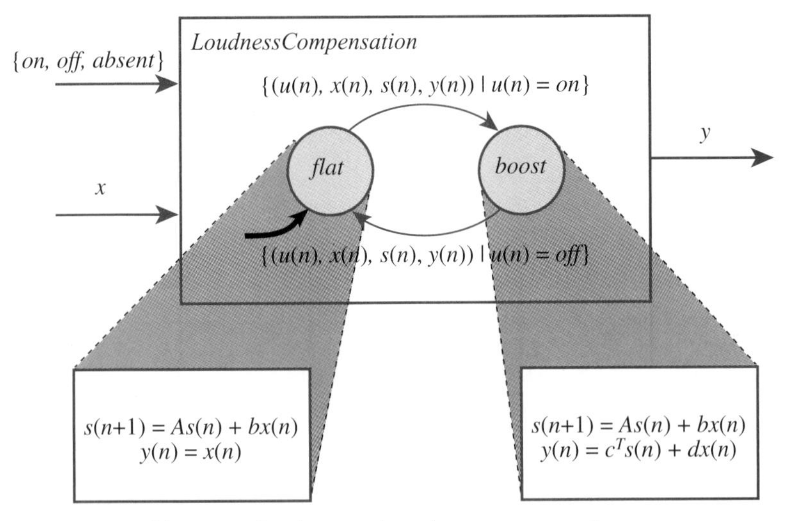

FIGURE 6.3: This system implements loundness compensation.

the two modes. The upper input is simply an event indicating the position of the control switch when it is thrown. The lower input x is a discrete-time signal, probably sampled at 44,100 samples/second, the compact disc rate. The *LoudnessCompensation* hybrid system has two modes. In the *flat* mode, the output y is simply set equal to the input x; that is, if $T_{flat} \subset Integers$ is the time indexes during which the machine is in the *flat* mode, then

$$\forall\, n \in T_{flat}, \quad y(n) = x(n).$$

This obviously does not boost low frequencies, inasmuch as the output is equal to the input.

When the *on* event occurs, the machine transitions to the *boost* mode, in which the filter is applied to the input x. This is done by using the state update and output equations

$$\forall\, n \in T_{boost}, \quad s(n+1) = As(n) + bx(n)$$

and

$$y(n) = c^T s(n) + dx(n),$$

where A, b, c, and d are chosen to boost the low frequencies (how to do that is explained in chapter 9).

Note that in the *flat* mode, even though the output equation does not depend on the state, the state update equation is still applied. This ensures that during a switch between states, no glitches are heard in the audio signal. The state of the *boost* refinement is maintained even when the mode is *flat*.

This loudness compensator is not very sophisticated. A more sophisticated version would have a set of compensation filters and would select among them according to the volume level. This is explored in exercise 1 at the end of this chapter. ❐

We consider a sequence of special cases of hybrid systems. Although the next few examples are all continuous-time models, it is easy to construct similar discrete-time models.

6.3 Timed automata

Timed automata are the simplest continuous-time hybrid systems. They are modal models in which the time-based refinements have very simple dynamics; all they do is measure the passage of time. Such refinements are called **clocks**. The resulting models are finite-state machines (automata) with clocks.

A clock is modeled by a first-order differential equation,

$$\forall\, t \in T_m, \quad \dot{s}(t) = a,$$

where s: *Reals* → *Reals* is a function, $s(t)$ is the value of the clock at time t, and $T_m \subset T$ is the subset of time during which the hybrid system is in mode m. The rate of the clock, a, is constant while the system is in this mode.

Example 6.4: Suppose we want to produce a sequence of output events called *tick* in which the time between two consecutive *ticks* alternates between one and two seconds; that is, we want to produce a *tick* at times $1, 3, 4, 6, 7, 9, \ldots$.

A hybrid system *tickGenerator* that does this is illustrated in figure 6.4. There are two modes, labeled *mode 1* and *mode 2*. The refinement state in each mode is the value of a clock at time t, denoted by $s \in$ *Reals*. Therefore, at any time t, the state of *tickGenerator* is the pair $(mode(t), s(t))$. The output is the event signal v and the time-based signal s. There is no input.

In both modes, s evolves according to the differential equation $\dot{s}(t) = 1$, where $\dot{s}(t)$ is the derivative of s with regard to time evaluated at some time t. Thus, s simply measures the passage of time; its value rises 1 second for every second of elapsed time.

The behavior of the system is shown in figure 6.5. At time 0, as indicated by the bold arrow in figure 6.4, the system enters *mode 1*. The bold arrow has an action, "$s(0) := 0$," which sets $s(0)$ to 0. The notation ":=" is used instead of "=" to emphasize that this is an assignment, not an assertion (see section A.1.1 in appendix A).

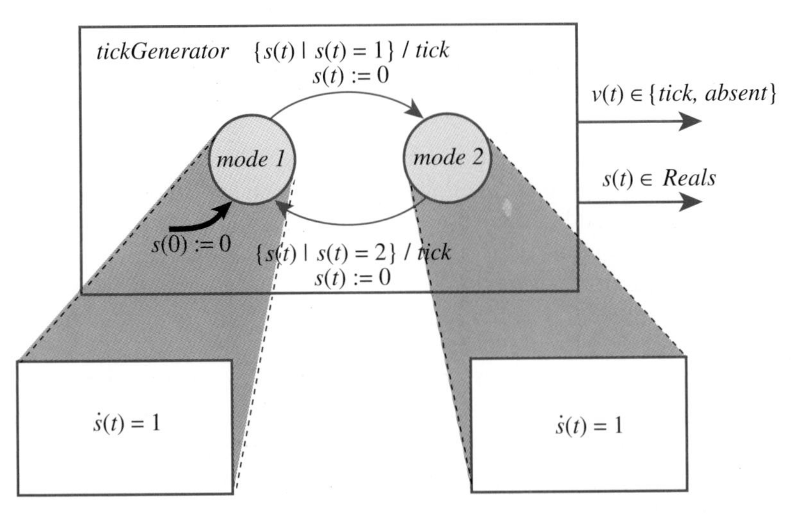

FIGURE 6.4: This hybrid system generates *tick* at time intervals alternating between one and two seconds. It is a timed automaton.

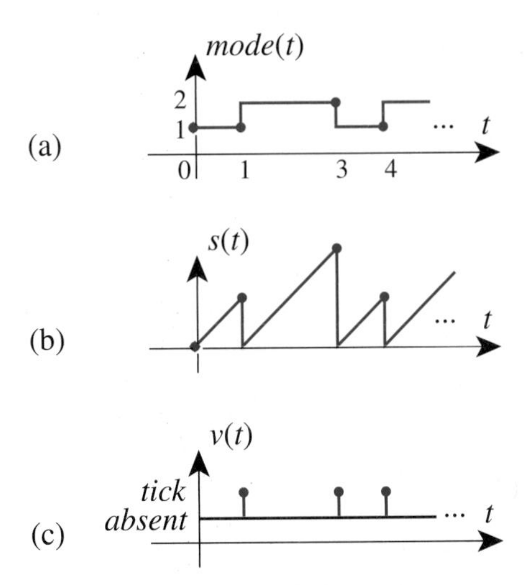

FIGURE 6.5: The modes of the hybrid system of figure 6.4 (a), the refinement state s (b), and the discrete event output v (c).

In this example, there is no input, and so a guard is a subset of the possible values (*Reals*) of the refinement states. The guard on the transition from *mode 1* to *mode 2* is

$$\{s(t) \mid s(t) = 1\},$$

which is satisfied one time unit after beginning. For all $t \in [0, 1]$, $s(t) = t$. At time $t = 1$, this guard is satisfied, the transition is taken, and the output event $v(1) = tick$ is produced. For all $t \in [0, 1)$, $v(t)$ has value *absent*.

This transition also has an action, "$s(t) := 0$," which resets s to zero. This action produces the initial condition for the refinement system of the destination mode. In our definition, at time $t = 1$, $s(t) = 1$, even though the action seems to contradict this. This is emphasized in figure 6.5 by showing with a boldface dot the value of s at each discontinuity. The action $s(t) := 0$ merely provides the initial conditions for the refinement of the destination mode. However, the destination mode is not active until $t > 1$, and so the action is setting $s(1+)$ to 0, where 1+ denotes a time infinitesimally larger than 1.

For $t \in (1, 3]$, the system remains in *mode 2*, evolving according to the differential equations

$$\dot{s}(t) = 1$$

and

$$s(1) = 0.$$

Therefore, for $1 < t \leq 3$,

$$s(t) = s(1) + \int_1^t 1dt = t - 1.$$

At time $t = 3$, the guard on the arc from *mode 2* to *mode 1* is satisfied, and so the transition is taken. Again, the output event *tick* is produced, and s is reset to 0. ❐

Notice in figure 6.5 that the output v is *absent* for all but a few discrete values of $t \in Reals$. This signal is called a **discrete event** signal for this reason. Of course, this signal can also be reinterpreted as a sequence of *tick* events with an arbitrary number of stuttering events in between. That signal could therefore be supplied as input to an ordinary state machine, enabling compositions of ordinary state machines with hybrid systems.

Notice also in figure 6.5 that the hybrid system evolves in alternating phases: There is a **time-passage phase**, in which the system stays in the same mode and its refinement state changes with the passage of time; this is followed by an instantaneous **discrete-event phase**, in which a mode transition occurs, an output event is produced, and the refinement state in the destination mode is initialized. In the figure, the time-passage phases are $(0, 1]$, $(1, 3]$, $(3, 4]$, ... and the discrete-event phases occur at $1, 3, 4, \ldots$.

Transitions between modes are associated with actions. Sometimes, it is useful to have transitions from one mode back to itself, just so that the action can be realized. This is illustrated in the next example.

Example 6.5: Figure 6.6 shows a hybrid system representation of the 60-minute parking meter considered in chapter 3. In the version in figure 3.6, the states of a state machine are used to measure the passage of time by counting ticks provided by the environment. In the hybrid version of figure 6.6, the passage of time is explicitly modeled by first-order differential equations.

There are two modes, *expired* and *safe*, and the refinement state at time t is $s(t) \in Reals$. At time $t = 0$, the initial mode is *expired*, and $s(0) = 0$. In the *expired* mode, s remains at 0. The input events *coin5* and *coin25* each cause

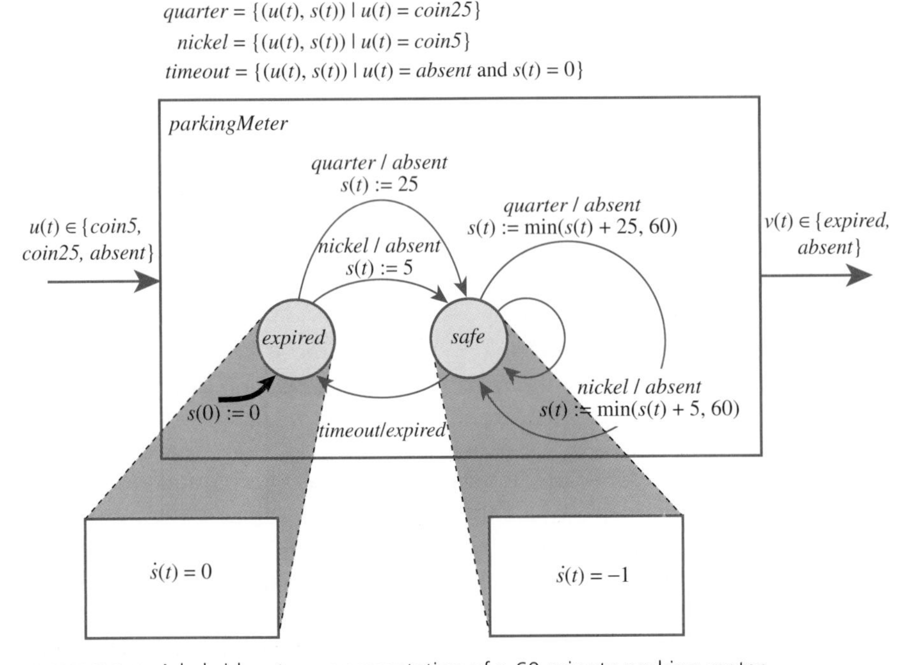

FIGURE 6.6: A hybrid system representation of a 60-minute parking meter.

one of two transitions from *expired* to *safe* to be taken. These transitions have guards that are named *nickel* and *quarter* and are defined by

$$nickel = \{(u(t), s(t)) \mid u(t) = coin5\}$$

and

$$quarter = \{(u(t), s(t)) \mid u(t) = coin25\}.$$

Using names for these guards in the figure makes it more readable. It would be cluttered if the guards were directly noted on the transitions.

The transitions from *expired* to *safe* produce *absent*. The actions on the transitions change the value of s to 5 or 25, depending on whether *coin5* or *coin25* is received.

In the *safe* mode, the refinement state decreases according to the differential equation of the clock:

$$\forall\, t \in T_{safe}, \quad \dot{s}(t) = -1.$$

There are three possible outgoing transitions from this mode. If the input event *coin5* or *coin25* occurs, then one of two self-loop transitions is taken, no output is produced, and the associated action increases s by setting $s(t) := \min(s(t)+5, 60)$ or $s(t) := \min(s(t)+25, 60)$. But if the guard *timeout* is satisfied, where

$$timeout = \{(u(t), s(t)) \mid u(t) = absent \text{ and } s(t) = 0\},$$

then there is a transition to *expired*, and the output event *expired* is produced. Note that this guard requires that $u(t) = absent$, so that if the parking meter expires at the very moment that a coin arrives, then the coin is properly registered.

In this system, the refinement state evolves differently in the two modes: In *expired*, s remains at 0 (because $\dot{s}(t) = 0$), but in *safe*, s obeys the differential equation $\dot{s}(t) = -1$. ❒

In the previous example, the transitions from *safe* back to *safe* were used for their actions, which occur in reaction to input events by setting the values of refinement states. This provides a clean way to model discontinuities in continuous-time signals, because the state trajectory is a continuous-time signal. A more extreme example, in which there is only one mode, is given next.

Example 6.6: We could also implement the parking meter as a cascade composition by using a timed automaton, *TickGenerator*, which has only one mode, *timer*, and produces a *tick* event every minute. This event serves as an input to the parking meter finite-state machine of figure 3.6. The cascade

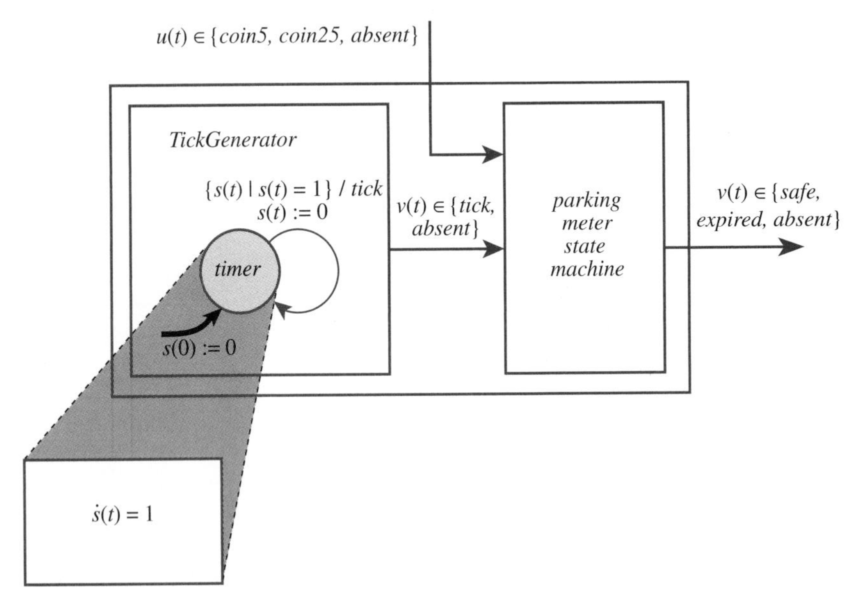

FIGURE 6.7: The 60-minute parking meter as a cascade composition of *tickGenerator* and an ordinary finite-state machine.

composition is shown in figure 6.7. The parking meter machine also accepts an additional (product-form) input event from {*coin5*, *coin25*} and produces the output event *safe* or *expired*. The difference between figures 3.6 and 6.7 is that in the former, *tick* was an input event from the environment, whereas in the latter, we explicitly construct a component—namely, *TickGenerator*—that produces a *tick* every minute. ❐

Timed automata are commonly used in modeling **communication protocols**, the logic used to achieve communication over a network. The following example models the transport layer of a sender of data on the Internet.

Example 6.7: Consider how an application such as an e-mail program sends a file over a communication network such as the Internet. There are two host computers, called *Sender* and *Receiver*. The file that *Sender* wants to send to *Receiver* is first divided into a sequence of finite bit strings called **packets**. For the purposes of this example, we do not care what the packets contain, and so we consider *packet* to be an event. We are interested in the fact that it needs to be transmitted, not in its contents.

The problem we address in this example is that the network is unreliable. Packets that are launched into it may never emerge. If the network is congested, packets get dropped. We design a protocol whereby the sender of a packet waits a certain amount of time for an acknowledgment. If it does not receive the acknowledgment in that time, then it retransmits the packet. This is an ideal application for timed automata.

The upper diagram in figure 6.8 shows the structure of the communication system. Everything begins when the sender produces a *packet* event. The *SenderProtocol* system reacts by producing a *transmit* event, which instructs its **network interface card** (**NIC**) to launch the packet into the Internet. The

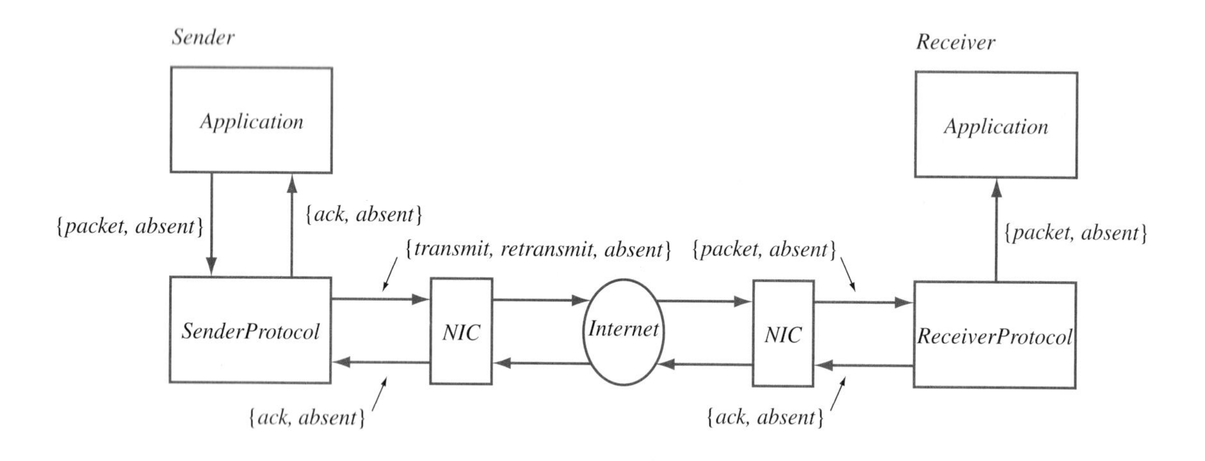

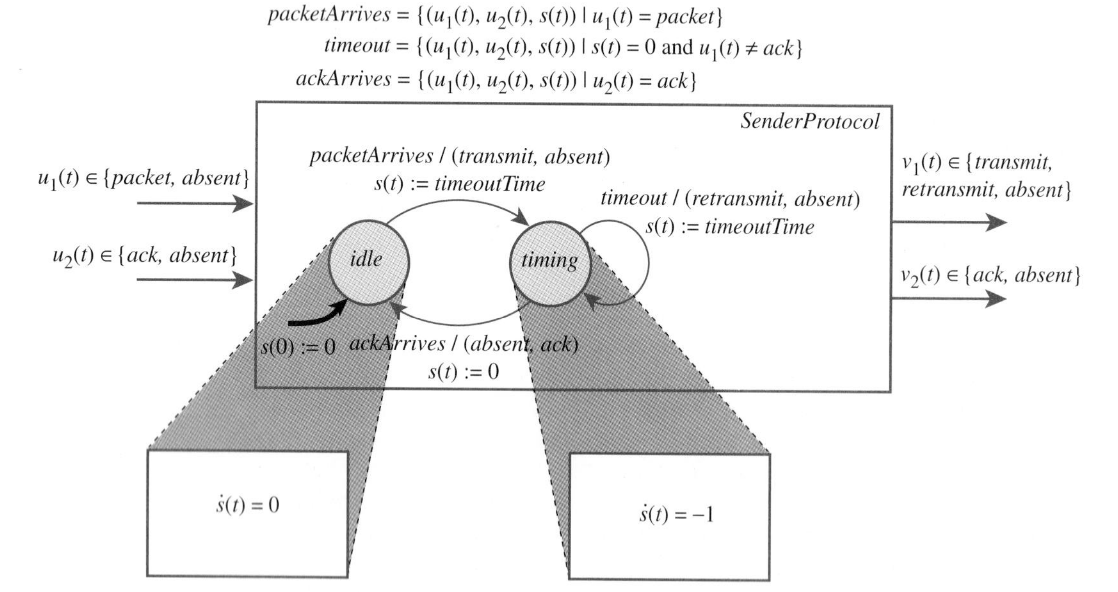

FIGURE 6.8: The top diagram describes the structure of a communication system. The lower diagram is the timed automaton that implements the sender protocol.

NIC is the physical device (such as the Ethernet card in a desktop computer) that converts the packet into the appropriate electrical signal that is transmitted through the network. The Internet transfers this signal to the NIC of the receiver. That NIC converts the signal back into the packet and forwards it to the *ReceiverProtocol* component. The *ReceiverProtocol* in turn forwards the packet to the e-mail application in the *Receiver* and simultaneously sends an acknowledgment packet, called *ack*, to its NIC.

The receiver's NIC sends the *ack* packet back through the network to the *Sender*. The sender's NIC receives this packet and forwards an *ack* to the *SenderProtocol*. The *SenderProtocol* notifies the application that the packet was indeed delivered. The application can now send the next *packet*, and the cycle is repeated until the entire file is delivered.

In reality, the network may drop the packet so that it is not delivered to the receiver, who therefore does not send the corresponding *ack*. The *SenderProtocol* system is designed to take care of this contingency. It is a timed automaton with two modes, *idle* and *timing*, and one refinement state, s, corresponding to a clock. Initially it is in the *idle* mode and $s(0) = 0$. In the idle mode, $\dot{s}(t) = 0$, and so the refinement state remains at zero. When *SenderProtocol* receives a *packet*, it makes a transition to the *timing* mode, sends the output event *transmit* to its NIC, and resets s to a timeout value, *timeoutTime*.

In the *timing* mode, there are two possible transitions. In the normal case, the input event *ack* is received before the guard *timeout* is satisfied. The transition to mode *idle* is taken, the output event *ack* is sent to the application, and the clock value $s(t)$ is reset to 0. The system waits for another *packet* from the application. In the second case, the guard *timeout* is satisfied (before event *ack*), and the self-loop transition is taken. In this case, the output event *retransmit* is sent to the NIC, and $s(t)$ is reset to *timeoutTime*.

Notice a feature of this design that may not be expected. If a packet arrives while the machine is in mode *timing*, the packet is ignored. What happens if a packet arrives simultaneously with an *ack* while the machine is mode *timing*?

Exercise 10 at the end of this chapter asks you to construct the corresponding receiver protocol, which is simpler.

In summary, the *SenderProtocol* machine repeatedly retransmits a packet every *timeoutTime* seconds until it receives an *ack*. This reveals a flaw in the protocol. If the network is for some reason unable ever to successfully transmit a packet to the receiver, the machine will continue retransmission forever. A better protocol would retransmit a packet a certain number of times, say five times, and if it is unsuccessful, it would return to *idle* and send a message *connectionFailed* to the application. The hybrid system of figure 6.9 incorporates this feature by adding another clock whose value is $r(t)$. ❒

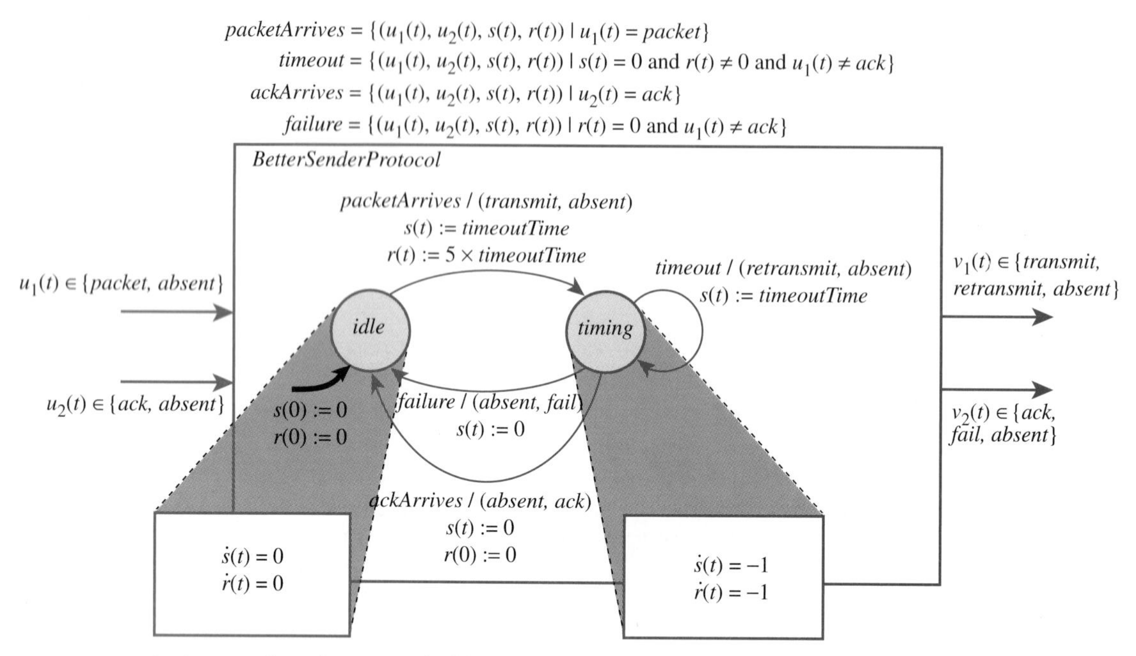

FIGURE 6.9: An improved sender protocol with two clocks, one of which detects a failed connection.

PROBING FURTHER

Internet protocols

Communication between two computers, *Sender* and *Receiver*, each connected to the Internet, is coordinated by a set of **protocols**. Each protocol can be modeled by a pair of hybrid systems, one in the *Sender* and the other in the *Receiver*. These protocols are arranged in a **protocol stack**, as shown in figure 6.10. Each layer in the stack performs a certain function and interacts with the corresponding layer in the other computer. The physical layer converts a bit stream into an electrical signal and vice versa and transfers the signal over one link of the network.

The network itself consists of many physical links connected by routers. The routers themselves act as computers but are missing the higher levels of the protocol stack. The **physical layer** transports bits over wires, optical fibers, or radio links. The **medium access layer** manages contention for the physical communication resource, preventing collisions among multiple users of the link. The **network layer** routes packets appropriately through the network. The **transport layer** ensures that the end-to-end transfer of packets is reliable, even if the network layer service is unreliable. The **application layer** converts whatever information is to be sent (such as an image or e-mail) into packets and then reassembles the packets into the appropriate information.

PROBING FURTHER

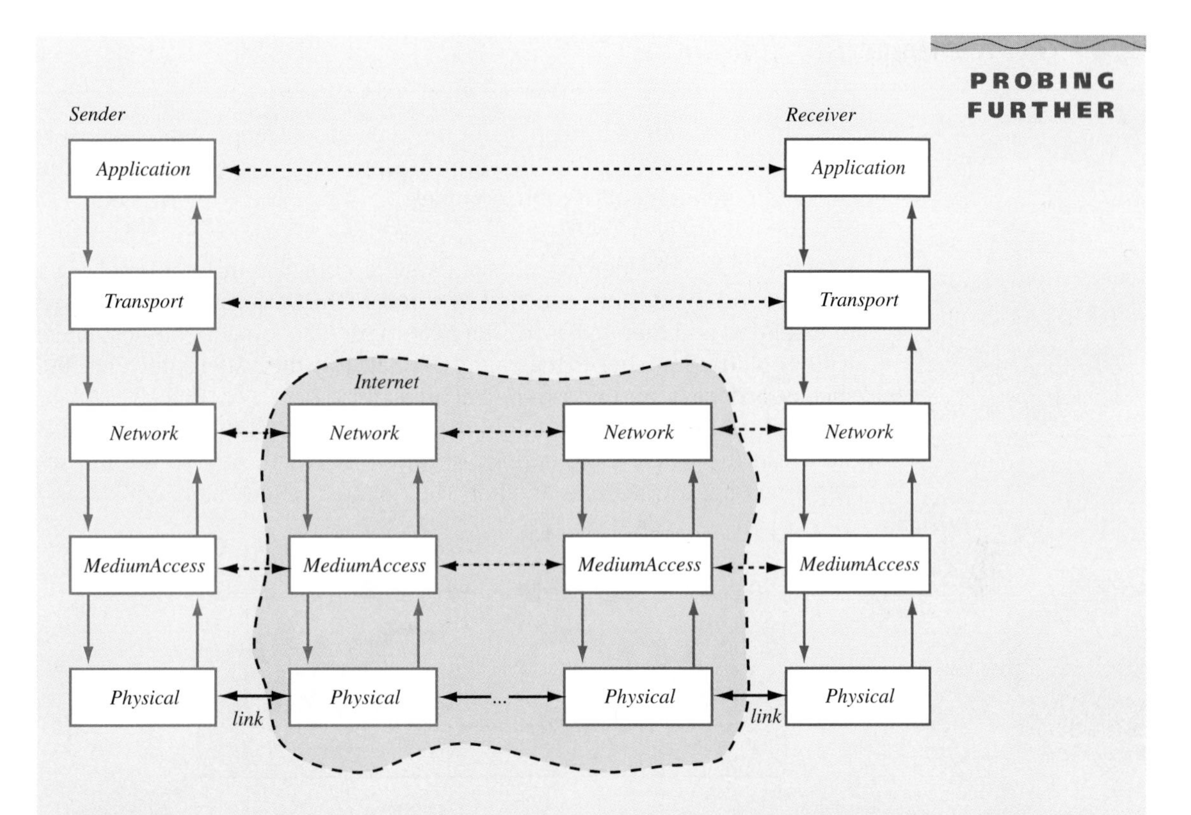

FIGURE 6.10: Network protocols are organized in a stack. Each protocol interacts with the corresponding layer in a remote computer. The dotted lines indicate conceptual interactions; the solid lines indicate physical interactions.

This layered approach provides an *abstraction* mechanism. Each layer conceptually interacts with the corresponding layer at a remote machine, as suggested by the dotted lines in figure 6.10. Each layer provides a "service" to the layer above it, using the service offered by the layer underneath. For example, the medium access layer offers as a service the transfer of a packet over a single link. The network layer uses this service to transfer a packet over a sequence of links between the end hosts. This abstraction mechanism permits the design of a single layer—say, the transport layer—assuming the service of the network layer, without regard to the layers below the network layer. The hybrid system in example 6.7, for instance, models only the transport layer.

The transport layer is an end-to-end protocol, and so it is implemented only at the end points in the connection, as shown in figure 6.10. The routers in the network need to implement only the lower layers.

As mentioned, each protocol layer is modeled as a pair of hybrid systems. Typically, these are timed automata, inasmuch as coordination between end hosts is achieved via several clocks, as in example 6.7. When a guard associated with a clock is satisfied, this signals some contingency in the communication, just as the timeout of the clock in figure 6.8 signals that a packet may be lost.

6.4 More interesting dynamics

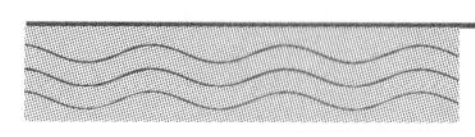

In timed automata, all that happens in the time-based refinement systems is that time passes. Hybrid systems, however, are much more interesting when the behavior of the refinements is more complex.

Example 6.8: Consider the physical system depicted in figure 6.11. Two sticky round masses are attached to springs. The springs are compressed or extended and then released. The masses oscillate on a frictionless table. If they collide, they stick together and oscillate together. After some time, the stickiness decays, and masses pull apart again.

A plot of the displacement of the two masses as a function of time is shown in the figure. Both springs begin compressed, and so the masses begin moving toward one another. They almost immediately collide and

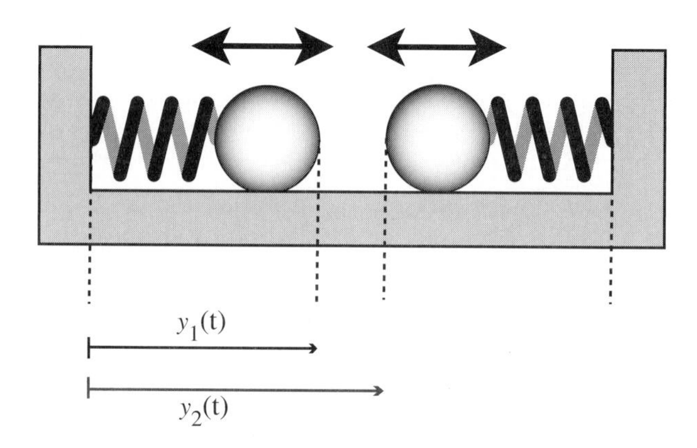

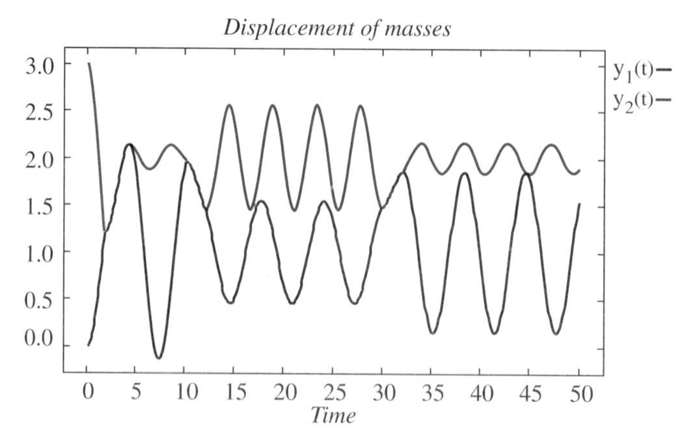

FIGURE 6.11: Sticky masses system.

then oscillate together for a brief period until they pull apart. In this plot, they collide two more times and almost collide a third time.

The physics of this problem is quite simple if we assume idealized springs. Let $y_1(t)$ denote the right edge of the left mass at time t, and let $y_2(t)$ denote the left edge of the right mass at time t, as shown in figure 6.11. Let p_1 and p_2 denote the neutral positions of the two masses (i.e., when the springs are neither extended nor compressed), so that the force is zero. For an ideal spring, the force at time t on the mass is proportional to $p_1 - y_1(t)$ (for the left mass) and $p_2 - y_2(t)$ (for the right mass). The force is positive to the right and negative to the left.

Let the spring constants be k_1 and k_2, respectively. Then the force on the left spring is $k_1(p_1 - y_1(t))$, and the force on the left spring is $k_2(p_2 - y_2(t))$. Let the masses be m_1 and m_2, respectively. Now we can use Newton's law, which relates force, mass, and acceleration:

$$f = ma.$$

The acceleration is the second derivative of the position with respect to time, which we write $\ddot{y}_1(t)$ and $\ddot{y}_2(t)$, respectively. Thus, as long as the masses are separate, their dynamics are given by

$$\ddot{y}_1(t) = k_1(p_1 - y_1(t))/m_1 \tag{6.1}$$

and

$$\ddot{y}_2(t) = k_2(p_2 - y_2(t))/m_2. \tag{6.2}$$

When the masses collide, however, the situation changes. When the masses are stuck together, they behave as a single object with mass $m_1 + m_2$. This single object is pulled in opposite directions by two springs. While the masses are stuck together, $y_1(t) = y_2(t)$. Let

$$y(t) = y_1(t) = y_2(t).$$

The dynamics are then given by

$$\ddot{y}(t) = \frac{k_1 p_1 + k_2 p_2 - (k_1 + k_2)y(t)}{m_1 + m_2}. \tag{6.3}$$

It is now easy to see how to construct a hybrid systems model for this physical system. The model is shown in figure 6.12. It has two modes, *apart* and *together*. The refinement of the *apart* mode is given by (6.1) and (6.2), whereas the refinement of the *together* mode is given by (6.3).

We still, however, have to label the transitions. In figure 6.12, the initial transition is shown entering the *apart* mode. Thus, we are assuming that the masses begin apart. Moreover, this transition is labeled with an action that

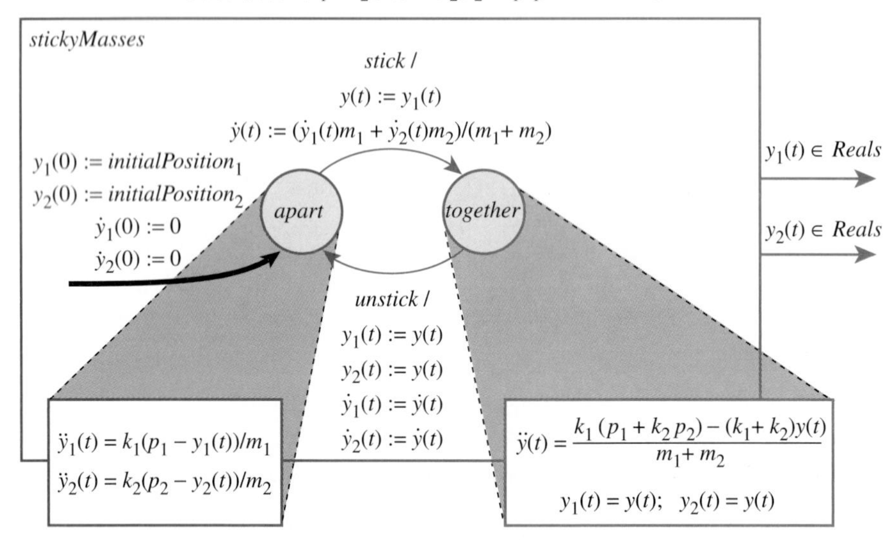

FIGURE 6.12: Hybrid system model for the sticky masses system.

sets the initial refinement state. Intuitively, the initial state of the masses is their positions and their initial velocities. In fact, we can define the refinement state to be

$$s(t) = \begin{bmatrix} y_1(t) \\ \dot{y}_1(t) \\ y_2(t) \\ \dot{y}_2(t) \end{bmatrix}.$$

It is then a simple matter to rewrite (6.1) and (6.2) in the form

$$\dot{s}(t) = g(s(t)) \tag{6.4}$$

for a suitably chosen function g (see exercise 12 at the end of this chapter). In figure 6.12, in the initial state, the masses are at some specified displacement, and the velocities are at zero.

The transition from *apart* to *together* has the guard

$$\textit{stick} = \{(y_1(t), \dot{y}_1(t), y_2(t), \dot{y}_2(t)) \mid y_1(t) = y_2(t)\}.$$

Thus, when the refinement state of *apart* satisfies this guard, the transition will be taken. No event output is produced, as indicated by the blank after the slash. However, an action is taken to set the initial refinement state of *together*. The refinement state of *together* could be the same $s(t)$ as previously, with the additional constraint that $y_1(t) = y_2(t)$ and $\dot{y}_1(t) = \dot{y}_2(t)$, because the masses are stuck together. More simply, we could define the state $z(t)$ of *together* to be the position $y(t)$ and velocity $\dot{y}(t)$, where $y(t) = y_1(t) = y_2(t)$:

$$z(t) = \begin{bmatrix} y(t) \\ \dot{y}(t) \end{bmatrix}.$$

The transition from *apart* to *together* sets $y(t)$ equal to $y_1(t)$ (it could equally well have chosen $y_2(t)$, because these are equal). It sets the velocity to conserve momentum. The momentum of the left mass is $\dot{y}_1(t)m_1$, the momentum of the right mass is $\dot{y}_2(t)m_2$, and the momentum of the combined masses is $\dot{y}(t)(m_1 + m_2)$. To make these equal, it sets

$$\dot{y}(t) = \frac{\dot{y}_1(t)m_1 + \dot{y}_2(t)m_2}{m_1 + m_2}.$$

The transition from *together* to *apart* has the more complicated guard

$$\mathit{unstick} = \{(y(t), \dot{y}(t)) \mid (k_1 - k_2)y(t) + (k_2p_2 - k_1p_1) > \mathit{stickiness}\}.$$

This guard is satisfied when the right-pulling force on the right mass exceeds the right-pulling force on the left mass by more than the stickiness. The right-pulling force on the right mass is simply

$$f_2(t) = k_2(p_2 - y(t)),$$

and the right-pulling force on the left mass is

$$f_1(t) = k_1(p_1 - y(t)).$$

Thus,

$$f_2(t) - f_1(t) = (k_1 - k_2)y(t) + (k_2p_2 - k_1p_1).$$

When this exceeds the stickiness, the masses pull apart.

An interesting elaboration on this example, considered in exercise 13 at the end of this chapter, modifies the *together* mode so that the stickiness is initialized to a starting value but then decays according to the differential equation

$$\dot{s}(t) = -as(t),$$

where $s(t)$ is the stickiness at time t and a is some positive constant. The dynamics of such an elaboration are plotted in figure 6.11. ❐

As in example 6.7, it is sometimes useful to have hybrid system models with only one state. The actions on one or more state transitions define the discrete event behavior that combines with the time-based behavior.

Example 6.9: Consider a bouncing ball. At time $t = 0$, the ball is dropped from a height $y(0) = \mathit{initialHeight}$ meters. It falls freely. At some later time t_1, it hits the ground with a velocity $\dot{y}(t_1) < 0$ meters/second. A *bump* event is produced when the ball hits the ground. The collision is inelastic, and the ball bounces back up with velocity $-a\dot{y}(t_1)$, where a is some constant in $(0, 1)$. The ball then rises to a certain height and falls back to the ground repeatedly.

The behavior of the bouncing ball can be described by the hybrid system of figure 6.13. There is only one mode, called *free*. We know that when the ball is not in contact with the ground, it follows the second-order differential equation

$$\ddot{y}(t) = -g, \tag{6.5}$$

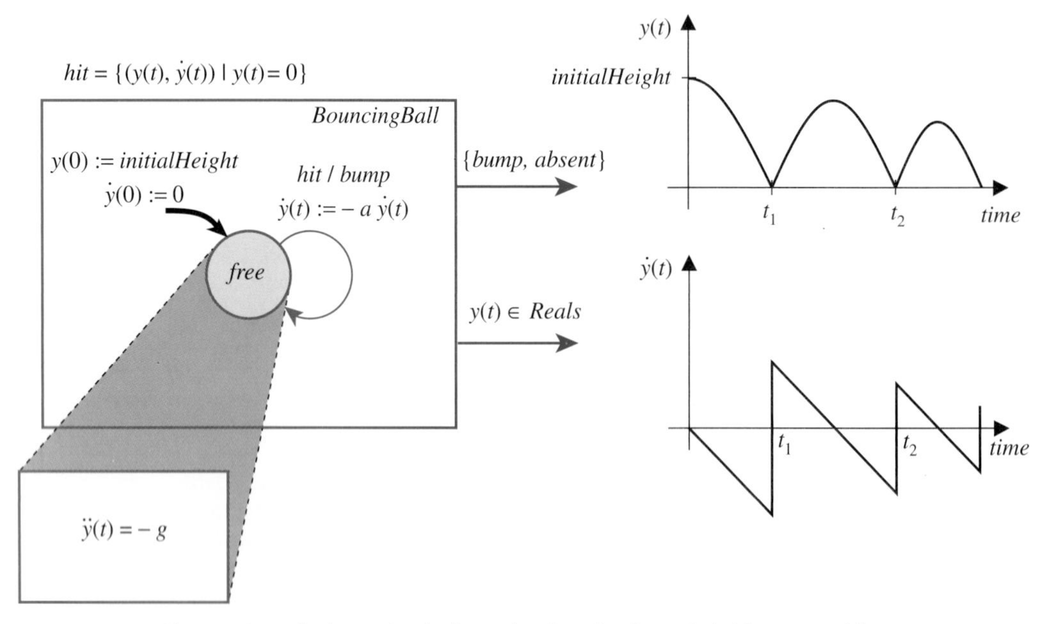

FIGURE 6.13: The motion of a bouncing ball may be described as a hybrid system with only one mode. The system outputs a *bump* each time the ball hits the ground and also outputs the position of the ball. The position and velocity are plotted versus time at the right.

where $g = 10$ m/sec^2 is the acceleration imposed by gravity. We can define the refinement state of the *free* mode to be

$$s(t) = \begin{bmatrix} y(t) \\ \dot{y}(t) \end{bmatrix}$$

with the initial conditions $y(0) = \mathit{initialHeight}$ and $\dot{y}(0) = 0$. It is then a simple matter to rewrite (6.5) as a first-order differential equation,

$$\dot{s}(t) = f(s(t)), \tag{6.6}$$

for a suitably chosen function f (see exercise 12 at the end of this chapter).

At the time t_1 when the ball first hits the ground, the guard

$$\mathit{hit} = \{(y(t), \dot{y}(t)) \mid y(t) = 0\}$$

is satisfied, and the self-loop transition is taken. The output *bump* is produced, and the action $\dot{y}(t) := -a\dot{y}(t)$ assigns $\dot{y}(t_1+) = -a\dot{y}(t_1)$. Here, $\dot{y}(t_1+)$ is the velocity after the bump, and $\dot{y}(t_1)$ is the velocity before the bump. Then (6.5) is followed again until the guard becomes true again.

By integrating (6.5), we get, for all $t \in (0, t_1)$,

$$\dot{y}(t) = -gt$$

and

$$y(t) = y(0) + \int_0^t \dot{y}(\tau)d\tau = \mathit{initialHeight} - \frac{1}{2}gt^2.$$

Therefore, $t_1 > 0$ is determined by $y(t_1) = 0$. It is the solution to the equation

$$\mathit{initialHeight} - \frac{1}{2}gt^2 = 0.$$

Thus,

$$t_1 = \sqrt{2\ \mathit{initialHeight}/g}.$$

Figure 6.13 plots the refinement state versus time. ❒

6.5 Supervisory control

We introduce supervisory control through a detailed example. A control system involves four components. A system called the **plant** comprises the physical process that is to be controlled, the environment in which the plant operates, the sensors that measure some variables of the plant and the environment, and the controller that determines the mode transition structure and selects the time-based inputs to the plant. The controller has two levels: the supervisory control that determines the mode transition structure, and the "low-level" control that selects the time-based inputs that control the behavior of the refinements. A complete design includes both levels of control, as in the following example.

Example 6.10: The plant is an **automated guided vehicle** (**AGV**) that moves along a closed track painted on a warehouse or factory floor. We design a controller to enable the vehicle to closely follow the track.

The vehicle has two degrees of freedom. At any time t, it can move forward along its body axis with speed $u(t)$, with the restriction that $0 \leq u(t) \leq 10$ mph. It can also rotate about its center of gravity with an angular speed $\omega(t)$ restricted to $-\pi \leq \omega(t) \leq \pi$ radians/second. We ignore the inertia of the vehicle.

Let $(x(t), y(t)) \in Reals^2$ be the position and $\phi(t) \in [-\pi, \pi]$ the angle (in radians) of the vehicle at time t in relation to some fixed coordinate frame, as shown on the left in figure 6.14. In terms of this coordinate frame, the motion of the vehicle is given by a system of three differential equations:

$$\begin{aligned} \dot{x}(t) &= u(t) \cos \phi(t), \\ \dot{y}(t) &= u(t) \sin \phi(t), \text{ and} \\ \dot{\phi}(t) &= \omega(t). \end{aligned} \tag{6.7}$$

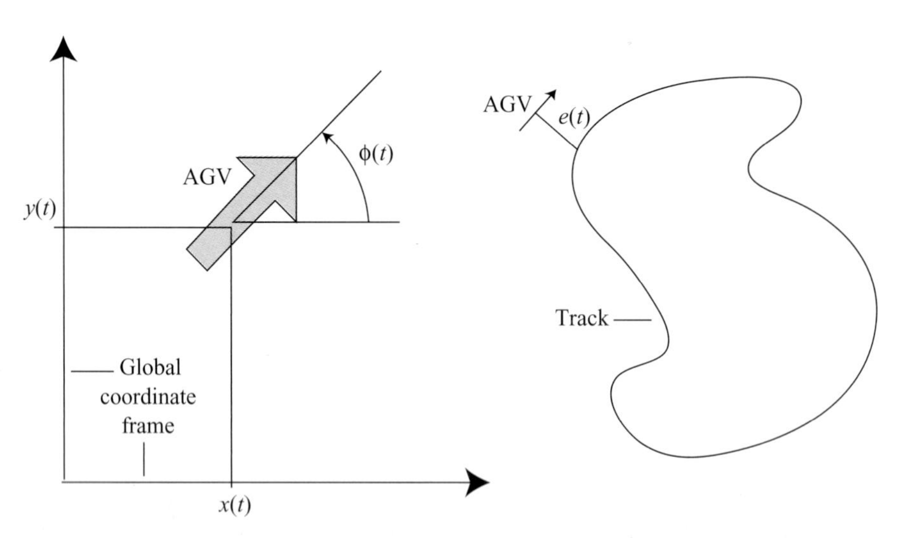

FIGURE 6.14: Illustration of the automated guided vehicle (AGV). The vehicle is shown as a large arrow on the left and as a small arrow on the right. On the right, the vehicle has deviated from a curved painted track by a distance $e(t)$. The coordinates of the vehicle at time t with respect to the global coordinate frame are $(x(t), y(t), \phi(t))$.

The track and the vehicle are shown on the right of figure 6.14. Equations (6.7) describe the plant. The environment is the closed painted track. It could be described by an equation. We will describe it indirectly by means of a sensor.

The two-level controller design is based on a simple idea. The vehicle always moves at its maximum speed of 10 mph. If the vehicle strays too far to the left of the track, the controller steers it toward the right; if it strays too far to the right of the track, the controller steers it toward the left. If the vehicle is close to the track, the controller maintains the vehicle in a straight direction. Thus, the controller guides the vehicle in four modes: *left*, *right*, *straight*, and *stop*. In *stop* mode, an operator brings the vehicle to a halt.

The following differential equations govern the AGV's motion in the refinement of the four modes. They describe the low-level controller—that is, the selection of the time-based inputs in each mode:

$$\begin{aligned}
\textit{straight}: \quad & \dot{x}(t) = 10\cos\phi(t) \\
& \dot{y}(t) = 10\sin\phi(t) \\
& \dot{\phi}(t) = 0 \\[1em]
\textit{left}: \quad & \dot{x}(t) = 10\cos\phi(t) \\
& \dot{y}(t) = 10\sin\phi(t) \\
& \dot{\phi}(t) = \pi \\[1em]
\textit{right}: \quad & \dot{x}(t) = 10\cos\phi(t) \\
& \dot{y}(t) = 10\sin\phi(t) \\
& \dot{\phi}(t) = -\pi \\[1em]
\textit{stop}: \quad & \dot{x}(t) = 0 \\
& \dot{y}(t) = 0 \\
& \dot{\phi}(t) = 0
\end{aligned}$$

In the *stop* mode, when the vehicle is stopped, $x(t)$, $y(t)$, and $\phi(t)$ are constant. In the *left* mode, $\phi(t)$ increases at the rate of π radians/second, and so from figure 6.14 we see that the vehicle moves to the left. In the *right* mode, it moves to the right. In the *straight* mode, $\phi(t)$ is constant, and the vehicle moves straight ahead with a constant heading. The refinements of the four modes are shown in the boxes of figure 6.15.

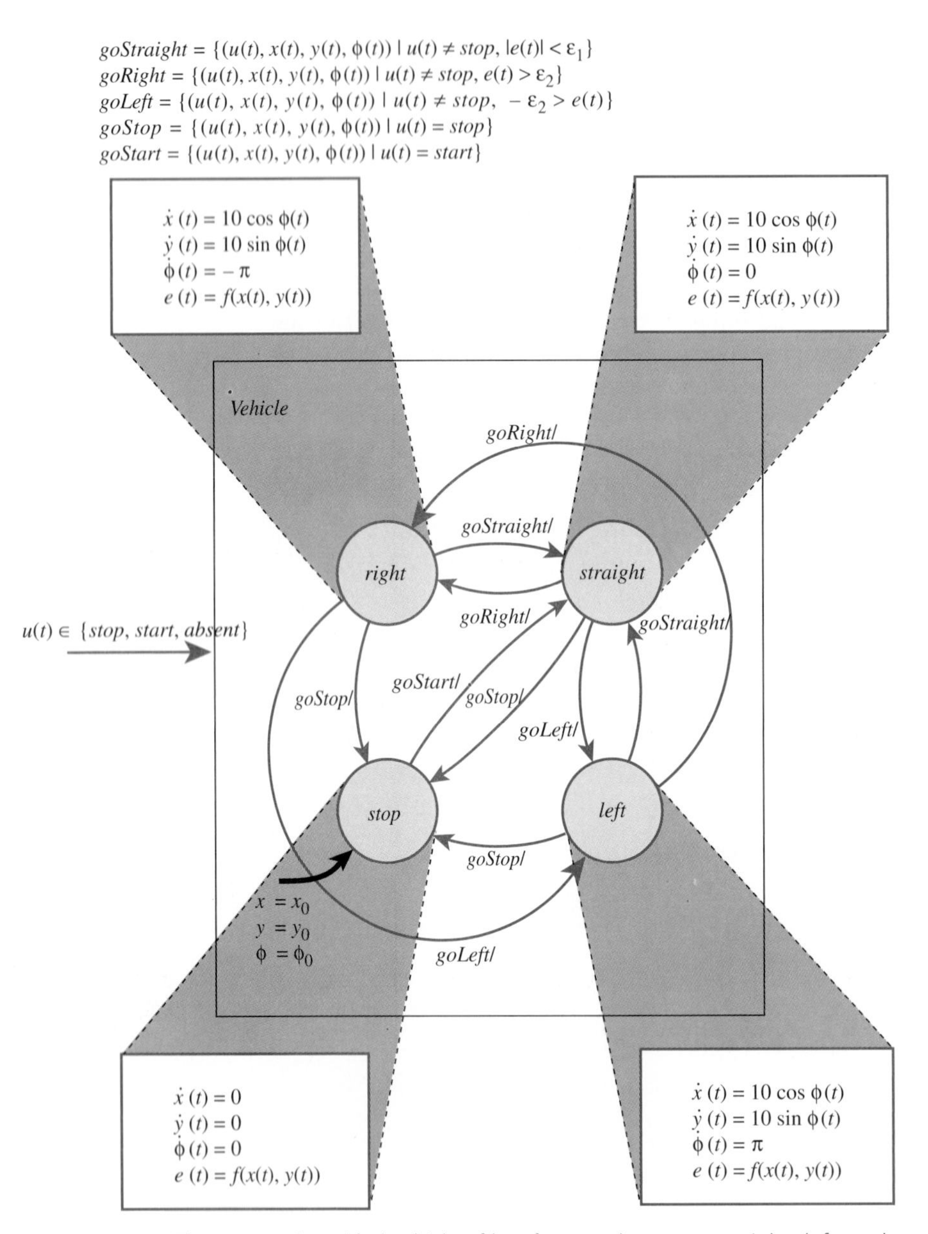

FIGURE 6.15: The automatic guided vehicle of has four modes: *stop*, *straight*, *left*, and *right*.

We design the supervisory control governing transitions between modes in such a way that the vehicle closely follows the track, using a sensor that determines how far the vehicle is to the left or right of the track. We can build such a sensor by using photodiodes. Suppose the track is painted with a light-reflecting color, whereas the floor is relatively dark. Underneath the AGV, we place an array of photodiodes, as shown in figure 6.16. The array is perpendicular to the AGV body axis. As the AGV passes over the track, the photodiode directly above the track generates more current than the other photodiodes. By comparing the magnitudes of the currents through the different photodiodes, the sensor gives the displacement $e(t)$ of the center of the array (hence, the center of the AGV) from the track. We adopt the convention that $e(t) < 0$ means that the AGV is to the right of the track and $e(t) > 0$ means that it is to the left. We model the sensor output as a function f of the AGV's position:

$$\forall t, \quad e(t) = f(x(t), y(t)).$$

The function f depends of course on the environment—the track. We now specify the supervisory controller precisely. We select two thresholds, $0 < \epsilon_1 < \epsilon_2$, as shown in figure 6.16. If the magnitude of the displacement is small, $|e(t)| < \epsilon_1$, we consider that the AGV is close enough to the track, and the AGV can move straight ahead, in *straight* mode. If $0 < \epsilon_2 < e(t)$ (where $e(t)$ is large and positive), the AGV has strayed too far to the left and must be steered to the right, by switching to *right* mode. If $0 > -\epsilon_2 > e(t)$ (where $e(t)$ is large and negative), the AGV has strayed too far to the right and must be steered to the left, by switching to *left* mode. This control logic is captured in the mode transitions of figure 6.15. The input events are *stop*, *start*, and *absent*. By selecting events *stop* and *start*, an operator can stop or start the

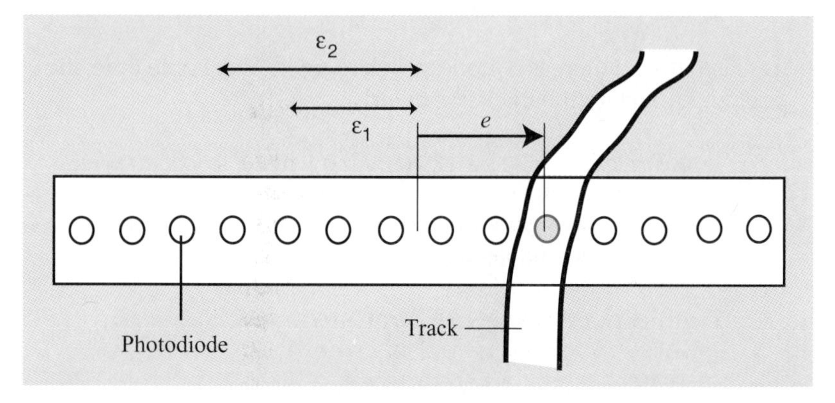

FIGURE 6.16: An array of photodiodes under the AGV is used to estimate the displacement e of the AGV relative to the track. The photodiode directly above the track generates more current than do the others.

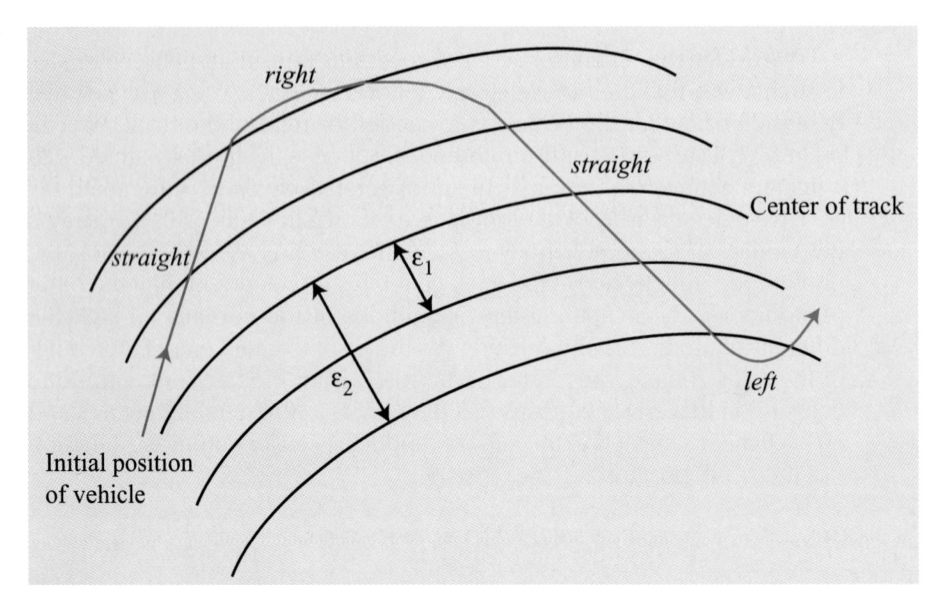

FIGURE 6.17: A trajectory of the automated guided vehicle (AGV), annotated with modes.

AGV. There is no time-based input. There is no external output. The initial mode is *stop*, and the initial values of its refinement are x_0, y_0, and ϕ_0.

We analyze how the AGV will move. Figure 6.17 is a sketch of one possible trajectory. Initially, the vehicle is within distance ϵ_1 of the track, and so it moves straight. At some later time, the vehicle goes too far to the left, the guard

$$goRight = \{(u(t), x(t), y(t), \phi(t)) \mid u(t) \neq stop, \epsilon_2 < e(t)\}$$

is satisfied, and there is a mode switch to *right*. After some time, the vehicle is close enough to the track, the guard

$$goStraight = \{(u(t), x(t), y(t), \phi(t)) \mid u(t) \neq stop, \mid e(t) \mid < \epsilon_1\}$$

is satisfied, and there is a mode switch to *straight*. Some time later, the vehicle is too far to the right, the guard

$$goLeft = \{(u(t), x(t), \phi(t)) \mid u(t) \neq stop \mid -\epsilon_2 > e(t)\}$$

is satisfied, there is a mode switch to *left*, and so on. ❐

The example illustrates the four components of a control system. The plant is described by the differential equations (6.7) that govern the evolution of the refinement state at time t, $(x(t), y(t), \phi(t))$, in terms of the time-based input, $(u(t), \omega(t))$. The second component is the environment—the closed track. The third component is the sensor, whose output at time t, $e(t) = f(x(t), y(t))$, gives the position of the AGV in relation to the track. The fourth component is the two-level controller. The supervisory controller comprises the four modes and the guards that determine when to switch between modes. The low-level controller specifies how the time-based inputs, u and ω, are selected in each mode.

6.6 Formal model

We develop a formal model of a hybrid system similar to the sets and functions model of section 3.1. A hybrid system *HybridSystem* is a five-tuple,

$$HybridSystem = (States, Inputs, Outputs, TransitionStructure, initalState),$$

where *States*, *Inputs*, and *Outputs* are sets and $initalState \in States$ is the initial state. *TransitionStructure* consists of several items that determine how the hybrid system evolves in time $t \in T$. T may be $Reals_+$ or $Naturals_0$. Here we assume $T = Reals_+$.

$States = Modes \times RefinementStates$ is the state space. *Modes* is the finite set of modes. *RefinementStates* is the state space of the refinements. If the current state at time t is $(m(t), s(t))$, we say that the system is in mode $m(t)$ and its refinement is in state $s(t)$.

$Inputs = InputEvents \times TimeBasedInputs$ is the set of input symbols. *InputEvents* is the finite alphabet of discrete input symbols, including a stuttering symbol, whereas *TimeBasedInputs* is the set of input values to which the refinement reacts. An input signal consists of a pair of functions (u, x) where $u\colon Reals_+ \to InputEvents$ and $x\colon Reals_+ \to TimeBasedInputs$. For all except a discrete set of times t, $u(t)$ is the stuttering symbol, *absent*.

$Outputs = OutputEvents \times TimeBasedOutputs$ is the set of output symbols. *OutputEvents* is the finite alphabet of discrete output symbols, including a stuttering output, *absent*, and *TimeBasedOutputs* is the set of continuous output values. An output signal consists of a pair of functions (v, y) where $v\colon Reals_+ \to OutputEvents$ and $y\colon Reals_+ \to TimeBasedOutputs$. For all except a discrete set of times, $v(t) = absent$.

The transition structure determines how a mode transition occurs and how the refinement state changes over time. Suppose the input signal is (u, x). Suppose at time t the mode is m and the refinement state is s. For each destination mode d, there is a guard,

$$G_{m,d} \subset InputEvents \times TimeBasedInputs \times RefinementStates.$$

There is also an output event—say, $\upsilon_{m,d}$—and an action $A_{m,d}$: *RefinementStates* $\to$ *RefinementStates* that assigns a (possibly new) value to each refinement state, (possibly) depending on the current value of the refinement state. If there is a match $(u(t), x(t), s(t)) \in G_{m,d}$, then there is a discrete transition at t; the mode after the transition is d, the output event $\upsilon(t) = \upsilon_{m,d}$ is produced, and the refinement state in mode d at time $t+$ immediately after the transition is set to $s(t+) = A_{m,d}(s(t))$.

If no guard is satisfied at time t, then the refinement state $s(t)$ and the time-based output $y(t)$ are determined by the time-based input signal x according to the equations governing the refinement dynamics. Here the setting must be concrete. In all the previous examples,

$$\begin{aligned} \mathit{RefinementStates} &= \mathit{Reals}^N, \\ \mathit{TimeBasedInputs} &= \mathit{Reals}^M, \text{ and} \\ \mathit{TimeBasedOutputs} &= \mathit{Reals}^K. \end{aligned}$$

In this concrete setting, the refinement dynamics are given as

$$\forall\, t \in T_m, \quad \dot{s}(t) = f_m(s(t), x(t)) \tag{6.8}$$

and

$$y(t) = g_m(s(t), x(t)), \tag{6.9}$$

where $T_m \subset T$ is the set of times t when the system is in mode m and the functions

$$f_m: \mathit{Reals}^N \times \mathit{Reals}^M \to \mathit{Reals}^N$$

and

$$g_m: \mathit{Reals}^N \times \mathit{Reals}^M \to \mathit{Reals}^K$$

characterize the behavior of the refinement system in mode m. The function

$$s: \mathit{Reals}_+ \to \mathit{RefinementStates}$$

is the trajectory of the refinement states.

We can now see how the hybrid system evolves over time. At time $t = 0$, the system starts in the initial state: say, $(m_0), s(0))$. It evolves in phases of time passage, $(t_0 = 0, t_1]$, $(t_1, t_2]$, ..., alternating with discrete transitions at $t_1, t_2, \ldots$. During the first interval $(t_0, t_1]$, no guard is satisfied and the system remains in mode $m(0)$; the refinement state $s(t)$ and time-based output $y(t)$ are determined by (6.8) and (6.9); and the discrete event output $\upsilon(t) = \mathit{absent}$.

At time t_1, the guard Gm_0m_1 for some destination mode m_1 is matched by $(u(t_1), x(t_1), s(t_1))$. There is a mode transition to m_1, the output event $v(t_1)$ is produced, and the continuous state is set to $s(t_1+) = Am_0m_1(s(t_1))$. The discrete transition phase is now over, and the system begins the time passage phase in the new mode m_1 and the continuous state $s(t_1+)$.

6.7 Summary

Hybrid systems provide a bridge between time-based models and state-machine models. The combination of the two families of models provides a rich framework for describing real-world systems. There are two key ideas. First, discrete events are embedded in a time base. Second, a hierarchical description is useful, particularly for when the system undergoes discrete transitions between different modes of operation. Associated with each mode of operation is a time-based system called the refinement of the mode. Mode transitions are made when guards that specify the combination of inputs and refinement states are satisfied. The action associated with a transition, in turn, sets the refinement state in the destination mode.

The behavior of a hybrid system is understood through the use of the tools of state machine analysis for mode transitions and the tools of time-based analysis for the refinement systems. The design of hybrid systems similarly proceeds on two levels: State machines are designed to achieve the appropriate logic of mode transitions, and refinement systems are designed to secure the desired time-based behavior in each mode.

EXERCISES

KEY: E = mechanical T = requires plan of attack C = more than 1 answer

C 1. Consider the loudness compensation of example 6.3. Suppose that instead of a switch on the front panel, the system automatically selects from among four compensation filters with state-space models $[A, b, c_1, d_1]$, $[A, b, c_2, d_2]$, $[A, b, c_3, d_3]$, and $[A, b, c_4, d_4]$. For all four, the A matrices are the same and the b vectors are the same. Which filter is used depends on a discrete-time v input, where at index n, $v(n)$ represents the current volume level. When the volume is high, above some threshold, filter 4 should be used. When it is low, filter 1 should be used. Design a hybrid system that does this.

E 2. Construct a timed automaton similar to that of figure 6.4 that produces *tick* at times $1, 2, 3, 5, 6, 7, 8, 10, 11, \ldots$, that is, ticks are produced with intervals between them of one second (three times) and two seconds (once).

E 3. The objectives of this problem are to understand a timed automaton and then to modify it as specified.

(a) For the timed automaton shown as follows, describe the output y. Avoid imprecise or sloppy notation.

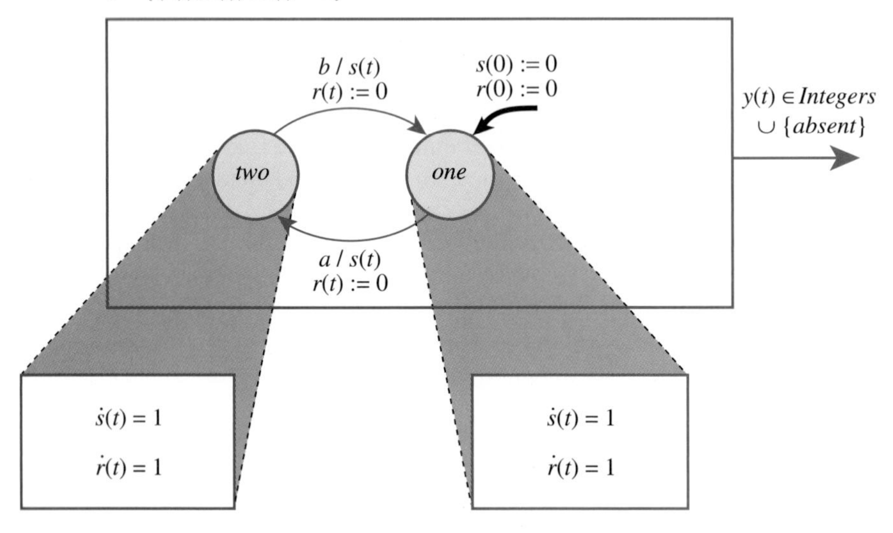

(b) Assume there is a new input u: *Reals* → *Inputs* with alphabet

$$Inputs = \{reset, absent\}$$

and that when the input has value *reset*, the hybrid system starts over, behaving as if it were starting at time 0 again. Modify the hybrid system from part (a) so that it behaves this way.

E 4. You have an analog source that produces a pure tone. You can switch the source on or off by the input event *on* or *off*. Construct a system that upon receiving an input event *ring*, produces an 80-msec-long sound consisting of three 20-msec-long bursts of the pure tone separated by two 10-msec intervals of silence. What does your system do if it receives two *ring* events that are 50-msec apart?

C 5. Automobiles today have the features listed as follows. Implement each feature as a timed automaton.

(a) The dome light is turned on as soon as any door is opened. It stays on for 30 seconds after all doors are shut. What sensors are needed?

(b) Once the engine is started, a beeper is sounded and a red light warning is indicated if there are passengers who have not buckled their seat belts. The beeper stops sounding after 30 seconds or as soon the seat belts are buckled, whichever is sooner. The warning light is on all the time the seat belt is unbuckled. Hint: Assume that the sensors provide a *warn* event when the ignition is turned on and a passenger is not buckled in or if the ignition is already on and a passenger sits in a seat without buckling the seat belt. Assume further that the sensors provide a *noWarn* event when a passenger departs from a seat, when the buckle is buckled, or when the ignition is turned off.

E 6. A programmable thermostat allows you to select four times, $0 \leq T_1 \leq \cdots \leq T_4 < 24$ (for a 24-hour cycle) and the corresponding temperatures $a_1, \ldots, a_4$. Construct a timed automaton that sends the event a_i to the heating systems controller. The controller maintains the temperature close to the value a_i until it receives the next event. How many timers and modes do you need?

E 7. Construct a parking meter similar to that in figure 6.6 that allows a maximum of 30 minutes (rather than 60 minutes) and accepts *coin5* and *coin25* as inputs. Then draw the state trajectories (both the mode and the clock state) and the output signal for the situation in which *coin5* occurs at time 0, *coin25* occurs at time 3, and there is no further input event for the next 35 minutes.

T 8. Consider the timed automaton of figure 6.6. Suppose we view the box as a discrete-event system with input alphabet $\{coin5, coin25, absent\}$ and output alphabet $\{expired, absent\}$. Does the box behave as a finite-state machine?

C 9. Figure 6.18 depicts the intersection of two one-way streets, called Main and Secondary. A light on each street controls its traffic. Each light goes through a cycle consisting of red (R), green (G), and yellow (Y) phases. It is a safety requirement that when one light is in its green or yellow phase, the other is in its red phase. The yellow phase is always 20 seconds long.

The traffic lights operate as follows, in one of two modes. In the normal mode, there is a five-minute-long cycle in which Main Street's light is green for 4 minutes and then yellow for 20 seconds—Secondary Street's light is red for these 4 minutes and 20 seconds—and 40 seconds of red, during which Secondary Street's light is green for 20 seconds and then yellow for 20 seconds.

The second, or interrupt, mode works as follows. Its purpose is to quickly give a right of way to the secondary road. A sensor in the secondary road detects when a vehicle has stopped at the light there. When this happens, the Main Street light aborts its green phase and immediately switches to its 20-second yellow phase, followed by the 40-second

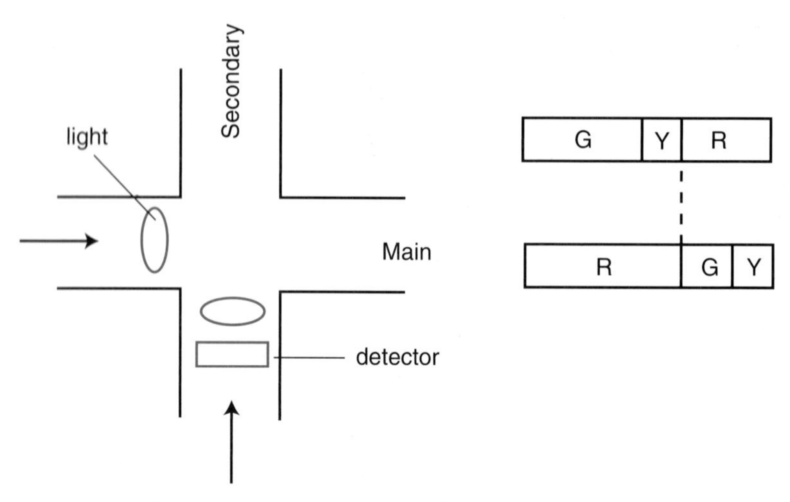

FIGURE 6.18: Traffic lights control the intersection of a main street and a secondary street. A detector senses when a vehicle crosses it. The red phase of one light must coincide with the green and yellow phases of the other light.

red phase. If the vehicle is detected while the Main Street light is yellow or red, the system continues in its normal mode.

Design a hybrid system that controls the lights. Let this hybrid system have discrete outputs that are pairs GG, GY, GR, and so forth, where the first letter denotes the color of the Main Street light and the second letter denotes the color of the Secondary Street light.

T 10. Design a *ReceiverProtocol* hybrid system that works together with the *SenderProtocol* of example 6.7.

E 11. For the bouncing ball of example 6.9, let t_n be the time when the ball hits the ground for the nth time, and let $v(n) = \dot{y}(t_n)$ be the velocity at that time.

(a) Find a relation between $v(n+1)$ and $v(n)$, and then calculate $v(n)$ in terms of $v(1)$.

(b) Obtain t_n in terms of $v(n)$.

(c) Calculate the maximum height reached by the ball after each bump.

E 12. Translate refinement systems that are described as second-order differential equations into first-order differential equations. Specifically,

(a) For the sticky masses system in example 6.8, find the function g such that (6.1) and (6.2) are represented as (6.4). Is this function linear?

(b) For the bouncing ball system in example 6.9, find the function f such that (6.5) is represented as (6.6). Is this function linear?

T 13. Elaborate the hybrid system model of figure 6.12 so that in the *together* mode, the stickiness decays according to the differential equation

$$\dot{s}(t) = -as(t),$$

where $s(t)$ is the stickiness at time t and a is some positive constant. On the transition into this mode, the stickiness should be initialized to some starting stickiness b.

T 14. Show that the trajectory of the AGV of figure 6.15 is a circle while it is in *left* or *right* mode. What is the radius of this circle, and how long does it take to complete a circle?

E 15. Express the hybrid system of figure 6.15 in terms of the formal model of section 6.6; that is, identify the sets *Inputs*, *Outputs* and the *TransitionStructure*.

INTERVIEW

P.R. Kumar

P.R. Kumar is a Franklin W. Woeltge Professor of Electrical and Computer Engineering, and a Research Professor in the Coordinated Science Laboratory, at the University of Illinois, Urbana–Champaign. He was the recipient of the Donald P. Eckman Award of the American Automatic Council and is a fellow of IEEE. His current research interests are in wireless networks from theory to protocol development, the convergence of control with communication and computing, the scheduling of wafer fabrication plants and other manufacturing systems, and machine learning. In addition, he likes to play table tennis.

What is your educational background, and how did you decide to study electrical engineering?

I obtained my Bachelor of Technology degree in electronics engineering from the Indian Institute of Technology, Madras in 1973. At that time, the IITs already had a fabulous reputation within India and I would have been glad to study anything at any IIT if it had come to that. Graduate studies in engineering were the last thing on my mind, and even less so was the thought of coming to the United States. However, a fellow student gave me an application form. So, sans GRE, TOEFL, and even application fees, I applied to Washington University in St. Louis.

Which person in the field has inspired you most? In what ways?

I do admire the work of Claude Shannon. His work serves as a role model in four ways. First, it is phenomenally broad and diverse. Second, his choice of areas to work on was characterized by an excellent strategic sense. Third, he was very creative; his ideas and tools were very innovative, and are still the basic workhorses of the field, fifty plus years later. Fourth, the theory that he created was what a perfect theory should be—his work shows that the best practical research that one can do is to develop a good theory. His work is of lasting value and guides the future development of technology.

How will EE, CE, and CS students benefit from learning signals and systems early on in the curriculum?

Even though we have all these labels of yesteryear, CS, CE, and EE, which serve to separate us, all of them have a large common core, and in turn fall under a larger umbrella that we may call "Information Technology." Signals and systems is the language of this larger discipline, and the sooner we learn to speak it, the more efficiently others will be able to communicate to us the concepts and facts of the larger field.

What is your vision for the future of electrical and computer engineering?

At the educational level, electrical engineering, computer engineering, and computer science are merging. This unification makes the whole more powerful than its parts. Students well versed in information theory, computational complexity, graph theory, and so forth, are beautifully poised to solve system's problems of the future.

At the research end, with the phenomenal advances in physical electronics, electromagnetics, and so forth, and the systems areas of communication, computing, control, circuits, and signal processing, it is not too much of an exaggeration to say that it is hard to think of an idea that can't actually be implemented. What limits us now is only our imagination. The future could well see orchestras of sensors, actuators, and computational nodes, all connected by wireless networks, and controlling our environment. The coming decades could thus be the age of building large-scale systems that pervade our lives—and EE, CS, and CE will be at the core.

Are there any special projects you are currently working on that you'd like to tell students about?

One of the two efforts I am excited about (the other is wireless networking—from theory to protocol development and implementation) anticipates what I believe is the next phase of the information technology revolution—the convergence of control with communication and competition. This will take place through the interaction of sensors, actuators, and computational capabilities, interconnected by wired, as well as wireless, networks. What I am currently investigating is the development of an architecture for such systems, one that is application and context independent. Just as the OSI hierarchy made possible the anarchic proliferation of communication networks, is there a universal architecture that will allow the anarchic proliferation of networked sensors, actuators, and computers that interact with the physical world, and control it? Currently, I am betting that there is, in fact, such a universal architecture, and am attempting to develop it by exploiting advances in software engineering and middleware.

CHAPTER 7

Frequency domain

We are interested in manipulating signals. We may wish to synthesize signals, as modems need to do in order to transmit a voice-like signal through the telephone channel. We may instead wish to analyze signals, as modems need to do in order to extract digital information from a received voice-like signal. In general, the field of **communications** is all about synthesizing signals with characteristics that match a channel and then analyzing signals, many of which have been corrupted by the channel, in order to extract the original information.

We may also wish to synthesize natural signals such as images or speech. Much of the field of **computer graphics** involves synthesizing natural-looking images. **Image processing** includes **image understanding**, which involves analyzing images to determine their content. The field of **signal processing** includes analysis and synthesis of speech and music signals.

We may wish to control a physical process. The physical process is sensed (through temperature, pressure, position, and speed sensors). The sensed signals are processed in order to estimate the internal state of the physical process. The physical process is controlled on the basis of the state estimate. Control system design includes the designs of **state estimators** and **controllers**.

In order to analyze or synthesize signals, we need models of those signals. Because a signal is a function, a model of the signal is a description or a definition of the function. We use two approaches. The first is a **declarative** (what is) approach. The second is an **imperative** (how to) approach. These two approaches are complementary. The situation determines which approach is better than the other.

Signals are functions. This chapter in particular deals with signals in which the domain is time (discrete or continuous). It introduces the concept of frequency-domain representation of these signals. The idea is that arbitrary signals can be described as sums of sinusoidal signals. This concept is first explained by referring to psychoacoustics, which is how humans hear sounds. Sinusoidal signals have particular psychoacoustic significance. But the real justification for the frequency domain approach is much broader. It turns out to be particularly easy to understand the effect that **linear time-invariant (LTI) systems** (discussed in chapter 5) have on sinusoidal signals. A powerful set of analysis and design techniques for arbitrary signals and the LTI systems that operate on them is then presented.

Although we know that few (if any) real-world systems are truly LTI, we can easily construct models in which the approximation is valid over some range of operation. Chapter 6 showed how modal models can be constructed to build realistic models over a broader range of operating conditions. Frequency domain methods are amenable to such hybrid system treatment.

7.1 Frequency decomposition

For some signals—particularly natural signals such as voice, music, and images—finding a concise and precise definition of the signal can be difficult. In such cases, we try to model signals as compositions of simpler signals that we can more easily model.

Psychoacoustics is the study of how humans hear sounds. Pure tones and their frequencies turn out to be very convenient concepts for describing sounds. Musical notes can be reasonably modeled accurately as combinations of relatively few pure tones (although subtle properties of musical sounds, such as the timbre of a sound, are harder to model accurately).

BASICS

Frequencies in Hertz and radians

A standard measure of frequency is **Hertz (Hz)**. It means **cycles per second**. A plot of one second of four sine waves of different frequencies is shown as follows:

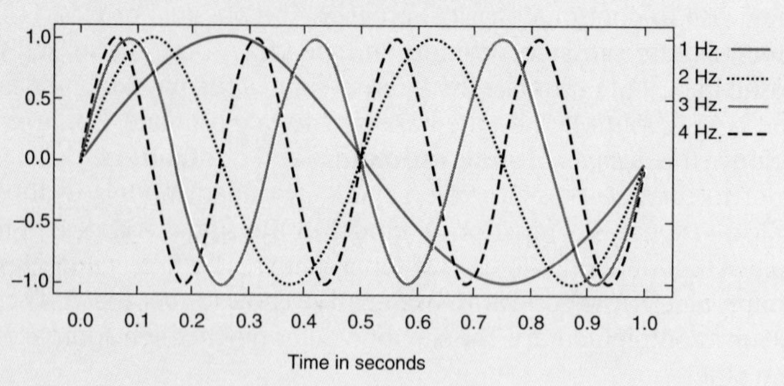

BASICS

For example, the frequencies in Hertz of the musical note A on the piano keyboard are as follows:

$$f_1 = 55, f_2 = 110, f_3 = 220, f_4 = 440, f_5 = 880, f_6 = 1{,}760, f_7 = 3{,}520, \text{ and } f_8 = 7{,}040.$$

A sinusoidal waveform x with frequency $f_4 = 440$ can be defined by

$$\forall t \in Reals, \quad x(t) = \sin(440 \times 2\pi t).$$

The factor 2π in this expression is a nuisance. An argument to a sine function has units of **radians**, and so 2π has units of radians per cycle. Explicitly showing all the units (in square brackets), we have

$$440[\text{cycles/second}] \times 2\pi\,[\text{radians/cycle}]t\,[\text{seconds}] = (440 \times 2\pi t)\,[\text{radians}].$$

To avoid having to keep track of the factor 2π everywhere, it is common to use the alternative unit for frequency, **radians per second**. The symbol ω is commonly used to denote frequencies in radians per second, and f is used for frequencies in Hertz. The relationship between Hertz and radians per second is simple:

$$\omega = 2\pi f,$$

as is easily confirmed by checking the units. Thus, in radians per second, the frequencies of the musical note A on the piano keyboard are as follows:

$$\omega_1 = 2\pi \times 55, \omega_2 = 2\pi \times 110, \omega_3 = 2\pi \times 220, \omega_4 = 2\pi \times 440, \omega_5 = 2\pi \times 880,$$

$$\omega_6 = 2\pi \times 1{,}760, \omega_7 = 2\pi \times 3{,}520, \text{ and } \omega_8 = 2\pi \times 7{,}040.$$

BASICS

Ranges of frequencies

Natural and synthetic signals contain an extremely wide range of frequencies. The following abbreviations are common:

- Hz (hertz): cycles per second.
- kHz (kilohertz): thousands of cycles per second.
- MHz (megahertz): millions of cycles per second.
- GHz (gigahertz): billions of cycles per second.
- THz (terahertz): trillions of cycles per second.

Audible sound signals are in the range of 20 Hz to 20 kHz. Frequencies above this range are called ultrasonic. Electromagnetic waves range from less than 1 Hz (used speculatively in seismology for earthquake prediction) through visible light (near 10^{15} Hz) to cosmic ray radiation (up to 10^{25} Hz).

In studying sounds, it is reasonable on psychoacoustic grounds to decompose the sounds into sums of sinusoids. It turns out that the motivation for doing this extends well beyond psychoacoustics. Pure tones have very convenient mathematical properties that make it useful to model other types of signals as sums of sinusoids, even when there is no psychoacoustic basis for doing so. For example, there is no psychoacoustic reason for modeling radio signals as sums of sinusoids.

Consider the range of frequencies covering one **octave**, ranging from 440 Hz to 880 Hz. "Octave" is the musical term for a factor of two in frequency. Both the frequencies 440 Hz and 880 Hz correspond to the musical note A, but one octave apart. The next higher A in the musical scale would have the frequency 1,760 Hz, twice 880 Hz. In the western musical scale, there are 12 notes in every octave. These notes are evenly distributed (geometrically), so the next note above A, which is B flat (B♭) has frequency $440 \times \sqrt[12]{2}$, where $\sqrt[12]{2} \approx 1.0595$. The next note above B♭, which is B, has frequency $440 \times \sqrt[12]{2} \times \sqrt[12]{2}$.

In table 7.1, the frequencies of the complete musical scale between middle A (A-440) and A-880 are shown. Each frequency is $\beta = \sqrt[12]{2}$ times the frequency below it.

Frequencies that are harmonically related tend to sound good together. Figure 7.1 shows the graph of a signal that is a major triad, a combination of the notes A, C sharp (C♯), and E. By "combination" we mean "sum." The A is a sinusoidal signal at 440 Hz. It is added to a C♯, which is a sinusoidal signal at 554 Hz. This sum is then added to an E, which is a sinusoidal signal at 659 Hz.

TABLE 7.1
Frequencies of notes over one octave of the western musical scale, in Hertz.

Note	Frequency
A	880
A♭	831
G	784
F♯	740
F	698
E	659
D♯	622
D	587
C♯	554
C	523
B	494
B♭	466
A	440

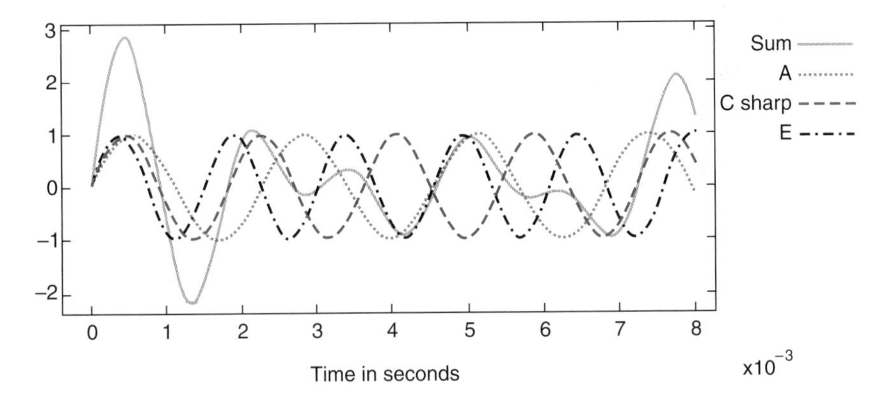

FIGURE 7.1: Graph of a major triad, showing its three sinusoidal components and their sum.

Each of the components is also shown, so you can verify graphically that at each point in time, the value of the solid signal is equal to the sum of values of the dashed signals at that time.

The stimulus presented to the ear is the solid waveform. What you hear, however, assuming a small amount of musical training, is the three sinusoidal components, which the human ear interprets as musical notes. The human ear decomposes the stimulus into its sinusoidal components.

PROBING FURTHER

Circle of fifths

The western musical scale is based on our perception of frequency and the harmonic relationships between frequencies. All the following frequencies correspond to the note A: 110, 220, 440, 880, 1,760, and 3,520. What about $440 \times 3 = 1{,}320$? Notice that $1{,}320/2 = 660$, which is almost exactly the E in table 7.1. Thus, 440×3 is (roughly) the note E, one octave above the E above A-440. E and A are closely harmonically related, and to most people, they sound good together. It is because

$$440 \times 3 \approx 659 \times 2.$$

The notes A, C♯, and E form a major triad. Where does the C♯ come from? Its frequency is 554 (see table 7.1). Notice that

$$440 \times 5 \approx 554 \times 4.$$

Among all the harmonic relationships in the scale, A, C♯, and E have one of the simplest. This is the reason for their sounding pleasing together.

continued on next page

PROBING FURTHER

For more arcane reasons, the interval between A and E, which is a frequency rise of approximately 3/2, is called a **fifth**. The note 3/2 (a fifth) above E has frequency 988, which is one octave above B-494. Another 3/2 above that is approximately F sharp (740 Hz). Continuing in this manner, multiplying frequencies by 3/2 and then possibly dividing by two, you can approximately trace the 12 notes of the scale. On the 13th, you return to A, approximately. This progression is called the **circle of fifths**. The notions of **key** and **scale** in music are based on this circle of fifths, as is the fact that there are 12 notes in the scale.

Table 7.1 is calculated by multiplying each frequency by $\sqrt[12]{2}$ to get the next higher frequency, not by using the circle of fifths. Indeed, the $\sqrt[12]{2}$ method applied 12 times yields a note that is *exactly* one octave higher than the starting point, whereas the circle of fifths yields only an approximation. The $\sqrt[12]{2}$ method yields the **well-tempered scale**. This scale was popularized by the composer J. S. Bach. It sounds much better than a scale based on the circle of fifths when the range of notes spans more than one octave.

Example 7.1: The major triad signal can be written as a sum of sinusoids:

$$s(t) = \sin(440 \times 2\pi t) + \sin(554 \times 2\pi t) + \sin(659 \times 2\pi t),$$

for all $t \in Reals$. The human ear hears as distinct tones the frequencies of these sinusoidal components. Musical sounds such as chords can be characterized as sums of pure tones. ❐

Purely sinusoidal signals, however, do not sound very good. Although they are recognizable as notes, they do not sound like any familiar musical instrument. Truly musical sounds are much more complex than a pure sinusoid. The characteristic sound of an instrument is its **timbre**, and as we shall see, some aspects of timbre can also be characterized as sums of sinusoids.

Timbre is caused in part by the fact that musical instruments do not produce purely sinusoidal sounds. Instead, to a first approximation, they produce sounds that consist of a fundamental sinusoidal component and harmonics. The fundamental is the frequency of the note being played, and the harmonics are multiples of that frequency. Figure 7.2 shows a waveform for a sound that is heard as an A-220 but has a much more interesting timbre than a sinusoidal signal with frequency 220 Hz. In fact, this waveform is generated by adding together sinusoidal signals with frequencies of 220 Hz, 440 Hz, 660 Hz, 880 Hz, 1,100 Hz, 1,320 Hz, and higher multiples, with varying weights. The 220-Hz component is called the **fundamental**, and the others are called **harmonics**. The **first harmonic** is the component at 440 Hz, the **second harmonic** is the component at 660 Hz, and so forth. The relative weights of the harmonics constitute a major part of what makes one musical instrument sound different from another.

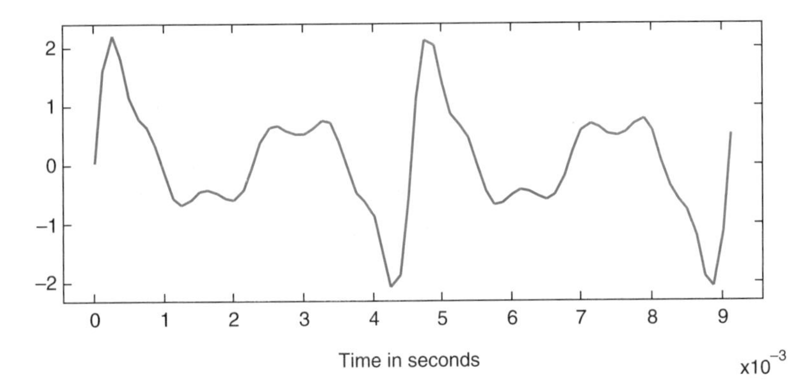

FIGURE 7.2: A sound waveform for an A-220 with more interesting timbre.

7.2 Phase

A sinusoidal sound has not just a frequency but also a **phase**. The phase may be thought of as the relative starting point of the waveform. Figure 7.3 shows five sinusoidal signals with the same frequency but five different phases. All these signals represent the sound A-440, and all sound identical. For a simple sinusoidal signal, phase obviously has no bearing on what the human ear hears.

Somewhat more surprising is that when two or more sinusoids are added together, the relative phase has a significant impact on the shape of the waveform but no impact on the perceived sound. The human ear is relatively insensitive to the phase of sinusoidal components of a signal, even though the phase of those components can strongly affect the shape of the waveform. If these waveforms represent something other than sound—stock prices, for example—the effect

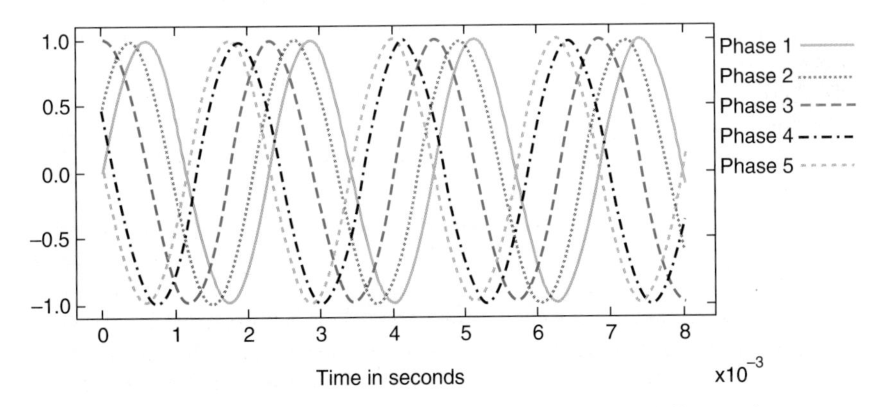

FIGURE 7.3: Five sinusoidal signals with the same frequency and different phases.

of phase could be quite significant. For a sinusoidal signal, the phase affects whether a particular point in time corresponds to a peak or a valley, for example. For stock prices, it makes a difference whether you sell at a high or a low.

There are certain circumstances in which the human ear is sensitive to phase. In particular, when two sinusoids of the same frequency combine, the relative phase has a big impact, because it affects the amplitude of the sum. For example, if the two sinusoids differ in phase by 180 degrees (π radians), then when they add, they exactly cancel, yielding a zero signal. The human brain can use the relative phase of a sound in the two ears to help spatially locate the origin of a sound. Also, audio systems with two speakers, which simulate spatially distributed sounds ("stereo"), can be significantly affected by the relative phase of the signal produced by the two speakers.

Phase is measured in either radians or degrees. An A-440 note can be represented by

$$g(t) = \sin(440 \times 2\pi t + \phi),$$

for all $t \in Reals$, where $\phi \in Reals$ is the phase. Regardless of the value of ϕ, this signal is still an A-440. If $\phi = \pi/2$, then

$$g(t) = \cos(440 \times 2\pi t).$$

7.3 Spatial frequency

Psychoacoustics provides a compelling motivation for decomposing audio signals as sums of sinusoids. In principle, images can also be similarly decomposed. However, the justification in this case is more mathematical than perceptual.

Figure 7.4 shows three images that are sinusoidal. Specifically, the **intensity** of the image (the amount of white light that is reflected by the page) varies spatially according to a sinusoidal function. In the left image, it varies only

FIGURE 7.4: Images that are sinusoidal horizontally, vertically, and both.

horizontally. There is no vertical variation in intensity. In the middle image, it varies only vertically. In the right image, it varies in both dimensions.

The sinusoidal image has **spatial frequency** rather than temporal frequency. Its units are cycles per unit distance. The images in figure 7.4 have frequencies of roughly 2.5 cycles/inch. Recall that a grayscale picture is represented by a function

$$Image: VerticalSpace \times HorizontalSpace \to Intensity.$$

An image that varies sinusoidally along the horizontal direction (with a spatial period of H inches) and is constant along the vertical direction is therefore represented by

$$\forall\, x \in VerticalSpace \text{ and } \forall\, y \in HorizontalSpace, \quad Image(x, y) = \sin(2\pi y/H).$$

Similarly, an image that varies sinusoidally along the vertical direction (with a spatial period of V inches) and is constant along the horizontal direction is represented by

$$\forall\, x \in VerticalSpace \text{ and } \forall\, y \in HorizontalSpace, \quad Image(x, y) = \sin(2\pi x/V).$$

An image that varies sinusoidally along both directions is represented by

$$\forall\, x \in VerticalSpace \text{ and } \forall\, y \in HorizontalSpace,$$

$$Image(x, y) = \sin(2\pi x/V) \times \sin(2\pi y/H).$$

These sinusoidal images have much less meaning than audio sinusoids, which we perceive as musical tones. Nonetheless, images can be described as sums of sinusoids, and such description is sometimes useful.

7.4 Periodic and finite signals

When the domain is continuous or discrete time, we can define a **periodic signal**. Assuming the domain is *Reals*, a periodic signal x with period $p \in Reals$ is one in which for all $t \in Reals$,

$$x(t) = x(t + p). \tag{7.1}$$

A signal with period p also has period $2p$, because

$$x(t) = x(t + p) = x(t + 2p).$$

In fact, it has period Kp, for any positive integer K. Usually, we define the period to be the *smallest* $p > 0$ such that

$$\boxed{\forall\, t \in Reals, \quad x(t) = x(t + p).}$$

Example 7.2: The sinusoidal signal x where, for all $t \in Reals$,

$$x(t) = \sin(\omega_0 t)$$

is a periodic signal with period $2\pi/\omega_0$, because for all $t \in Reals$,

$$\sin(\omega_0(t + 2\pi/\omega_0)) = \sin(\omega_0 t). \ \square$$

A periodic signal is defined over an infinite interval. If the domain is instead a subset $[a, b] \subset Reals$, for some finite a and b, then we call this a **finite signal**.

Example 7.3: The signal y, where for all $t \in [0, 2\pi/\omega_0]$

$$y(t) = \sin(\omega_0 t),$$

is a finite signal with duration $2\pi/\omega_0$. This interval spans exactly one cycle of the sine wave. ❐

A finite signal with duration p can be used to define a periodic signal with period p. All that is needed is to periodically repeat the finite signal. Formally, given a finite signal $y\colon [a, b] \to Reals$, we can define a signal $y'\colon Reals \to Reals$ by

$$\forall\, t \in Reals, \quad y'(t) = \begin{cases} y(t) & \text{if } t \in [a, b] \\ 0 & \text{otherwise.} \end{cases} \tag{7.2}$$

In other words, $y'(t)$ is simply $y(t)$ inside its domain and zero elsewhere. Then the periodic signal can be given by*

$$\boxed{x(t) = \sum_{m=-\infty}^{\infty} y'(t - mp),} \tag{7.3}$$

where $p = b - a$. This is called a **shift-and-add summation**, illustrated in figure 7.5. The periodic signal is a sum of versions of $y'(t)$ that have been shifted in time by multiples of p. Do not let the infinite sum intimidate: All but one of the terms of the summation are zero for any fixed t. Thus, a periodic signal can be defined

* If this notation is unfamiliar, see Basics: Summations box on page 69.

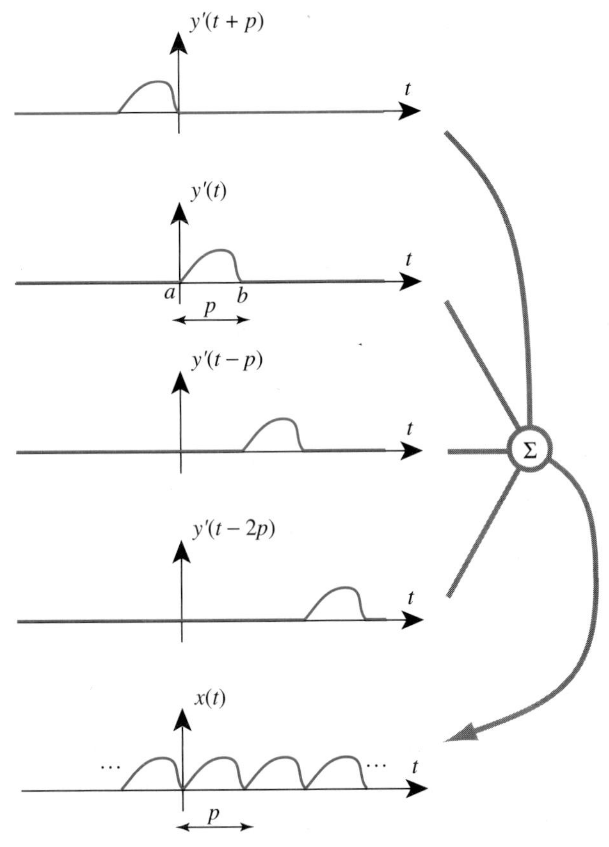

FIGURE 7.5: By "Shift-and-add" summation of the finite signal y we can obtain a periodic signal x.

in terms of a finite signal, which represents one period. Conversely, a finite signal can be defined in terms of a periodic signal (by taking one period).

We can check that x given by (7.3) is indeed periodic with period p,

$$\begin{aligned} x(t+p) &= \sum_{m=-\infty}^{\infty} y'(t+p-mp) \\ &= \sum_{m=-\infty}^{\infty} y'(t-(m-1)p) \\ &= \sum_{k=-\infty}^{\infty} y'(t-kp) = x(t), \end{aligned}$$

by using a change of variables, $k = m - 1$.

It is also important to note that the periodic signal x agrees with y in the finite domain $[a, b]$ of y, because

$$\begin{aligned}\forall\, t \in [a, b], \quad x(t) &= \sum_{m=-\infty}^{\infty} y'(t - mp) \\ &= y'(t) + \sum_{m \neq 0} y'(t - mp) \\ &= y(t),\end{aligned}$$

because, according to (7.2), for $t \in [a, b]$, $y'(t) = y(t)$ and $y'(t - mp) = 0$ if $m \neq 0$.

Any periodic signal, and hence any finite signal, can be described as a sum of sinusoidal signals. This result, known as the Fourier series, is one of the fundamental tools in the study of signals and systems.

7.5 Fourier series

A remarkable result, obtained by Joseph Fourier (1768–1830), is that a periodic signal x: *Reals* → *Reals* with period $p \in$ *Reals* can usually be described as a constant term plus a sum of sinusoids:

$$\boxed{x(t) = A_0 + \sum_{k=1}^{\infty} A_k \cos(k\omega_0 t + \phi_k).} \tag{7.4}$$

This representation of x is called its **Fourier series**. The Fourier series is widely used for signal analysis. Each term in the summation is a cosine with amplitude A_k and phase ϕ_k. The particular values of A_k and ϕ_k depend on x, of course. The frequency ω_0, which is measured in units of radians per second (assuming the domain of x is in seconds), is called the **fundamental frequency** and is related to the period p by

$$\boxed{\omega_0 = 2\pi/p.}$$

In other words, a signal with fundamental frequency ω_0 has period $p = 2\pi/\omega$. The constant term A_0 is sometimes called the **DC** term, a reference to the early applications of this theory in electrical circuit analysis in which "DC" stands for **direct current**. The terms in which $k \geq 2$ are called **harmonics**.

Equation (7.4) is often called the **Fourier series expansion** for x because it expands x in terms of its sinusoidal components.

If we had a facility for generating individual sinusoids, we could use the Fourier series representation (7.4) to synthesize any periodic signal. However, using the Fourier series expansion for synthesis of periodic signals is problematic because of the infinite summation. But for most practical signals, the coefficients A_k become very small (or even zero) for large k, and so a finite summation can be used as an approximation. A **finite Fourier series approximation** with $K + 1$ terms has the form

$$\tilde{x}(t) = A_0 + \sum_{k=1}^{K} A_k \cos(k\omega_0 t + \phi_k). \tag{7.5}$$

The infinite summation of (7.4) is, in fact, the limit of (7.5) as K goes to infinity. We therefore need to be concerned with whether this limit exists. The Fourier series expansion is valid only if the limit exists. There are some technical mathematical conditions on x that, if satisfied, ensure that the limit exists (see boxes on pages 262 and 263). Fortunately, these conditions are met almost always by practical, real-world time-domain signals.

Example 7.4: Figure 7.6(a) shows a square wave with a period of eight milliseconds and some finite Fourier series approximations to the square wave. Only one period of the square wave is shown. The method for constructing these approximations is covered in detail in chapter 10. Here, we merely observe the general structure of the approximations.

Notice in figure 7.6(a) that the $K = 1$ approximation consists only of the DC term (which is zero in this case) and a sinusoid with an amplitude slightly larger than that of the square wave. Its amplitude is depicted in figure 7.6(b) as the height of the largest bar. The horizontal position of the bar corresponds to the frequency of the sinusoid, 125 Hz, which is (one cycle)/(8 milliseconds), the fundamental frequency. The $K = 3$ waveform is the sum of the $K = 1$ waveform and one additional sinusoid with a frequency of 375 Hz and an amplitude equal to the height of the second largest bar in figure 7.6(b). ❐

A plot like that in figure 7.6(b) is called a **frequency domain** representation of the square wave, because it depicts the square wave by the amplitude and frequency of its sinusoidal components. Actually, a complete frequency domain representation also needs to specify the phase of each sinusoidal component.

Notice in figure 7.6(b) that all even terms of the Fourier series approximation have zero amplitude. For example, there is no component at 250 Hz. This is a consequence of the symmetry of the square wave, as will become clear in chapter 8.

Notice also that as the number of terms in the summation increases, the approximation more closely resembles a square wave but the amount of its overshoot does not appear to decrease. This is known as **Gibb's phenomenon**. In fact, the maximum difference between the finite Fourier series approximation

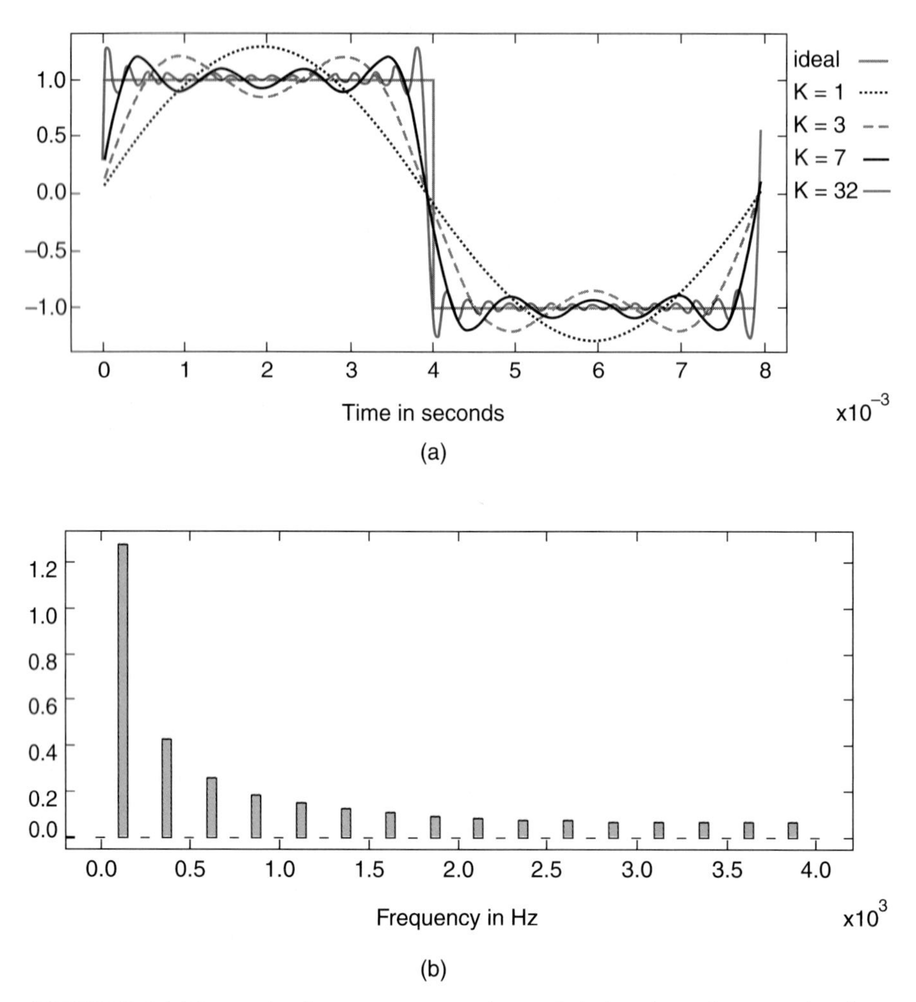

FIGURE 7.6: (a) One cycle of a square wave and some finite Fourier series approximations. (b) The amplitudes of the Fourier series terms for the square wave.

and the square wave does not converge to zero as the number of terms in the summation increases. In this sense, the square wave cannot be exactly described with a Fourier series (see Probing further: Uniform convergence box). Intuitively, the problem results from the abrupt discontinuity in the square wave in its transitions between its high value and its low value. In another sense, however, the square wave *is* accurately described by a Fourier series. Although the maximum difference between the approximation and the square wave does not converge to zero, the **mean square error** does converge to zero (see Probing further: Mean square convergence box). For practical purposes, the mean square error is an adequate criterion for convergence, and so we can work with the Fourier series expansion of the square wave.

Example 7.5: Figure 7.7 shows some finite Fourier series approximations for a triangle wave. This waveform has no discontinuities; therefore, the maximum error in the finite Fourier series approximation converges to zero (see the following box). Notice that its Fourier series components decrease in amplitude much more rapidly than do those of the square wave. Moreover, the time-domain approximations appear to be more accurate with fewer terms in the finite summation. ❐

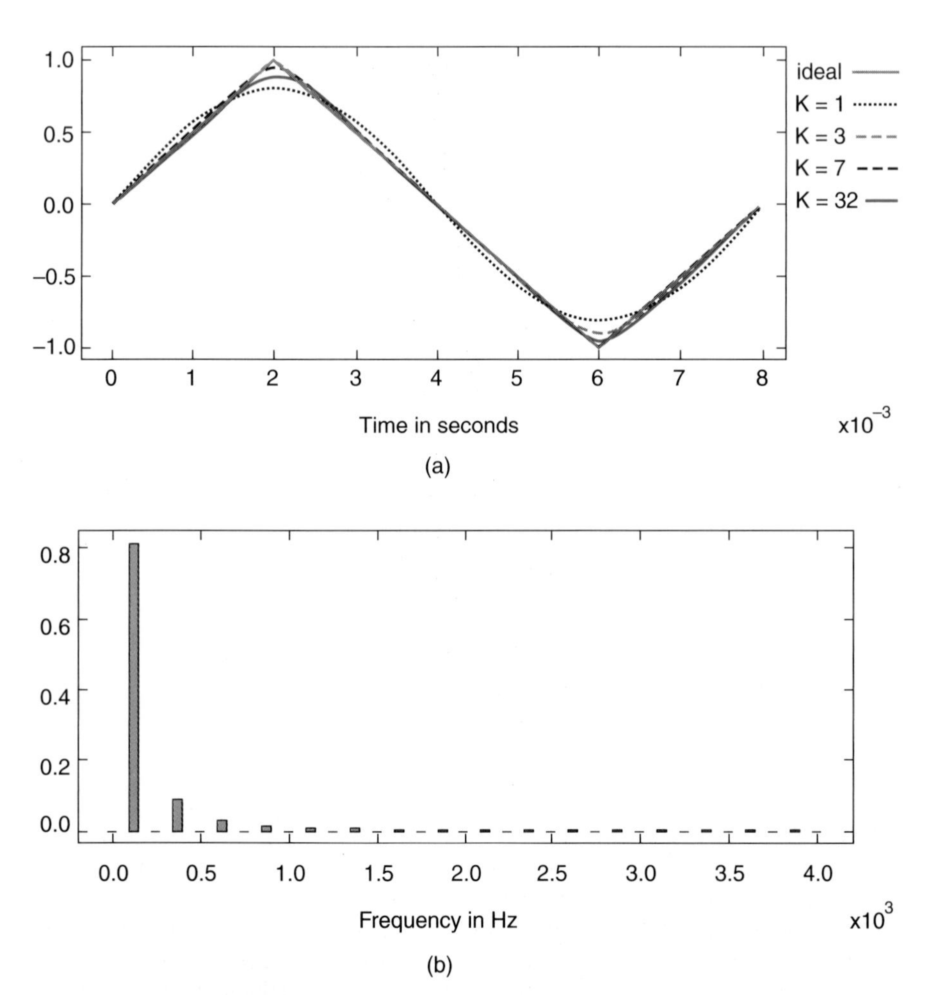

FIGURE 7.7: (a) One cycle of a triangle wave and some finite Fourier series approximations. (b) The amplitudes of the Fourier series terms for the triangle wave.

Many practical, real-world signals, such as audio signals, do not have discontinuities and thus do not exhibit the sort of convergence problems exhibited by the square wave (Gibb's phenomenon). Other signals, such as images, however, are full of discontinuities. A (spatial) discontinuity in an image is simply an edge. Most images have edges. Nonetheless, a Fourier series representation for such a signal is almost always still valid, in a mean square error sense (see Probing further: Mean square convergence box). This is sufficient for almost all engineering purposes.

PROBING FURTHER

Uniform convergence of the Fourier series

The Fourier series representation of a periodic signal x is a limit of a sequence of functions x_N for $N = 1, 2, \ldots,$ where

$$\forall\, t \in Reals, \quad x_N(t) = A_0 + \sum_{k=1}^{N} A_k \cos(k\omega_0 t + \phi_k).$$

Specifically, for the Fourier series representation to be valid, we must have that for all $t \in Reals$,

$$x(t) = \lim_{N\to\infty} x_N(t).$$

A strong criterion for validity of the Fourier series is **uniform convergence** of this limit, in which for each real number $\epsilon > 0$, there exists a positive integer M such that for all $t \in Reals$ and for all $N > M$,

$$|x(t) - x_N(t)| < \epsilon.$$

A sufficient condition for uniform convergence is that the signal x be continuous and that its first derivative be piecewise continuous.

A square wave, for example, is not continuous and hence does not satisfy this sufficient condition. Indeed, the Fourier series does not converge uniformly, as you can see in figure 7.6 by observing that the peak difference between $x(t)$ and $x_K(t)$ does not decrease to zero. A triangle wave, however, is continuous and has a piecewise continuous first derivative. Thus, it does satisfy the sufficient condition. Its Fourier series approximation therefore converges uniformly, as suggested in figure 7.7. A weaker, but still useful, criterion for validity of the Fourier series is considered in the next sidebar. That criterion is met by the square wave.

See, for example, R. G. Bartle, *The Elements of Real Analysis*, Second Edition; New York, John Wiley & Sons, 1976, p. 117 (for uniform convergence) and p. 337 (for this sufficient condition).

PROBING FURTHER

Mean square convergence of the Fourier series

The Fourier series representation of a periodic signal x with period p is a limit of a sequence of functions x_N for $N = 1, 2, \ldots$, where

$$\forall\, t \in \mathit{Reals}, \quad x_N(t) = A_0 + \sum_{k=1}^{N} A_k \cos(k\omega_0 t + \phi_k).$$

Specifically, for the Fourier series representation to be valid, we must have that for all $t \in \mathit{Reals}$,

$$x(t) = \lim_{N\to\infty} x_N(t).$$

For some practical signals, such as the square wave of figure 7.6, this statement is not quite true for all $t \in \mathit{Reals}$. For practical purposes, however, this need not be true. A weaker condition for validity of the Fourier series is that the total **energy** in the error over one period be zero. Specifically, we say that $x_N(t)$ **converges in mean square** to $x(t)$ if

$$\lim_{N\to\infty} \int_0^p |x(t) - x_N(t)|^2 dt = 0.$$

For certain physical signals the integral here is the energy in the error $x(t) - x_N(t)$ over one period. It turns out that if x itself has finite energy over one period, then $x_N(t)$ converges in mean square to $x(t)$; that is, all we need is that

$$\int_0^p |x(t)|^2 dt < \infty.$$

Virtually all signals with any engineering importance satisfy this criterion. Note that convergence in mean square does not guarantee that at any particular $t \in \mathit{Reals}$, $x(t) = \lim_{N\to\infty} x_N(t)$. For a condition that (almost) ensures this for all practical signals, see the following box.

See, for example, R. V. Churchill, *Fourier Series and Boundary Value Problems*, Third Edition; New York, McGraw-Hill Book Company, 1978.

PROBING FURTHER

Dirichlet conditions for validity of the Fourier series

The Fourier series representation of a periodic signal x with period p is a limit of a sequence of finite Fourier series approximations x_N for $N = 1, 2, \ldots$. We have seen in the "Probing further: Uniform convergence" box a strong condition that ensures that $\forall\, t \in \mathit{Reals}$,

$$x(t) = \lim_{N\to\infty} x_N(t). \tag{7.6}$$

continued on next page

PROBING FURTHER

We have seen in the preceding sidebar a weaker condition that does not guarantee this but guarantees instead that the energy in the error over one period is zero. It turns that for almost all signals of interest, we can assert that (7.6) holds for *almost all* $t \in Reals$. In particular, if the **Dirichlet conditions** are satisfied, then (7.6) holds for all t except where x is discontinuous. There are three Dirichlet conditions:

- Over one period, x is **absolutely integrable**, meaning that

$$\int_0^p |x(t)| dt < \infty.$$

- Over one period, x is of **bounded variation**, meaning that there are no more than a finite number of maxima or minima; that is, if the signal is oscillating between high and low values, it can oscillate only a finite number of times in each period.
- Over one period, x is continuous at all but a finite number of points.

These conditions are satisfied by the square wave and, indeed, by any physical signal of practical engineering importance.

Example 7.6: Consider an audio signal given by

$$s(t) = \sin(440 \times 2\pi t) + \sin(550 \times 2\pi t) + \sin(660 \times 2\pi t).$$

This is a major triad in a non–well-tempered scale. The first tone is A-440. The third is approximately E, with a frequency 3/2 that of A-440. The middle term is approximately C♯, with a frequency 5/4 that of A-440. These simple frequency relationships result in a pleasant sound. We choose the non–well-tempered scale because it makes it much easier to construct a Fourier series expansion for this waveform. The more difficult problem of finding the Fourier series coefficients for a well-tempered major triad is explored in exercise 5 at the end of this chapter.

To construct the Fourier series expansion, we can follow these steps:

- Find p, the period. The period is the smallest number $p > 0$ such that $s(t) = s(t - p)$ for all t in the domain. To do this, note that

$$\sin(2\pi ft) = \sin(2\pi f(t - p))$$

if fp is an integer. Thus, we want to find the smallest p such that $440p$, $550p$, and $660p$ are all integers. Equivalently, we want to find the largest fundamental frequency $f_0 = 1/p$ such that $440/f_0$, $550/f_0$, and $660/f_0$ are all integers. Such an f_0 is called the **greatest common divisor** of 440, 550, and 660. This can be computed by using the `gcd` function in

MATLAB.®* In this case, however, we can do it mentally, observing that $f_0 = 110$.

- Find A_0, the constant term. Inspection reveals that there is no constant component in $s(t)$, only sinusoidal components, and so $A_0 = 0$.
- Find A_1, the fundamental term. Inspection reveals that there is no component at 110 Hz, and so $A_1 = 0$. Because $A_1 = 0$, ϕ_1 is immaterial.
- Find A_2, the first harmonic. Inspection reveals that there is no component at 220 Hz, and so $A_2 = 0$.
- Find A_3. Inspection reveals that there is no component at 330 Hz, and so $A_3 = 0$.
- Find A_4. There is a component at 440 Hz, $\sin(440 \times 2\pi t)$. We need to find A_4 and ϕ_4 such that

$$A_4 \cos(440 \times 2\pi t) + \phi_4 = \sin(440 \times 2\pi t).$$

  Inspection reveals that $\phi_4 = -\pi/2$ and $A_4 = 1$.
- Similarly determine that $A_5 = A_6 = 1$, that $\phi_5 = \phi_6 = -\pi/2$, and that all other terms are zero.

Putting this all together, we can write the Fourier series expansion

$$s(t) = \sum_{k=4}^{6} \cos(k\omega_0 t - \pi/2),$$

where $\omega_0 = 2\pi f_0 = 220\pi$. ❒

The method used in this example for determining the Fourier series coefficients is clearly tedious and error prone and will work only for simple signals. Much better techniques are presented in chapters 8 and 10.

7.5.1 *Uniqueness of the Fourier series*

If x: *Reals* → *Reals* is a periodic function with period p, then the Fourier series expansion is unique. In other words, if it is true that both

$$x(t) = A_0 + \sum_{k=1}^{\infty} A_k \cos(k\omega_0 t + \phi_k)$$

and

$$x(t) = B_0 + \sum_{k=1}^{\infty} B_k \cos(k\omega_0 t + \theta_k),$$

* MATLAB is a registered trademark of The MathWorks, Inc.

where $\omega_0 = 2\pi/p$, then it must also be true that

$$\forall k \geq 0, \quad A_k = B_k \text{ and } \phi_k \bmod 2\pi = \theta_k \bmod 2\pi .$$

(The modulo operation is necessary because of the nonuniqueness of phase.) Thus, when we talk about *the* frequency content of a signal, we are talking about something that is unique and well defined. For a suggestion about how to prove this uniqueness, see exercise 11 at the end of this chapter.

7.5.2 *Periodic, finite, and aperiodic signals*

We have seen in section 7.4 that periodic signals and finite signals have much in common. One can be defined in terms of the other. Thus, a Fourier series can be used to describe a finite signal as well as a periodic one. The "period" is simply the extent of the finite signal. Thus, if the domain of the signal is $[a, b] \subset Reals$, then $p = b - a$. The fundamental frequency is therefore just $\omega_0 = 2\pi/(b - a)$.

An aperiodic signal, like an audio signal, can be partitioned into finite segments, and a Fourier series can be constructed from each segment.

Example 7.7: Consider the train whistle shown in figure 7.8(a). Figure 7.8(b) shows a 16-msec segment. Notice that within this segment, the sound clearly has a somewhat periodic structure. It is not hard to envision how it could be described as sums of sinusoids. The magnitudes of the A_k Fourier series coefficients for this 16-msec segment are shown in figure 7.8(c). These are calculated on a computer by techniques we discuss later, rather than being calculated mentally as in example 7.6. Notice that there are three dominant frequency components that give the train whistle its tonality and timbre. ❐

7.5.3 *Fourier series approximations to images*

Images are invariably finite signals. Given any image, it is possible to construct a periodic image by just tiling a plane with the image. Thus, there is again a close relationship between a periodic image and a finite one.

We have seen sinusoidal images (figure 7.4), and so it follows that it ought to be possible to construct a Fourier series representation of an image. The only difficulty is that images have a two-dimensional domain and thus are finite in two distinct dimensions. The sinusoidal images in figure 7.4 have a vertical frequency, a horizontal frequency, or both.

Suppose that the domain of an image is $[a, b] \times [c, d] \subset Reals \times Reals$. Let $p_H = b - a$ and $p_V = d - c$ represent the horizontal and vertical "periods" for the

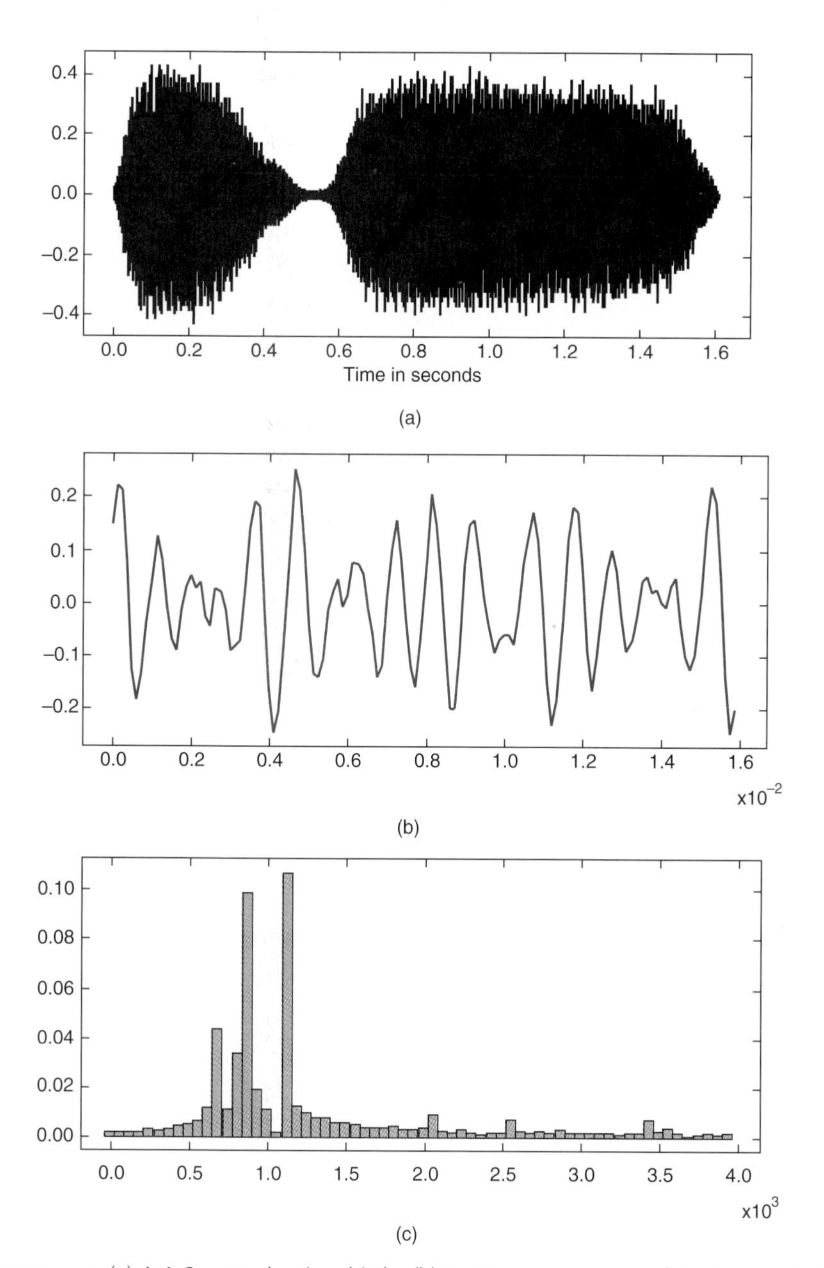

FIGURE 7.8: (a) A 1.6-second train whistle. (b) A 16-msec segment of the train whistle. (c) The Fourier series coefficients A_k for the 16-msec segment.

equivalent periodic image. For constructing a Fourier series representation, we can define the horizontal and vertical fundamental frequencies as follows:

$$\omega_H = 2\pi / p_H$$

and

$$\omega_V = 2\pi / p_V.$$

The Fourier series representation of *Image*: $[a, b] \times [c, d] \to$ *Intensity* is

$$Image(x, y) = \sum_{k=0}^{\infty} \sum_{m=0}^{\infty} A_{k,m} \cos(k\omega_H x + \phi_k) \cos(m\omega_V y + \varphi_m).$$

For convenience, we have included the constant term $A_{0,0}$ in the summation, so we assume that $\phi_0 = \varphi_0 = 0$. (Recall that $\cos(0) = 1$).

7.6 Discrete-time signals

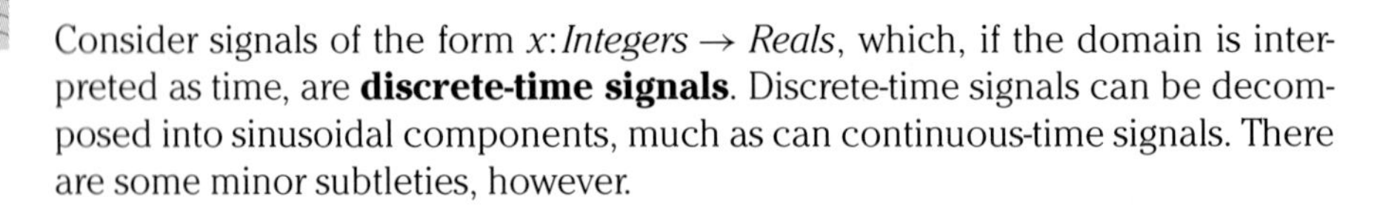

Consider signals of the form x: *Integers* $\to$ *Reals*, which, if the domain is interpreted as time, are **discrete-time signals**. Discrete-time signals can be decomposed into sinusoidal components, much as can continuous-time signals. There are some minor subtleties, however.

7.6.1 *Periodicity*

A discrete-time signal is **periodic** if there is a nonzero integer $p > 0$ such that

$$\forall n \in Integers, \quad x(n + p) = x(n).$$

Note, somewhat counterintuitively, that not all sinusoidal discrete-time signals are periodic. Consider

$$x(n) = cos(2\pi fn). \tag{7.7}$$

For this to be periodic, we must be able to find a nonzero integer p such that for all integers n,

$$x(n + p) = cos(2\pi fn + 2\pi fp) = cos(2\pi fn) = x(n).$$

This can be true only if $(2\pi fp)$ is an integer multiple of 2π—that is, if there is some integer m such that

$$2\pi fp = 2\pi m.$$

Dividing both sides by $2\pi p$, we see that this signal is periodic only if we can find nonzero integers p and m such that

$$f = m/p.$$

In other words, f must be rational. Only if f is rational is this signal periodic.

Discrete-time frequencies

BASICS

When the domain of a signal is *Integers*, then the units of frequency are cycles per sample. Consider, for example, the discrete-time signal given by

$$\forall n \in Integers, \quad x(n) = \cos(2\pi fn).$$

Suppose this represents an audio signal that is sampled at 8,000 samples/second. Then to convert f to Hertz, just observe the units:

$$f[cycles/sample] \times 8{,}000[samples/second] = 8{,}000f[cycles/second].$$

The frequency could have been given equally appropriately in units of radians per sample, as ω in

$$x(n) = \cos(\omega n).$$

for all $n \in Integers$. To convert ω to Hertz,

$$\omega[radians/sample] \times 8{,}000[samples/second] \times (1/2\pi)[cycles/radian]$$
$$= (8{,}000\omega/2\pi)[cycles/second].$$

Example 7.8: Consider a discrete-time sinusoid x given by

$$\forall n \in Integers, \quad x(n) = \cos(4\pi n/5).$$

Putting this into the form of (7.7),

$$x(n) = \cos(2\pi(2/5)n),$$

we see that it has frequency $f = 2/5$ cycles/sample. If this were a continuous-time sinusoid, we could invert this to get the period. However, the inverse is not an integer, and so it cannot possibly be the period. Noting that the inverse is 5/2 samples/cycle, we need to find the smallest multiple of this that is an integer. Multiplying by 2, we get (5 samples)/(2 cycles). Therefore, the period is $p = 5$ samples. ❐

In general, for a discrete sinusoid with frequency of f cycles/sample, the period is $p = K/f$, where $K > 0$ is the smallest integer such that K/f is an integer.

7.6.2 *The discrete-time Fourier series*

Assume that we are given a periodic discrete-time signal x with period p. Just as with continuous-time signals, this signal can be described as a sum of sinusoids, called the **discrete-time Fourier series** (**DFS**) expansion,

$$\boxed{x(n) = A_0 + \sum_{k=1}^{K} A_k \cos(k\omega_0 n + \phi_k),} \tag{7.8}$$

where

$$K = \begin{cases} (p-1)/2 & \text{if } p \text{ is odd} \\ p/2 & \text{if } p \text{ is even.} \end{cases}$$

In contrast to the continuous-time case, this sum is finite. This is because discrete-time signals cannot represent frequencies above a certain value. We examine this phenomenon in more detail in chapter 11, but for now, it proves extremely convenient. Mathematically, the relation just given is much simpler than the continuous-time Fourier series. All computations are finite. There is a finite number of signal values in one period of the waveform. There is a finite number of terms in the Fourier series representation for each signal value. In contrast to the continuous-time case, it is easy for computers to manage this representation. Given a finite set of values, $A_0, \ldots, A_K$, a computer can calculate $x(n)$. Moreover, the representation is exact for any periodic signal. No approximation is needed, and there is no question of convergence. In the continuous-time case, the Fourier series representation is accurate only for certain signals. For the discrete-time case, it is always accurate.

The DFS can be calculated efficiently on a computer by using an algorithm called the **fast Fourier transform** (**FFT**). All of the Fourier series examples that are plotted in this text were calculated with the use of the FFT algorithm.

7.7 *Summary*

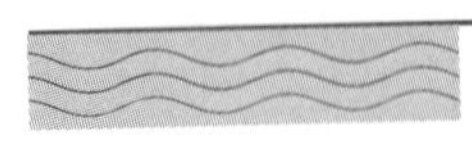

A time-based signal can be described as a sum of sinusoids. This sum is called a Fourier series. The magnitudes and phases of the sinusoids, taken together as a function of frequency, are called the frequency-domain representation of the signal. For audio signals, this frequency-domain representation has a direct psychoacoustic significance. But we see in chapter 8 that this representation has significance for all signals when LTI systems are used to process the signals. We see that the effect that an LTI system has on a signal is particularly easy to understand in the frequency domain.

EXERCISES

KEY: E = mechanical T = requires plan of attack C = more than 1 answer

E 1. In (7.1) we defined periodic for continuous-time signals.

(a) Define **finite** and **periodic** for discrete-time signals, where the domain is *Integers*.

(b) Define **finite** and **periodic** for images.

E 2. Which of the following signals is periodic with a period greater than zero, and what is that period? All functions are of the form $x\colon Reals \to Complex$. The domain is time, measured in seconds, and so the period is in seconds.

(a) $\forall\, t \in Reals, \quad x(t) = 10\sin(2\pi t) + (10 + 2i)\cos(2\pi t)$.

(b) $\forall\, t \in Reals, \quad x(t) = \sin(2\pi t) + \sin(\sqrt{2}\pi t)$.

(c) $\forall\, t \in Reals, \quad x(t) = \sin(2\sqrt{2}\pi t) + \sin(\sqrt{2}\pi t)$.

E 3. Consider the discrete-time signal x where

$$\forall\, n \in Integers, \quad x(n) = 1 + \cos(4\pi n/9).$$

(a) Find the period p, where $p > 0$.

(b) Give the fundamental frequency corresponding to the period in (a). Give the units.

(c) Give the coefficients $A_0, A_1, A_2, \ldots$ and $\phi_1, \phi_2, \ldots$ of the Fourier series expansion for this signal.

E 4. Consider a continuous-time signal $x\colon Reals \to Reals$ defined by

$$\forall\, t \in Reals, \quad x(t) = \cos(\omega_1 t) + \cos(\omega_2 t),$$

where $\omega_1 = 2\pi$ and $\omega_2 = 3\pi$ radians/second.

(a) Find the smallest period $p \in Reals_+$, where $p > 0$.

(b) Give the fundamental frequency corresponding to the period in (a). Give the units.

(c) Give the coefficients $A_0, A_1, A_2, \ldots$ and $\phi_1, \phi_2, \ldots$ of the Fourier series expansion for x.

T 5. Determine the fundamental frequency and the Fourier series coefficients for the following well-tempered major triad:

$$s(t) = \sin(440 \times 2\pi t) + \sin(554 \times 2\pi t) + \sin(659 \times 2\pi t).$$

E 6. Define $x\colon Reals \to Reals$ by

$$\forall\, t \in Reals, \quad x(t) = 5\cos(\omega_0 t + \pi/2) + 5\cos(\omega_0 t - \pi/6) + 5\cos(\omega_0 t - 2\pi/3).$$

Find A and ϕ so that

$$\forall\, t \in Reals, \quad x(t) = A\cos(\omega_0 t + \phi).$$

Hint: Appendix B may be useful.

T 7. In this problem, we examine a practical application of the mathematical result in exercise 6. In particular, we consider multipath interference, a common problem with wireless systems in which multiple paths from a transmitter to a receiver can result in destructive interference of a signal.

When a transmitter sends a radio signal to a receiver, the received signal consists of the direct path plus several reflected paths. In figure 7.9, the transmitter is on a tower at the left of the figure, the receiver is the telephone in the foreground, and there are three paths: the direct path is l_0 meters long, the path reflected from the hill is l_1 meters long, and the path reflected from the building is l_2 meters long.

Suppose the transmitted signal is an f-Hz sinusoid, $x\colon Reals \to Reals$,

$$\forall\, t \in Reals, \quad x(t) = A\cos(2\pi ft).$$

Thus, the received signal is y such that $\forall\, t \in Reals$,

$$y(t) = \alpha_0 A\cos(2\pi f(t - \frac{l_0}{c})) + \alpha_1 A\cos(2\pi f(t - \frac{l_1}{c})) + \alpha_2 A\cos(2\pi f(t - \frac{l_2}{c})). \tag{7.9}$$

Here, $0 \le \alpha_i \le 1$ are numbers that represent the attenuation (or reduction in signal amplitude) of the signal, and $c = 3 \times 10^8$ m/sec is the speed of light in a vacuum.* Answer the following questions.

* In reality, the reflections are more complicated than the model here.

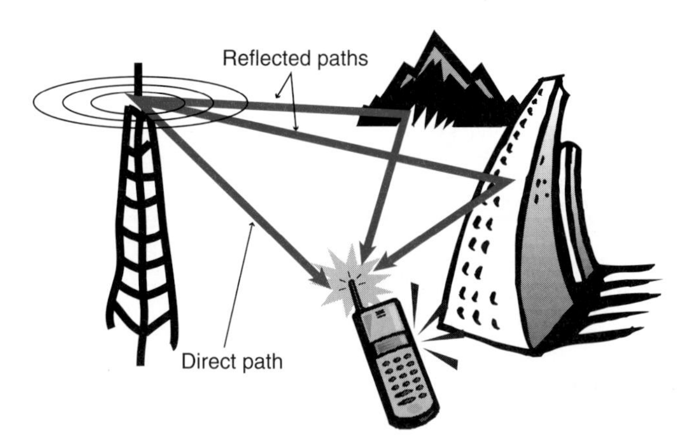

FIGURE 7.9: One direct path and two reflected paths from transmitter to receiver.

(a) Explain why the description of y given in (7.9) is a reasonable model of the received signal.

(b) What would be the description if, instead of the three paths as shown in figure 7.9, there were 10 paths (one direct and nine reflected)?

(c) The signals received over the different paths cause different phase shifts, ϕ_i, and so the signal y (with three paths) can also be written as

$$\forall\, t \in Reals, \quad y(t) = \sum_{k=0}^{2} \alpha_k A \cos(2\pi f t - \phi_k).$$

What are the ϕ_k? Give an expression in terms of f, l_k, and c.

(d) Let $\Phi = \max\{\phi_1 - \phi_0, \phi_2 - \phi_0\}$ be the largest difference in the phase of the received signals, and let $L = \max\{l_1 - l_0, l_2 - l_0\}$ be the maximum path length difference. What is the relationship among Φ, L, and f?

(e) Suppose for simplicity that there is only one reflected path of distance l_1—that is, let $\alpha_2 = 0$ in the earlier expressions in this exercise. Then $\Phi = \phi_1 - \phi_0$. When $\Phi = \pi$, the reflected signal is said to *destroy* the direct signal. Explain why the term "destroy" is appropriate. (This phenomenon is called destructive interference.)

(f) In the context of mobile radio shown in figure 7.9, typically $L \leq 500$ meters. For what values of f is $\Phi \leq \pi/10$? (Note that if $\Phi \leq \pi/10$, the signals will not interact destructively by much.)

(g) For the two-path case, derive an expression that relates the frequencies f that interfere destructively to the path length difference $L = l_1 - l_0$.

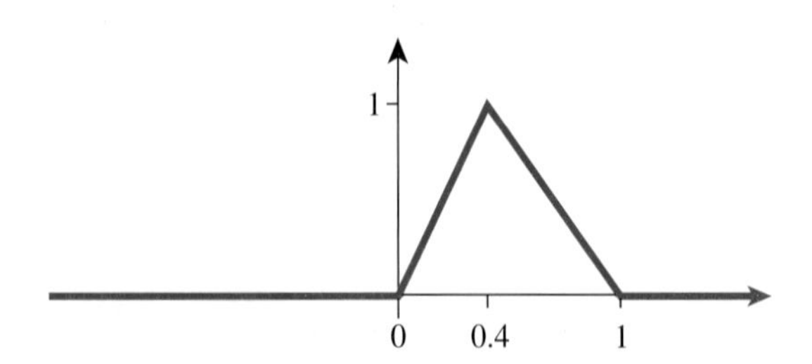

FIGURE 7.10: The graph of x.

T 8. The function x: *Reals* $\to$ *Reals* is given by its graph shown in figure 7.10. Note that $\forall\, t \notin [0, 1],\ x(t) = 0$, and $x(0.4) = 1$. Define y by

$$\forall\, t \in Reals, \quad y(t) = \sum_{k=-\infty}^{\infty} x(t - kp),$$

where p is the period.

(a) Prove that y is periodic with period p; that is,

$$\forall\, t \in Reals,\ y(t) = y(t + p).$$

(b) Plot y for $p = 1$.

(c) Plot y for $p = 2$.

(d) Plot y for $p = 0.8$.

(e) Plot y for $p = 0.5$.

(f) Suppose the function z is obtained by advancing x by 0.4; that is,

$$\forall\, t \in Reals, \quad z(t) = x(t + 0.4).$$

Define w by

$$\forall\, t \in Reals, \quad w(t) = \sum_{k=-\infty}^{\infty} z(t - kp).$$

What is the relation between w and y. Use this relation to plot w for $p = 1$.

T 9. Suppose x: *Reals* $\to$ *Reals* is a periodic signal with period p; that is,

$$\forall\, t \in Reals, \quad x(t) = x(t + p).$$

Let $f\colon Reals \to Reals$ be any function, and define the signal $y\colon Reals \to Reals$ by $y = f \circ x$; that is,

$$\forall\, t \in Reals, \quad y(t) = f(x(t)).$$

(a) Prove that y is periodic with period p.

(b) Suppose $\forall\, t \in Reals, \quad x(t) = \sin(2\pi t)$. Suppose f is the sign function:

$$\forall\, a \in Reals, f(a) = \begin{cases} 1 & \text{if } a \geq 0 \\ -1 & \text{if } a < 0. \end{cases}$$

Plot x and y.

(c) Suppose $\forall\, t \in Reals, \quad x(t) = \sin(2\pi t)$. Suppose f is the square function, $\forall\, x \in Reals, f(x) = x^2$. Plot y.

C 10. Suppose the periodic square wave shown on the left in Figure 7.11 has the Fourier series representation

$$A_0 + \sum_{k=1}^{\infty} A_k \cos(2\pi kt/p + \phi_k).$$

Use this to obtain a Fourier series representation of the two-dimensional pattern of rectangles on the right, which is an image.

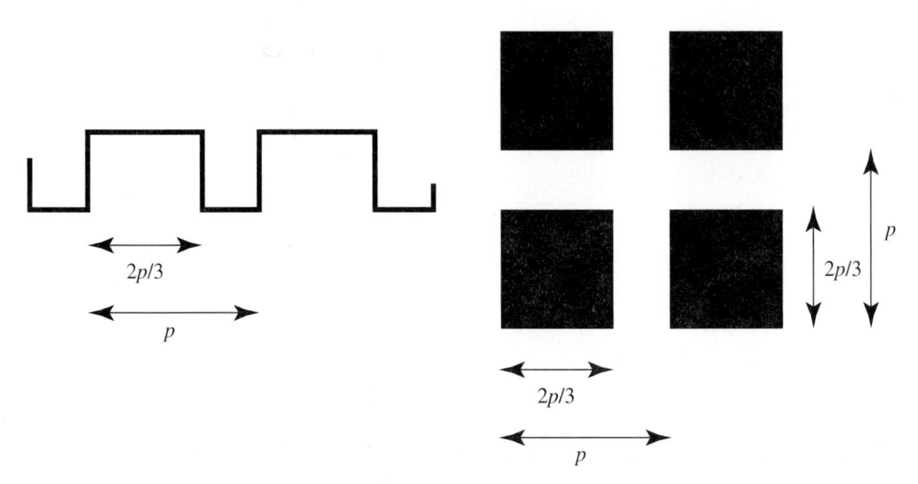

FIGURE 7.11: A periodic square wave (left) and a periodic pattern (right).

T 11. Suppose $A_k \in Complex$, $\omega_k \in Reals$, and $k = 1, 2$, such that

$$\forall\, t \in Reals, \quad A_1 e^{i\omega_1 t} = A_2 e^{i\omega_2 t}. \tag{7.10}$$

Show that $A_1 = A_2$ and $\omega_1 = \omega_2$. Hint: Evaluate both sides of (7.10) at $t = 0$, and evaluate their derivatives at $t = 0$.

Discussion: This result shows that in order for two complex exponential signals to be equal, their frequencies, phases, and amplitudes must be equal. More interestingly, this result can be used to show that if a signal can be described as a sum of complex exponential signals, then that description is unique. There is no other sum of complex exponentials (one involving different frequencies, phases, or amplitudes) that will also describe the signal. In particular, the Fourier series representation of a periodic signal is unique, as stated previously in section 7.5.1.

CHAPTER 8

Frequency response

A class of systems that yield to sophisticated analysis techniques is the class of **linear time-invariant (LTI) systems**, discussed in chapter 5. LTI systems have a key property: Given a sinusoidal input, the output is a sinusoidal signal with the same frequency but possibly a different amplitude and a different phase. If an input is a sum of sinusoids, the output is a sum of the same sinusoids, each with its amplitude and phase possibly modified.

We can justify describing audio signals as sums of sinusoids on purely psychoacoustic grounds. However, because of this property of LTI systems, it is often convenient to describe any signal as a sum of sinusoids, regardless of whether there is a psychoacoustic justification. The real value in this mathematical device is that by using the theory of LTI systems, we can design systems that operate more or less independently on the sinusoidal components of a signal. For example, abrupt changes in the signal value require higher frequency components. Thus, we can enhance or suppress these abrupt changes by enhancing or suppressing the higher frequency components. Such an operation is called **filtering** because it filters frequency components. We design systems by crafting their **frequency response**, which is their response to sinusoidal inputs. An audio equalizer, for example, is a filter that enhances or suppresses certain frequency components. Images can also be filtered. Enhancing the higher frequency components will sharpen the image, whereas suppressing the higher frequency components will blur the image.

State-space models described in previous chapters are precise and concise but, in a sense, not as powerful as a frequency response. For an LTI system, a given frequency response can reveal a great deal about the relationship between an input signal and an output signal. Fewer assertions are practical in general with state-space models.

LTI systems can, in fact, also be described with state-space models, with the use of difference equations and differential equations, as explored in chapter 5. But state-space models can also describe systems that are not LTI. Thus, state-space models are more general. It should come as no surprise that the price we pay for this increased generality is fewer analysis and design techniques. In this chapter, we explore the (very powerful) analysis and design techniques that apply to the special case of LTI systems.

8.1 LTI systems

LTI systems have received a great deal of intellectual attention for two reasons. First, they are relatively easy to understand. Their behavior is predictable and can be fully characterized in fairly simple terms, on the basis of the frequency domain representation of signals that we introduced in chapter 7. Second, many physical systems can be reasonably approximated by LTI systems. Few physical systems fit the model perfectly, but many fit very well within a certain operating range. Hybrid system models can be used when we wish to model more than one operating range, each with a different LTI system.

8.1.1 *Time invariance*

Consider the set of signals whose domain is interpreted as time. Such signals are functions of time, sometimes called **time-domain signals**. The domain might be *Reals*, for **continuous-time signals**, or *Integers*, for **discrete-time signals**. Physical audio signals, for example, are continuous-time signals, whereas a digital audio file is a discrete-time signal. Systems with continuous-time input and output signals are called **continuous-time systems**. Systems with discrete-time input and output signals are called **discrete-time systems**.

A simple continuous-time system is the **delay** system D_τ, where if the input is x, then the output $y = D_\tau(x)$ is given by

$$\forall t \in Reals, \quad y(t) = x(t - \tau). \tag{8.1}$$

Positive values of τ result in positive delays (despite the subtraction in $x(t - \tau)$). Any delay results in a leftward or rightward shifting of the graph of a signal, as shown in figure 8.1.

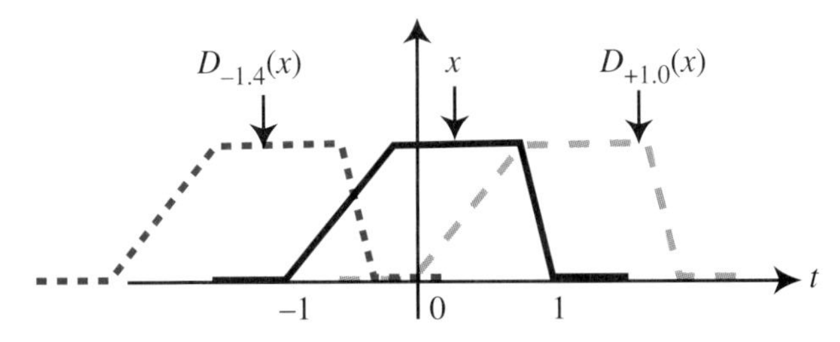

FIGURE 8.1: Illustration of the delay system D_τ. $D_{-1.4}(x)$ is the signal x to the left by 1.4, and $D_{+1.0}(x)$ is x moved to the right by 1.0.

Example 8.1: Consider a continuous-time signal x given by

$$\forall\, t \in Reals, \quad x(t) = \cos(2\pi t).$$

Let $y = D_{0.5}(x)$. Then

$$\forall\, t \in Reals, \quad y(t) = \cos(2\pi(t - 0.5)) = \cos(2\pi t - \pi).$$

The delay by 0.5 seconds is equivalent to a phase shift of $-\pi$ radians. For a sinusoidal signal, and only for a sinusoidal signal, time delay and phase changes are equivalent, except for the fact that phase is measured in radians (or degrees) rather than in seconds. In addition, a phase change of q is equivalent to a phase change of $q + K2\pi$ for any integer K. Phase applies to sinusoidal signals, whereas delay applies to any signal that is a function of time. ❐

Intuitively, a **time-invariant system** is one whose response to inputs does not change with time. More precisely, a continuous-time system S is said to be time invariant if

$$\boxed{\forall\, \tau \in Reals, \quad S \circ D_\tau = D_\tau \circ S.} \tag{8.2}$$

Figure 8.2 illustrates this equivalence, where the left side, $S \circ D_\tau$, is shown on top and the right side, $D_\tau \circ S$, is shown on the bottom. Time invariance implies that the upper and lower systems in figure 8.2 have identical behavior.

Equivalently, S is time invariant if for all x and τ,

$$S(D_\tau(x)) = D_\tau(S(x)).$$

Because both sides of this equation are functions, this equation means that

$$\forall\, t \in Reals, \quad S(D_\tau(x))(t) = D_\tau(S(x))(t).$$

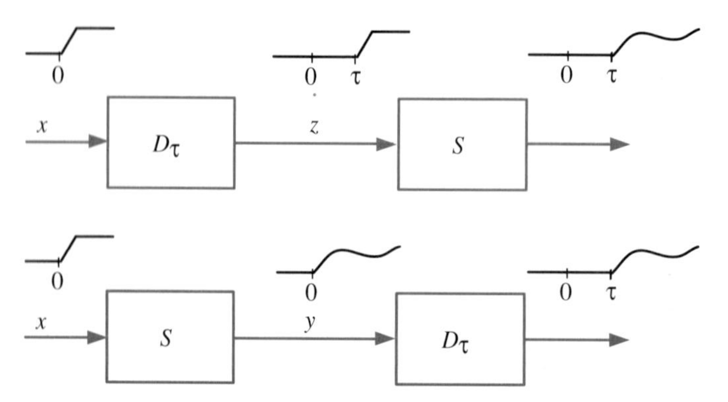

FIGURE 8.2: Time invariance implies that the top and bottom systems produce the same output signal for the same input signal.

Thus, a less compact version of (8.2) is*

$$\forall\, \tau, t, x, \quad S(D_\tau(x))(t) = D_\tau(S(x))(t).$$

This is interpreted as follows:

> A system S is time invariant if, for any input x that produces output y, a delayed input $D_\tau(x)$ produces the delayed output $D_\tau(y)$.

Similarly, the discrete-time M-sample delay is written D_M. The signal $y = D_M(x)$ is given by

$$\forall\, n \in Integers, \quad y(n) = x(n - M), \tag{8.3}$$

where $M \in Integers$. A discrete-time system S is time invariant if

$$\boxed{\forall\, M, x, \quad S(D_M(x)) = D_M(S(x)).} \tag{8.4}$$

Example 8.2: Consider a discrete-time system S, where

$$S\colon [Integers \to Reals] \to [Integers \to Reals].$$

* We use the shorthand "$\forall\, x$" instead of "$\forall\, x \in [Reals \to Reals]$" when the set is understood. Similarly, we can write "$\forall\, \tau, t, x$" instead of "$\forall\, \tau \in Reals, \forall\, t \in Reals, \forall\, x \in [Reals \to Reals]$."

Suppose that any input x produces output y, where

$$\forall\, n \in \mathit{Integers}, \quad y(n) = x(n) + 0.9x(n-1). \tag{8.5}$$

This system is time invariant. To show this, consider a delayed input $\hat{x} = D_M(x)$ for some integer M; that is,

$$\forall\, n, \quad \hat{x}(n) = x(n-M). \tag{8.6}$$

Suppose that this input produces output $\hat{y} = S(\hat{x})$. Then according to the relation (8.5) between an input and an output,

$$\forall\, n, \quad \hat{y}(n) = \hat{x}(n) + 0.9\hat{x}(n-1).$$

Substituting (8.6), we see that

$$\forall\, n, \quad \hat{y}(n) = x(n-M) + 0.9x(n-M-1) = y(n-M).$$

Because $\hat{y} = D_M(y)$, the system is time invariant. ❐

Example 8.3: Consider the system *DelayAndSquare* (*DS*), where

$$DS\colon [\mathit{Reals} \to \mathit{Reals}] \to [\mathit{Reals} \to \mathit{Reals}].$$

Suppose that any input x produces output y, where

$$\forall\, t \in \mathit{Reals}, \quad y(t) = (x(t-1))^2. \tag{8.7}$$

This system is time invariant. To show this, consider a delayed input $\hat{x} = D_\tau(x)$, for some real number τ; that is,

$$\forall\, t, \quad \hat{x}(t) = x(t-\tau). \tag{8.8}$$

Suppose that this input produces output $\hat{y} = S(\hat{x})$. Then according to the relation (8.7) between an input and an output,

$$\forall\, t, \quad \hat{y}(t) = (\hat{x}(t-1))^2.$$

Substituting (8.8), we see that

$$\forall\, t, \quad \hat{y}(t) = (x(t-1-\tau))^2 = y(t-\tau).$$

Because $\hat{y} = D_\tau(y)$, the system is time invariant. ❐

Example 8.4: Consider a system *ReverseTime* (*RT*), where

$$RT\colon [Reals \to Reals] \to [Reals \to Reals],$$

for which any input x produces output y related by

$$\forall\, t \in Reals, \quad y(t) = x(-t). \tag{8.9}$$

This system is *not* time invariant. To show this, consider a delayed input $\hat{x} = D_\tau(x)$, for some real number τ; that is,

$$\forall\, t, \quad \hat{x}(t) = x(t - \tau). \tag{8.10}$$

Suppose that this input produces output $\hat{y} = S(\hat{x})$. Then according to the relation (8.9) between an input and an output,

$$\forall\, t, \quad \hat{y}(t) = \hat{x}(-t).$$

Substituting (8.10), we see that

$$\forall\, t, \quad \hat{y}(t) = x(-t - \tau) \neq y(t - \tau) = x(-t + \tau).$$

Because $\hat{y} \neq D_\tau(y)$, the system is *not* time invariant.

To be completely convinced that these two signals are different in general, consider a particular signal x such that $\forall\, t,\ x(t) = t$, and let $\tau = 1$. Then $\hat{y}(0) = -1$, but $(D_\tau(y))(0) = 1$. ❒

Time invariance is a mathematical idealization. No electronic system is time invariant in the strict sense. For one thing, such a system is turned on at some time. Its behavior before it is turned on is clearly not the same as its behavior after it is turned on. Nevertheless, time invariance proves to be a very convenient mathematical fiction, and it is a reasonable approximation for many systems if their behavior is constant over a relatively long period of time (in relation to whatever phenomenon we are studying). For example, an audio amplifier is not a time-invariant system. Its behavior changes drastically when it is turned on or off and changes less drastically when the volume is raised or lowered. However, for the duration of a compact disc, if the volume is left fixed, the system can be reasonably assured to be time invariant.

Some systems have a similar property even though they operate on signals whose domain is not time. For example, the domain of an image is a region of a plane. The output of an image processing system may not depend significantly on where in the plane the input image is placed. Shifting the input image shifts the output image only by the same amount. This property, which generalizes time invariance and holds for some image processing systems, is called **shift invariance** (see exercise 5 at the end of this chapter).

8.1.2 *Linearity*

Consider the set of signals whose range is *Reals* or *Complex*. Such signals are real-valued functions or complex-valued functions. Because real-valued functions are a subset of complex-valued functions, we need to talk only about complex-valued functions. It does not matter (for now) whether they are continuous-time signals or discrete-time signals. The domain could be *Reals* or *Integers*.

Suppose x is a complex-valued function and a is a complex constant. Then we can define a new complex-valued function ax such that for all t in the domain of x,

$$(ax)(t) = a(x(t)).$$

In other words, the new function, which we call ax, is simply scaled by the constant a.

Similarly, given two complex-valued functions x_1 and x_2 with the same domain and range, we can define a new function $(x_1 + x_2)$ such that for all t in the domain,

$$(x_1 + x_2)(t) = x_1(t) + x_2(t).$$

Consider the set of all systems that map complex-valued functions to complex-valued functions. Such systems are called **complex systems**. Again, it does not matter (for now) whether they are discrete-time systems or continuous-time systems. Suppose that S is a complex system. S is said to be **linear** if for all $a \in Complex$ and for all complex signals x,

$$\boxed{S(ax) = aS(x),} \tag{8.11}$$

and for all complex signals x_1 and x_2 in the domain of S,

$$\boxed{S(x_1 + x_2) = S(x_1) + S(x_2).} \tag{8.12}$$

According to the first of these—called the **homogeneity** property—if you scale the input, the output is scaled. According to the second one—called the **additivity** property—if the input is described as the sum of two component signals, then the output can be described as the sum of two signals that would result from the components alone. Recall that linear functions were introduced in section 5.2. A **linear system** is one whose function relating the output to the input is linear.

In pictures, the first property holds that the two systems in figure 8.3 are equivalent if S is linear. Here, the triangle represents the scaling operation. The second property holds that the two systems in figure 8.4 are equivalent.

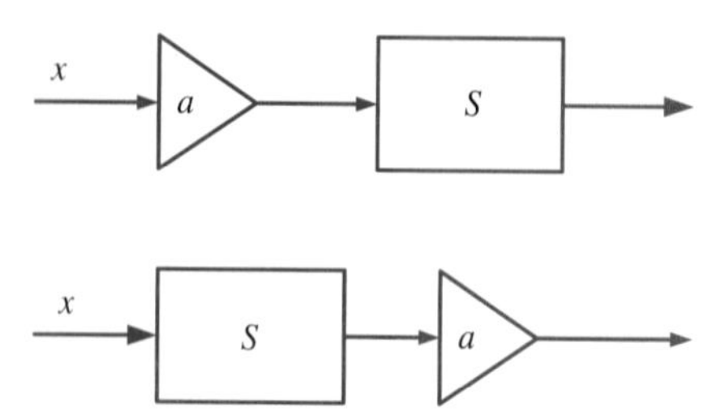

FIGURE 8.3: If S is linear, then these two systems are equivalent. The triangle represents a system that scales a signal by some complex constant a.

Example 8.5: Consider the same discrete-time system S of example 8.2, where input x produces output y such that

$$\forall n \in Integers, \quad y(n) = x(n) + 0.9x(n-1).$$

This system is linear. To show this, we must show that (8.11) and (8.12) hold. Suppose input $\hat{x} = ax$ produces output $\hat{y} = S(\hat{x})$. Then

$$\begin{aligned} \forall n, \quad \hat{y}(n) &= \hat{x}(n) + 0.9\hat{x}(n-1) \\ &= ax(n) + 0.9ax(n-1) \\ &= a(x(n) + 0.9x(n-1)) \\ &= ay(n). \end{aligned}$$

Thus, $S(ax) = aS(x)$, establishing (8.11).

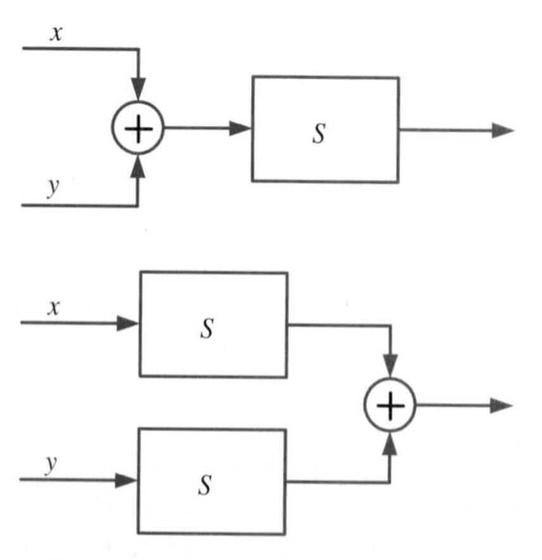

FIGURE 8.4: If S is linear, then these two systems are equivalent.

To check (8.12), suppose input x_1 produces output y_1 and x_2 produces y_2. Let $x = x_1 + x_2$ produce y. We must show that $y = y_1 + y_2$. We leave this as an exercise. ❐

Example 8.6: In the continuous-time system *DelayAndSquare* (*DS*) of example 8.3, an input x produces output y where

$$\forall\, t \in Reals, \quad y(t) = (x(t-1))^2.$$

This system is not linear. To show this, we must show that neither (8.11) nor (8.12) holds for *DS*. To show that (8.11) does not hold, consider a scaled input $\hat{x} = ax$, for some complex number a. Suppose that this input produces output $\hat{y} = S(\hat{x})$. Then

$$\forall\, t, \quad \hat{y}(t) = (\hat{x}(t-1))^2.$$

Because $\hat{x} = ax$,

$$\forall\, t, \quad \hat{y}(t) = (ax(t-1))^2 = a^2(x(t-1))^2 \neq ay(t) = a(x(t-1))^2.$$

In particular, take, for example, $a = 2$, $t = 0$, and x such that for all t, $x(t) = 1$. These values result in

$$\hat{y}(t) = 4 \neq ay(t) = 2.$$ ❐

Example 8.7: In the time-reversal system *RT* of example 8.4,

$$RT\colon [Reals \to Reals] \to [Reals \to Reals],$$

any input x produces output y related by

$$\forall\, t \in Reals, \quad y(t) = x(-t).$$

This system is linear. To show this, we must show that both (8.11) and (8.12) hold. We first show that (8.11) holds; that is, for all $a \in Reals, x \in [Reals \to Reals]$, and $t \in Reals$,

$$RT(ax)(t) = (aRT(x))(t).$$

But this is certainly true, because both the left and right sides are equal to $ax(-t)$. A similar argument is used to show that (8.12) holds. ❐

Linearity is a mathematical idealization. No electronic system is linear in the strict sense. A system is designed to work with a range of input signals, and arbitrary scaling of the input does not translate into arbitrary scaling of the output. If you provide an input to an audio amplifier that is a higher voltage than it is designed for, then it is not likely to merely produce louder sounds:

Its input circuits get overloaded, and signal distortion results. Nonetheless, as a mathematical idealization, linearity is extremely convenient. It allows us to decompose the inputs to a system and study the effect of the system on the individual components.

8.1.3 *Linearity and time invariance*

For time-domain systems, time invariance is a useful (if fictional) property. For complex or real systems, linearity is a useful (if fictional) property. For complex or real time-domain systems, the combination of these properties is extremely useful. LTI systems turn out to have particularly simple behavior with sinusoidal inputs:

> Given a sinusoid at the input, the output of an LTI system is a sinusoid with the same frequency but possibly with different phase and amplitude.

It then follows that

> Given an input that is described as a sum of sinusoids of certain frequencies, the output can be described as a sum of sinusoids with the same frequencies but with possible phase and amplitude changes at each frequency.

To show in a straightforward way that LTI systems have these properties, start by considering **complex exponentials** (for a review of complex numbers, see appendix B). A continuous-time complex exponential is a signal $x \in [Reals \to Complex]$ where

$$\boxed{\forall\, t \in Reals, \quad x(t) = e^{i\omega t} = \cos(\omega t) + i\sin(\omega t).}$$

Complex exponential functions have an interesting property that will prove useful to us. Specifically,

$$\forall\, t \in Reals \text{ and } \tau \in Reals, \quad x(t-\tau) = e^{i\omega(t-\tau)} = e^{-i\omega\tau}e^{i\omega t}.$$

This follows from the multiplication property of exponentials,

$$e^{b+c} = e^b e^c.$$

Because $D_\tau(x)(t) = x(t - \tau)$, we find that for the complex exponential x,

$$D_\tau(x) = ax,$$

where

$$a = e^{-i\omega\tau}. \tag{8.13}$$

In words, a delayed complex exponential is a scaled complex exponential, where the scaling constant is the complex number $a = e^{-i\omega\tau}$.

We now show that if the input to a continuous-time LTI system is $e^{i\omega t}$, then the output is $H(\omega)e^{i\omega t}$, where $H(\omega)$ is a constant (not a function of time) that depends on the frequency ω of the complex exponential. In other words, the output is only a scaled version of the input.

When the output of a system is only a scaled version of the input, the input is called an **eigenfunction**; this term is derived from the German word for "same." The output is (almost) the same as the input.

Complex exponentials are eigenfunctions of LTI systems, as we now show. This is the most important reason for the focus on complex exponentials in the study of signals and systems. This single property underlies much of the discipline of signal processing and is used extensively in circuit analysis, communication systems, and control systems.

Given an LTI system $S\colon [Reals \to Complex] \to [Reals \to Complex]$, let x be an input signal where

$$\forall\, t \in Reals, \quad x(t) = e^{i\omega t}.$$

Recall that S is time invariant if for all $\tau \in Reals$,

$$S \circ D_\tau = D_\tau \circ S;$$

thus,

$$S(D_\tau(x)) = D_\tau(S(x)).$$

From (8.13),

$$S(D_\tau(x)) = S(ax),$$

where $a = e^{-i\omega\tau}$, and from the linearity property,

$$S(ax) = aS(x);$$

therefore,

$$aS(x) = D_\tau(S(x)). \tag{8.14}$$

Let $y = S(x)$ be the corresponding output signal. Substituting into (8.14), we get

$$ay = D_\tau(y).$$

In other words,

$$\forall\, t, \tau \in Reals, \quad e^{-i\omega\tau} y(t) = y(t - \tau).$$

In particular, this is true for $t = 0$; thus, letting $t = 0$, we get

$$\forall\, \tau \in Reals, \quad y(-\tau) = e^{-i\omega\tau} y(0).$$

Changing variables, letting $t = -\tau$, we note that this implies that

$$\forall\, t \in Reals, \quad y(t) = e^{i\omega t} y(0).$$

Recall that $y(0)$ is the output evaluated at 0 when the input is $e^{i\omega t}$. It is a constant, in that it does not depend on t, which establishes that the output is a complex exponential, just like the input, except that it is scaled by $y(0)$. However, $y(0)$ does, in general, depend on ω, and so we define the function H: *Reals* → *Complex* by

$$\boxed{\forall\, \omega \in Reals, \quad H(\omega) = y(0) = (S(x))(0),}$$

where

$$\forall\, t \in Reals, \quad x(t) = e^{i\omega t}; \tag{8.15}$$

that is, $H(\omega)$ is the output at time zero when the input is a complex exponential with frequency ω.

Using this notation, we write the output y as

$$\boxed{\forall\, t \in Reals, \quad y(t) = H(\omega) e^{i\omega t}}$$

when the input is $e^{i\omega t}$. Note that $H(\omega)$ is a function of $\omega \in Reals$, the frequency of the complex exponential input.

The function H:*Reals* → *Complex* is called the **frequency response**. It defines the response of the LTI system to a complex exponential input at any given frequency. It gives the scaling factor that the system imposes on that complex exponential.

For discrete-time systems, the situation is similar. By reasoning identical to that previously, for an LTI system, if the input is a **discrete complex exponential**,

$$\boxed{\forall\, n \in Integers, \quad x(n) = e^{i\omega n},}$$

then the output is the same complex exponential scaled by a constant (a complex number that does not depend on time):

$$\boxed{\forall\, n \in Integers, \quad y(n) = H(\omega) e^{i\omega n}.}$$

H is once again called the frequency response, and because it is a function of ω and is possibly complex valued, it has the form H: *Reals* $\to$ *Complex*.

There is one key difference, however, between discrete-time systems and continuous-time systems. Because n is an integer, notice that

$$e^{i\omega n} = e^{i(\omega+2\pi)n} = e^{i(\omega+4\pi)n}$$

and so on; that is, a discrete complex exponential with frequency ω is identical to a discrete complex exponential with frequency $\omega + 2K\pi$, for any integer K. The frequency response must therefore be identical at these frequencies, because the inputs are identical. In other words,

$$\forall\, \omega \in Reals, \quad H(\omega) = H(\omega + 2K\pi)$$

for any integer K; that is, a discrete-time frequency response is periodic with period 2π.

8.2 Finding and using the frequency response

We have seen that if the input to an LTI system is a complex exponential signal $x \in [Reals \to Complex]$ where

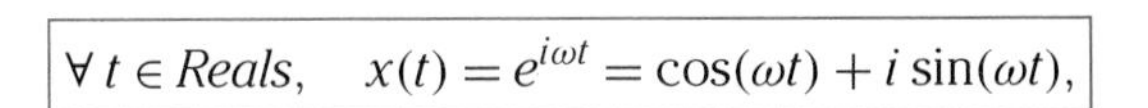

$$\forall\, t \in Reals, \quad x(t) = e^{i\omega t} = \cos(\omega t) + i \sin(\omega t),$$

then the output is

$$\forall\, t \in Reals, \quad y(t) = H(\omega) e^{i\omega t}, \tag{8.16}$$

where $H(\omega)$ is a possibly complex-valued number that is a property of the system. $H(\omega)$ is called the frequency response at frequency ω.

Example 8.8: Consider a delay system $S = D_T$, for some $T \in Reals$. It is an LTI system that is easy to verify by checking that (8.2), (8.11), and (8.12) are satisfied. Suppose the input to the delay system is the complex exponential x given by

$$\forall\, t \in Reals, \quad x(t) = e^{i\omega t}.$$

Then the output y satisfies

$$\forall\, t \in Reals, \quad y(t) = e^{i\omega(t-T)} = e^{-i\omega T} e^{i\omega t}.$$

Comparing this with (8.16), we see that the frequency response of the delay is

$$\forall \omega \in Reals, \quad H(\omega) = e^{-i\omega T}. \square$$

Example 8.9: Consider a discrete-time M-sample delay system $S = D_M$. If $y = S(x)$, then y is given by

$$\forall n \in Integers, \quad y(n) = x(n - M). \tag{8.17}$$

This is an LTI system, which is easy to verify. We could find the frequency response exactly the same way as in the previous example, but instead we use a slightly different method. Because the system is LTI, we know that if the input is x such that for all $n \in Integers$, $x(n) = e^{i\omega n}$, then the output is $H(\omega)e^{i\omega n}$, where H is the frequency response. By plugging this input and output into (8.17), we get

$$H(\omega)e^{i\omega n} = e^{i\omega(n-M)} = e^{i\omega n}e^{-i\omega M}.$$

Dividing both sides by $e^{i\omega n}$, we get

$$\forall \omega \in Reals, \quad H(\omega) = e^{-i\omega M}. \square$$

The techniques in the preceding examples can be used to find the frequency response of more complicated systems. Simply replace the input x in a difference equation like (8.17) with $e^{i\omega n}$, replace the output y with $H(\omega)e^{i\omega n}$, and then solve for $H(\omega)$.

Example 8.10: Consider a discrete-time, two-point moving average, given by the difference equation

$$\forall n \in Integers, \quad y(n) = (x(n) + x(n - 1))/2,$$

where x is the input and y is the output. When the input for all n is $e^{i\omega n}$, this becomes

$$H(\omega)e^{i\omega n} = (e^{i\omega n} + e^{i\omega(n-1)})/2.$$

Solving for $H(\omega)$, we find that the frequency response is

$$\forall \omega \in Reals, \quad H(\omega) = (1 + e^{-i\omega})/2. \square$$

A similar approach can be used to find the frequency response of a continuous-time system described by a differential equation.

Example 8.11: Consider a continuous-time system with input x and output y related by the differential equation

$$\forall\, t \in Reals, \quad RC\frac{dy}{dt}(t) + y(t) = x(t), \tag{8.18}$$

where R and C are real-valued constants. This differential equation describes the RC circuit of figure 8.5, which consists of an R-ohm resistance in series with a C-farad capacitor. The circuit has input voltage x, provided by a voltage source, shown as a circle on the left. The output is the voltage y across the capacitor. Kirchhoff's voltage law provides the differential equation. It is easy to verify that this differential equation describes an LTI system.

We can determine the frequency response of this system by assuming that the input x is given by, $\forall\, t \in Reals$, $x(t) = e^{i\omega t}$ and finding the output. Because the system is LTI, the output has the form $y(t) = H(\omega)e^{i\omega t}$. Plugging these values for x and y into (8.18) yields

$$RC(i\omega)H(\omega)e^{i\omega t} + H(\omega)e^{i\omega t} = e^{i\omega t}, \tag{8.19}$$

because

$$\frac{dy}{dt}(t) = i\omega H(\omega)e^{i\omega t}.$$

Dividing both sides of (8.19) by $e^{i\omega t}$ yields the frequency response of this circuit:

$$\forall\, \omega \in Reals, \quad H(\omega) = \frac{1}{1 + iRC\omega}. \quad \square$$

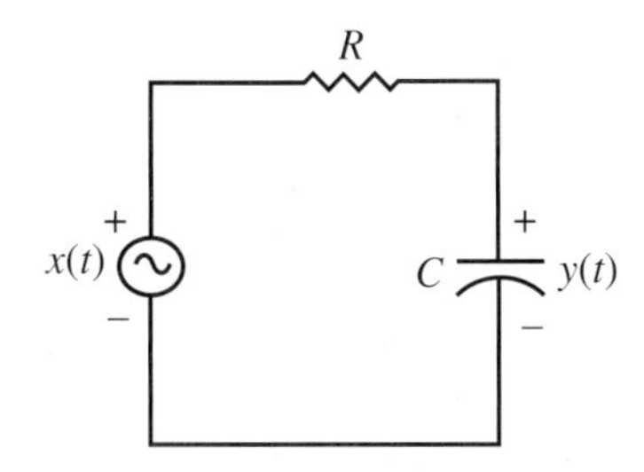

FIGURE 8.5: An RC circuit.

8.2.1 Linear difference and differential equations

The procedure of example 8.10 can be used to write down, after inspection, the frequency response of any high-order linear difference equation of the following form: $\forall\, n \in Integers$,

$$\begin{aligned} a_0 y(n) + a_1 y(n-1) + \cdots + a_N y(n-N) \\ = b_0 x(n) + b_1 x(n-1) + \cdots + b_M x(n-M). \end{aligned} \tag{8.20}$$

The coefficients of this difference equation, $a_0, \ldots, a_N$ and $b_0, \ldots, b_M$, are real constants. (They could also be complex.) This describes an LTI system, and a good way to recognize that a discrete-time system is LTI is to write it in this form. If the input for all n is $x(n) = e^{i\omega n}$, then the output for all n is $y(n) = H(\omega)e^{i\omega n}$. Plugging these values of input and output into (8.20) yields

$$\begin{aligned} a_0 H(\omega)e^{i\omega n} + a_1 H(\omega)e^{i\omega(n-1)} + \cdots + a_N H(\omega)e^{i\omega(n-N)} \\ = b_0 e^{i\omega n} + b_1 e^{i\omega(n-1)} + \cdots + b_M e^{i\omega(n-M)}. \end{aligned}$$

Recognizing that $e^{i\omega(n-m)} = e^{-im\omega}e^{i\omega n}$, we can divide both sides by $e^{i\omega n}$ and solve for $H(\omega)$ to get

$$\boxed{\forall\, \omega \in Reals, \quad H(\omega) = \frac{b_0 + b_1 e^{-i\omega} + \cdots + b_M e^{-iM\omega}}{a_0 + a_1 e^{-i\omega} + \cdots + a_N e^{-iN\omega}}.} \tag{8.21}$$

Observe that the frequency response (8.21) is a ratio of two polynomials in $e^{-i\omega}$.

Example 8.12: Consider an LTI system given by the difference equation

$$\forall\, n \in Integers, \quad y(n) - y(n-3) = x(n) + 2x(n-1) + x(n-2).$$

Its frequency response is

$$\forall\, \omega \in Reals, \quad H(\omega) = \frac{1 + 2e^{-i\omega} + e^{-i2\omega}}{1 - e^{-i3\omega}}. \quad \square$$

In a similar way, we can obtain the frequency response of any linear differential equation of the following form:

$$\forall t \in Reals, \quad a_N \frac{d^N y}{dt^N}(t) + \cdots + a_1 \frac{dy}{dt}(t) + a_0 y(t) = b_M \frac{d^M x}{dt^M}(t) + \cdots + b_1 \frac{dx}{dt}(t) + b_0 x(t), \tag{8.22}$$

where the coefficients $a_N, \ldots, a_0$ and $b_M, \ldots, b_0$, are real (or complex) constants. This describes an LTI system, and a good way to recognize that a continuous-time system is LTI is to write it in this form. If the input for all t is $x(t) = e^{i\omega t}$, the output for all t is $y(t) = H(\omega)e^{i\omega t}$. Plugging these values of input and output into (8.22), and recognizing that

$$\frac{d^k}{dt^k} e^{i\omega t} = (i\omega)^k e^{i\omega t},$$

yields

$$a_N (i\omega)^N H(\omega) e^{i\omega t} + \cdots + a_1 (i\omega) H(\omega) e^{i\omega t} + a_0 H(\omega) e^{i\omega t} = b_M (i\omega)^M e^{i\omega t} + \cdots + b_1 (i\omega) e^{i\omega t} + b_0 e^{i\omega t}.$$

We can divide both sides by $e^{i\omega t}$ to get

$$\boxed{\forall \omega \in Reals, \quad H(\omega) = \frac{b_M (i\omega)^M + \cdots + b_1 (i\omega) + b_0}{a_N (i\omega)^N + \cdots + a_1 (i\omega) + a_0}.} \tag{8.23}$$

Observe that the frequency response (8.23) is a ratio of two polynomials in $i\omega$.

Example 8.13: Consider the continuous-time system given by the differential equation

$$\forall t \in Reals, \quad \frac{d^2 y}{dt^2}(t) - 3\frac{dy}{dt}(t) + 2y(t) = \frac{dx}{dt} + x(t),$$

where x is the input and y is the output. It has frequency response

$$\forall \omega \in Reals, \quad H(\omega) = \frac{i\omega + 1}{(i\omega)^2 - 3i\omega + 2}. \square$$

BASICS

Sinusoids in terms of complex exponentials

Euler's formula states that

$$e^{i\theta} = \cos(\theta) + i\sin(\theta).$$

The complex conjugate is

$$e^{-i\theta} = \cos(\theta) - i\sin(\theta).$$

Summing these yields

$$e^{i\theta} + e^{-i\theta} = 2\cos(\theta)$$

or

$$\boxed{\cos(\theta) = (e^{i\theta} + e^{-i\theta})/2.}$$

Thus, for example,

$$\cos(\omega t) = (e^{i\omega t} + e^{-i\omega t})/2.$$

Similarly,

$$\boxed{\sin(\theta) = -i(e^{i\theta} - e^{-i\theta})/2.}$$

In appendix B we show that many useful trigonometric identities are easily derived from these simple relations.

TIPS AND TRICKS

Phasors

Consider a general continuous-time **sinusoidal signal**,

$$\forall t \in Reals, \quad x(t) = A\cos(\omega t + \phi).$$

Here A is the **amplitude**, ϕ is the **phase**, and ω is the **frequency** of the sine wave. (We call this a sine wave, even though we are using cosine to describe it.) The units of ϕ are radians. The units of ω are radians per second, assuming t is in seconds. This can be written

$$x(t) = Re\{Ae^{i(\omega t+\phi)}\} = Re\{Ae^{i\phi}e^{i\omega t}\} = Re\{Xe^{i\omega t}\}$$

TIPS AND TRICKS

where $X = Ae^{i\phi}$ is called the **complex amplitude** or **phasor**. The representation

$$\boxed{x(t) = Re\{Xe^{i\omega t}\}} \tag{8.24}$$

is called the phasor representation of x. It can be convenient.

Example 8.14: Consider summing two sinusoids with the same frequency:

$$x(t) = A_1 \cos(\omega t + \phi_1) + A_2 \cos(\omega t + \phi_2).$$

This is particularly easy to do with the use of phasors, because

$$\begin{aligned} x(t) &= Re\{(X_1 + X_2)e^{i\omega t}\} \\ &= |X_1 + X_2| \cos(\omega t + \angle(X_1 + X_2)), \end{aligned}$$

where $X_1 = A_1 e^{i\phi_1}$ and $X_2 = A_2 e^{i\phi_2}$. Thus, addition of the sinusoids reduces to addition of two complex numbers. ❒

The exponential $Xe^{i\omega t}$ in (8.24) is complex valued. In a two-dimensional plane, as represented in figure 8.6, it rotates in a counterclockwise direction as t increases. The

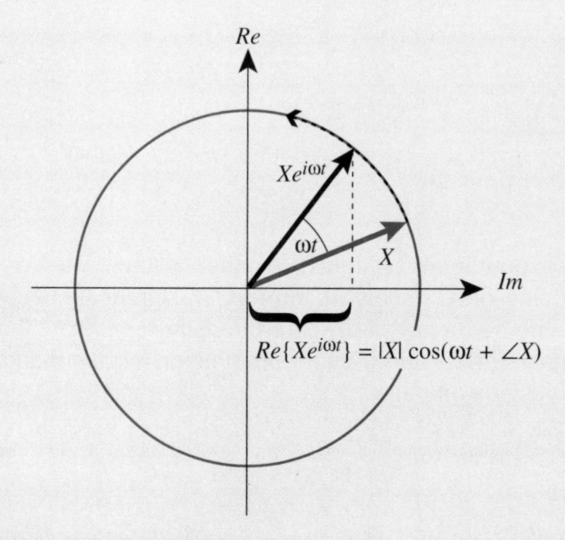

FIGURE 8.6: Phasor representation of a sinusoid.

continued on next page

TIPS AND TRICKS

frequency of rotation is ω radians per second. At time 0 it is X, shown in gray. The real-valued sine wave $x(t)$ is the projection of $Xe^{i\omega t}$ on the real axis: Namely,

$$Re\{Xe^{i\omega t}\} = |X| \cos(\omega t + \angle X).$$

The sum of two sinusoids with the same frequency is similarly depicted in figure 8.7. The two phasors, X_1 and X_2, are put head to tail and then rotated together. A similar method can be used to add sinusoids of different frequencies, but then the two vectors rotate at different rates.

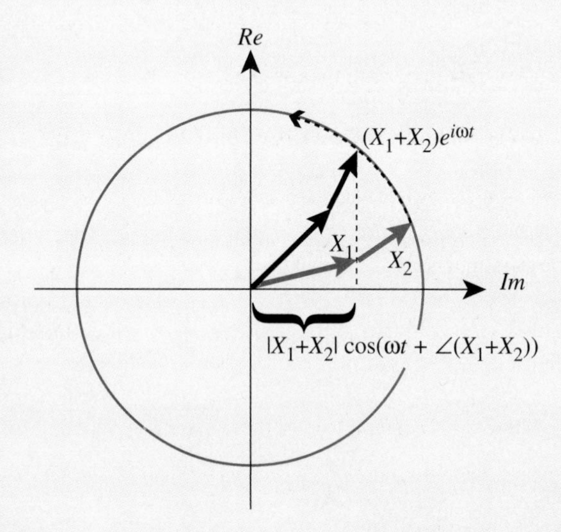

FIGURE 8.7: Phasor representation of the sum of two sinusoids with the same frequency.

Complex exponentials as inputs are rather abstract. We have seen that with audio signals, sinusoidal signals are intrinsically significant because the human ear interprets the frequency of the sinusoid as its tone. Note that a real-valued sinusoidal signal can be given as a combination of exponential signals (see Basics: Sinusoids box on page 294),

$$\cos(\omega t) = (e^{i\omega t} + e^{-i\omega t})/2.$$

Thus, if this is the input to an LTI system with frequency response H, then the output is

$$y(t) = (H(\omega)e^{i\omega t} + H(-\omega)e^{-i\omega t})/2. \tag{8.25}$$

Many (or most) LTI systems are not capable of producing complex-valued outputs when the input is real, and so for such systems, this $y(t)$ must be real. This implies that

$$H(\omega) = H^*(-\omega). \tag{8.26}$$

To see why this is so, note that if $y(t)$ is real, then so is $H(\omega)e^{i\omega t} + H(-\omega)e^{-i\omega t}$. But for this to be real, the imaginary parts of the two terms must cancel:

$$Im\{H(\omega)e^{i\omega t}\} = -Im\{H(-\omega)e^{-i\omega t}\}. \tag{8.27}$$

Note that

$$Im\{H(\omega)e^{i\omega t}\} = Re\{H(\omega)\}\sin(\omega t) + Im\{H(\omega)\}\cos(\omega t).$$

Using a similar fact for the right side of (8.27), we get

$$\begin{aligned} Re\{H(\omega)\}\sin(\omega t) + Im\{H(\omega)\}\cos(\omega t) = & -Re\{H(-\omega)\}\sin(-\omega t) \\ & - Im\{H(-\omega)\}\cos(-\omega t). \end{aligned}$$

If we evaluate this at $t = 0$, we get

$$Im\{H(\omega)\} = -Im\{H(-\omega)\},$$

and if we evaluate it at $t = \pi/(2\omega)$, we get

$$Re\{H(\omega)\} = Re\{H(-\omega)\};$$

these two equations together imply (8.26).

Property (8.26) is called **conjugate symmetry**. The frequency response of a real system (one whose input and output signals are real valued) is **conjugate symmetric**. Thus, combining (8.25) and (8.26) when the input is $x(t) = \cos(\omega t)$ yields the output

$$\boxed{\forall t \in Reals, \quad y(t) = Re\{H(\omega)e^{i\omega t}\}.}$$

If we write $H(\omega)$ in polar form,

$$H(\omega) = |H(\omega)|e^{i\angle H(\omega)},$$

then when the input is $\cos(\omega t)$, the output is

$$\boxed{\forall t \in Reals, \quad y(t) = |H(\omega)|\cos(\omega t + \angle H(\omega)).}$$

Thus, $H(\omega)$ consists of the **gain** $|H(\omega)|$ and the **phase shift** $\angle H(\omega)$ that a sinusoidal input with frequency ω experiences. $|H(\omega)|$ is called the **magnitude response** of the system, and $\angle H(\omega)$ is called the **phase response**.

Example 8.15: The delay system $S = D_T$ of example 8.8 has frequency response

$$H(\omega) = e^{-i\omega T}.$$

The magnitude response is

$$|H(\omega)| = 1.$$

Thus, any cosine input into a delay obviously yields a cosine output with the same amplitude. A filter with a constant unity magnitude response is called an **allpass filter**, because it passes all frequencies equally. A delay is a particularly simple form of an allpass filter.

The phase response is

$$\angle H(\omega) = -\omega T.$$

Thus, any cosine input with frequency ω yields a cosine output with the same frequency but with phase shift $-\omega T$. ❒

Example 8.16: The M-sample discrete-time delay of example 8.9 is also an allpass filter. Its phase response is

$$\angle H(\omega) = -\omega M. \; ❒$$

Example 8.17: The magnitude response of the two-point moving average considered in example 8.10 is

$$|H(\omega)| = |(1 + e^{-i\omega})/2|.$$

We can plot this by using the following MATLAB®* code, which (after adjusting the labels) results in figure 8.8:

```
omega = [0:pi/250:pi];
H = (1 + exp(-i*omega))/2;
plot(omega, abs(H)).
```

Notice that at frequency zero (a cosine with zero frequency has constant value), the magnitude response is 1; that is, a constant signal gets through

* MATLAB is a registered trademark of The MathWorks, Inc.

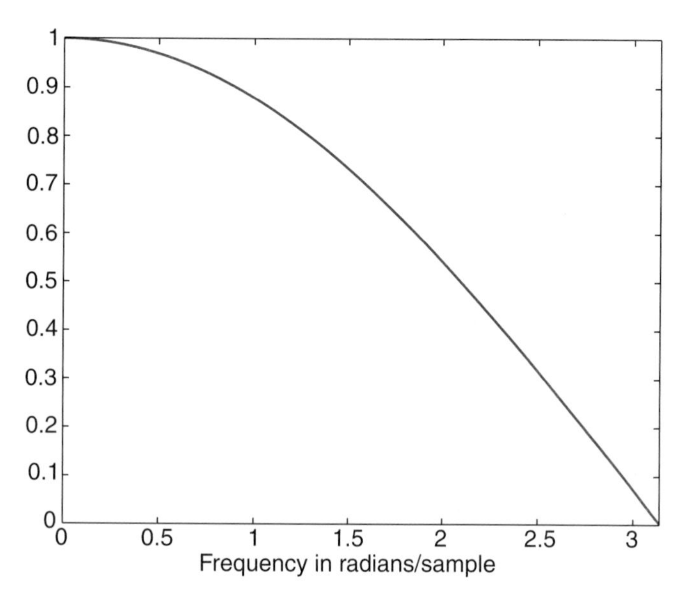

FIGURE 8.8: The magnitude response of a two-point moving average.

the filter without any reduction in amplitude. This is expected, because the average of two neighboring samples of a constant signal is simply the value of the constant signal. Notice that the magnitude response decreases as the frequency increases. Thus, the amplitudes of higher frequency signals are reduced more by the filter than are those of lower frequency signals. Such a filter is called a **lowpass** filter because it passes lower frequencies better than higher frequencies. ❐

Example 8.18: The RC circuit of example 8.11 has frequency response given by

$$\forall\, \omega \in Reals, \quad H(\omega) = \frac{1}{1 + iRC\omega}.$$

We can express this in polar form, $H(\omega) = |H(\omega)| e^{i\angle H(\omega)}$, to get the magnitude and phase responses:

$$|H(\omega)| = \frac{1}{\sqrt{1 + (RC\omega)^2}}, \qquad \angle H(\omega) = -\tan^{-1}(RC\omega).$$

If for all t, $x(t) = \cos(\omega t)$, then $y(t) = |H(\omega)| \cos(\omega t - \angle H(\omega))$. Because $|H(\omega)| \to 0$ as $\omega \to \infty$, this RC circuit is a lowpass filter, albeit not a very good one. ❐

Often, we are given a frequency response rather than some other description of an LTI system such as a difference equation. The frequency response, in fact, tells us everything we need to know about the system. The next example begins the exploration of that idea.

Example 8.19: Suppose that the frequency response H of a discrete-time LTI system *Filter* is given by

$$\forall \omega \in Reals, \quad H(\omega) = \cos(2\omega),$$

where ω is in units of radians per sample. Suppose that the input signal x: *Integers* $\rightarrow$ *Reals* is such that for all $n \in Integers$,

$$x(n) = \begin{cases} +1 & \text{if } n \text{ even} \\ -1 & \text{if } n \text{ odd.} \end{cases}$$

We can determine the output. All we have to do is notice that the input can be written as

$$x(n) = \cos(\pi n).$$

Thus, the input is a cosine with a frequency of π radians/sample. Hence, the output is

$$y(n) = |H(\pi)| \cos(\pi n + \angle H(\pi)) = \cos(\pi n) = x(n).$$

This input is passed unchanged by the system.

Suppose instead that the input is given by

$$x(n) = 5.$$

Once again, the input is a cosine but this time with zero frequency:

$$x(n) = 5\cos(0n).$$

Hence, the output is

$$y(n) = |H(0)| 5\cos(0n + \angle H(0)) = 5 = x(n).$$

This input is also passed unchanged by the system.

Suppose instead that the input is

$$x(n) = \cos(\pi n/2).$$

This input is given explicitly as a cosine, which makes our task easier. The output is

$$\begin{aligned} y(n) &= |H(\pi/2)| \cos(\pi n/2 + \angle H(\pi/2)) \\ &= \cos(\pi n/2 + \pi) \\ &= -\cos(\pi n/2) \\ &= -x(n). \end{aligned}$$

This input is inverted by the system.

Finally, suppose that the input is

$$x(n) = \cos(\pi n/4).$$

The output is

$$y(n) = |H(\pi/4)| \cos(\pi n/4 + \angle H(\pi/4)) = 0.$$

This input is filtered out by the system. ❐

8.2.2 *The Fourier series with complex exponentials*

The Fourier series for a continuous-time, periodic signal $x: Reals \to Reals$ with period $p = 2\pi/\omega_0$ can be written (see (7.4)) as

$$\forall t \in Reals, \quad x(t) = A_0 + \sum_{k=1}^{\infty} A_k \cos(k\omega_0 t + \phi_k).$$

For reasons that we can now understand, the Fourier series is usually written in terms of complex exponentials rather than cosines. Because complex exponentials are eigenfunctions of LTI systems, this form of the Fourier series decomposes a signal into components that, when processed by the system, are only scaled.

Each term of the Fourier series expansion has the form

$$A_k \cos(k\omega_0 t + \phi_k),$$

which we can write (see Basics: Sinusoids box on page 294)

$$A_k \cos(k\omega_0 t + \phi_k) = A_k (e^{i(k\omega_0 t + \phi_k)} + e^{-i(k\omega_0 t + \phi_k)})/2.$$

Therefore, the Fourier series can also be written

$$x(t) = A_0 + \sum_{k=1}^{\infty} \frac{A_k}{2}(e^{i(k\omega_0 t+\phi_k)} + e^{-i(k\omega_0 t+\phi_k)}).$$

Observe that

$$e^{i(k\omega_0 t+\phi_k)} = e^{ik\omega_0 t}e^{i\phi_k},$$

and let

$$X_k = \begin{cases} A_0 & \text{if } k = 0 \\ 0.5A_k e^{i\phi_k} & \text{if } k > 0 \\ 0.5A_{-k} e^{-i\phi_{-k}} & \text{if } k < 0. \end{cases} \tag{8.28}$$

Then the Fourier series becomes

$$\boxed{\forall\, t \in Reals, \quad x(t) = \sum_{k=-\infty}^{\infty} X_k e^{ik\omega_0 t}.} \tag{8.29}$$

This is the form in which the Fourier series usually appears. Notice from (8.28) that the Fourier series coefficients are **conjugate symmetric**,

$$X_k = X^*_{-k}.$$

Of course, since (8.29) is an infinite sum, we need to worry about convergence (see box on page 262).

The **discrete-time Fourier series** (**DFS**) can be similarly written. If x: *Integers* $\to$ *Reals* is a periodic signal with period $p = 2\pi/\omega_0$, then we can write

$$\boxed{\forall\, n \in Integers, \quad x(n) = \sum_{k=0}^{p-1} X_k e^{ik\omega_0 n}} \tag{8.30}$$

for suitably defined coefficients X_k. Relating the coefficients X_k to the coefficients A_k and ϕ_k is a bit more difficult in the discrete-time case than in the continuous-time case (see the following box).

There are two differences between (8.29) and (8.30). First, the sum in the discrete-time case is finite, which makes calculations manageable by computer. Second, it is exact for any periodic waveform. There are no mathematically tricky cases, and no approximation is needed.

PROBING FURTHER

Relating DFS coefficients

We have two DFS expansions (see (8.30) and (7.8)):

$$x(n) = \sum_{k=0}^{p-1} X_k e^{ik\omega_0 n} \tag{8.31}$$

and

$$x(n) = A_0 + \sum_{k=1}^{K} A_k \cos(k\omega_0 n + \phi_k), \quad K = \begin{cases} (p-1)/2 & \text{if } p \text{ is odd} \\ p/2 & \text{if } p \text{ is even.} \end{cases}$$

There is a relationship among the coefficients A_k, ϕ_k, and X_k, but the relationship is more complicated than in the continuous-time case, given by (8.28). To develop that relationship, begin with the second of these expansions and write

$$x(n) = A_0 + \sum_{k=1}^{K} \frac{A_k}{2}(e^{i(k\omega_0 n+\phi_k)} + e^{-i(k\omega_0 n+\phi_k)}).$$

Note that $\omega_0 = 2\pi/p$; therefore, for all integers n, $e^{i\omega_0 pn} = 1$, and so

$$e^{-i(k\omega_0 n+\phi_k)} = e^{-i(k\omega_0 n+\phi_k)} e^{i\omega_0 pn} = e^{i(\omega_0(p-k)n-\phi_k)}.$$

Thus,

$$\begin{aligned} x(n) &= A_0 + \sum_{k=1}^{K} \frac{A_k}{2} e^{i\phi_k} e^{ik\omega_0 n} + \sum_{k=1}^{K} \frac{A_k}{2} e^{-i\phi_k} e^{i\omega_0(p-k)n} \\ &= A_0 + \sum_{k=1}^{K} \frac{A_k}{2} e^{i\phi_k} e^{ik\omega_0 n} + \sum_{m=K}^{p-1} \frac{A_{p-m}}{2} e^{-i\phi_{p-m}} e^{i\omega_0 mn}, \end{aligned}$$

by change of variables. Comparing this with (8.31), we find

$$X_k = \begin{cases} A_0 & \text{if } k = 0 \\ A_k e^{i\phi_k}/2 & \text{if } k \in \{1, \ldots, K-1\} \\ A_k e^{i\phi_k}/2 + A_k e^{-i\phi_k}/2 = A_k \cos(\phi_k) & \text{if } k = K \\ A_{p-k} e^{-i\phi_{p-k}}/2 & \text{if } k \in \{K+1, \ldots, p-1\}. \end{cases}$$

This relationship is more complicated than (8.28). Fortunately, it is rare that we need to use both forms of the DFS, and so we can usually pick just one of the two forms and work only with its set of coefficients.

8.2.3 Examples

The Fourier series coefficients A_k of a square wave are shown in figure 7.6(b). The magnitudes of the corresponding coefficients X_k for the Fourier series expansion of (8.29) are shown in figure 8.9(b). Because each cosine is composed of two complex exponentials, there are twice as many coefficients.

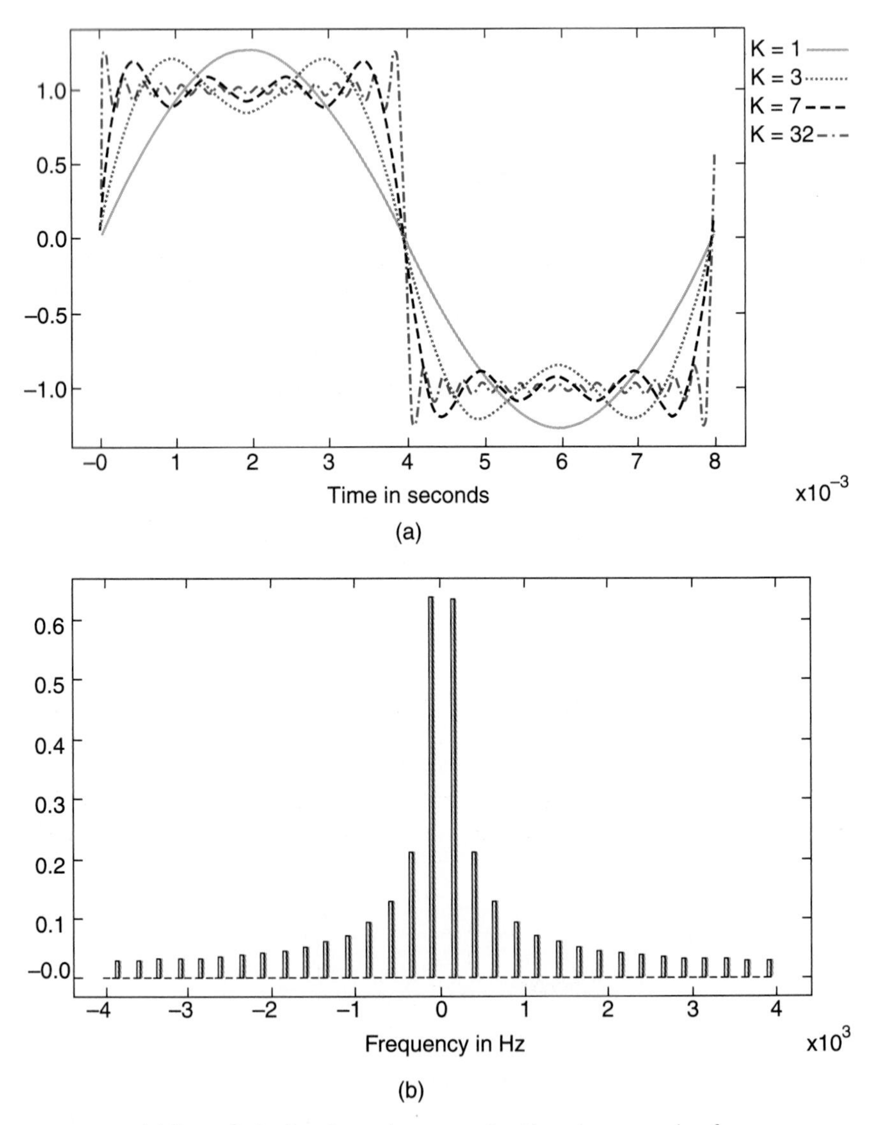

FIGURE 8.9: (a) Some finite Fourier series approximations to one cycle of a square wave. The number of Fourier series terms that are included in the approximation is $2K + 1$, and so K is the magnitude of the largest index of the terms. (b) The magnitude of the complex Fourier series coefficients, shown as a function of frequency.

Notice the symmetry in figure 8.9. There are frequency components shown at both positive and negative frequencies. Notice also that the amplitude of the components is half that in figure 7.6, $|X_k| = |A_k|/2$. This is because there are now two components, one at negative frequencies and one at positive frequencies, that contribute.

8.3 Determining the Fourier series coefficients

We have seen in chapter 7 that determining the Fourier series coefficients by directly attempting to determine the amplitude of individual frequency components can be difficult, even when the individual frequency components are known. Usually, however, they are not known. A general formula for computing the coefficients for a continuous-time periodic signal is

$$\forall m \in Integers, \quad X_m = \frac{1}{p} \int_0^p x(t) e^{-im\omega_0 t} dt. \tag{8.32}$$

The mth Fourier series coefficient is obtained by multiplying x by a complex exponential with frequency $-m\omega_0$ and averaging. The validity of this equation is demonstrated in the Probing further box on page 306.

The discrete-time case is somewhat simpler. For a discrete-time periodic signal x with period $p \in Integers$, its Fourier series coefficients are represented by

$$\forall k \in Integers, \quad X_k = \frac{1}{p} \sum_{m=0}^{p-1} x(m) e^{-imk\omega_0}. \tag{8.33}$$

This can be shown by manipulations similar to those in the Probing further box on the next page. The practical importance of computing is much greater than that of the Fourier series for continuous-time signals. Because this sum is finite, the DFS coefficients can be easily computed precisely on a computer.

PROBING FURTHER

Formula for Fourier series coefficients

To see that (8.32) is valid, try substituting for $x(t)$ its Fourier series expansion,

$$x(t) = \sum_{k=-\infty}^{\infty} X_k e^{ik\omega_0 t}$$

to get

$$X_m = \frac{1}{p} \int_0^p \sum_{k=-\infty}^{\infty} X_k e^{ik\omega_0 t} e^{-im\omega_0 t} dt.$$

Exchange the integral and summation (assuming this is valid; see the Probing further box on page 307) to get

$$X_m = \frac{1}{p} \sum_{k=-\infty}^{\infty} X_k \int_0^p e^{ik\omega_0 t} e^{-im\omega_0 t} dt.$$

The exponentials can be combined to get

$$X_m = \frac{1}{p} \sum_{k=-\infty}^{\infty} X_k \int_0^p e^{i(k-m)\omega_0 t} dt.$$

In the summation, in which k varies over all integers, there is exactly one term of the summation for which $k = m$. In that term, the integral evaluates to p. For the rest of the terms, $k \neq m$. Separating these two situations, we can write

$$X_m = X_m + \frac{1}{p} \sum_{k=-\infty, k\neq m}^{\infty} X_k \int_0^p e^{i(k-m)\omega_0 t} dt,$$

where the first term, X_m, is the value of the term in the summation in which $k = m$. For each remaining term of the summation, the integral evaluates to zero, thus establishing our result. To show that the integral evaluates to zero, let $n = k - m$, and note that $n \neq 0$. Then

$$\int_0^p e^{in\omega_0 t} dt = \int_0^p \cos(n\omega_0 t) dt + i \int_0^p \sin(n\omega_0 t) dt.$$

Because $\omega_0 = 2\pi/p$, these two integrals exactly span one or more complete cycles of the cosine or sine and hence integrate to zero.

PROBING FURTHER

Exchanging integrals and summations

The demonstration of the validity of the formula for the Fourier series coefficients in the preceding Probing further box relies on the ability to exchange an integral and an infinite summation. The infinite summation can be given as a limit of a sequence of functions

$$x_N(t) = \sum_{k=-N}^{N} X_k e^{ik\omega_0 t}.$$

Thus, we wish to exchange the integral and limit in

$$X_m = \frac{1}{p} \int_0^p (\lim_{N\to\infty} x_N(t))dt.$$

A sufficient condition for being able to perform the exchange is that the limit converges uniformly in the interval [0, p]. A sufficient condition for uniform convergence is that x is continuous and that its first derivative is piecewise continuous.

See R. G. Bartle, *The Elements of Real Analysis*, Second Edition; New York, John Wiley & Sons, 1976, p. 241.

8.3.1 *Negative frequencies*

The Fourier series expansion for a periodic signal $x(t)$ is

$$x(t) = \sum_{k=-\infty}^{\infty} X_k e^{ik\omega_0 t}.$$

This includes Fourier series coefficients for the constant term (when $k = 0$, $e^{ik\omega_0 t} = 1$) and for the fundamental and harmonics ($k \geq 1$). But it also includes terms that seem to correspond to negative frequencies. When $k \leq -1$, the frequency $k\omega_0$ is negative. These negative frequencies balance the positive ones so that the resulting sum is real valued.

8.4 Frequency response and the Fourier series

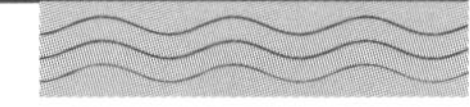

Recall that if the input to an LTI system S is a complex exponential signal $x \in [Reals \to Complex]$ where

$$\forall\, t \in Reals, \quad x(t) = e^{i\omega_0 t} = \cos(\omega_0 t) + i \sin(\omega_0 t),$$

then the output for all t is

$$y(t) = H(\omega_0) e^{i\omega_0 t},$$

where the complex number $H(\omega_0)$ is the frequency response of the system at the fundamental frequency ω_0 of the periodic input. It is equal to the output at time zero, $y(0)$, when the input is $e^{i\omega_0 t}$. H itself is a function $H: Reals \to Complex$ that in principle can be evaluated for any frequency $\omega \in Reals$, including negative frequencies.

Recall further that if an input x to the system S is a periodic signal with period p, then it can be represented as a Fourier series,

$$\forall t \in Reals, \quad x(t) = \sum_{k=-\infty}^{\infty} X_k e^{ik\omega_0 t},$$

where $\omega_0 = 2\pi/p$. According to the linearity and time invariance properties of S, the output $y = S(x)$ for this periodic input x is represented by

$$\forall t \in Reals, \quad y(t) = \sum_{k=-\infty}^{\infty} H(k\omega_0) X_k e^{ik\omega_0 t}.$$

Thus, according to linearity, if the input is decomposed into a sum of components, then the output can be decomposed into a sum of components in which each component is the response of the system to a single input component. Linearity together with time invariance tells us that each component, which is a complex exponential, is simply scaled. Thus, the output is given by a Fourier series with coefficients $X_k H(k\omega_0)$.

This major result is summarized as follows:

For an LTI system, if the input is given by a sum of complex exponentials, then the output can be given by a sum of the same complex exponentials, each one scaled by the frequency response evaluated at the corresponding frequency.

Among other things, this result tells us that

- All frequency components in the output are in the input. The output consists of the same frequency components as the input but with each component individually scaled.
- LTI systems can be used to enhance or suppress certain frequency components. Such operations are called **filtering**.
- The frequency response function characterizes which frequencies are enhanced or suppressed and also which phase shifts might be imposed on individual components by the system.

Chapter 9 contains many examples of filtering.

8.5 Frequency response of composite systems

In section 2.3.4 we studied several ways of composing systems (with the use of block diagrams) to obtain more complex, composite systems. This section shows that when each block is an LTI system, the resulting composite system is also LTI. Moreover, we can easily obtain the frequency response of the composite system from the frequency response of each block. This provides a useful way to construct interesting and complex systems by assembling simpler components. This tool works equally well with discrete and continuous systems.

8.5.1 *Cascade connection*

Consider the composite system S obtained by the cascade connection of systems S_1 and S_2 in figure 8.10. Suppose S_1 and S_2 are LTI. We first show that $S = S_2 \circ S_1$ is LTI. To show that S is time invariant, we must show that for any τ, $S \circ D_\tau = D_\tau \circ S$:

$$
\begin{aligned}
S \circ D_\tau &= S_2 \circ S_1 \circ D_\tau \\
&= S_2 \circ D_\tau \circ S_1, \text{ because } S_1 \text{ is time invariant} \\
&= D_\tau \circ S_2 \circ S_1, \text{ because } S_2 \text{ is time invariant} \\
&= D_\tau \circ S,
\end{aligned}
$$

as required.

We now show that S is linear. Let x be any input signal and a any complex number. Then

$$
\begin{aligned}
S(ax) &= (S_2 \circ S_1)(ax) \\
&= S_2(S_1(ax)) \\
&= S_2(aS_1(x)), \text{ because } S_1 \text{ is linear} \\
&= S_2(S_1(x)), \text{ because } S_2 \text{ is linear} \\
&= aS(x).
\end{aligned}
$$

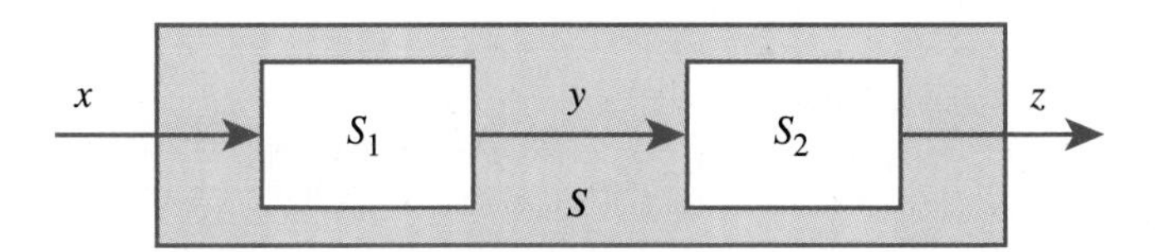

FIGURE 8.10: The cascade connection of the two LTI systems is the system $S = S_2 \circ S_1$. The frequency response is related by $\forall\, \omega, H(\omega) = H_1(\omega)H_2(\omega)$.

Last, if x_1 and x_2 are two input signals, then

$$\begin{aligned} S(x_1 + x_2) &= (S_2 \circ S_1)(x_1 + x_2) \\ &= S_2(S_1(x_1 + x_2)) \\ &= S_2(S_1(x_1) + S_1(x_2)), \text{ because } S_1 \text{ is linear} \\ &= S_2(S_1(x_1)) + S_2(S_1(x_2)), \text{ because } S_2 \text{ is linear} \\ &= S(x_1) + S(x_2). \end{aligned}$$

This shows that S is linear.

We now compute the frequency response of S. Let $H_1(\omega)$, $H_2(\omega)$, and $H(\omega)$ be the frequency responses of S_1, S_2, and S, respectively, at the frequency ω. Consider the complex exponential input x given by

$$\forall\, t \in Reals, \quad x(t) = e^{i\omega t}.$$

Then the signal $y = S_1(x)$ is a multiple of x: Namely, $y = H_1(\omega)x$. In particular, y is a (scaled) complex exponential, and so $z = S_2(y)$ is given by

$$z = H_2(\omega)y = H_2(\omega)H_1(\omega)x.$$

However, because $H(\omega)$ is the frequency response of S at the frequency ω, we also have

$$z = S(x) = H(\omega)x,$$

and so we obtain

$$\boxed{\forall\, \omega \in Reals, \quad H(\omega) = H_2(\omega)H_1(\omega).} \tag{8.34}$$

The frequency response of the cascade composition is the product of the frequency responses of the components. Exactly the same formula applies in the discrete-time case. This is a remarkable result. First, suppose that the cascade connection of figure 8.10 is reversed; that is, consider the system $\tilde{S} = S_1 \circ S_2$. Then the frequency response of $\tilde{S}$ is

$$\tilde{H}(\omega) = H_1(\omega)H_2(\omega) = H_2(\omega)H_1(\omega) = H(\omega);$$

that is, $\tilde{S}$ and S have the same frequency response! This implies, in fact, that S and $\tilde{S}$ are equivalent; they yield the same output for the same input. Hence,

> In any cascade connection of LTI systems, the order in which the systems are composed does not matter.

8.5.2 *Feedback connection*

The feedback arrangement shown in figure 8.11 is fundamental to the design of control systems. Typically, S_1 is some physical system, and S_2 is a controller that we design to get the physical system to do our bidding. The overall system S is called a **closed-loop** system. We first show that S is LTI if S_1 and S_2 are LTI, and we then calculate its frequency response.

Suppose x is the input signal, and define the signals u, z, and y as shown in figure 8.11. The circle with the plus sign represents the relationship $u = x + z$. The signals are then related by

$$
\begin{aligned}
y &= S_1(u) \\
&= S_1(x + z) \\
&= S_1(x) + S_1(z), \text{ because } S_1 \text{ is linear} \\
&= S_1(x) + S_1(S_2(y)).
\end{aligned}
$$

Note that this equation relates the input and output but that the output appears on both sides. We can rewrite this as

$$y - S_1(S_2(y)) = S_1(x). \tag{8.35}$$

Thus, given the input signal x, the output signal y is obtained by solving this equation. We assume that for any signal x, (8.35) has a unique solution y. Then, of course, $y = S(x)$. We can use (8.35) and methods similar to the ones we used for the cascade example to show that S is LTI (see the following Probing further box).

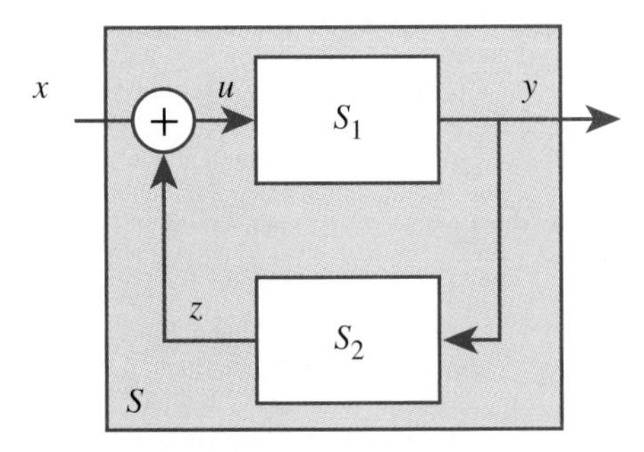

FIGURE 8.11: The feedback connection of the two LTI systems S_1 and S_2 is the LTI system S. The frequency response is related by $\forall\, \omega \in Reals,\ H(\omega) = H_1(\omega)/[1 - H_1(\omega)H_2(\omega)]$.

PROBING FURTHER

Feedback systems are LTI

To show that S in figure 8.11 is time invariant, we must show that for any $\tau \in Reals$,

$$S(D_\tau(x)) = D_\tau(S(x)) = D_\tau(y); \tag{8.36}$$

that is, we must show that $D_\tau(y)$ is the output of S when the input is $D_\tau(x)$. Now the left side of (8.35) with y replaced by $D_\tau(y)$ is

$$\begin{aligned} D_\tau(y) - S_1(S_2(D_\tau(y))) &= D_\tau(y) - D_\tau(S_1(S_2(y))), \\ &\quad \text{because } S_1 \text{ and } S_2 \text{ are time invariant} \\ &= D_\tau(y - S_1(S_2(y))), \text{ because } D_\tau \text{ is linear} \\ &= D_\tau(S_1(x)), \text{ according to (8.35)} \\ &= S_1(D_\tau(x)), \text{ because } S_1 \text{ is time invariant,} \end{aligned}$$

so that $D_\tau(y)$ is indeed the solution of (8.35) when the input is $D_\tau(x)$. This proves (8.36).

Linearity is shown in a similar manner. Let a be any complex number. To show that ay is the output when the input is ax, we evaluate the left side of (8.35) at ay:

$$\begin{aligned} ay - S_1(S_2(ay)) &= ay - aS_1(S_2(y)), \text{ because } S_2 \text{ and } S_1 \text{ are linear} \\ &= a[y - S_1(S_2(y))] \\ &= aS_1(x), \text{ according to (8.35)} \\ &= S_1(ax), \text{ because } S_1 \text{ is linear,} \end{aligned}$$

which shows that $S(ax) = aS(x)$.

Now suppose w is another input and $z = S(w)$ is the corresponding output; that is,

$$z - S_1(S_2(z)) = S_1(w). \tag{8.37}$$

We evaluate the left side of (8.35) at $y + z$,

$$\begin{aligned} (y+z) - S_1(S_2(y+z)) &= [y - S_1(S_2(y))] + [z - S_1(S_2(z))], d \text{ because } S_2 \text{ and } S_1 \text{ are linear} \\ &= S_1(x) + S_1(w), \text{ according to (8.35) and (8.37)} \\ &= S_1(x+w), \text{ because } S_1 \text{ is linear,} \end{aligned}$$

and so $S(x + w) = y + z = S(x) + S(z)$.

We now compute the frequency response $H(\omega)$ of S at frequency ω. Let the frequency response of S_1 be H_1, and let that of S_2 be H_2. Suppose the input signal is the complex exponential

$$\forall\, t \in Reals, \quad x(t) = e^{i\omega t}.$$

For this input, we know that $S_1(x) = H_1(\omega)x$ and $S_2(x) = H_2(\omega)x$. Because S is LTI, we know that the output signal y is given by

$$y = H(\omega)x.$$

Using this relation for y in (8.35), we get

$$\begin{aligned}
H(\omega)x - S_1(S_2(H(\omega)x)) &= H(\omega)[x - S_1(S_2(x))], \text{ because } S_2, S_1 \text{ are linear} \\
&= H(\omega)[x - H_1(\omega)H_2(\omega)x] \\
&= H(\omega)[1 - H_1(\omega)H_2(\omega)]x \\
&= S_1(x) \\
&= H_1(\omega)x, \text{ according to (8.35),}
\end{aligned}$$

from which we get the frequency response of the feedback system,

$$\boxed{H(\omega) = \frac{H_1(\omega)}{1 - H_1(\omega)H_2(\omega)}.} \tag{8.38}$$

This relation is at the foundation of linear feedback control design.

Example 8.20: Consider a discrete-time feedback system as in figure 8.11, where S_1 simply scales the input by 0.9; that is, for all discrete-time input signals u, $S_1(u) = 0.9u$. Suppose further that S_2 is a one-sample delay; that is, $S_2 = D_1$. According to example 8.9, the frequency response of S_2 is given by

$$\forall\, \omega \in Reals, \quad H_2(\omega) = e^{-i\omega}.$$

The frequency response of S_1 is (trivially) given by

$$\forall\, \omega \in Reals, \quad H_1(\omega) = 0.9.$$

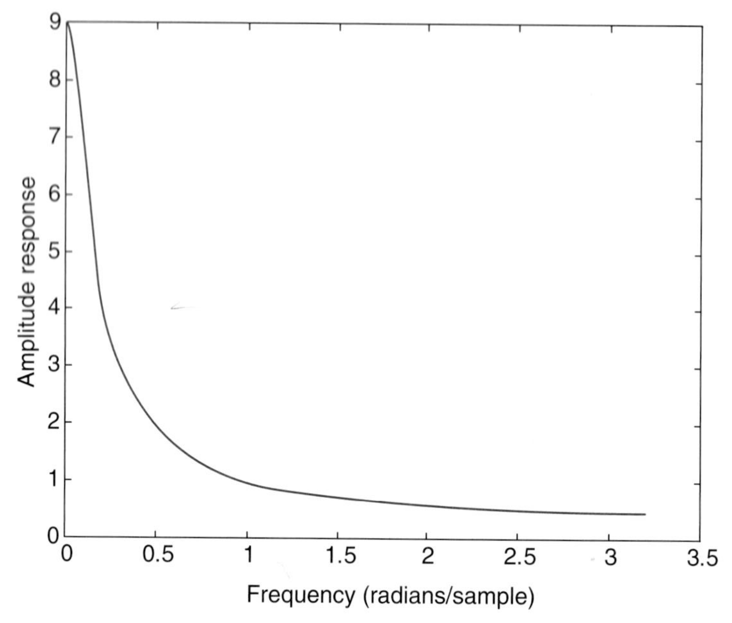

FIGURE 8.12: The magnitude response of the feedback composition.

Thus, the frequency response of the feedback composition is

$$\forall\, \omega \in Reals, \quad H(\omega) = \frac{0.9}{1 - 0.9e^{-i\omega}}.$$

We can plot the magnitude of this by using the following MATLAB code, which (after adjusting the labels) results in figure 8.12:

```
omega = [0:pi/250:pi];
H = 0.9./(1 - 0.9.*exp(-i*omega));
plot(omega, abs(H)).
```

Notice that at zero frequency, the gain is 9 and it rapidly drops off at higher frequencies. Thus, this system behaves as a lowpass filter. ❐

Example 8.21: We can use formula (8.38) to obtain the frequency response of more complex compositions, such as the one shown in figure 8.13. To find the frequency response of the composition S in the upper left of the figure, we first express as the composition in the upper right. S_3 is given in

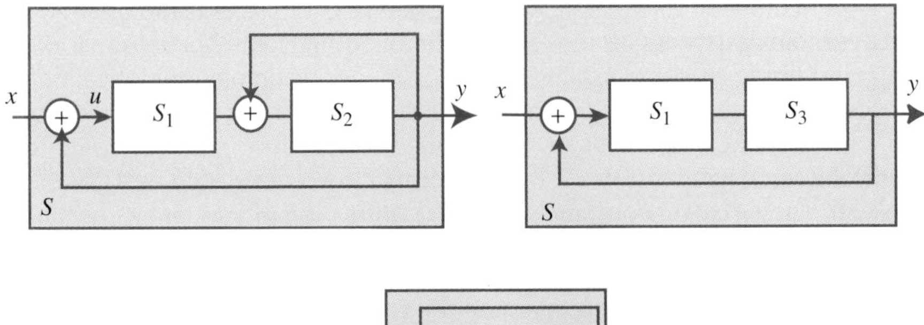

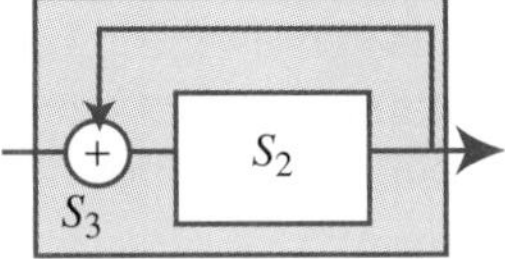

FIGURE 8.13: The composition on the upper left can be expressed as the one on the upper right by first considering the subsystem at the bottom.

the bottom part of the figure. The frequency response H of S can now be obtained from formula (8.38) as

$$H(\omega) = \frac{H_1(\omega)H_3(\omega)}{1 - H_1(\omega)H_3(\omega)}.$$

The same formula also gives the frequency response H_3 of S_3,

$$H_3(\omega) = \frac{H_2(\omega)}{1 - H_2(\omega)},$$

which, upon substitution in the previous expression, yields

$$H(\omega) = \frac{H_1(\omega)H_2(\omega)}{1 - H_2(\omega) - H_1(\omega)H_2(\omega)}. \square$$

8.6 Summary

LTI systems have a particularly useful property: If the input is a sinusoid, then the output is a sinusoid of the same frequency. Moreover, if the input is a sum of two sinusoids, then the output is a sum of two sinusoids with the same frequencies as the input sinusoids. Each sinusoid is scaled and shifted in phase by the system. The scaling and phase shift, as a function of frequency, is called the frequency response of the system.

It turns out that, mathematically, this phenomenon is easiest to analyze by using complex exponentials instead of real-valued sinusoids. The reason for this is that the phase shift and scaling together amount to simple multiplication of the complex exponential by a complex constant. If an input is represented as a sum of complex exponentials (a form of the Fourier series), then the output is simply the same Fourier series with each term of the series scaled by a complex constant. These scaling constants, viewed as a function of frequency, constitute the frequency response of the system.

Composing LTI systems is particularly easy. The frequency response of the cascade of two LTI systems is simply the product of the frequency responses of the individual systems. This simple fact can be used to quickly ascertain the frequency response of complicated compositions.

EXERCISES

KEY: E = mechanical T = requires plan of attack C = more than 1 answer

E 1. Find $A \in Complex$ so that

$$\forall\, t \in Reals, \quad Ae^{i\omega t} + A^* e^{-i\omega t} = \cos(\omega t + \pi/4),$$

where A^* is the complex conjugate of A.

E 2. Plot the function $s\colon Reals \to Reals$ given by

$$\forall\, x \in Reals, \quad s(x) = Im\{e^{(-x+i2\pi x)}\}.$$

You are free to choose a reasonable interval for your plot, but be sure it includes $x = 0$.

E 3. This exercise explores the fact that a delay in a sine wave causes a phase shift; that is, for any real numbers τ and ω, there is a phase shift $\phi \in Reals$ such that for all $t \in Reals$,

$$\sin(\omega(t - \tau)) = \sin(\omega t - \phi).$$

Give ϕ in terms of τ and ω. What are the units of τ, ω, and ϕ? Hint: The argument to the sine function has units of radians.

E 4. Let $x\colon Reals \to Reals$. Show that x is periodic with period p if and only if $D_p(x) = x$. Now show that if S is a time-invariant system and x is a periodic signal, then $S(x)$ is also periodic with period p.

E 5. Analogously to D_τ and D_M in section 8.1.1, define formally the following variants:

(a) A shift operator $S_{v,h}$ that shifts an image v units vertically and h units horizontally, where $v \in Reals$ and $h \in Reals$.

(b) A shift operator $S_{m,n}$ that shifts a discrete image m units vertically and n units horizontally, where $m \in Integers$ and $n \in Integers$.

E 6. Consider a discrete-time system D: [*Integers* → *Reals*] → [*Integers* → *Reals*] where, if $y = D(x)$, then

$$\forall\, n \in Integers, \quad y(n) = x(n-1).$$

(a) Is D linear? Justify your answer.

(b) Is D time invariant? Justify your answer.

E 7. Consider a continuous-time system *TimeScale*: [*Reals* → *Reals*] → [*Reals* → *Reals*] where, if $y = TimeScale(x)$, then

$$\forall\, t \in Reals, \quad y(t) = x(2t).$$

(a) Is *TimeScale* linear? Justify your answer.

(b) Is *TimeScale* time invariant? Justify your answer.

E 8. Consider the continuous-time signal x where

$$\forall\, t \in Reals, \quad x(t) = 1 + \cos(\pi t) + \cos(2\pi t).$$

Suppose that x is the input to an LTI system with frequency response given by

$$\forall\, \omega \in Reals, \quad H(\omega) = \begin{cases} e^{i\omega} & \text{if } |\omega| < 4 \text{ radians/second} \\ 0 & \text{otherwise.} \end{cases}$$

Find the output y of the system.

T 9. Suppose that the continuous-time signal x: *Reals* → *Reals* is periodic with period p. Let the fundamental frequency be $\omega_0 = 2\pi/p$. Suppose that the Fourier series coefficients for this signal are known constants $A_0, A_1, A_2, \ldots$ and $\phi_1, \phi_2, \ldots$. Give the Fourier series coefficients $A'_0, A'_1, A'_2, \ldots$ and $\phi'_1, \phi'_2, \ldots$ for each of the following signals:

(a) ax, where $a \in Reals$ is a constant;

(b) $D_\tau(x)$, where $\tau \in Reals$ is a constant and D_τ is the delay system; and

(c) $S(x)$, where S is an LTI system with frequency response H given by

$$\forall\, \omega \in \mathit{Reals}, \quad H(\omega) = \begin{cases} 1 & \text{if } \omega = 0 \\ 0 & \text{otherwise.} \end{cases}$$

(Note that this is a highly unrealistic frequency response.)

(d) Let $y\colon \mathit{Reals} \to \mathit{Reals}$ be another periodic signal with period p. Suppose y has Fourier series coefficients $A''_0, A''_1, A''_2, \ldots$ and $\phi''_1, \phi''_2, \ldots$. Give the Fourier series coefficients of $x + y$.

E 10. Consider discrete-time systems with input $x\colon \mathit{Integers} \to \mathit{Reals}$ and output $y\colon \mathit{Integers} \to \mathit{Reals}$. Each of the following defines such a system. For each, indicate whether it is linear (L), time invariant (TI), both (LTI), or neither (N).

(a) $\forall\, n \in \mathit{Integers}, \quad y(n) = x(n) + 0.9y(n-1)$.

(b) $\forall\, n \in \mathit{Integers}, \quad y(n) = \cos(2\pi n)x(n)$.

(c) $\forall\, n \in \mathit{Integers}, \quad y(n) = \cos(2\pi n/9)x(n)$.

(d) $\forall\, n \in \mathit{Integers}, \quad y(n) = \cos(2\pi n/9)(x(n) + x(n-1))$.

(e) $\forall\, n \in \mathit{Integers}, \quad y(n) = x(n) + 0.1(x(n))^2$.

(f) $\forall\, n \in \mathit{Integers}, \quad y(n) = x(n) + 0.1(x(n-1))^2$.

E 11. Suppose that the frequency response of a discrete-time LTI system S is given by

$$H(\omega) = |\sin(\omega)|,$$

where ω has units of radians per sample. Suppose the input is the discrete-time signal x given by $\forall\, n \in \mathit{Integers},\ x(n) = 1$. Give a *simple* expression for $y = S(x)$.

T 12. Find the smallest positive integer n such that

$$\sum_{k=0}^{n} e^{i5k\pi/6} = 0.$$

Hint: Note that the term being summed is a periodic function of k. What is its period? What is the sum of a complex exponential over one period?

T 13. Consider a continuous-time periodic signal x with fundamental frequency $\omega_0 = 1$ radian/second. Suppose that the Fourier series coefficients (see (7.4)) are

$$A_k = \begin{cases} 1 & \text{if } k = 0, 1, \text{ or } 2 \\ 0 & \text{otherwise} \end{cases}$$

and for all $k \in Naturals_0$, $\phi_k = 0$.

(a) Find the Fourier series coefficients X_k for all $k \in Integers$ (see (8.29)).

(b) Consider a continuous-time LTI system $Filter\colon [Reals \to Reals] \to [Reals \to Reals]$ with frequency response

$$H(\omega) = \cos(\pi\omega/2).$$

Find $y = Filter(x)$; that is, give a simple expression for $y(t)$ that is valid for all $t \in Reals$.

(c) For y calculated in (b), find the fundamental frequency in radians per second; that is, find the largest $\omega_0' > 0$ such that

$$\forall\, t \in Reals, \quad y(t) = y(t + 2\pi/\omega_0').$$

T 14. Consider a continuous-time LTI system S. Suppose that when the input x is given by

$$\forall\, t \in Reals, \quad x(t) = \begin{cases} 1 & \text{if } 0 \le t < 1 \\ 0 & \text{otherwise,} \end{cases}$$

then the output $y = S(x)$ is given by

$$\forall\, t \in Reals, \quad y(t) = \begin{cases} 1 & \text{if } 0 \le t < 2 \\ 0 & \text{otherwise.} \end{cases}$$

Give an expression and a sketch for the output of the same system if the input is

(a)

$$\forall\, t \in Reals, \quad x'(t) = \begin{cases} 1 & \text{if } 0 \le t < 1 \\ -1 & \text{if } 1 \le t < 2 \\ 0 & \text{otherwise.} \end{cases}$$

(b)

$$\forall\, t \in Reals, \quad x'(t) = \begin{cases} 1 & \text{if } 0 \le t < 1/2 \\ 0 & \text{otherwise.} \end{cases}$$

T 15. Suppose that the frequency response H of a discrete-time LTI system *Filter* is given by

$$\forall\, \omega \in [-\pi, \pi], \quad H(\omega) = |\omega|.$$

where ω has units of radians/sample. Note that because a discrete-time frequency response is periodic with period 2π, this definition implicitly describes $H(\omega)$ for all $\omega \in \mathit{Reals}$. Give simple expressions for the output y when the input signal $x\colon \mathit{Integers} \to \mathit{Reals}$ is such that $\forall\, n \in \mathit{Integers}$, each of the following is true:

(a) $x(n) = \cos(\pi n/2)$.

(b) $x(n) = 5$.

(c) $x(n) = \begin{cases} +1 & \text{if } n \text{ even} \\ -1 & \text{if } n \text{ odd.} \end{cases}$

T 16. Consider a continuous-time LTI system S. Suppose that when the input is given by

$$x(t) = \begin{cases} \sin(\pi t) & \text{if } 0 \le t < 1 \\ 0 & \text{otherwise,} \end{cases}$$

then the output $y = S(x)$ is given by

$$y(t) = \begin{cases} \sin(\pi t) & \text{if } 0 \le t < 1 \\ \sin(\pi(t-1)) & \text{if } 1 \le t < 2 \\ 0 & \text{otherwise} \end{cases}$$

for all $t \in \mathit{Reals}$.

(a) Carefully sketch these two signals.

(b) Give an expression and a sketch for the output of the same system if the input is

$$x(t) = \begin{cases} \sin(\pi t) & \text{if } 0 \le t < 1 \\ -\sin(\pi(t-1)) & \text{if } 1 \le t < 2 \\ 0 & \text{otherwise.} \end{cases}$$

T 17. Suppose you are given the following building blocks for building block diagrams:

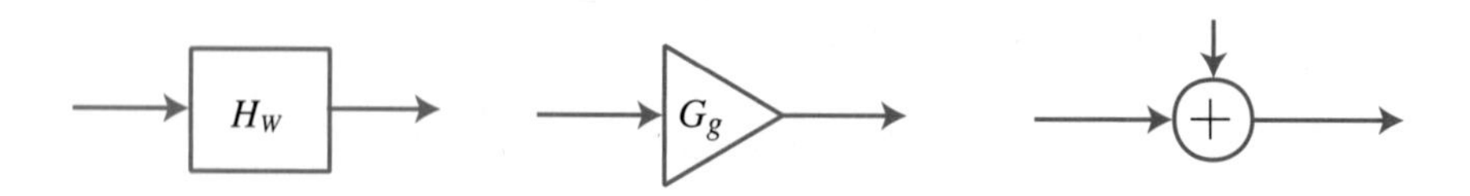

These blocks are defined as follows:

(a) An LTI system H_W: $[Reals \to Reals] \to [Reals \to Reals]$ that has a rectangular frequency response given by

$$\forall\, \omega \in Reals, \quad H(\omega) = \begin{cases} 1 & \text{if } -W < \omega < W \\ 0 & \text{otherwise,} \end{cases}$$

where W is a parameter you can set.

(b) A gain block G_g: $[Reals \to Reals] \to [Reals \to Reals]$ where, if $y = g(x)$, then

$$\forall\, t \in Reals, \quad y(t) = gx(t),$$

where $g \in Reals$ is a parameter you can set.

(c) An adder, which can add two continuous-time signals; specifically, Add: $[Reals \to Reals] \times [Reals \to Reals] \to [Reals \to Reals]$ such that if $y = Add(x_1, x_2)$, then

$$\forall\, t \in Reals, \quad y(t) = x_1(t) + x_2(t).$$

Use these building blocks to construct a system with the following frequency response:

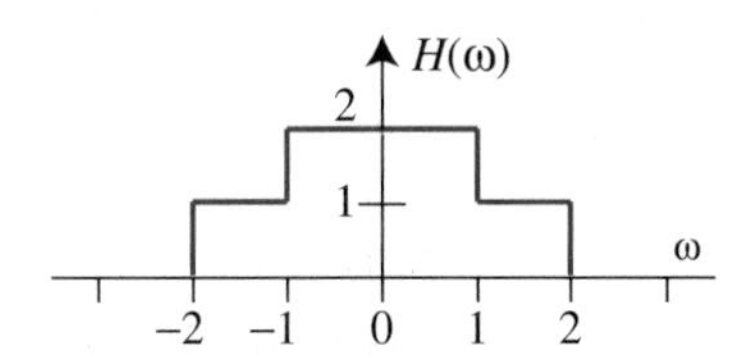

T 18. Let u be a discrete-time signal given by

$$\forall\, n \in Integers, \quad u(n) = \begin{cases} 1 & \text{if } 0 \le n \\ 0 & \text{otherwise.} \end{cases}$$

This is the **unit step** signal, which we saw before in (2.15). Suppose that a discrete-time system H that is known to be LTI is such that if the input is u, the output is $y = H(u)$ given by

$$\forall\, n \in Integers, \quad y(n) = nu(n).$$

This is called the **step response** of the system. Find a simple expression for the output $w = H(p)$ when the input is p given by

$$\forall\, n \in \mathit{Integers}, \quad p(n) = \begin{cases} 2 & \text{if } 0 \le n < 8 \\ 0 & \text{otherwise.} \end{cases}$$

Sketch w.

T 19. Suppose you are given the Fourier series coefficients $\ldots X_{-1}, X_0, X_1, X_2, \ldots$ for a periodic signal $x\colon \mathit{Reals} \to \mathit{Reals}$ with period p. Find the fundamental frequency and the Fourier series coefficients of the following signals in terms of those of x:

(a) y such that $\forall\, t \in \mathit{Reals},\ y(t) = x(at)$, for some positive real number a.

(b) w such that $\forall\, t \in \mathit{Reals},\ w(t) = x(t)e^{i\omega_0 t}$, where $\omega_0 = 2\pi/p$.

(c) z such that $\forall\, t \in \mathit{Reals},\ z(t) = x(t)\cos(\omega_0 t)$, where $\omega_0 = 2\pi/p$.

E 20. Analogously to the Probing further box on page 306, show that the formula (8.33) for the discrete Fourier series coefficients is valid.

C 21. Consider a system $\mathit{Squarer}\colon [\mathit{Reals} \to \mathit{Reals}] \to [\mathit{Reals} \to \mathit{Reals}]$ where, if $y = \mathit{Squarer}(x)$, then

$$\forall\, t \in \mathit{Reals}, \quad y(t) = (x(t))^2.$$

(a) Show that this system is memoryless.

(b) Show that this system is not linear.

(c) Show that this system is time invariant.

(d) Suppose that the input x is given by

$$\forall\, t \in \mathit{Reals}, \quad x(t) = \cos(\omega t),$$

for some fixed ω. Show that the output y contains a component at frequency 2ω.

INTERVIEW
Dawn Tilbury

Dawn Tilbury is an Associate Professor in Mechanical Engineering at the University of Michigan in Ann Arbor. Her research interests lie in the area of control systems, including nonlinear control, logic control, and hybrid control. In conjunction with teaching, she developed a set of web-based tutorials to introduce students to the use of MATLAB. These tutorials received an Undergraduate Computational Science Award from the U.S. Department of Energy through the Ames Laboratory, and the Educom Medal, nominated by the ASME.

How did you decide to study electrical engineering?
My father was an electrical engineer, he got his degree from the University of Minnesota, and he wanted me to follow in his footsteps. I wanted to study French, but he convinced me that I could keep my options open by taking physics and calculus along with French my first year. It didn't take me long to decide that at the college level, physics was more interesting than French.

What was your first job in the industry?
I worked as a summer intern at Honeywell in Minneapolis during my undergraduate years. Although I really liked school, I didn't like those jobs very much (although they paid well), so when I got a fellowship I decided to go to graduate school and keep studying.

Which person in the field has inspired you most? In what ways?
My father inspired me to study engineering and was convinced that I was good at it. My research adviser at Berkeley was inspiring for his constant enthusiasm to attack new, unsolved problems.

What is your vision for the future of electrical and computer engineering?
Computers and electrical engineering are becoming more pervasive in our lives. As things get more complex, EE and CE designers need to think about the user interfaces and how users will work with them and how the things will work with each other. Engineers should harness the power of technology to make people's lives better and easier.

The design of feedback control is based on a frequency-domain analysis of linear systems. What new concepts, analysis and design tools, and examples are needed to deal with the design of control "logic" and networked control systems? What is the current status of this area of design?

Most of the frequency-domain analysis of linear systems has been developed for single-input, single-output systems. Logic and networked control systems have many inputs and outputs (possibly even hundreds or thousands). To manage the complexity, hierarchical and modular frameworks for analysis are needed. The SISO loops are at the bottom and are the foundation stones upon which the complex controllers are built.

Are there any special projects you are currently working on that you'd like to tell students about?

I'm currently working with IBM on a project to control computing systems such as mail servers and web servers. Large software systems typically have many different parameters that can be set for admission control, session length, priorities, and so forth. We are using feedback control techniques to manage and improve the performance of these systems, to simplify the user interface for system administrators, and to adapt the servers automatically to the current workload that they are experiencing. This is part of a larger effort in "autonomic computing," which means that computer systems should be able to administer and manage themselves.

CHAPTER 9
Filtering

A property of linear time-invariant (LTI) systems is that if the input is described as a sum of sinusoids, then the output is a sum of sinusoids of the same frequency. Each sinusoidal component is typically scaled differently, and each is subjected to a phase change, but the output contains only sinusoidal components that are also present in the input. For this reason, an LTI system is often called a **filter**. It can filter out frequency components of the input, and it can also enhance other components, but it cannot introduce components that are not already present in the input. It merely changes the relative amplitudes and phases of the frequency components that are present in the inputs.

LTI systems arise in two circumstances in an engineering context. First, they may be used as a model of a physical system. Many physical systems are accurately modeled as LTI systems. Second, they may represent an ideal for an engineered system. For example, they may specify the behavior that an electronic system is expected to exhibit.

Consider, for example, an audio system. The normal human ear hears frequencies in the range of about 30 to 20,000 Hz, and so a specification for a high-fidelity audio system is typically that the frequency response be constant (in magnitude) over this range. The human ear is relatively insensitive to phase, and so the same specification may include nothing about the phase response (the argument, or angle of the frequency response). An audio system is free to filter out frequencies outside this range.

Consider an acoustic environment: a physical context such as a lecture hall where sounds are heard. The hall itself alters the sound. The sound heard by a listener's ear is not identical to the sound created by the lecturer. The room introduces echoes, caused by reflections of the sound by the walls. These echoes tend to occur more for the lower frequency components in the sound than for the higher frequency components because the walls and objects in the room tend to absorb higher frequency sounds better. Thus, the lower frequency sounds bounce around in the room, reinforcing each other, whereas the higher frequency sounds, which are more quickly absorbed, become relatively less pronounced. In an extreme circumstance—in a room in which the walls are lined with shag carpeting, for example—the higher frequency sounds are absorbed so effectively that the sound gets muffled.

Such a room can be modeled by an LTI system in which the frequency response $H(\omega)$ is smaller in magnitude for large ω than for small ω. This is a simple form of **distortion** introduced by a **channel** (the room), which in this case carries a sound from its transmitter (the lecturer) to its receiver (the listener). This form of distortion is called **linear distortion**, a shorthand expression for linear, time-invariant distortion (the time invariance is left implicit).

A public address system in a lecture hall may compensate for the acoustics of the room by boosting the high-frequency content in the sound. Such a compensator is called an **equalizer** because it corrects for distortion in the channel so that all frequencies are received equally well by the receiver.

In a communications system, a channel may be a physical medium, such as a pair of wires, that carries an electrical signal. That physical medium distorts the signal, and this distortion is often reasonably well approximated as linear and time invariant. An equalizer in the receiver compensates for this distortion. In contrast to the audio example, however, such an equalizer often needs to compensate for the phase response, not just the magnitude response. Because the human ear is relatively insensitive to phase distortion, a public address system equalizer need not compensate for phase distortion. But the wire pair may be carrying a signal that is not an audio signal. It may be, for example, a modem signal.

Images may also be processed by LTI systems. Consider the three images shown in figure 9.1. The top image is the original, undistorted image. The lower left image is blurred, as might result, for example, from unfocused optics. The lower right image is, in a sense, the opposite of the blurred image. Whereas the blurred image deemphasizes the patterns in the outfit, for example, the right image deemphasizes the regions of constant value, changing them all to a neutral gray.

For images, time is not the critical variable. Its role is replaced by two spatial variables, one in the horizontal direction and one in the vertical direction. Thus, instead of LTI, we may talk about an image processing system being a **linear space-invariant** (**LSI**) system. The blurred image is constructed from the original by an LSI system that eliminates high (spatial) frequencies, passing

FIGURE 9.1: An image (top) and two versions that have been distorted by a linear, space-invariant system (bottom).

the low frequencies unaltered. Such a system is called a **lowpass** filter. The lower right image is constructed from the original by an LSI system that eliminates low frequencies, passing the high frequencies unaltered. Such a system is called a **highpass** system. Both images were created with the use of Adobe Photoshop, although the blurred image could have been just as easily created by a defocused lens.

9.1 Convolution

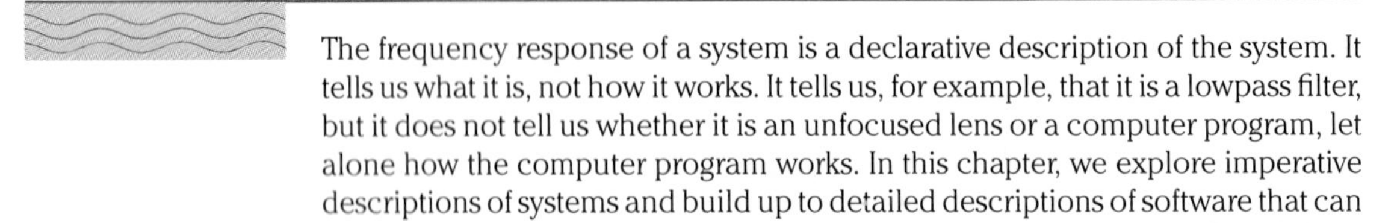

The frequency response of a system is a declarative description of the system. It tells us what it is, not how it works. It tells us, for example, that it is a lowpass filter, but it does not tell us whether it is an unfocused lens or a computer program, let alone how the computer program works. In this chapter, we explore imperative descriptions of systems and build up to detailed descriptions of software that can implement certain kinds of LTI (or LSI) systems. These imperative descriptions are based on **convolution**.

9.1.1 Convolution sum and integral

For discrete-time signals, the convolution operator is called the convolution sum, and for continuous-time signals, it is called the convolution integral. We define these two operators now and note some important properties.

Let $x, y \in [Integers \to Reals]$ be two discrete-time signals. The convolution of x and y is the discrete-time signal, denoted $x * y$, expressed by the **convolution sum**

$$\boxed{\forall\, n \in Integers, \quad (x * y)(n) = \sum_{k=-\infty}^{\infty} x(k)y(n-k).} \tag{9.1}$$

We note two properties. First, the order in the convolution does not matter: $x * y = y * x$. Indeed, if in (9.1) we change the variables in the summation, letting $m = n - k$, we get

$$\begin{aligned} \forall\, n \in Integers, \quad (x * y)(n) &= \sum_{k=-\infty}^{\infty} x(k)y(n-k) \\ &= \sum_{m=-\infty}^{\infty} x(n-m)y(m). \end{aligned}$$

Thus,

$$\boxed{(x * y)(n) = (y * x)(n).} \tag{9.2}$$

This property is called **commutativity** of the convolution operator.*

* Matrix multiplication is an example of an operator that is not commutative, whereas matrix addition is. Because the sum of two matrices M and N (of the same size) does not depend on the order (i.e., $M + N = N + M$), the matrix sum is a commutative operator. However, the product of two matrices depends on the order; that is, it is not always true that $M \times N = N \times M$, and so the matrix product is not commutative.

Another property of convolution is **linearity**; that is, if x, y_1, and y_2 are three signals and a_1 and a_2 are real numbers, then

$$\boxed{x * (a_1 y_1 + a_2 y_2) = a_1(x * y_1) + a_2(x * y_2),} \tag{9.3}$$

which may be checked directly by applying definition (9.1) to both sides.

We now use the convolution sum to define some LTI systems. Fix a discrete-time signal h, and define the system

$$S: [Integers \to Reals] \to [Integers \to Reals]$$

by

$$\forall\, x \in [Integers \to Reals], \quad S(x) = h * x.$$

The output signal $y = S(x)$ corresponding to the input signal x is thus represented by

$$\forall\, n \in Integers, \quad y(n) = \sum_{k=-\infty}^{\infty} h(k)x(n-k).$$

We now show that S is LTI. The linearity of S follows immediately from the linearity property of the convolution. To show time invariance, we must show that for any integer M and any input signal x,

$$D_M(h * x) = h * (D_M(x)),$$

where D_M is a delay by M. But this is easy to see, because for all n,

$$\begin{aligned}
(D_M(h * x))(n) &= (h * x)(n - M) \\
&= \sum_{k=-\infty}^{\infty} h(k)x(n - M - k), \text{ according to definition (9.1)} \\
&= \sum_{k=-\infty}^{\infty} h(k)z(n - k), \text{ where } z = D_M(x) \\
&= (h * z)(n).
\end{aligned}$$

Thus, every discrete-time signal h defines an LTI system through convolution. The next section shows the converse result: Every LTI system is defined by convolution with some signal h.

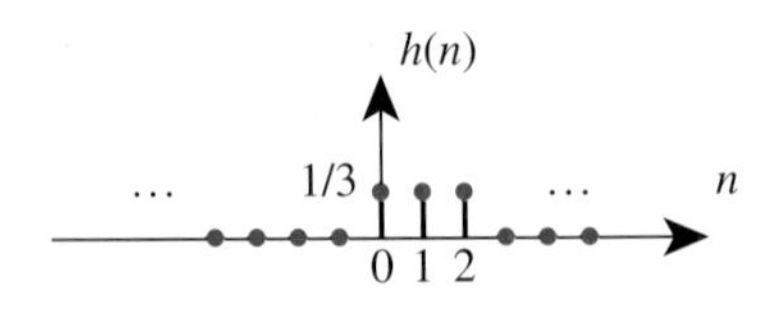

FIGURE 9.2: A signal.

Example 9.1: Consider a discrete-time signal h defined by

$$\forall\, n \in \mathit{Integers}, \quad h(n) = \begin{cases} 1/3 & \text{if } n \in \{0, 1, 2\} \\ 0 & \text{otherwise.} \end{cases}$$

This is shown in figure 9.2. Define a system S as follows. If the input is x, then the output is

$$y = S(x) = h * x;$$

that is,

$$\begin{aligned} \forall\, n \in \mathit{Integers}, \quad y(n) &= \sum_{k=-\infty}^{\infty} h(k)x(n-k) \\ &= \sum_{k=0}^{2} (1/3)x(n-k) \\ &= (x(n) + x(n-1) + x(n-2))/3. \end{aligned} \qquad (9.4)$$

This system calculates the three-point moving average! ❐

We now turn to the continuous-time case. Let $x, y \in [\mathit{Reals} \to \mathit{Reals}]$ be two continuous-time signals. The convolution of x and y is the continuous-time signal, denoted $x * y$, defined by the following **convolution integral**:

$$\boxed{\forall\, t \in \mathit{Reals}, \quad (x * y)(t) = \int_{-\infty}^{\infty} x(\tau)y(t-\tau)d\tau.} \qquad (9.5)$$

By a change of variable in the integral, we can check that convolution again is commutative,

$$\boxed{\forall\, t \in \mathit{Reals}, \quad (x * y)(t) = (y * x)(t),} \qquad (9.6)$$

and that it is linear; that is, if x, y_1, and y_2 are three continuous-time signals and a_1 and a_2 are real numbers, then

$$\boxed{x * (a_1 y_1 + a_2 y_2) = a_1(x * y_1) + a_2(x * y_2).} \tag{9.7}$$

Again, fix $h \in [Reals \to Reals]$, and define the system

$$S\colon [Reals \to Reals] \to [Reals \to Reals]$$

by

$$\forall\, x \in [Reals \to Reals], \quad S(x) = h * x.$$

Then in exactly the same way as for the discrete-time case, we can show that S is LTI.

Example 9.2: Consider a continuous-time signal h defined by

$$\forall\, t \in Reals, \quad h(t) = \begin{cases} 1/3 & \text{if } t \in [0, 3] \\ 0 & \text{otherwise.} \end{cases}$$

This is shown in figure 9.3. Define a system S as follows. If the input is x, then the output is

$$y = S(x) = h * x;$$

that is,

$$\begin{aligned} \forall\, t \in Reals, \quad y(t) &= \int_{-\infty}^{\infty} h(\tau) x(t - \tau) d\tau \\ &= \frac{1}{3} \int_{0}^{3} x(t - \tau) d\tau. \end{aligned} \tag{9.8}$$

This system is the length-three continuous-time moving average! ❐

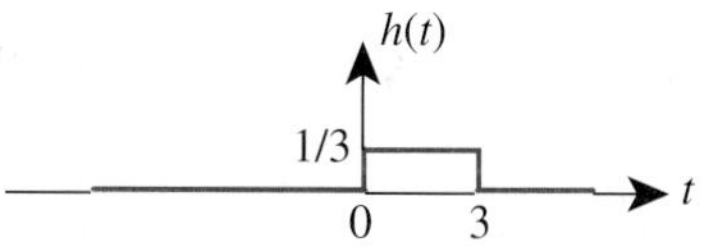

FIGURE 9.3: A signal.

Note that we are using the symbol "*" for both the convolution sum and the convolution integral. The context determines which operator is intended.

9.1.2 *Impulses*

Intuitively, an impulse is a signal that is zero everywhere except at time zero. In the discrete-time case, the **Kronecker delta function**,

$$\delta\colon Integers \to Reals,$$

is defined by

$$\boxed{\forall\, n \in Integers, \quad \delta(n) = \begin{cases} 1 & \text{if } n = 0 \\ 0 & \text{otherwise.} \end{cases}} \tag{9.9}$$

Its graph is shown in figure 9.4.

The continuous-time case, which is called the **Dirac delta function**, is mathematically much more difficult to work with. Like the Kronecker delta function, it is zero everywhere except at zero. In contrast to the Kronecker delta function, however, its value is infinite at zero. We do not concentrate on its subtleties; rather, we just introduce it and assert some results without fully demonstrating their validity. The Dirac delta function is defined to be

$$\delta\colon Reals \to Reals_{++},$$

where $Reals_{++} = Reals \cup \{\infty, -\infty\}$, and

$$\forall\, t \in Reals \text{ where } t \neq 0, \quad \delta(t) = 0$$

and where the following property is satisfied for any $\epsilon > 0$ in $Reals_+$:

$$\int_{-\epsilon}^{\epsilon} \delta(t)dt = 1.$$

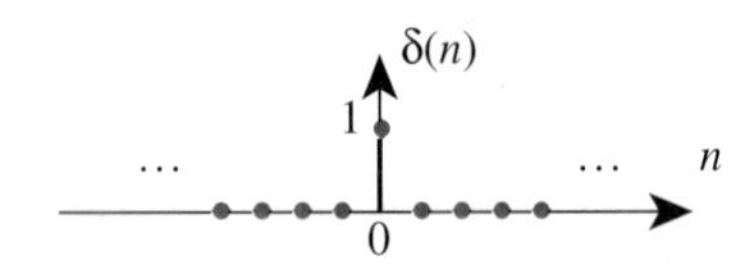

FIGURE 9.4: The Kronecker delta function, a discrete-time impulse.

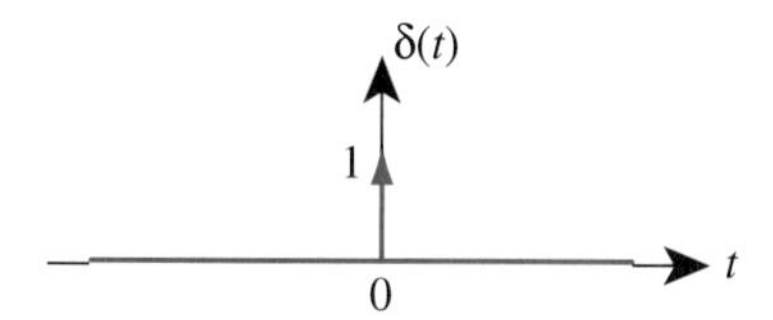

FIGURE 9.5: The Dirac delta function, a continuous-time impulse.

For the latter property to be satisfied, clearly no finite value at $t = 0$ would suffice. This is why the value must be infinite at $t = 0$. Notice that the Kronecker delta function has a similar property,

$$\sum_{n=-a}^{a} \delta(n) = 1$$

for any integer $a > 0$, but in this case, the property is trivial. There is no mathematical subtlety.

The Dirac delta function is usually depicted as in figure 9.5. In any figure, of course, the infinite value at $t = 0$ cannot be shown explicitly. The infinite value is suggested by a vertical arrow. Next to the arrow is the **weight** of the delta function, "1" in this case. In general, a Dirac delta function can be multiplied by any real constant a. Of course, this does not change its value at $t = 0$, which is infinite, nor does it change its value at $t \neq 0$, which is zero. What it does change is its integral:

$$\int_{-\epsilon}^{\epsilon} a\delta(t)dt = a.$$

Thus, although the impulse is still infinitely narrow and infinitely high, the area under the impulse has been scaled by a.

9.1.3 *Signals as sums of weighted delta functions*

Any discrete-time signal x: *Integers* → *Reals* can be expressed as a sum of weighted Kronecker delta functions:

$$\forall n \in Integers, \quad x(n) = \sum_{k=-\infty}^{\infty} x(k)\delta(n-k). \tag{9.10}$$

The kth term in the sum is $x(k)\delta(n-k)$. This term, by itself, defines a signal that is zero everywhere except at $n = k$, where it has value $x(k)$. This signal is called a **weighted delta function** because it is a (time-shifted) delta function with a specified weight. Thus, any discrete-time signal is a sum of weighted delta functions, much the way that the Fourier series describes a signal as a sum of weighted complex exponential functions.

Example 9.3: The signal h in example 9.1 can be written in terms of Kronecker delta functions:

$$\forall n \in Integers, \quad h(n) = \frac{1}{3}(\delta(n) + \delta(n-1) + \delta(n-2)).$$

This has the form of (9.10) and is illustrated in figure 9.6. It is described as a sum of signals in which each signal contains only a single weighted impulse. ❐

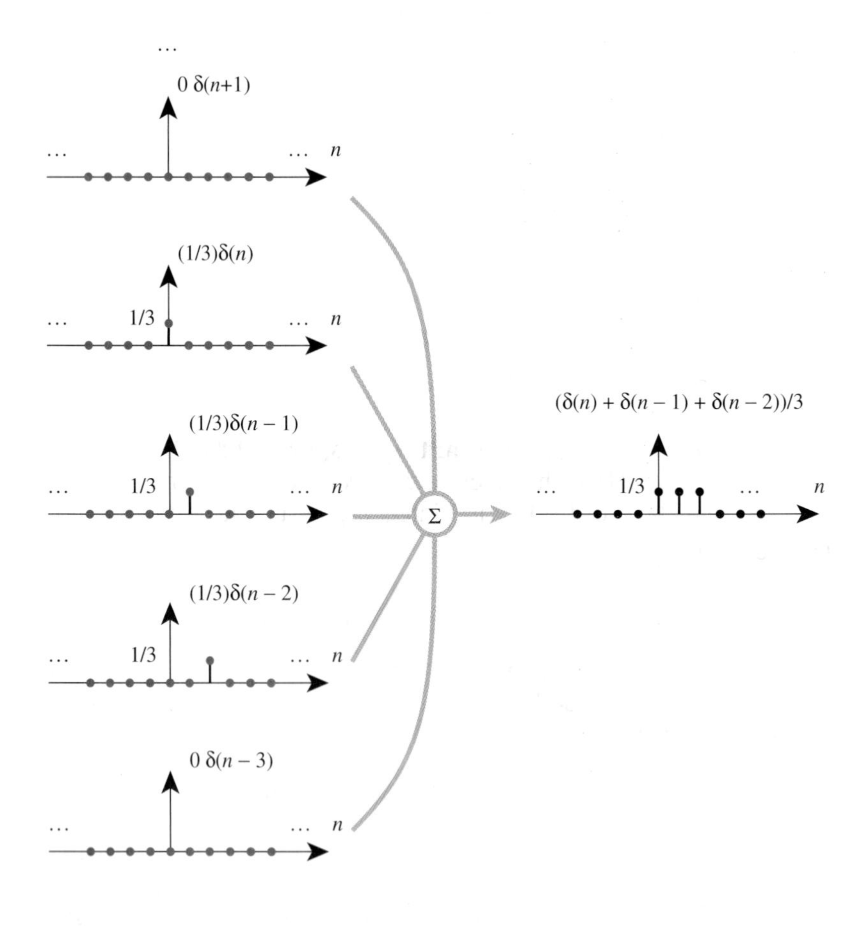

FIGURE 9.6: A discrete-time signal is a sum of weighted delta functions.

Equation (9.10) is sometimes called the **sifting property** of the Kronecker delta function because it "sifts out" the value of a function x at some integer n; that is, the infinite sum reduces to a single number. This property can often be used to eliminate infinite summations in more complicated expressions.

The continuous-time version of this is similar, except that the sum becomes an integral (integration, after all, is just summation over a continuum). Given any signal $x\colon Reals \to Reals$,

$$\forall\, t \in Reals, \quad x(t) = \int_{-\infty}^{\infty} x(\tau)\delta(t-\tau)d\tau. \tag{9.11}$$

Although this is mathematically much more subtle than the discrete-time case, it is very similar in structure. It describes a signal x as a sum (or, more precisely, an integral) of weighted Dirac delta functions.

Example 9.4: The signal h in figure 9.3 and example 9.2 can be written as a sum (an integral, actually) of weighted Dirac delta functions:

$$\forall\, t \in Reals, \quad h(t) = \int_{-\infty}^{\infty} h(\tau)\delta(t-\tau)d\tau$$

$$= \int_{0}^{3} (1/3)\delta(t-\tau)d\tau.$$

This has the form of (9.11). ❐

Equation (9.11) is also called the **sifting property** of the Dirac delta function, because it sifts out from the function x the value at a given time t. The sifting property can often be used to eliminate integrals, because it replaces an integral with a single value.

9.1.4 *Impulse response and convolution*

Consider a discrete-time LTI system $S\colon [Integers \to Reals] \to [Integers \to Reals]$. Define its **impulse response** h to be the output signal when the input signal is the Kronecker delta function (an impulse), $h = S(\delta)$; that is,

$$\forall\, n \in Integers, \quad h(n) = (S(\delta))(n).$$

Now let x be any input signal, and let $y = S(x)$ be the corresponding output signal. In (9.10), x is given as sum of components, in which each component is a weighted delta function. Because S is LTI, the output can be given as a sum of the responses to these components. Each component is a signal $x(k)\delta(n-k)$ for fixed k, and the response to this signal is $x(k)h(n-k)$. The response to a scaled and delayed impulse is a scaled and delayed impulse response. The overall output is therefore

$$\forall n \in Integers, \quad y(n) = \sum_{k=-\infty}^{\infty} x(k)h(n-k) = (x * h)(n) = (h * x)(n).$$

Thus,

the output of any discrete-time LTI system is given by the convolution of the input signal and the impulse response.

Example 9.5: The three-point moving average system S of example 9.1 has impulse response

$$\forall n \in Integers, \quad h(n) = \frac{1}{3}(\delta(n) + \delta(n-1) + \delta(n-2)).$$

This can be determined from (9.4) by just letting the input be $x = \delta$. The impulse response, after all, is defined to be the output when the input is an impulse. The impulse response is shown in figure 9.2. ❐

Consider now a continuous-time LTI system $S: [Reals \to Reals] \to [Reals \to Reals]$. Define its impulse response as the output signal h when the input signal is the Dirac delta function, $h = S(\delta)$; that is,

$$\forall t \in Reals, \quad h(t) = (S(\delta))(t).$$

Now let x be any input signal and let $y = S(x)$ be the corresponding output signal. According to the sifting property, we can express x as the sum (integral) of weighted delta functions:

$$x(t) = \int_{-\infty}^{\infty} x(\tau)\delta(t-\tau)d\tau.$$

Because S is LTI, the output is a sum (integral) of the responses to each of the components (the integrand for fixed τ), or

$$\forall t \in Reals, \quad y(t) = \int_{-\infty}^{\infty} x(\tau)h(t-\tau)d\tau = (x * h)(t) = (h * x)(t). \tag{9.12}$$

Thus,

The output of any continuous-time LTI system is given by the convolution of the input signal and the impulse response.

Example 9.6: The length-three continuous-time moving average system S of example 9.2 has impulse response

$$\forall\, t \in Reals, \quad h(t) = \begin{cases} 1/3 & \text{if } t \in [0, 3] \\ 0 & \text{otherwise.} \end{cases}$$

This can be determined from (9.8) by just letting the input be $x = \delta$ and then using the sifting property of the Dirac delta function. The impulse response is, after all, defined to be the output when the input is an impulse. The impulse response is shown in figure 9.3. In general, a moving average has an impulse response with this rectangular shape. ❒

Example 9.7: Consider an M-step discrete-time delay system, as in example 8.9, where the output y is represented in terms of the input x by the difference equation

$$\forall\, n \in Integers, \quad y(n) = x(n - M). \tag{9.13}$$

The impulse response can be found by letting $x = \delta$, to get

$$h(n) = \delta(n - M).$$

The output can be represented as a convolution of the input and the impulse response,

$$y(n) = \sum_{k=-\infty}^{\infty} x(k)\delta(n - M - k) = x(n - M),$$

by using the sifting property. Of course, this agrees with (9.13). ❒

Example 9.8: Consider a T-second continuous-time delay system, where the output y is defined in terms of the input x by the equation

$$\forall\, t \in Reals, \quad y(t) = x(t - T). \tag{9.14}$$

The impulse response can be found by letting $x = \delta$, to get

$$h(t) = \delta(t - T).$$

The output can be represented as a convolution of the input and the impulse response,

$$y(t) = \int_{-\infty}^{\infty} x(\tau)\delta(t - T - \tau)d\tau = x(t - T),$$

by using the sifting property. Of course, this agrees with (9.14). ❒

Example 9.9: Suppose that we have two LTI systems (discrete or continuous time) with impulse responses h_1 and h_2 and that we connect them in a cascade structure as shown in figure 9.7. We can find the impulse response of the cascade composition by letting the input be an impulse, $x = \delta$. Then the output of the first system is its impulse response, $w = h_1$. This provides the input to the second system, and so its output is $y = h_1 * h_2$. Thus, the overall impulse response is the convolution of the two impulse responses:

$$h = h_1 * h_2.$$

We also know from chapter 8 that the frequency responses relate in a very simple way: namely,

$$H(\omega) = H_1(\omega)H_2(\omega).$$

In general, convolution in the time domain is equivalent to multiplication in the frequency domain. ❒

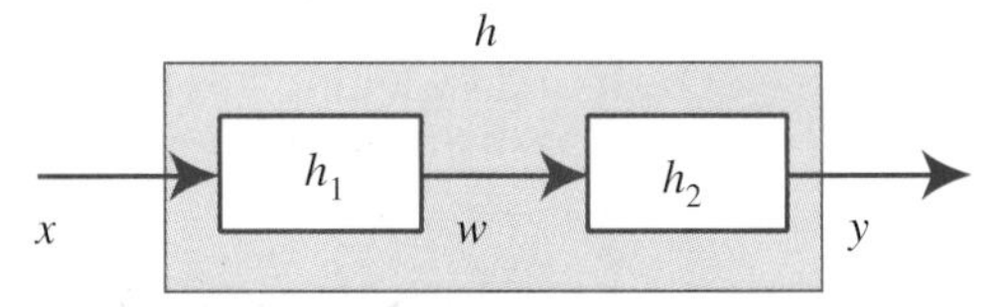

FIGURE 9.7: A cascade combination of two discrete-time LTI systems.

9.2 Frequency response and impulse response

If a discrete-time LTI system has impulse response h, then the output signal y corresponding to the input signal x is represented by the convolution sum

$$\forall n \in Integers, \quad y(n) = \sum_{m=-\infty}^{\infty} h(m)x(n - m).$$

In particular, suppose the input is the complex exponential function

$$\forall n \in Integers, \quad x(n) = e^{i\omega n},$$

for some real ω. Then the output signal is

$$\forall n \in Integers, \quad y(n) = \sum_{m=-\infty}^{\infty} h(m)e^{i\omega(n-m)} = e^{i\omega n} \sum_{m=-\infty}^{\infty} h(m)e^{-i\omega m}.$$

Recall further that when the input is a complex exponential with frequency ω, then the output is represented by

$$\forall\, n \in Integers, \quad y(n) = H(\omega)e^{i\omega n},$$

where $H(\omega)$ is the frequency response. Comparing these two expressions for the output, we see that the frequency response is related to the impulse response by

$$\boxed{\forall\, \omega \in Reals, \quad H(\omega) = \sum_{m=-\infty}^{\infty} h(m)e^{-i\omega m}.} \tag{9.15}$$

This expression allows us, in principle, to calculate the frequency response from the impulse response. Equation (9.15) gives us a way to transform h, a time-domain function, into H, a frequency-domain function. Equation (9.15) is called a **discrete-time Fourier transform** (**DTFT**). Equivalently, we say that H is the DTFT of h. Therefore, the frequency response of a discrete-time system is the DTFT of its impulse response.

Example 9.10: Consider the M-step delay from example 9.7. Its impulse response is

$$\forall\, n \in Integers, \quad h(n) = \delta(n - M).$$

We can find the frequency response by calculating the DTFT:

$$\begin{aligned} \forall\, \omega \in Reals, \quad H(\omega) &= \sum_{m=-\infty}^{\infty} h(m)e^{-i\omega m} \\ &= \sum_{m=-\infty}^{\infty} \delta(m - M)e^{-i\omega m} \\ &= e^{-i\omega M}, \end{aligned}$$

where the last step follows from the sifting property. This same result was obtained more directly in example 8.9. Note that the magnitude response is particularly simple:

$$|H(\omega)| = 1.$$

This is intuitive. An M-step delay does not change the magnitude of any complex exponential input. It only shifts its phase. ❒

Notice from (9.15) that

$$\boxed{\text{if } h \text{ is real-valued, then } H^*(-\omega) = H(\omega)} \tag{9.16}$$

(just conjugate both sides of (9.15) and evaluate at $-\omega$). This property is called **conjugate symmetry**. It implies that

$$|H(-\omega)| = |H(\omega)|. \tag{9.17}$$

This says that for any LTI system with a real-valued impulse response, a complex exponential with frequency ω undergoes the same amplitude change as a complex exponential with frequency $-\omega$.

Notice further from (9.15) that

$$\boxed{\forall\, \omega \in Reals, \quad H(\omega + 2\pi) = H(\omega);} \tag{9.18}$$

that is, the DTFT is periodic with period 2π. This says that a complex exponential with frequency ω undergoes the same amplitude and phase changes as a complex exponential with frequency $\omega + 2\pi$. This should not be surprising, inasmuch as the two complex exponentials are in fact identical:

$$\forall\, n \in Integers, \quad e^{i(\omega+2\pi)n} = e^{i\omega n} e^{i2\pi n} = e^{i\omega n},$$

because $e^{i2\pi n} = 1$ for any integer n.

The continuous-time version proceeds in the same way. Let S be a continuous-time system with impulse response h. Then the output signal y corresponding to an input signal x is expressed by

$$\forall\, t \in Reals, \quad y(t) = \int_{-\infty}^{\infty} x(t-\tau)h(\tau)d\tau.$$

In particular, if the input signal is the complex exponential

$$\forall\, t \in Reals, \quad x(t) = e^{i\omega t},$$

then the output signal is

$$y(t) = \int_{-\infty}^{\infty} e^{i\omega(t-\tau)}h(\tau)d\tau = e^{i\omega t}\int_{-\infty}^{\infty} e^{-i\omega\tau}h(\tau)d\tau.$$

The output is also given by $y(t) = H(\omega)e^{i\omega t}$, where $H(\omega)$ is the frequency response, and so we have

$$H(\omega) = \int_{-\infty}^{\infty} h(t)e^{-i\omega t}dt. \tag{9.19}$$

Therefore, given its impulse response, we can calculate the frequency response of a continuous-time LTI system by evaluating the integral (9.19). Like the DTFT, this integral transforms a time-domain signal h into a frequency-domain signal H. It is called the **continuous-time Fourier transform** (**CTFT**) or, more commonly, simply the **Fourier transform** (**FT**). Thus, the frequency response H of a continuous-time LTI system is just the CTFT of its impulse response h.

Example 9.11: Consider the T-second delay from example 9.8. Its impulse response is

$$h(t) = \delta(t - T).$$

We can find the frequency response by calculating the CTFT:

$$\begin{aligned} H(\omega) &= \int_{-\infty}^{\infty} h(t)e^{-i\omega t}dt \\ &= \int_{-\infty}^{\infty} \delta(t - T)e^{-i\omega t}dt \\ &= e^{-i\omega T}, \end{aligned}$$

where the last step follows from the sifting property. Note that the magnitude response is particularly simple:

$$|H(\omega)| = 1.$$

This is intuitive. A T-second delay does not change the magnitude of any complex exponential input. It only shifts its phase. ❐

Notice from (9.19) that the CTFT is also conjugate symmetric if h is real:

$$H^*(-\omega) = H(\omega) \tag{9.20}$$

and

$$|H(-\omega)| = |H(\omega)|. \tag{9.21}$$

9.3 Causality

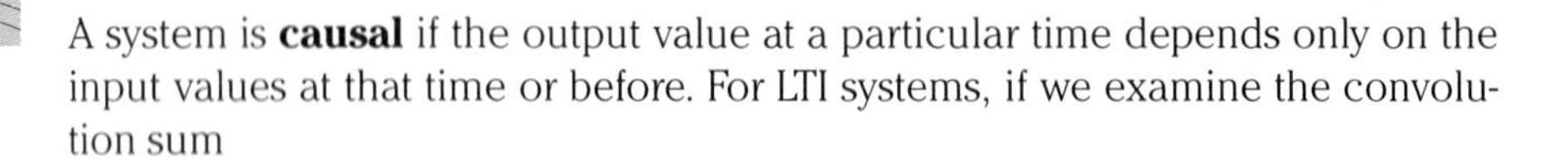

A system is **causal** if the output value at a particular time depends only on the input values at that time or before. For LTI systems, if we examine the convolution sum

$$y(n) = \sum_{m=-\infty}^{\infty} h(m)x(n-m),$$

it must be true for a causal system that $h(m) = 0$ for all $m < 0$. Were this not true, there would be nonzero terms in the sum with $m < 0$, and those terms would involve a future sample of the input, $x(n-m)$. Conversely, if $h(m) = 0$ for all $m < 0$, then the system is causal because the sum for $y(n)$ involves only the previous input values, $x(n), x(n-1), x(n-2), \ldots$.

Causality is an important practical property of a system that receives its data in **real time** (physical time). Such systems cannot possibly evaluate the future, at least not until someone invents a time machine. However, there are many situations in which causality is irrelevant. A system that processes stored data, such as digital audio properties of a compact disc or audio files in a computer, has no difficulty looking ahead in "time."

PROBING FURTHER

Causality

We can give a much more general and formal definition of causality that does not require a system to be LTI. Consider a system $S: [A \to B] \to [A \to B]$ that operates on signals of type $[A \to B]$. Assume that A is an ordered set and that B is any ordinary set. An ordered set is one with relations "$<$" and "$>$" in which, for any two elements a and b, one of the following assertions must be true:

$$a = b, \quad a > b, \text{ or } \quad a < b.$$

Examples of ordered sets are *Integers* and *Reals*.

An example of a set that is not an ordered set is *Integers* × *Integers*, in which we define the ordering relation "$<$" so that $(a, b) < (c, d)$ if $a < c$ and $b < c$, and we define "$>$" similarly. According to these definitions, for the elements (1,2) and (2,1), none of the preceding assertions is true, and so the set is not ordered.

However, we could define the ordering relation "$<$" so that $(a, b) < (c, d)$ if one of the following is true:

$$a < c$$

or

$$a = c \text{ and } b < c.$$

PROBING FURTHER

The relation ">" could be defined similarly. According to these definitions, the set is ordered.

Define a function $Prefix_t$: $[A \to B] \to [A_t \to B]$ that extracts a portion of a signal up to $t \in A$. Formally, $A_t \subset A$ such that $a \in A_t$ if $a \in A$ and $a \leq t$. Then for all $x \in [A \to B]$ and for all $t \in A_t$, $(Prefix_t(x))(t) = x(t)$.

A system S is causal if, for all $t \in A$ and for all $x, y \in [A \to B]$,

$$Prefix_t(x) = Prefix_t(y) \Rightarrow Prefix_t(S(x)) = Prefix_t(S(y)).$$

The symbol "$\Rightarrow$" reads "implies." In words, the system is causal if, when two input signals have the same prefix up to time t, it follows that the two corresponding output signals have the same prefix up to time t.

9.4 Finite impulse response filters

Consider an LTI system S: $[Integers \to Reals] \to [Integers \to Reals]$ with impulse response h: $Integers \to Reals$ that has the properties

$$h(n) = 0 \text{ if } n < 0 \quad \text{and} \quad h(n) = 0 \text{ if } n \geq L,$$

where L is some positive integer. Such a system is called a **finite impulse response** (**FIR**) system because the interesting part (the nonzero part) of the impulse response is finite in extent. Because of that property, the convolution sum becomes a finite sum:

$$y(n) = \sum_{m=-\infty}^{\infty} x(n-m)h(m) = \sum_{m=0}^{L-1} x(n-m)h(m). \qquad (9.22)$$

L is the length of the impulse response.

This sum, because it is finite, is convenient to work with. It can be used to define a procedure for computing the output of an FIR system when given its input. This makes it easy to implement on a computer. It is also reasonably straightforward to implement certain **infinite impulse response** (**IIR**) filters on a computer.

A continuous-time finite impulse response could be defined, but there is not as much motivation for doing so because, in practice, continuous-time systems rarely have finite impulse responses. Moreover, even though the convolution integral acquires finite limits, this does not make it any more computable. Computing integrals on a computer is a difficult proposition whether the limits are finite or not.

Example 9.12: We have seen in example 9.1 a three-point moving average system. An L-point moving average system with input x has output y as expressed by

$$y(n) = \frac{1}{L} \sum_{m=0}^{L-1} x(n-m).$$

The output at index n is the average of the most recent L inputs. Such a filter is widely used in stock markets to try to detect trends in stock prices. We can study its properties. First, it is easy to see that it is an LTI system. In fact, it is an FIR system with impulse response

$$h(n) = \begin{cases} 1/L & \text{if } 0 \leq n < L \\ 0 & \text{otherwise.} \end{cases}$$

To see this, just let $x = \delta$. The frequency response is the DTFT of this impulse response:

$$\begin{aligned} H(\omega) &= \sum_{m=-\infty}^{\infty} h(m) e^{-i\omega m} \\ &= \frac{1}{L} \sum_{m=0}^{L-1} e^{-i\omega m}. \end{aligned}$$

With the help of the useful identity*

$$\boxed{\sum_{m=0}^{L-1} a^m = \frac{1-a^L}{1-a},} \tag{9.23}$$

we can write the frequency response as

$$H(\omega) = \frac{1}{L} \left(\frac{1-e^{-i\omega L}}{1-e^{-i\omega}} \right),$$

* You can verify the identity by multiplying both sides by $1-a$.

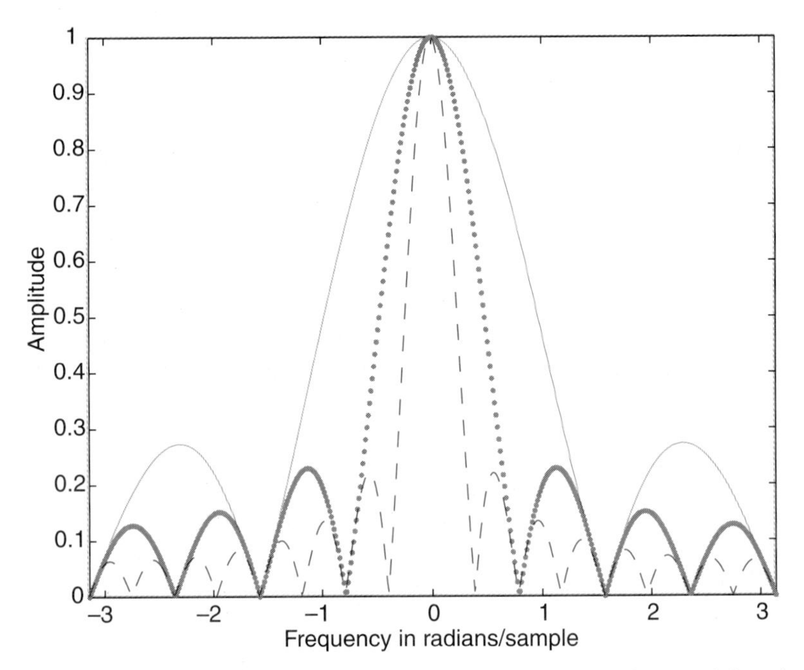

FIGURE 9.8: The magnitude response of the moving average filter with lengths $L = 4$ (color line), $L = 8$ (gray line), and $L = 16$ (dashed color line).

where $a = e^{-i\omega}$. We can plot the magnitude of the frequency response by using MATLAB®* as follows (for $L = 4$):†

```
L = 4;
omega = [-pi:pi/250:pi];
H = (1/L)*(1-exp(-i*omega*L))./(1-exp(-i*omega));
plot(omega, abs(H));
xlabel('frequency in radians/sample')
```

The plot is shown in figure 9.8, together with plots for $L = 8$ and $L = 16$. Notice that the plot shows only the frequency range $-\pi < \omega < \pi$. Because the DTFT is periodic with period 2π, this plot simply repeats outside this range.

Notice that in all three cases shown in figure 9.8, the frequency response has a lowpass characteristic. A constant component (zero frequency) in the input passes through the filter unattenuated. Certain higher frequencies,

* MATLAB is a registered trademark of The MathWorks, Inc.

† This issues a “divide-by-zero” warning but yields a correct plot nonetheless. The magnitude of the frequency response at $\omega = 0$ is 1, as you can verify by using L’Hôpital’s rule.

such as $\pi/2$, are completely eliminated by the filter. However, if the intent was to design a lowpass filter, then we have not done very well. Some of the higher frequencies are attenuated only by a factor of about 1/10 (for the 16-point moving average) or 1/3 (for the four-point moving average). We can do better than that with a more intelligently designed filter. ❒

9.4.1 *Design of FIR filters*

The moving average system in example 9.12 exhibits a lowpass frequency response but not a particularly good lowpass frequency response. A great deal of intellectual energy has historically gone into finding ways to choose the impulse response of an FIR filter. The subject is quite deep. Fortunately, much of the work that has been done is readily available in the form of easy-to-use software, and so the researcher does not need to be particularly knowledgeable about these techniques to be able to design good FIR filters.

Example 9.13: MATLAB's filter design facilities in its digital signal processing (DSP) toolbox provide some well-studied algorithms for filter design. For example,

```
>> h = remez(7,[0,0.125,0.25,1],[1,1,0,0])
h =
  Columns 1 through 6
    0.0849    0.1712    0.1384    0.1912    0.1912    0.1384
  Columns 7 through 8
    0.1712    0.0849
```

This returns the impulse response of a length eight-FIR lowpass filter. The arguments to the **remez** function specify the filter by outlining approximately the desired frequency response. You can use MATLAB's online help to get the details, but in brief, the arguments just given define a **passband** (a region of frequency in which the gain is approximately 1) and a **stopband** (a region of frequency in which the gain is approximately 0). The first argument, 7, specifies that the length of the filter should be 8 (The MathWorks may explain why this is off by one). The second argument, `[0,0.125,0.25,1]`, specifies that the first frequency band begins at 0 and extends to 0.125π radians/sample and that the second band begins at 0.25π and extends to π radians/sample. (The π factor is omitted.) The unspecified band, from 0.125π to 0.25π, is a "don't care" region, a **transition band** that allows for a gradual transition between the passband and the stopband.

The last argument, `[1,1,0,0]`, specifies that the first band should have a magnitude frequency response of approximately 1 and that the second band should have a magnitude frequency response of approximately 0.

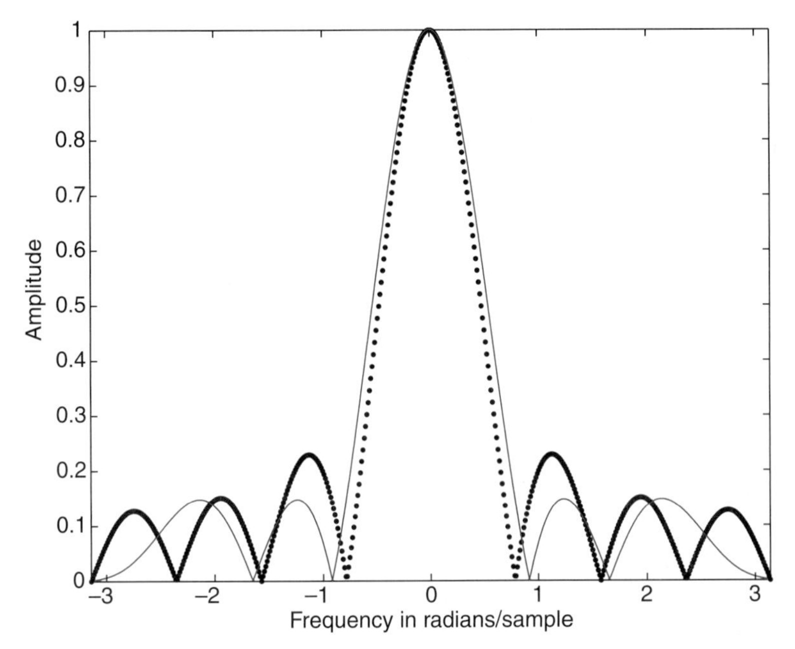

FIGURE 9.9: Magnitude response of an FIR filter (color line) designed with the Parks-McClellan algorithm (the MATLAB `remez` function), in comparison with an ordinary moving average (black dotted line).

The frequency response of this filter can be directly calculated and plotted with the following (rather brute force) MATLAB code:

```
omega = [-pi:pi/250:pi];
H = h(1) + h(2)*exp(-i*omega) + h(3)*exp(-i*2*omega) +
    h(4)*exp(-i*3*omega) + h(5)*exp(-i*4*omega) +
    h(6)*exp(-i*5*omega) + h(7)*exp(-i*6*omega) +
    h(8)*exp(-i*7*omega);
plot(omega, abs(H));
```

The result is shown in figure 9.9, where it is plotted together with the magnitude response of a moving average filter with the same length. Notice that the attenuation in the stopband is better for the filter designed with the `remez` function than for the moving average filter. Also notice that it is not 0, despite our request that it be 0, and that the passband gain is not 1, despite our request that it be 1. ❒

The `remez` function used in this example applies an optimization algorithm called the Parks-McClellan algorithm, which is based on the Remez exchange algorithm (hence the name of the function). This algorithm ensures that the sidelobes (the humps in the stopband) are of equal size. This turns out to minimize their maximum size. The Parks-McClellan algorithm is widely used for designing FIR filters.

In this example, the `remez` function is unable to deliver a filter that meets our request. The gain in the stopband is not 0, and the gain in the passband is not 1. In general, this is the problem with filter design (and much of the rest of life): We cannot have what we want. We can obtain an approximation by using more resources (also as in much of life). In this case, that means designing a filter with a longer impulse response. Because the impulse response is longer, the filter is more costly to implement. This is because the finite summation in (9.22) has more terms.

Example 9.14: Consider, for example, the impulse response returned by

```
h = remez(63,[0,0.125,0.25,1],[1,1,0,0]);
```

This has length 64. The magnitude frequency response of this can be calculated by using the following MATLAB code (see lab 10 in the Lab Manual for an explanation):

```
H = fft(h, 1024);
magnitude = abs([H(513:1024),H(1:512)]);
plot([-pi:pi/512:pi-pi/1024], magnitude)
```

This is plotted in figure 9.10. ❐

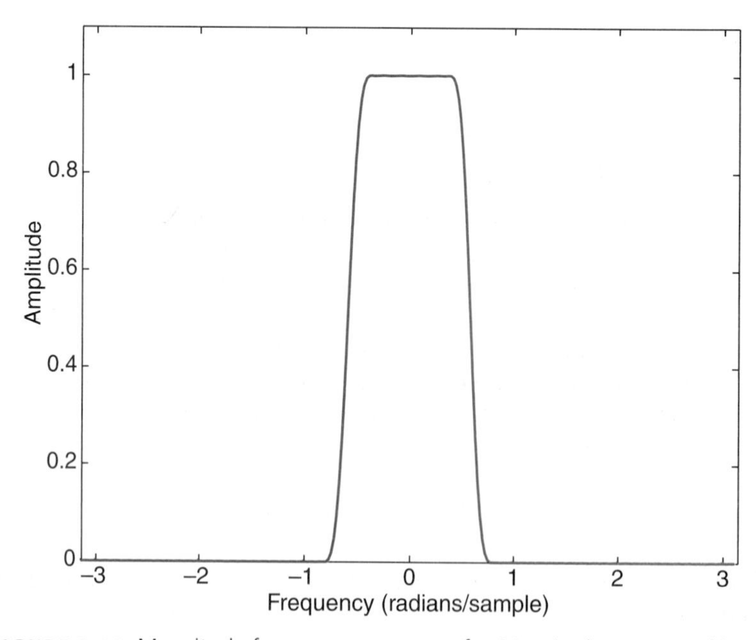

FIGURE 9.10: Magnitude frequency response of a 64-point lowpass FIR filter designed with the Parks-McClellan algorithm.

The frequency response achieved in this example appears in figure 9.10 to exactly match our requirements. However, this plot is somewhat deceptive. It suggests that the magnitude frequency response in the stopband is in fact 0. In fact, it is not, except at certain discrete frequencies. If you zoom in sufficiently on the plot, you will see that. Instead of zooming in on such plots, engineers usually construct them by using a logarithmic vertical axis, as explained in the next subsection.

9.4.2 *Decibels*

The amplitude response of a filter is the magnitude of the frequency response. It specifies the **gain** experienced by a complex exponential at that frequency. This gain is simply the ratio of the output amplitude to the input amplitude. Because it is the ratio of two amplitudes with same units, the gain itself is unitless.

It is customary in engineering to describe gains in a logarithmic scale. Specifically, if the gain of a filter at frequency ω is $|H(\omega)|$, then we write instead

$$\boxed{G(\omega) = 20\log_{10}(|H(\omega)|).} \tag{9.24}$$

This has units of **decibels**, written **dB**. The multiplication by 20 and the use of base 10 logarithms is by convention (see Probing further: Decibels box).

Example 9.15: The plot in figure 9.10 is redone with the following MATLAB commands:

```
H = fft(h, 1024);
dB = 20*log10(abs([H(513:1024),H(1:512)]));
plot([-pi:pi/512:pi-pi/1024], dB)
```

After some adjustment of the axes, this yields the plot in figure 9.11. Notice that in the passband, the gain is 0 dB. This is because $\log_{10}(1) = 0$. Zero decibels corresponds to a gain of unity. In the stopband, the gain is almost -70 dB, a very respectable attenuation. If this were an audio signal, then frequency components in the stopband would probably not be audible as long as there is some signal in the passband to mask them. They are 70 dB weaker than the components in the passband, which translates into a factor of 3,162 smaller in amplitude, because

$$20\log_{10}(1/3{,}162) \approx -70.$$

We can find the gain given decibels by solving (9.24) for $|H(\omega)|$ in terms of $G(\omega)$ to get

$$|H(\omega)| = 10^{G(\omega)/20}.$$

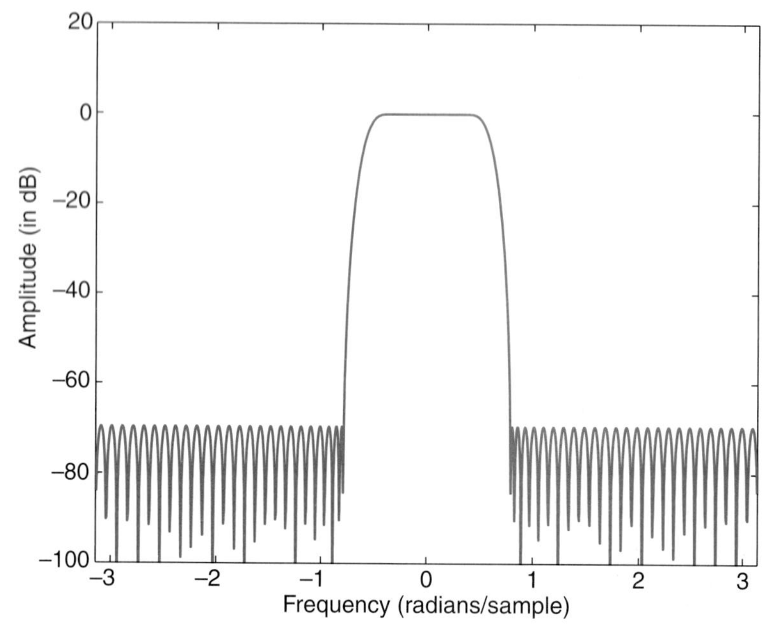

FIGURE 9.11: Magnitude frequency response in decibels of a 64-point lowpass FIR filter designed with the Parks-McClellan algorithm.

In this case, in the stopband,

$$|H(\omega)| = 10^{-70/20} \approx 1/3{,}162.$$

Notice that $20\log_{10}(0) = -\infty$. Thus, whenever the gain is zero, the gain in decibels is $-\infty$. The downward spikes in figure 9.11 occur at frequencies at which the gain is 0, or $-\infty$ dB. The spikes do not necessarily show this precisely, because the plot does not necessarily evaluate the gain at precisely the frequency that yields 0; hence the ragged bottoms. ❐

PROBING FURTHER

Decibels

The term "decibel" literally means one tenth of a **bel**, which is named after Alexander Graham Bell (1847–1922). This unit of measure was originally developed by telephone engineers at Bell Telephone Labs to designate the ratio of the **power** of two signals.

Power is a measure of energy dissipation (work done) per unit time. It is measured in **watts** for electronic systems. One bel is defined to be a factor of 10 in power. Thus, a 1,000-watt hair dryer dissipates 1 bel, or 10 dB, more power than that used by a 100-watt

PROBING FURTHER

lightbulb. Let $p_1 = 1{,}000$ watts be the power of the hair dryer and $p_2 = 100$ be the power of the lightbulb. Then the ratio is

$$\log_{10}(p_1/p_2) = 1 \text{ bel}.$$

In decibels, this becomes

$$10\log_{10}(p_1/p_2) = 10 \text{ dB}.$$

Comparing this with (9.24), we notice a discrepancy: there, the multiplying factor is 20, not 10. That is because the ratio in (9.24) is a ratio of amplitudes, not powers. In electronic circuits, if an amplitude represents the voltage across a resistor, then the power dissipated by the resistor is proportional to the *square* of the amplitude. Let a_1 and a_2 be two such amplitudes. Then the ratio of their powers is

$$10\log_{10}(a_1^2/a_2^2) = 20\log_{10}(a_1/a_2).$$

Hence the multiplying factor of 20 instead of 10.

A 3-dB power ratio amounts to a factor of 2 in power. In amplitudes, this is a ratio of $\sqrt{2}$. The edge of the passband of a bandpass filter is often defined to be the frequency at which the power drops to half, which is the frequency at which the gain is 3 dB below the passband gain. The magnitude frequency response here is $1/\sqrt{2}$ times the passband gain.

In audio technology, decibels are used to measure sound pressure. You might hear statements such as "A jet engine at 10 meters produces 120 dB of sound." But decibels are measures of power ratios, not absolute power, so what does such a statement mean? By convention, sound pressure is measured in relation to a defined reference of 20 micropascals, where 1 pascal is a pressure of 1 newton per square meter. For most people, this is approximately the threshold of hearing at 1 kHz. Thus, a sound at 0 dB is barely audible. A sound at 10 dB has 10 times the power. A sound at 100 dB has 10^{10} times the power. You would therefore not be surprised to learn that the jet engine just described would probably make you deaf without ear protection.

9.5 Infinite impulse response (IIR) filters

The equation for an FIR filter (9.22) is a difference equation relating the output at index n to the inputs at indices $n - L + 1$ through n. A more general form of this difference equation includes more indices of the output,

$$\forall n \in Integers, \quad y(n) = \sum_{m=0}^{L-1} x(n-m)b(m) + \sum_{m=1}^{M-1} y(n-m)a(m), \quad (9.25)$$

where L and M are positive integers and $b(m)$ and $a(m)$ are **filter coefficients**, which are usually real valued. If all the a coefficients are zero, then this reduces to an FIR filter with impulse response $b = h$. However, if any a coefficient is nonzero, then the impulse response of this filter never completely dies out, inasmuch as the output keeps getting recycled to affect future outputs. Such a filter is called an infinite impulse response (IIR) filter or, sometimes, a **recursive filter**.

Example 9.16: Consider a causal LTI system defined by the difference equation

$$\forall n \in Integers, \quad y(n) = x(n) + 0.9y(n-1).$$

We can find the impulse response by letting $x = \delta$, the Kronecker delta function. Because the system is causal, we know that $h(n) = 0$ for $n < 0$ (see section 9.3). Thus,

$$\begin{aligned} y(n) &= 0, \quad \text{if } n < 0 \\ y(0) &= 1 \\ y(1) &= 0.9 \\ y(2) &= 0.9^2 \\ y(3) &= 0.9^3. \end{aligned}$$

Noticing the pattern, we conclude that

$$y(n) = (0.9)^n u(n),$$

where u is the **unit step**, seen before in (2.15):

$$u(n) = \begin{cases} 1 & \text{if } n \geq 0 \\ 0 & \text{otherwise.} \end{cases} \qquad (9.26)$$

The output y and the magnitude of the frequency response (in decibels) are plotted in figure 9.12. Notice that this filter has about 20 dB of gain at DC or zero frequency, dropping to about −6 dB at π radians/sample. ❒

9.5.1 *Designing IIR filters*

Designing IIR filters amounts to choosing the a and b coefficients for (9.25). As with FIR filters, how to choose these coefficients is a well-studied subject, and the results of this study are (for the most part) available in easy-to-use software. There are four widely used methods for calculating these coefficients; these methods

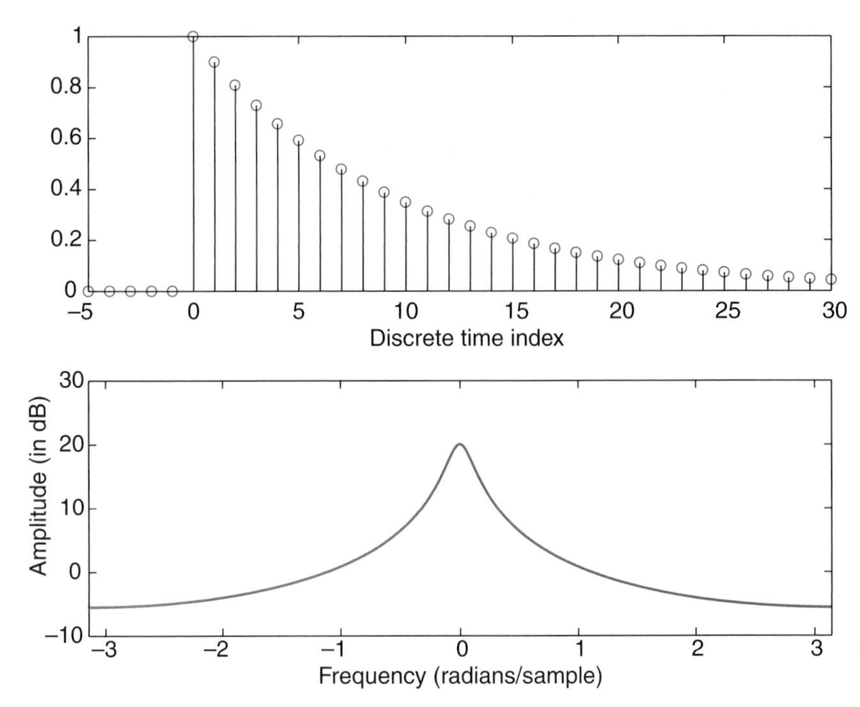

FIGURE 9.12: Impulse response (top) and magnitude frequency response in decibels (bottom) of a simple causal IIR filter.

result in the creation of four types of filters: **Butterworth**, **Chebyshev 1**, **Chebyshev 2**, and **elliptic** filters.

Example 9.17: We can design one of each of the four types of filters by using the MATLAB commands `butter`, `cheby1`, `cheby2`, and `ellip`. The arguments for each of these specify either a **cutoff frequency**, which is the frequency at which the filter achieves -3 dB gain, or, in the case of `cheby2`, the edge of the stopband. For example, to get lowpass filters with a gain of about 1 from 0 to $\pi/8$ radians/sample and a stopband at higher frequencies, we can use the following commands:

```
N = 5;
Wn = 0.125;
[B1, A1] = butter(N, Wn);
[B2, A2] = cheby1(N,1,Wn);
[B3, A3] = cheby2(N,70,0.25);
[B4, A4] = ellip(N,1,70,Wn);
```

The returned values are vectors containing the a and b coefficients of (9.25). The magnitude frequency responses of the resulting filters are shown in figure 9.13. In that figure, we show only positive frequencies, because the magnitude frequency response is symmetric. We also show the frequency range only from 0 to π, because the frequency response is periodic with period 2π. Notice that the Butterworth filter has the most gradual **rolloff** from the passband to the stopband. Thus, for a given filter order, a Butterworth filter yields the weakest lowpass characteristic of the four. The elliptic filter has the sharpest rolloff. The Butterworth filter, on the other hand, has the smoothest characteristic. The Chebyshev 1 filter has **ripple** in the passband, the Chebyshev 2 filter has ripple in the stopband, and the elliptic filter has ripple in both.

In the preceding MATLAB commands, `N` is the filter **order**, equal to L and M in (9.25). It is a constraint of these filter design techniques that $L = M$ in (9.25). `Wn` is the cutoff frequency, as a fraction of π radians/sample. A cutoff

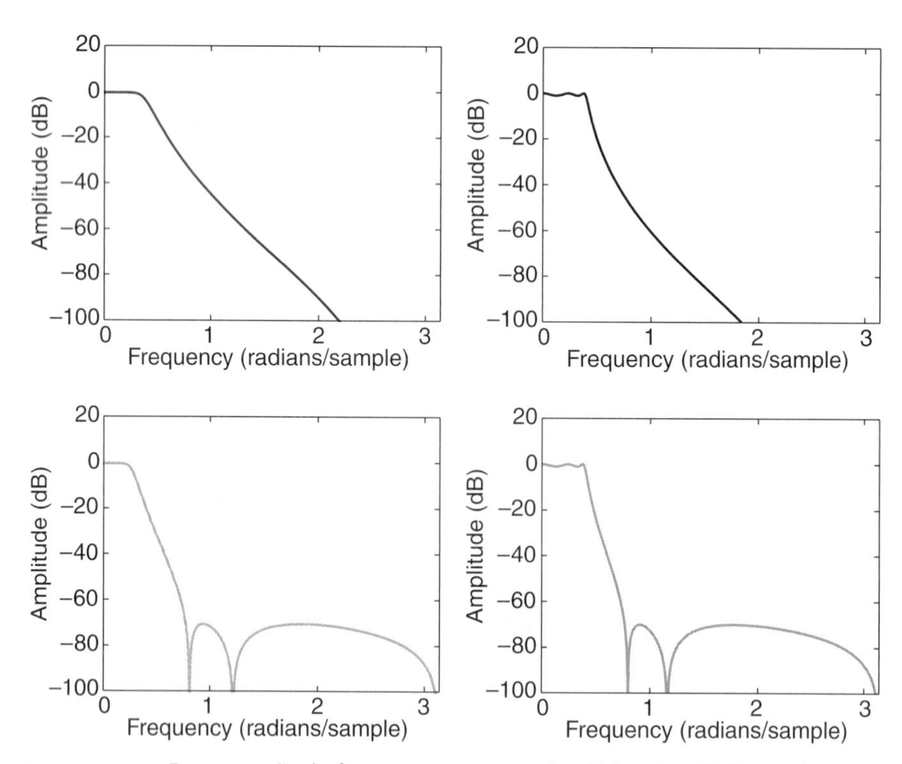

FIGURE 9.13: Four magnitude frequency responses for fifth-order IIR filters of types Butterworth (upper left), Chebyshev 1 (upper right), Chebyshev 2 (lower left), and elliptic (lower right).

frequency of 0.125 means $0.125\pi = \pi/8$ radians/sample. The "1" argument in the `cheby1` command specifies the amount of passband ripple that we are willing to tolerate (in decibels). The 70 in the `cheby2` and `ellip` commands specifies the amount of stopband attenuation that we want (in decibels). Finally, the 0.25 in the `cheby2` line specifies the edge of the stopband. ❐

9.6 Implementation of filters

We now have several ways to describe an LTI system (a filter). We can give a state-space description, a frequency response, an impulse response, or a difference equation such as (9.25) or (9.22). All are useful. The state-space description and the difference equations prove the most useful when the filter is constructed.

A realization of a filter in hardware or software is called an **implementation**. Do not confuse **filter design** with **filter implementation**. The term "filter design" is used in the engineering community to refer to the choice of frequency response, impulse response, or coefficients for a difference equation, not to how the frequency response is implemented in hardware or software. In this section, we talk about implementation.

The output y of an FIR filter with impulse response h and input x is given by (9.22). The output of an IIR filter with coefficients a and b is given by (9.25). Each of these difference equations defines a procedure for calculating the output given the input. We discuss various ways of implementing these procedures.

9.6.1 *MATLAB implementation*

If x is finite and we can interpret it as an infinite signal that is zero outside the specified range, then we can compute the output of an FIR filter by using MATLAB's `conv` function, and we can compute the output of an IIR filter by using `filter`.

Example 9.18: Consider an FIR filter with an impulse response of length L. If `x` is a vector containing the P values of the input and `h` is a vector containing the L values of the impulse response, then

```
y = conv(x, h);
```

yields a vector containing $L + P - 1$ values of the output. ❐

PROBING FURTHER

Java implementation of an FIR filter

The following Java class implements an FIR filter:

```
1  class FIR {
2    private int length;
3    private double[] delayLine;
4    private double[] impResp;
5    private int count = 0;
6    FIR(double[] coefs) {
7      length = coefs.length;
8      impResp = coefs;
9      delayLine = new double[length];
10   }
11   double getOutputSample(double inputSample) {
12     delayLine[count] = inputSample;
13     double result = 0.0;
14     int index = count;
15     for (int i=0; i<length; i++) {
16       result += impResp[i] * delayLine[index--];
17       if (index < 0) index = length-1;
18     }
19     if (++count >= length) count = 0;
20     return result;
21   }
22 }
```

A class in Java (and in any object-oriented language) has both data members and methods. The methods are procedures that operate on the data members and may or may not take arguments or return values. In this case, there are two procedures: "FIR" and "getOutputSample." The first, lines 6 to 10, is a constructor, which is a procedure that is called to create an FIR filter. It takes one argument, an array of double-precision floating-point numbers that specify the impulse response of the filter. The second, lines 11 to 22, takes a new input sample value as an argument and returns a new output sample. It also updates the delay line by using a strategy called circular buffering; that is, the "count" member is used to keep track of where each new input sample should go. It is increased incrementally (line 19) each time the `getOutputSample()` method is called. When it exceeds the length of the buffer, it is reset to zero. Thus, at all times, it contains the L most recently received input data samples. The most recent one is at index `count` in the buffer. The second most recent is at `count - 1` or, if that is negative, at `length - 1`. Line 17 makes sure that the variable `index` remains within the confines of the buffer as we iterate through the loop.

PROBING FURTHER

Programmable DSP implementation of an FIR filter

The following section of code is the assembly language for a programmable DSP, which is a specialized microprocessor designed to implement signal processing functions efficiently in embedded systems (such as cellular telephones, digital cordless telephones, and digital audio systems). This particular code is for the Motorola DSP56000 family of processors.

```
1    fir    movep           x:input,x:(r0)
2           clr a           x:(r0)-,x0        y:(r4)+,y0
3           rep m0
4           mac x0,y0,a     x:(r0)-,x0        y:(r4)+,y0
5           macr x0,y0,a    (r0)+
6           movep a,x:output
7           jmp fir
```

This processor has two memory banks, called `x` and `y`. The code assumes that each input sample can be successively read from a memory location called `input` and that the impulse response is stored in `y` memory beginning at an address stored in register `r4`. Moreover, it assumes that register `r0` contains the address of a section of `x` memory to use for storing input samples (the delay line) and that this register has been set up to perform modulo addressing. Modulo addressing means that if it increases or decreases beyond the range of its buffer, then the address wraps around to the other end of the buffer. Finally, it assumes that the register `m0` contains an integer specifying the number of samples in the impulse response minus one.

The key line (the one that does most of the work) is line 4. It follows a `rep` instruction, which causes that one line to be repeatedly executed the number of times specified by register `m0`. Line 4 multiplies the data in register `x0` by that in `y0` and adds the result to `a` (the accumulator). Such an operation is called a "multiply and accumulate," or mac, operation. At the same time, it loads `x0` with an input sample from the delay line, in `x` memory at a location given by `r0`. It also loads register `y0` with a sample from the impulse response, stored in `y` memory at a location given by `r4`. At the same time, it increases `r0` and decreases `r4`. This one-line instruction, which carries out several operations in parallel, is obviously highly tuned to FIR filtering. Indeed, FIR filtering is a major function of processors of this type.

For IIR examples involving `filter`, see lab 10 (in the Lab Manual). This strategy, of course, works only for finite input data, inasmuch as MATLAB requires that the input be available in a finite vector.

Discrete-time filters can be implemented by using standard programming languages and using assembly languages on specialized processors (see the following boxes). These implementations do not have the limitation that the input be finite.

9.6.2 *Signal flow graphs*

We can describe the computations in a discrete-time filter by using a block diagram with three visual elements: a unit delay, a multiplier, and an adder. In the convolution sum for an FIR filter,

$$y(n) = \sum_{m=0}^{L-1} h(m)x(n-m),$$

notice that at each n, we need access to $x(n), x(n-1), x(n-2), \ldots, x(n-L+1)$. We can maintain this set of values by cascading a set of **unit delay** elements to form a **delay line**, as shown in figure 9.14.

For each integer n, the output sample is the values in the delay line scaled by $h(0), h(1), \ldots, h(L-1)$. To obtain the values in the delay line, we simply tap the delay line, as shown in figure 9.15. The triangular boxes denote **multipliers** that multiply by a constant ($h(m)$, in this case). The circles with the plus signs denote **adders**. The structure in figure 9.15 is called a **tapped delay line** description of an FIR filter.

Diagrams of the type shown in figure 9.15 are called **signal flow graphs** because they describe computations by focusing on the flow of signals between operators. Signal flow graphs can be quite literal descriptions of digital hardware that implements a filter. But what they really describe is the computation, regardless of whether the implementation is in hardware or software.

An IIR filter can also be described with the use of a signal flow graph. Consider the difference equation in (9.25). A signal flow graph description of this equation is shown in figure 9.16. Notice that the left side is structurally equivalent

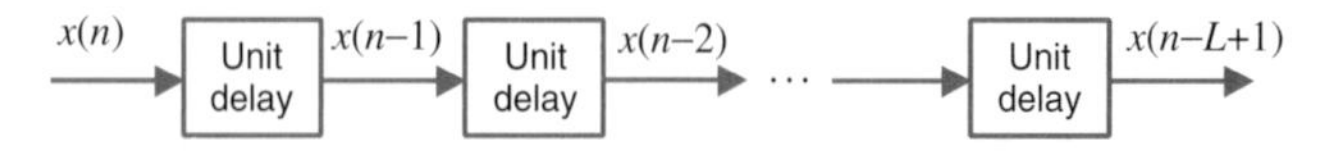

FIGURE 9.14: A delay line.

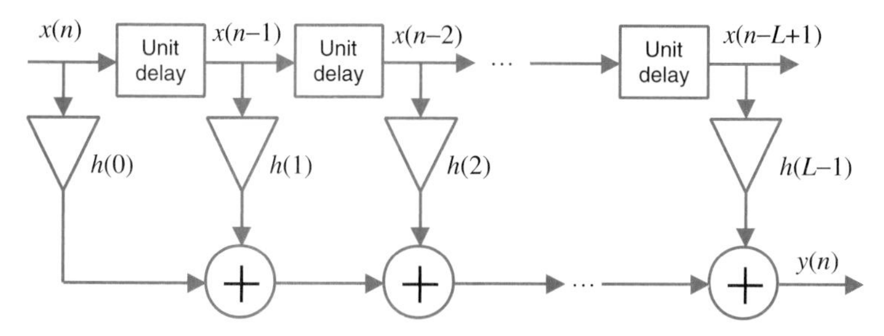

FIGURE 9.15: A tapped delay line realization of an FIR filter, described as a signal flow graph.

to the tapped delay line but stood on end. The right side represents the **feedback** in the filter, or the **recursion**, that makes it an IIR filter rather than an FIR filter.

The filter structure in figure 9.16 can be simplified somewhat. Observe that because the left and right sides are LTI systems, their order can be reversed without changing the output. Once the order is reversed, the two delay lines can be combined into one, as shown in figure 9.17. There are many other structures that can be used to realize IIR filters.

The relative advantages of one structure over another is a fairly deep topic, depending primarily on an understanding of the effects of finite precision arithmetic. It is beyond the scope of this text.

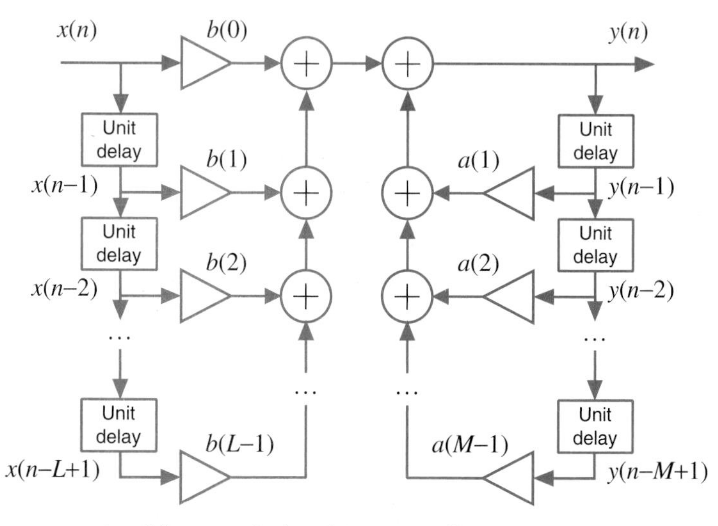

FIGURE 9.16: A signal flow graph describing an IIR filter. This is called a direct form 1 filter structure.

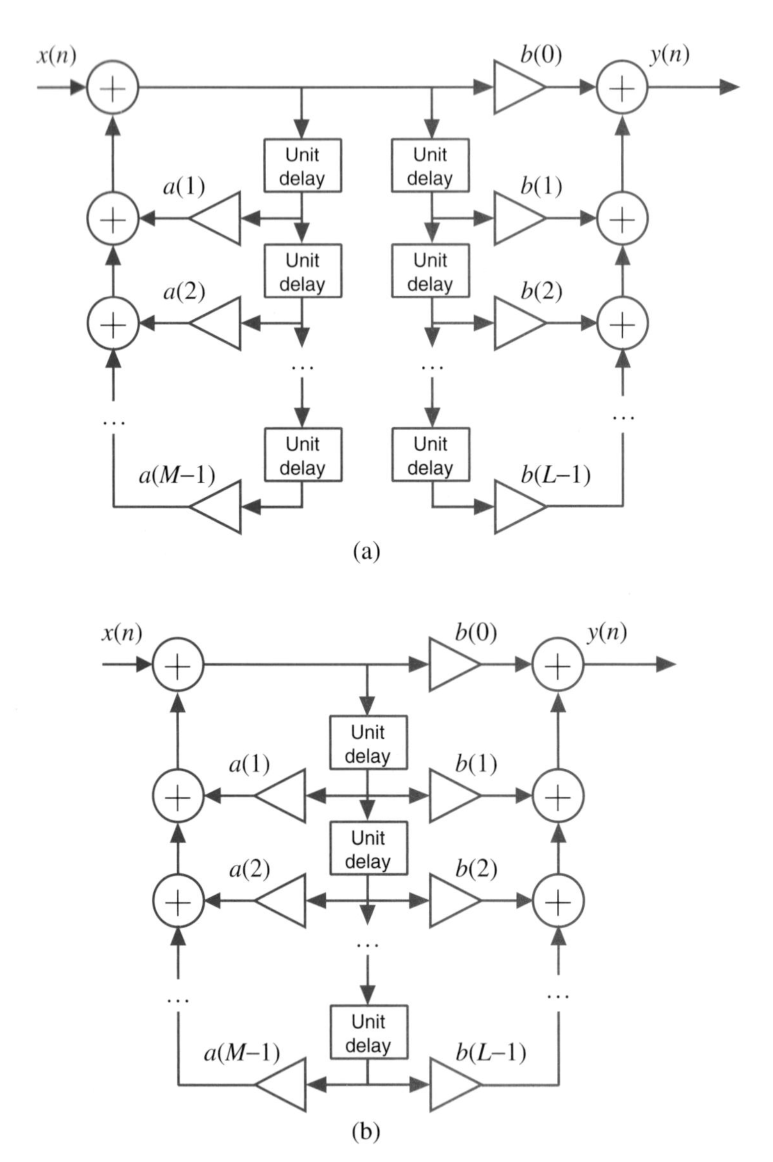

FIGURE 9.17: (a) A signal flow graph equivalent to that in figure 9.16, obtained by reversing the left and right components. (b) A signal flow graph equivalent to that in (a) obtained by merging the two delay lines into one. This is called a direct form 2 filter structure.

9.7 Summary

The output of an LTI system contains only those frequencies that are also present in the input. Because of this property, LTI systems are often called filters. This chapter considers how filters may be implemented. The convolution sum (for discrete-time systems) and convolution integral (for continuous-time systems) provide descriptions of how to calculate an output when given an input signal. This description highlights the usefulness of the response of a system to an idealized impulse, because the output is given by the convolution of the input and the impulse response. The frequency response, moreover, is shown to be the Fourier transform of the impulse response. The Fourier transform is a generalization of the Fourier series considered in previous chapters. Fourier transforms are considered in much more depth in chapter 10.

This chapter also considers the design of filters, by which we mean the selection of a suitable frequency response. Engineers are, as usual, forced to compromise to get an acceptable response at an acceptable computational cost. The response is assessed by considering how effectively it performs the desired filtering operation, and the cost is assessed by considering the arithmetic calculation that must be realized to implement the filter. The chapter closes with a discussion of alternative ways of realizing the arithmetic calculation.

EXERCISES

KEY: E = mechanical T = requires plan of attack C = more than 1 answer

E 1. Consider an LTI discrete-time system *Filter* with impulse response

$$h(n) = \sum_{k=0}^{7} \delta(n-k),$$

where δ is the Kronecker delta function.

(a) Sketch $h(n)$.

(b) Suppose the input signal $x\colon Integers \to Reals$ is such that $\forall\, n \in Integers$, $x(n) = \cos(\omega n)$, where $\omega = \pi/4$ radians/sample. Give a simple expression for $y = Filter(x)$.

(c) Give the value of $H(\omega)$ for $\omega = \pi/4$, where H is the frequency response.

E 2. Consider the continuous-time moving average system S, whose impulse response is shown in figure 9.3. Find its frequency response. The following fact from calculus may be useful:

$$\int_a^b e^{c\omega} c\, d\omega = e^{cb} - e^{ca}$$

for real a and b and complex c. Use MATLAB to plot the magnitude of this frequency response over the range −5 Hz to 5 Hz. Note the symmetry of the magnitude response, as required by (9.21).

E 3. Consider a continuous-time LTI system with impulse response given by

$$\forall\, t \in Reals, \quad h(t) = \delta(t-1) + \delta(t-2),$$

where δ is the Dirac delta function.

(a) Find a simple equation relating the input x and output y of this system.

(b) Find the frequency response of this system.

(c) Use MATLAB to plot the magnitude frequency response of this system in the range −5 to 5 Hz.

E 4. Consider a discrete-time LTI system with impulse response h given by

$$\forall\, n \in Integers, \quad h(n) = \delta(n) + 2\delta(n-1).$$

(a) Plot the impulse response.

(b) Find and sketch the output when the input is u, the unit step, defined by (9.26).

(c) Find and sketch the output when the input is a ramp:

$$r(n) = \begin{cases} n & \text{if } n \geq 0 \\ 0 & \text{otherwise.} \end{cases}$$

(d) Find the frequency response.

(e) Show that the frequency response is periodic with period 2π.

(f) Show that the frequency response is conjugate symmetric.

(g) Give a simplified expression for the magnitude response.

(h) Give a simplified expression for the phase response.

(i) Suppose that the input x is given by

$$\forall\, n \in Integers, \quad x(n) = \cos(\pi n/2 + \pi/6) + \sin(\pi n + \pi/3).$$

Find the output y.

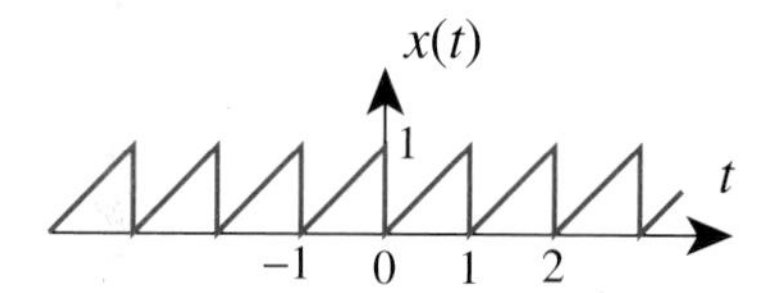

FIGURE 9.18: A sawtooth signal.

E 5. Consider the sawtooth signal shown in figure 9.18. This is a periodic, continuous-time signal. Suppose it is filtered by an LTI system with frequency response

$$H(\omega) = \begin{cases} 1 & \text{if } |\omega| \le 2.5 \text{ radians/second} \\ 0 & \text{otherwise.} \end{cases}$$

What is the output?

E 6. Suppose that the following difference equation relates the input x and output y of a discrete-time, causal LTI system S:

$$y(n) + \alpha y(n-1) = x(n) + x(n-1)$$

for some constant α.

(a) Find the impulse response h.

(b) Find the frequency response H.

(c) Find a sinusoidal input with nonzero amplitude such that the output is zero.

(d) Use MATLAB to create a plot of the magnitude of the frequency response, assuming $\alpha = -0.9$.

(e) Find a state-space description for this system (define the state s and find A, b, c^T, d).

(f) Suppose $\alpha = 1$. Find the impulse response and frequency response. Make sure your answer makes sense (check it against the original difference equation).

T 7. Each of the following statements refers to a discrete-time system S with input x and output y. Determine whether the statement is true or false. The signal u described in parts (c) and (d) is the unit step, defined by (9.26). The signal δ described in parts (e) and (f) is the Kronecker delta function.

(a) Suppose you know that if x is a sinusoid, then y is a sinusoid. Then you can conclude that S is LTI.

(b) Suppose you know that S is LTI and that if $x(n) = \cos(\pi n/2)$, then $y(n) = 2\cos(\pi n/2)$. Then you have enough information to determine the frequency response.

(c) Suppose you know that S is LTI and that if $x(n) = \delta(n)$, then

$$y(n) = (0.9)^n u(n).$$

Then you have enough information to determine the frequency response.

(d) Suppose you know that S is LTI and that if $x(n) = u(n)$, then $y(n) = (0.9)^n u(n)$. Then you have enough information to determine the frequency response.

(e) Suppose you know that S is causal, that input $x(n) = \delta(n)$ produces output $y(n) = \delta(n) + \delta(n-1)$, and that input $x'(n) = \delta(n-2)$ produces output $y'(n) = 2\delta(n-2) + \delta(n-3)$. Then you can conclude that S is not LTI.

(f) Suppose you know that S is causal and that if $x(n) = \delta(n) + \delta(n-2)$, then $y(n) = \delta(n) + \delta(n-1) + 2\delta(n-2) + \delta(n-3)$. Then you can conclude that S is not LTI.

T 8. Consider the continuous-time systems S_k given by, $\forall\, t \in Reals$,

$$(S_1(x))(t) = x(t-2),$$

$$(S_2(x))(t) = x(t+2),$$

$$(S_3(x))(t) = x(t) - 2,$$

$$(S_4(x))(t) = x(2-t),$$

$$(S_5(x))(t) = x(2t),$$

$$(S_6(x))(t) = t^2 x(t).$$

(a) Which of these systems are linear?

(b) Which of these systems are time invariant?

(c) Which of these systems are causal?

T 9. Consider an LTI discrete-time system *Filter* with impulse response

$$\forall\, n \in Integers, \quad h(n) = \delta(n) + \delta(n-2),$$

where δ is the Kronecker delta function.

(a) Sketch h.

(b) Find the output when the input is u, the unit step, represented by (9.26).

(c) Find the output when the input is a ramp:

$$r(n) = \begin{cases} n & \text{if } n \geq 0 \\ 0 & \text{otherwise.} \end{cases}$$

(d) Suppose the input signal x is such that

$$\forall\, n \in Integers, \quad x(n) = \cos(\omega n),$$

where $\omega = \pi/2$ radians/sample. Give a simple expression for $y = Filter(x)$.

(e) Give an expression for $H(\omega)$ that is valid for all ω, where H is the frequency response.

(f) Sketch the magnitude of the frequency response. Can you explain your answer in part (d)?

(g) Is there any other frequency at which a sinusoidal input with a nonzero amplitude yields an output that is zero?

T 10. Consider an LTI discrete-time system *Filter* with impulse response

$$h(n) = \delta(n) - \delta(n-1),$$

where δ is the Kronecker delta function.

(a) Sketch $h(n)$.

(b) Suppose the input signal $x\colon Integers \to Reals$ is such that $\forall\, n \in Integers$, $x(n) = 1$. Give a simple expression for $y = Filter(x)$.

(c) Give an expression for $H(\omega)$ that is valid for all ω, where H is the frequency response.

(d) Sketch the magnitude of the frequency response. Can you explain your answer in part (b)?

T 11. Consider a discrete-time LTI system with impulse response h given by

$$\forall\, n \in Integers, \quad h(n) = \frac{1}{2}\delta(n-1) + \frac{1}{2}\delta(n+1),$$

and consider the periodic discrete-time signal given by

$$\forall\, n \in Integers, \quad x(n) = 2 + \sin(\pi n/2) + \cos(\pi n).$$

(a) Is the system causal?

(b) Find the frequency response of the system. Check that your answer is periodic with period 2π.

(c) For the given signal x, find the fundamental frequency ω_0 and the Fourier series coefficients X_k in the Fourier series expansion

$$x(n) = \sum_{k=-\infty}^{\infty} X_k e^{i\omega_0 kn}.$$

Give the units of the fundamental frequency.

(d) Assuming the input to the system is x as given, find the output.

T 12. Consider a continuous-time LTI system S. Suppose that when the input is the continuous-time **unit step**, defined by

$$u(t) = \begin{cases} 1 & \text{if } t \geq 0 \\ 0 & \text{if } t < 0, \end{cases} \tag{9.27}$$

then the output $y = S(u)$ is given by

$$y(t) = \begin{cases} 1 & \text{if } 0 \leq t \leq 1 \\ 0 & \text{otherwise.} \end{cases}$$

This output y is called the **step response** because it is the response to a unit step.

(a) Express y in terms of sums and differences of u and $D_1(u)$, where D_1 is the delay operator.

(b) Sketch a signal flow graph that produces the result $y = S(u)$ when the input is u. Note: We know that if two LTI systems have the same impulse response, then they are the same system. It is a fact, albeit a nontrivial one to demonstrate, that if two LTI systems have the same step response, then they are also the same system. Thus, your signal flow graph implements S.

(c) Use your signal flow graph to determine what the output y' of S is when the input is

$$x(t) = \begin{cases} 1 & \text{if } 0 \leq t \leq 1 \\ 0 & \text{otherwise.} \end{cases}$$

Plot your answer.

(d) What is the frequency response $H(\omega)$ of S?

T 13. Suppose a discrete-time LTI system S has impulse response

$$h(n) = u(n)/2^n = \begin{cases} 1/2^n & \text{if } n \geq 0 \\ 0 & \text{otherwise,} \end{cases}$$

where u is the unit step function represented by (9.26).

(a) What is the step response of this system? The step response is defined to be the output when the input is the unit step. Hint: The identity (9.23) may be helpful.

(b) What is the frequency response? Plot the magnitude and phase response (you may use MATLAB or do it manually). Hint: The following variant of the identity (9.23) may be useful: If $|a| < 1$, then

$$\sum_{m=0}^{\infty} a^m = \frac{1}{1-a}.$$

This follows immediately from (9.23) by letting L go to infinity.

(c) Suppose S is put in cascade with another identical system. What is the frequency response of the cascade composition?

(d) Suppose S is arranged in a feedback composition as shown in figure 9.19. What is the frequency response of the feedback composition?

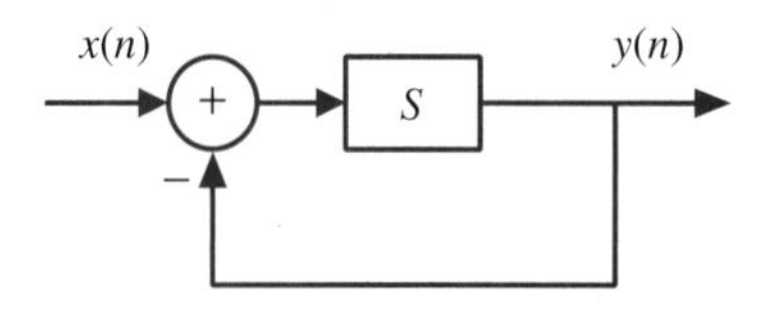

FIGURE 9.19: Feedback composition.

CHAPTER 10

The four Fourier transforms

In chapter 7 we saw that the Fourier series describes a periodic signal as a sum of complex exponentials. In chapter 8 we saw that if the input to a linear time-invariant (LTI) system is a sum of complex exponentials, then the frequency response of the LTI system describes its response to each of the component exponentials. Thus, we can calculate the system response to any periodic input signal by combining the responses to the individual components.

In chapter 9 we saw that the response of an LTI system to any input signal can also be obtained as the convolution of the input signal and the impulse response. The impulse response and the frequency response give us the same information about the system but in different forms. The impulse response and the frequency response are related by the Fourier transform, where in chapter 9 we saw both discrete-time and continuous-time versions.

This chapter shows that for discrete-time systems, the frequency response can be described as a sum of weighted complex exponentials (the discrete-time Fourier transform [DTFT]), in which the weights turn out to be the impulse response samples. We show that the impulse response is, in fact, a Fourier series representation of the frequency response, with the roles of time and frequency reversed from those in the uses of the Fourier series that we have seen so far.

This reappearance of the Fourier series is not a coincidence. In this chapter, we explore this pattern by showing that the Fourier series is a special case of a family of representations of signals that are collectively called **Fourier transforms**. The Fourier series applies specifically to continuous-time, periodic signals. The discrete Fourier series applies to discrete-time, periodic signals. We complete the description with a discussion of the continuous-time Fourier transform (CTFT), which applies to continuous-time signals that are not periodic, and the DTFT, which applies to discrete-time signals that are not periodic.

10.1 Notation

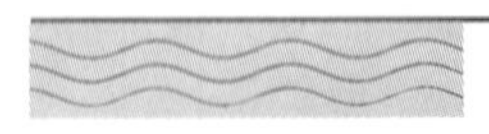

We define the following four sets of signals:

1. *ContSignals* = [*Reals* → *Complex*]. Because *Reals* is included in *Complex*, *ContSignals* includes continuous-time signals whose range is *Reals*, and so we do not need to consider these separately. *ContSignals* includes continuous-time signals, but the domain need not be interpreted as time. Indeed, sometimes the domain could be interpreted as space. It is sometimes useful to interpret the domain as frequency.
2. *DiscSignals* = [*Integers* → *Complex*]. This includes discrete-time signals whose domain is time or sample number, but again the domain need not be interpreted as time.
3. $ContPeriodic_p \subset ContSignals$. This set is defined to contain all continuous signals that are periodic with period p, where p is a real number.
4. $DiscPeriodic_p \subset DiscSignals$. This set is defined to contain all discrete signals that are periodic with period p, where p is an integer.

Note that whenever we discuss periodic signals we could equally well discuss finite signals, for which the domain is that of one cycle of the periodic signal.

10.2 The Fourier series

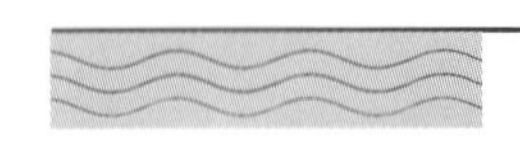

The **continuous-time Fourier series** representation of a periodic signal $x \in ContPeriodic_p$ is

$$\forall\, t \in Reals, \quad x(t) = \sum_{k=-\infty}^{\infty} X_k e^{ik\omega_0 t}, \tag{10.1}$$

where $\omega_0 = 2\pi/p$ (radians/second). The Fourier series coefficients are given by

$$\boxed{\forall\, m \in Integers, \quad X_m = \frac{1}{p}\int_0^p x(t)e^{-im\omega_0 t}dt.} \qquad (10.2)$$

Observe that the sequence of Fourier series coefficients given by (10.2) can be regarded as a signal $X \in DiscSignals$, where

$$\forall\, m \in Integers, \quad X(m) = X_m.$$

Therefore, we can define a system $FourierSeries_p$ with domain $ContPeriodic_p$ and range *DiscSignals* such that if the input is the periodic signal x, the output is its Fourier series coefficients, X; that is,

$$FourierSeries_p\colon ContPeriodic_p \to DiscSignals.$$

This system is the first of four forms of the Fourier transform and is depicted graphically in figure 10.1(a). Its inverse is a system

$$InverseFourierSeries_p\colon DiscSignals \to ContPeriodic_p,$$

depicted in figure 10.1(b).

The two systems, $FourierSeries_p$ and $InverseFourierSeries_p$, are **inverses** of each other, because (see the following box)

$$\forall\, x \in ContPeriodic_p, \quad (InverseFourierSeries_p \circ FourierSeries_p)(x) = x \qquad (10.3)$$

and

$$\forall\, X \in DiscSignals, \quad (FourierSeries_p \circ InverseFourierSeries_p)(X) = X. \qquad (10.4)$$

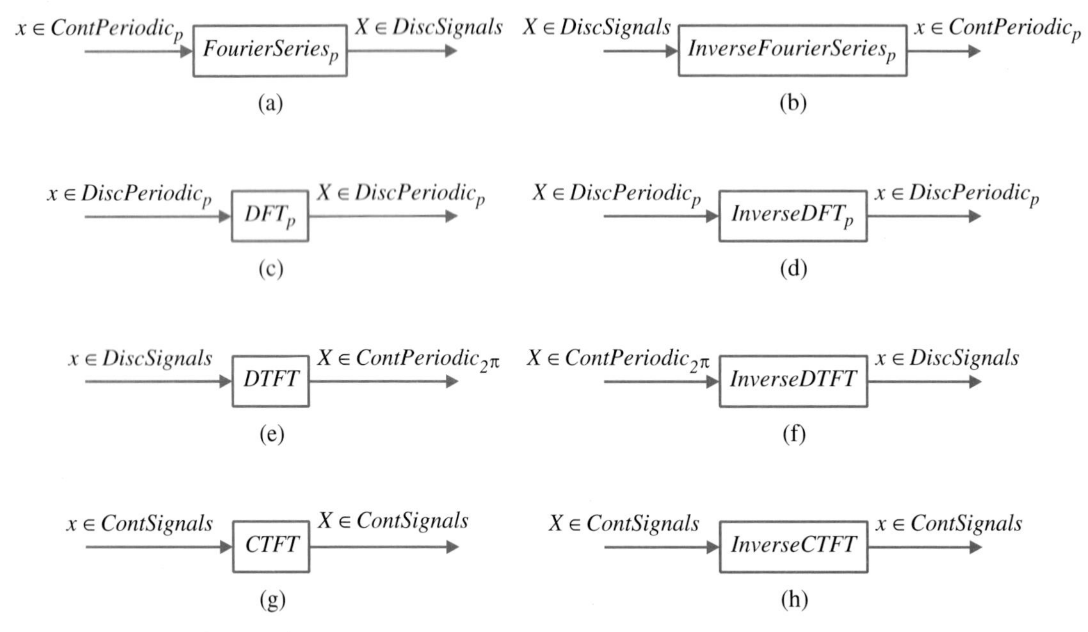

FIGURE 10.1: Fourier transforms as systems.

PROBING FURTHER

Showing inverse relations

In this and previous chapters, we have given formulas that describe time-domain functions in terms of frequency-domain functions and vice versa. By convention, a formula that gives the frequency-domain function in terms of the time-domain function is called a Fourier transform, and the formula that gives the time-domain function in terms of the frequency-domain function is called an **inverse Fourier transform**. As shown in figure 10.1, these transforms can be viewed as systems that take as inputs signals in one domain and return output signals in the other. We discuss four distinct Fourier transforms and their corresponding inverses. In each case, it is possible to show that the Fourier transform and its inverse are in fact inverses of one another, as stated in (10.3) and (10.4). We prove the second relation, (10.4), to illustrate how this is done. Similar proofs can be carried out for all four types of Fourier transforms, although sometimes these proofs require you to be adept at manipulating Dirac delta functions, which requires significant mathematical skill.

Let $X \in DiscSignals$ be the Fourier series for some $x \in ContPeriodic_p$; that is, $x = InverseFourierSeries_p(X)$. Let $Y = FourierSeries_p(x)$. We now show that $Y = X$—that is, that $Y_m = X_m$ for all m:

$$Y_m = \frac{1}{p}\int_0^p x(t)e^{-im\omega_0 t}dt, \text{ according to (10.2)}$$

PROBING FURTHER

$$= \frac{1}{p} \int_0^p [\sum_{k=-\infty}^{\infty} X_k e^{ik\omega_0 t}] e^{-im\omega_0 t} dt, \text{ according to (10.1)}$$

$$= \sum_{k=-\infty}^{\infty} \frac{1}{p} X_k \int_0^p e^{i(k-m)\omega_0 t} dt$$

$$= \frac{1}{p} X_m \int_0^p dt + \sum_{k \neq m} \frac{1}{p} X_k \int_0^p e^{i(k-m)\omega_0 t} dt$$

$$= X_m,$$

because for $k \neq m$,

$$\int_0^p e^{i(k-m)\omega_0 t} dt = 0.$$

Example 10.1: Consider a **square wave**, $x \in ContPeriodic_2$, shown in figure 10.2. It is periodic with period $p = 2$, and so its fundamental frequency is $\omega_0 = 2\pi/2 = \pi$ radians/second. We can find its continuous-time Fourier series coefficients by using (10.2) as follows:

$$X_m = \frac{1}{2} \int_0^2 x(t) e^{-im\pi t} dt$$

$$= \frac{1}{2} \int_0^1 e^{-im\pi t} dt.$$

Notice that when $m = 0$, this integral is easy to evaluate because $e^0 = 1$, and so

$$X_0 = 1/2.$$

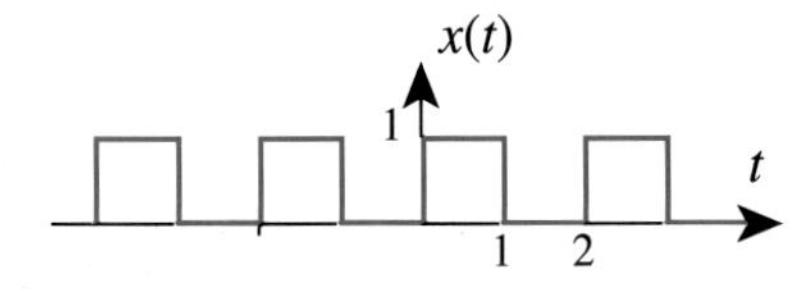

FIGURE 10.2: A square wave.

The following fact from calculus helps us solve the integral when $m \neq 0$:

$$\int_a^b e^{c\omega} c \, d\omega = e^{cb} - e^{ca} \tag{10.5}$$

for real a and b and complex c. Letting $c = -im\pi$, the integral becomes

$$\begin{aligned} X_m &= \frac{1}{2c} \int_0^1 e^{ct} c \, dt \\ &= \frac{-1}{2im\pi}(e^{-im\pi} - e^0) \\ &= \frac{i}{2m\pi}(e^{-im\pi} - 1), \end{aligned}$$

where the last step follows from multiplying top and bottom by i, with the observations that $i^2 = -1$ and that $e^0 = 1$. Notice that

$$e^{-im\pi} = \begin{cases} 1 & \text{if } m \text{ is even} \\ -1 & \text{if } m \text{ is odd.} \end{cases}$$

Thus, when $m \neq 0$,

$$X_m = \begin{cases} 0 & \text{if } m \text{ is even} \\ -i/m\pi & \text{if } m \text{ is odd.} \end{cases}$$

The magnitudes of these Fourier series coefficients are plotted in figure 10.3. Notice that the magnitudes of the coefficients decay rather slowly as m gets large (they decay as $1/m$). This accounts for the persistent overshoot (Gibb's phenomenon) seen in figure 7.6. Finite Fourier series approximations of this periodic square wave are not very accurate because the size of the Fourier series coefficients decays slowly. ❐

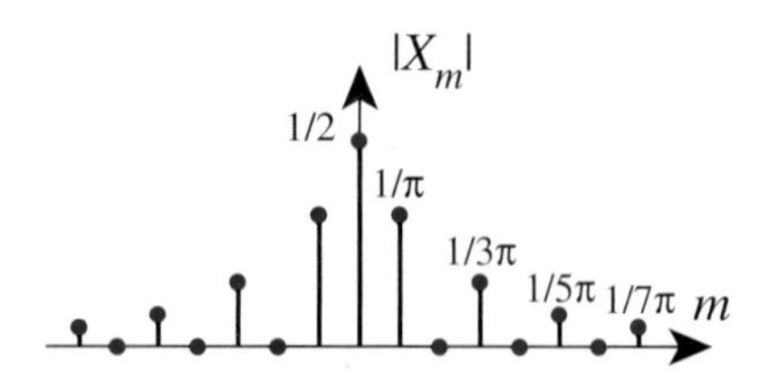

FIGURE 10.3: Magnitudes of the Fourier series coefficients of a square wave.

Other examples of Fourier series coefficients for various signals are given in table 10.1.

TABLE 10.1
Fourier series coefficients of periodic signals.*

Signal	FS	Reference
$\forall\, t \in Reals$, $x(t) = e^{i\omega_0 t}$, where $\omega_0 \neq 0$.	$\forall\, m \in Integers$, $X_m = \begin{cases} 1 & \text{if } m = 1 \\ 0 & \text{otherwise} \end{cases}$	Exercise 2
$\forall\, t \in Reals$, $x(t) = \cos(\omega_0 t)$, where $\omega_0 \neq 0$.	$\forall\, m \in Integers$, $X_m = \begin{cases} 1/2 & \text{if } \|m\| = 1 \\ 0 & \text{otherwise} \end{cases}$	Exercise 2
$\forall\, t \in Reals$, $x(t) = \sin(\omega_0 t)$, where $\omega_0 \neq 0$.	$\forall\, m \in Integers$, $X_m = \begin{cases} 1/2i & \text{if } m = 1 \\ -1/2i & \text{if } m = -1 \\ 0 & \text{otherwise} \end{cases}$	Exercise 2
$\forall\, t \in Reals$, $x(t) = 1$	$\forall\, m \in Integers$, $X_m = \begin{cases} 1 & \text{if } m = 0 \\ 0 & \text{otherwise} \end{cases}$	Exercise 2
Square wave: $\forall\, t \in [0, p]$, $x(t) = \begin{cases} 1 & \text{if } t < T \text{ or } t > p - T \\ 0 & \text{otherwise} \end{cases}$	$\forall\, m \in Integers$, $X_m = \sin(m_{\omega_0} T)/(m\pi)$	Exercise 3
Impulse train: $\forall\, t \in Reals$, $x(t) = \sum_{n=-\infty}^{\infty} \delta(t - np)$	$\forall\, m \in Integers$, $X_m = 1/p$	Exercise 4

* In all cases, $\omega_0 = 2\pi/p$, where p is the period. We can obtain Fourier transforms for each of these signals by using (10.16).

10.3 The discrete Fourier transform

The **discrete-time Fourier series** (**DFS**) expansion for $x \in DiscPeriodic_p$ (see (8.30)) is

$$\forall n \in Integers, \quad x(n) = \sum_{k=0}^{p-1} X_k e^{ik\omega_0 n}, \tag{10.6}$$

where $\omega_0 = 2\pi/p$ radians/sample. The Fourier series coefficients can be found by using the formula

$$\forall k \in Integers, \quad X_k = \frac{1}{p} \sum_{m=0}^{p-1} x(m) e^{-imk\omega_0}. \tag{10.7}$$

For historical reasons, the **discrete Fourier transform** (**DFT**) is the DFS with slightly different scaling. It is defined by

$$\boxed{\forall n \in Integers, \quad x(n) = \frac{1}{p} \sum_{k=0}^{p-1} X'_k e^{ik\omega_0 n}} \tag{10.8}$$

and

$$\boxed{\forall k \in Integers, \quad X'_k = \sum_{m=0}^{p-1} x(m) e^{-imk\omega_0}.} \tag{10.9}$$

Obviously, the DFT coefficients are related to the DFS coefficients by

$$X'_k = pX_k.$$

This scaling is somewhat unfortunate, because it means that the DFT coefficients do not have the same units as the signal x, but the scaling is firmly established in the literature, and so we stick to it. We omit the prime when it is clear that we are talking about the discrete Fourier transform instead of the Fourier series.

Observe that $X' = DFT_p(x)$ is a discrete signal that is itself periodic with period p. To verify this, note that for any integer N and for all integers k,

$$
\begin{aligned}
X'_{k+Np} &= \sum_{m=0}^{p-1} x(m) e^{-im(k+Np)\omega_0} \\
&= \sum_{m=0}^{p-1} x(m) e^{-imk\omega_0} e^{-imNp\omega_0} \\
&= \sum_{m=0}^{p-1} x(m) e^{-imk\omega_0} e^{-imN2\pi}, \text{ because } \omega_0 = 2\pi/p \\
&= \sum_{m=0}^{p-1} x(m) e^{-imk\omega_0} \\
&= X'_k.
\end{aligned}
$$

Note that (10.9) looks like a discrete Fourier series expansion of the periodic function X'; that is, the periodic function is described as a sum of complex exponentials. The only substantial difference from (10.6) is the sign of the exponent. This sign difference can be easily removed by changing variables. In doing so, you can discover that $x(-m)$ is the mth Fourier series coefficient for the function X'. The DFT is therefore rather special in that both the time- and frequency-domain representations of a function are Fourier series and are discrete and periodic.

The DFT therefore is the function

$$DFT_p\colon DiscPeriodic_p \to DiscPeriodic_p,$$

defined by (10.9). The inverse DFT is the function

$$InverseDFT_p\colon DiscPeriodic_p \to DiscPeriodic_p,$$

defined by (10.8). As in (10.3) and (10.4), DFT_p and $InverseDFT_p$ are inverses of each other. This can be verified by using methods similar to those in the preceding box. The DFT and its inverse are computed by systems that we can represent as shown in figure 10.1(c) and (d).

The DFT is the most useful form of the Fourier transform for computation, because the system DFT_p is easily implemented on a computer. Both summations (10.8) and (10.9) are finite. Moreover, there is an algorithm called the **fast Fourier transform** (**FFT**) that calculates these summations with far fewer arithmetic operations than the most direct method. Moreover, because the sums are finite, the DFT always exists. There are no mathematical problems with convergence.

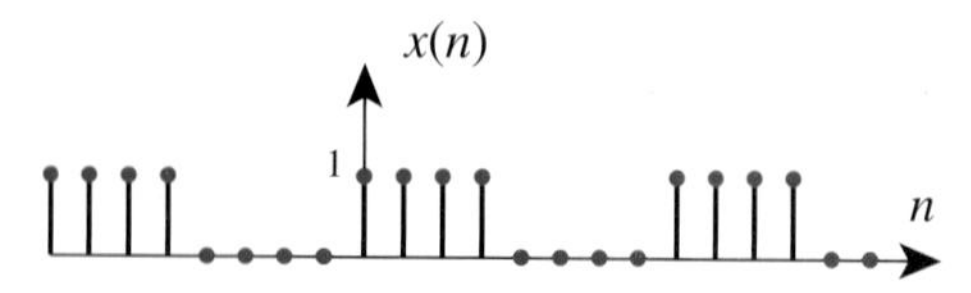

FIGURE 10.4: A discrete square wave.

Example 10.2: Consider a **discrete square wave**, $x \in DiscPeriodic_8$, shown in figure 10.4. It is periodic with period $p = 8$, and so its fundamental frequency is $\omega_0 = 2\pi/8 = \pi/4$ radians/sample. We can find its DFT by using (10.9) as follows:

$$X'_k = \sum_{m=0}^{7} x(m)e^{-imk\omega_0}$$
$$= \sum_{m=0}^{3} e^{-imk\pi/4}.$$

Notice that when $k = 0$, this sum is easy to evaluate because $e^0 = 1$, and so

$$X'_0 = 4.$$

Moreover, when k is any multiple of 8, $X'_k = 4$, as it should, because the DFT is periodic with period 8. The following identity helps us simplify the summation when $k \neq 0$ or any multiple of 8:

$$\boxed{\sum_{m=0}^{N} a^m = \frac{1-a^{N+1}}{1-a}.} \tag{10.10}$$

(To demonstrate the validity of this identity, just multiply both sides by $1 - a$.) If $N = 3$ and $a = e^{-ik\pi/4}$, then for $k \neq 0$ or any multiple of 8,

$$X'_k = \frac{1 - e^{-ik\pi}}{1 - e^{-ik\pi/4}}.$$

Notice that the numerator is particularly simple, because

$$1 - e^{-ik\pi} = \begin{cases} 0 & \text{if } k \text{ is even} \\ 2 & \text{if } k \text{ is odd.} \end{cases}$$

The magnitudes of these DFT coefficients are plotted in figure 10.5. ❐

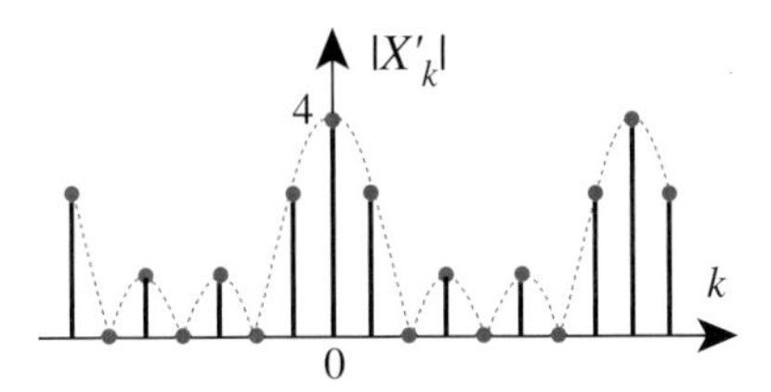

FIGURE 10.5: Magnitude of the DFT of a discrete square wave.

Other examples of discrete Fourier transform coefficients for various signals are given in table 10.2.

TABLE 10.2
Discrete Fourier transform of periodic signals.*

Signal	DFT	Reference
$\forall n \in Integers$, $x(n) = e^{i2\pi fn}$, where $f \neq 0$.	$\forall k \in Integers$, $X'_k = \begin{cases} p & \text{if } k \in A \\ 0 & \text{otherwise} \end{cases}$	Exercise 5
$\forall n \in Integers$, $x(n) = \cos(2\pi fn)$, where $f \neq 0$.	$\forall k \in Integers$, $X'_k = \begin{cases} p/2 & \text{if } k \in A \\ p/2 & \text{if } k \in B \\ 0 & \text{otherwise} \end{cases}$	Exercise 5
$\forall n \in Integers$, $x(n) = \sin(i2\pi fn)$, where $f \neq 0$.	$\forall k \in Integers$, $X'_k = \begin{cases} p/2i & \text{if } k \in A \\ -p/2i & \text{if } k \in B \\ 0 & \text{otherwise} \end{cases}$	Exercise 5
$\forall n \in Integers$, $x(n) = 1$	$\forall k \in Integers$, $X'_k = \begin{cases} p & \text{if } k \in C \\ 0 & \text{otherwise} \end{cases}$	Exercise 5
Square wave: $\forall n \in \{0, 1, \ldots, p-1\}$, $x(n) = \begin{cases} 1 & \text{if } n \leq M \text{ or } n \geq p - M \\ 0 & \text{otherwise} \end{cases}$	$\forall k \in Integers$, $X'_k = \sin(k(M + 0.5)\omega_0)/\sin(k\omega_0/2)$	Exercise 6
Impulse train: $\forall n \in Integers$, $x(n) = \sum_{k=-\infty}^{\infty} \delta(n - kp)$	$\forall k \in Integers$, $X'_k = 1$	Exercise 7

* The fundamental frequency is $\omega_0 = 2\pi/p$, where p is the period. For the complex exponential and sinusoidal signals, the frequency f must be rational, and is related to the period p by $f = m/p$ for some integer m (see section 7.6.1). The following sets are used in this definition: $A = \{\ldots m - 2p, m - p, m, m + p, m + 2p, \ldots\}$, $B = \{\ldots - m - 2p, -m - p, -m, -m + p, -m + 2p, \ldots\}$, and $C = \{\ldots - 2p, -p, 0, p, 2p, \ldots\}$.

10.4 The discrete-time Fourier transform

We have shown in section 9.2 that the frequency response H of an LTI system is related to the impulse response h by

$$\forall\, \omega \in Reals, \quad H(\omega) = \sum_{m=-\infty}^{\infty} h(m)e^{-i\omega m}.$$

H is called the **discrete-time Fourier transform** (**DTFT**) of h. For any $x \in$ *DiscSignals* (not just an impulse response), its DTFT is defined to be

$$\boxed{\forall\, \omega \in Reals, \quad X(\omega) = \sum_{m=-\infty}^{\infty} x(m)e^{-i\omega m}.} \tag{10.11}$$

Of course, this definition is valid only for those x and those ω for which the sum converges (it is not trivial mathematically to characterize this).

Notice that the function X is periodic with period 2π; that is, $X(\omega) = X(\omega + 2\pi N)$ for any integer N, because

$$e^{-i\omega t} = e^{-i(\omega+2\pi N)t}$$

for any integer N. Thus, $X \in ContPeriodic_{2\pi}$.

Note that (10.11) looks like a DFS expansion of the periodic function X; that is, the periodic function is described as a sum of complex exponentials. The only substantial difference from (10.1) is the sign of the exponent. This sign difference can be easily removed by changing variables. In doing so, you can discover that $x(-n)$ is the nth Fourier series coefficient for the function X.

The DTFT (10.11) has similar structure to the DFT (10.9). In fact, the DTFT can be viewed as a generalization of the DFT to signals that are neither periodic nor finite. In other words, as p approaches infinity, ω_0 approaches zero, and so instead of having a discrete set of frequencies spaced by ω_0, we have a continuum.

The DTFT is a system

$$DTFT\colon DiscSignals \to ContPeriodic_{2\pi},$$

and its inverse is

$$InverseDTFT\colon ContPeriodic_{2\pi} \to DiscSignals.$$

The inverse is given by

$$\forall\, n \in Integers, \quad x(n) = \frac{1}{2\pi}\int_{0}^{2\pi} X(\omega)e^{i\omega n}d\omega. \qquad (10.12)$$

Notice that because X is periodic, this can be written equivalently as

$$\forall\, n \in Integers, \quad x(n) = \frac{1}{2\pi}\int_{-\pi}^{\pi} X(\omega)e^{i\omega n}d\omega.$$

We integrate over one cycle of the periodic function, and so it does not matter where we start. The system and its inverse are depicted graphically in figure 10.1(e) and (f).

DTFT and *InverseDTFT* are inverses of each other. This follows from the fact that $FourierSeries_p$ and $InverseFourierSeries_p$ are inverses of each other.

Example 10.3: Consider a **discrete rectangle**, $x \in DiscSignals$, shown in figure 10.6. We can find its DTFT by using (10.11) as follows:

$$X(\omega) = \sum_{m=-\infty}^{\infty} x(m)e^{-i\omega m}$$

$$= \sum_{m=0}^{3} e^{-i\omega m}.$$

Notice that when $\omega = 0$, this sum is easy to evaluate because $e^0 = 1$, and so

$$X(0) = 4.$$

Moreover, when ω is any multiple of 2π, $X(\omega) = 4$, as it should, because the DTFT is periodic with period 2π. We can again use the identity (10.10). Letting $N = 3$ and $a = e^{-i\omega}$, we get

$$X(\omega) = \frac{1 - e^{-i4\omega}}{1 - e^{-i\omega}}.$$

The magnitude of this function is plotted in figure 10.7. ❐

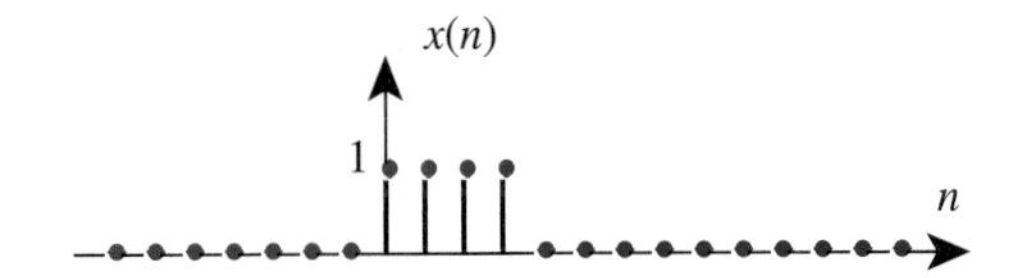

FIGURE 10.6: A discrete rectangle signal.

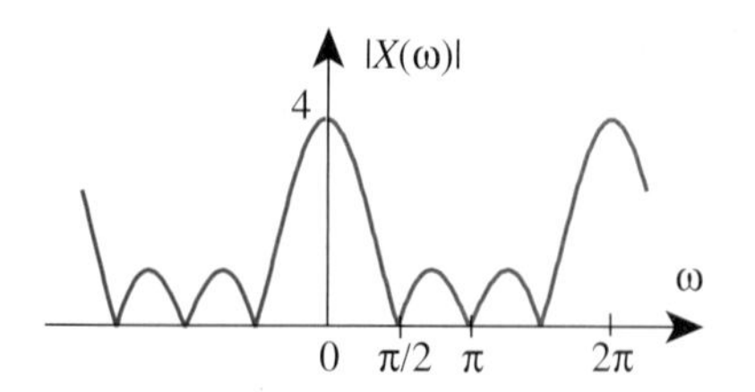

FIGURE 10.7: Magnitude of the DTFT of a discrete rectangle.

Other examples of DTFTs for various signals are given in table 10.3.

TABLE 10.3
Discrete time Fourier transforms of key signals.*

Signal	DTFT	Reference
$\forall n \in Integers,$ $x(n) = \delta(n)$	$\forall \omega \in Reals,$ $X(\omega) = 1$	Example 10.8
$\forall n \in Integers,$ $x(n) = \delta(n - N)$	$\forall \omega \in Reals,$ $X(\omega) = e^{-i\omega N}$	Example 10.8
$\forall n \in Integers,$ $x(n) = K$	$\forall \omega \in Reals,$ $X(\omega) = 2\pi K \sum_{k=-\infty}^{\infty} \delta(\omega - k2\pi)$	Section 10.7.5
$\forall n \in Integers,$ $x(n) = a^n u(n), \quad \lvert a \rvert < 1$	$\forall \omega \in Reals,$ $X(\omega) = 1/1 - ae^{-i\omega}$	Exercise 18
$\forall n \in Integers,$ $x(n) = \begin{cases} 1 & \text{if } \lvert n \rvert \le M \\ 0 & \text{otherwise} \end{cases}$	$\forall \omega \in Reals,$ $X(\omega) = \sin(\omega(M + 0.5))/\sin(\omega/2)$	Exercise 8
$\forall n \in Integers,$ $x(n) = \sin(Wn)/\pi n, \quad 0 < W < \pi$	$\forall \omega \in [-\pi, \pi],$ $X(\omega) = \begin{cases} 1 & \text{if } \lvert \omega \rvert \le W \\ 0 & \text{otherwise} \end{cases}$	—

* The function u is the unit step, given by (2.15).

10.5 The continuous-time Fourier transform

The frequency response and impulse response of a continuous-time LTI system are related by the **continuous-time Fourier transform** (**CTFT**), more commonly called simply the **Fourier transform** (**FT**):

$$\forall\, \omega \in \mathit{Reals}, \quad X(\omega) = \int_{-\infty}^{\infty} x(t) e^{-i\omega t} dt. \tag{10.13}$$

The CTFT can be defined for any function $x \in \mathit{ContSignals}$ where the integral exists. It need not be the impulse response of any system, and it need not be periodic or finite. The inverse relation is

$$\forall\, t \in \mathit{Reals}, \quad x(t) = \frac{1}{2\pi} \int_{-\infty}^{\infty} X(\omega) e^{i\omega t} d\omega. \tag{10.14}$$

It is true, but difficult to prove, that the CTFT and the inverse CTFT are indeed inverses.

The CTFT can be viewed as a generalization of both the FS and DTFT in which neither the frequency domain nor the time domain signal needs to be periodic. Alternatively, it can be viewed as the last remaining Fourier transform, in which neither time nor frequency is discrete. Graphically, the CTFT is a system as shown in figure 10.1(g) and (h).

Example 10.4: Consider a **rectangular pulse** $x \in \mathit{ContSignals}$, shown in figure 10.8. We can find its continuous-time Fourier transform by using (10.13) as follows:

$$\begin{aligned} X(\omega) &= \int_{-\infty}^{\infty} x(t) e^{-i\omega t} dt \\ &= \int_{0}^{1} e^{-i\omega t} dt. \end{aligned}$$

Notice that when $\omega = 0$, this integral is easy to evaluate because $e^0 = 1$, and so

$$X(0) = 1.$$

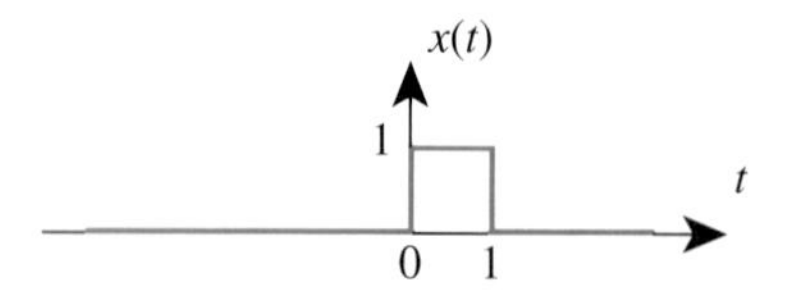

FIGURE 10.8: A rectangular pulse.

We can use (10.5) to solve the integral in general:

$$X(\omega) = \frac{i}{\omega}\left[e^{-i\omega} - 1\right].$$

We can get better intuition about this by manipulating it as follows:

$$\begin{aligned} X(\omega) &= \frac{i}{\omega}e^{-i\omega/2}\left[e^{-i\omega/2} - e^{i\omega/2}\right] \\ &= e^{-i\omega/2}\left[\frac{\sin(\omega/2)}{\omega/2}\right]. \end{aligned}$$

The leading factor has unit magnitude, and so we see that the magnitude can be written

$$|X(\omega)| = \left|\frac{\sin(\omega/2)}{\omega/2}\right|.$$

This magnitude is plotted in figure 10.9. Notice that the magnitude decays rather slowly as ω gets large (it decays as $1/\omega$). ❐

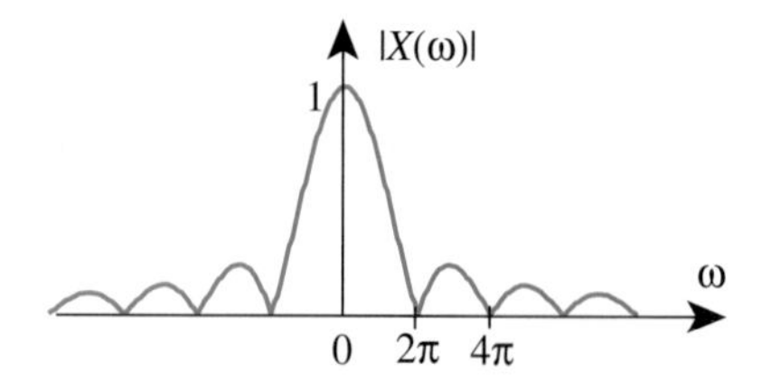

FIGURE 10.9: Magnitude of the Fourier transform of the rectangular pulse in figure 10.8.

Other examples of Fourier transforms for various signals are given in table 10.4. The four Fourier transforms are summarized in table 10.5.

TABLE 10.4
Continuous time Fourier transforms of key signals.

Signal	CTFT	Reference
$\forall\, t \in Reals,$ $x(t) = \delta(t)$	$\forall\, \omega \in Reals,$ $X(\omega) = 1$	Example 10.7
$\forall\, t \in Reals,$ $x(t) = \delta(t-\tau),\ \tau \in Reals$	$\forall\, \omega \in Reals,$ $X(\omega) = e^{-i\omega\tau}$	Example 10.7
$\forall\, t \in Reals,$ $x(t) = K$	$\forall\, \omega \in Reals,$ $X(\omega) = 2\pi K \delta(\omega)$	Section 10.7.5
$\forall\, t \in Reals,$ $x(t) = a^t u(t),\quad 0 < a < 1$	$\forall\, \omega \in Reals,$ $X(\omega) = 1/j\omega - \ln(a)$	—
$\forall\, t \in Reals,$ $x(t) = \begin{cases} \pi/a & \text{if } \lvert t\rvert \le a \\ 0 & \text{otherwise} \end{cases}$	$\forall\, \omega \in Reals,$ $X(\omega) = 2\pi \sin(a\omega)/a\omega$	Exercise 23
$\forall\, t \in Reals,$ $x(t) = \sin(\pi t/T)/\pi t/T$	$\forall\, \omega \in Reals,$ $X(\omega) = \begin{cases} T & \text{if } \lvert\omega\rvert \le \pi/T \\ 0 & \text{otherwise} \end{cases}$	Exercise 21

* The function u is the unit step, given by (9.27).

10.6 Fourier transforms versus Fourier series

For each continuous and discrete-time signal, we have a Fourier transform and a Fourier series (see table 10.5). The Fourier series applies only to periodic signals, whereas the Fourier transforms apply to any signal. The Fourier transforms must therefore also be applicable to periodic signals. Why do we need the Fourier series? Moreover, if we have a finite signal, then we can find its Fourier series by considering it to be a periodic signal, or we can find its Fourier transform by considering it to be zero outside its finite domain. Which should we do?

10.6.1 *Fourier transforms of finite signals*

Consider a discrete-time signal y that is finite. Suppose it is defined for the domain $[0, p-1] \subset Integers$. As we did in section 7.4, we can define a related periodic signal x as

$$\forall\, n \in Integers, \quad x(n) = \sum_{m=-\infty}^{\infty} y'(n - mp),$$

TABLE 10.5
The four Fourier transforms summarized.*

	Aperiodic Time, Continuous Frequency	**Periodic Time, Discrete Frequency**
Aperiodic Frequency, Continuous Time	*CTFT*: *ContSignals* $\to$ *ContSignals* $X(\omega) = \int_{-\infty}^{\infty} x(t)e^{-i\omega t}dt$ *InverseCTFT*: *ContSignals* $\to$ *ContSignals* $x(t) = \frac{1}{2\pi}\int_{-\infty}^{\infty} X(\omega)e^{i\omega t}d\omega$	*FourierSeries*$_p$: *ContPeriodic*$_p$ $\to$ *DiscSignals* $X_m = \frac{1}{p}\int_{0}^{p} x(t)e^{-im\omega_0 t}dt$ *InverseFourierSeries*$_p$: *DiscSignals* $\to$ *ContPeriodic*$_p$ $x(t) = \sum_{k=-\infty}^{\infty} X_k e^{ik\omega_0 t}$
Periodic Frequency, Discrete Time	*DTFT*: *DiscSignals* $\to$ *ContPeriodic*$_{2\pi}$ $X(\omega) = \sum_{n=-\infty}^{\infty} x(n)e^{-in\omega}$ *InverseDTFT*: *ContPeriodic*$_{2\pi}$ $\to$ *DiscSignals* $x(n) = \frac{1}{2\pi}\int_{-\pi}^{\pi} X(\omega)e^{i\omega n}d\omega$	*DFT*$_p$: *DiscPeriodic*$_p$ $\to$ *DiscPeriodic*$_p$ $X_k = \sum_{n=0}^{p-1} x(n)e^{-ink\omega_0}$ *InverseDFT*$_p$: *DiscPeriodic*$_p$ $\to$ *DiscPeriodic*$_p$ $x(n) = \frac{1}{p}\sum_{k=0}^{p-1} X_k e^{ik\omega_0 n}$

* The column and row titles tell you when to use the specified Fourier transform and its inverse. For example, the first row applies to continuous-time signals, and the second column applies to periodic signals. Thus, if you have a continuous-time periodic signal, you should use the Fourier series at the upper right.

where

$$\forall\, n \in Integers, \quad y'(n) = \begin{cases} y(n) & \text{if } n \in [0, p-1] \\ 0 & \text{otherwise.} \end{cases}$$

The signal y', of course, is an ordinary discrete-time signal and hence possesses a DTFT. The signal x is a periodic signal and therefore possesses a DFT. The DTFT and DFT are closely related. The DTFT is given by

$$\forall\, \omega \in Reals, \quad Y'(\omega) = \sum_{n=-\infty}^{\infty} y'(n)e^{-in\omega}$$

$$= \sum_{n=0}^{p-1} y(n)e^{-in\omega}.$$

The DFT is

$$\forall\, k \in Integers, \quad X_k = \sum_{n=0}^{p-1} x(n)e^{-ink\omega_0}$$

$$= \sum_{n=0}^{p-1} y(n)e^{-ink\omega_0}.$$

Comparing the DTFT and DFT, we see that

$$\boxed{\forall\, k \in Integers, \quad X_k = Y'(k\omega_0).}$$

In words, the DFT of the periodic signal is equal to samples of the DTFT of the finite signal with sampling interval $\omega_0 = 2\pi/p$. This fact proves extremely useful in Fourier analysis of real-world signals, as we see in the next subsection.

10.6.2 *Fourier analysis of a speech signal*

Consider the speech signal shown in figure 10.10. It is two seconds of sound sampled at 8 kHz, for a total of 16,000 samples. This is a finite signal, and so we can find a DFT or DTFT, as explained in the previous section. However, this would not be all that useful. As is evident from the figure, the character of the signal changes several times in the two seconds. The speech sound is one of the authors saying, "This is the sound of my voice." Usually, it is more interesting to perform Fourier analysis on a single **phoneme**, or elemental sound. Such analysis may be, for example, the first stage of a speech recognition system, which attempts to automatically determine what was said.

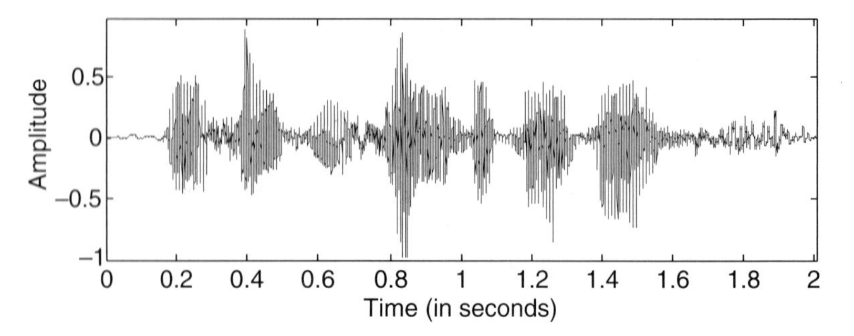

FIGURE 10.10: Two seconds of speech.

Consider a much smaller segment of speech whose waveform is shown at the top of figure 10.11. This is approximately 64 msec of speech, or 512 samples at a rate of 8,000 samples/second. This particular segment is the vowel sound in the word "sound." Notice that the signal has quite a bit of structure. Fourier analysis can reveal a great deal about that structure.

The first issue is to decide which of the four Fourier transforms to apply. This is easy, because only one of the four, the DFT, is computable. Because we do not have an analytic expression for the speech signal, we cannot algebraically

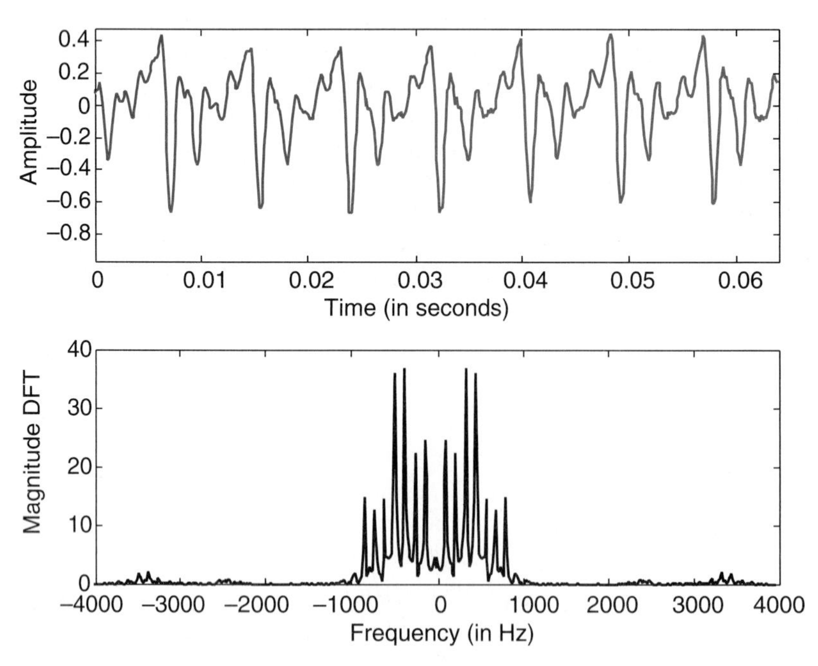

FIGURE 10.11: A voiced segment of speech (top) and one cycle of the magnitude of its DFT (bottom).

determine the DTFT. However, we know from the previous section that the DFT yields samples of the DTFT of a finite signal, and so calculating the DFT reveals the structure of the DTFT.

The DFT of these 512 samples is shown at the bottom of figure 10.11, which shows one cycle of the periodic DFT. The horizontal axis is labeled with the frequencies in Hertz rather than the index k of the DFT coefficient. Recall that the kth coefficient of the DFT represents a complex exponential with frequency $k\omega_0 = k2\pi/p$ radians/sample. To convert this to Hertz, we manipulate the units,

$$(k2\pi/p)[\text{radians/sample}] \times (1/T)[\text{samples/second}] \\ \times (1/2\pi)[\text{cycles/radian}] = k/(pT)[\text{cycles/second}],$$

where $T = 1/8{,}000$ is the sampling interval.

Although this plot shows the DFT, the plot can equally well be interpreted as a plot of the DTFT. It shows 512 values of the DFT (one cycle of the periodic signal), and instead of showing each individual value, it runs them together as a continuous curve. In fact, because the DFT comprises samples of the DTFT, that continuous curve is a pretty good estimate of the shape of the DTFT. Of course, the samples might not be close together enough to accurately represent the DTFT, but they probably are. This issue is examined in considerable detail in chapter 11.

Notice that in figure 10.11, most of the signal is concentrated below 1 kHz. In figure 10.12, we have zoomed in to this region of the spectrum. Notice that the DFT is strongly peaked and the lowest frequency peak occurs at about 120 Hz. Indeed, this is not surprising, because the time-domain signal at the top of figure 10.11 looks somewhat periodic, with a period of about 80 msec, approximately the inverse of 120 Hz; that is, it looks like a periodic signal with a fundamental frequency of 120 Hz and various harmonics. The weights of the harmonics are proportional to the heights of the peaks in figure 10.12.

The vowel in the word "sound" is created by vibrating the human vocal chords and then shaping the mouth into an acoustic cavity that alters the sound

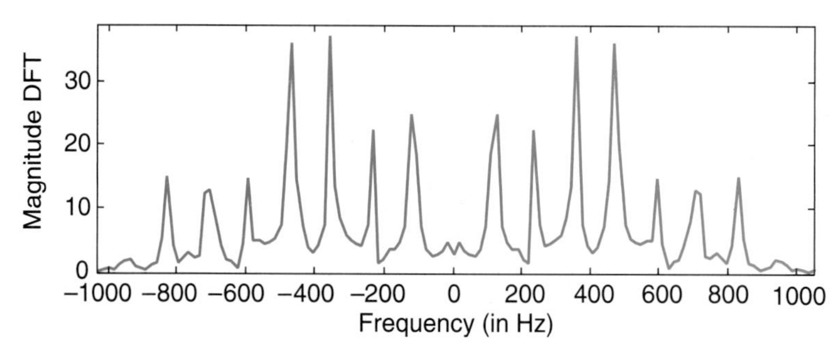

FIGURE 10.12: DFT of the voiced segment of speech, shown over a narrower frequency range.

produced by the vocal chords. A sound that entails the use of the vocal chords is called by linguists a **voiced sound**. In contrast, the "s" in "this" is an **unvoiced sound**. It is created without the help of the vocal chords by forcing air through a narrow passage in the mouth to create turbulence.

The waveform of the "s" in "this" is shown in figure 10.13. It is very different from the waveform of the vowel sound in figure 10.11. In particular, it looks much less regular. There is no evident periodicity in the signal, and so the DFT does not reveal a fundamental frequency and harmonics. It is a noise-like signal, for which the DFT reveals that much of the noise is at low frequencies, below 500 Hz, but there are also significant components all the way up to 4 kHz.

This analysis of the segments of speech is a typical use of Fourier analysis. A long (even infinite) signal is divided into short segments, such as the two 512-sample segments that we examined previously. These segments are studied by calculating their DFT. Because the DFT is a finite summation, it is easy to realize on a computer. Also, as mentioned before, there is a highly efficient algorithm called the fast Fourier transform (FFT) for calculating the DFT. The `fft` function in MATLAB®* was used to calculate the DFTs in figures 10.11 and 10.13. The DFT comprises samples of the DTFT of the finite segment of signal. This idea is explored in much more detail in lab 7 (see Lab Manual), in which it is shown how to repeatedly calculate DFTs of short segments to create a **spectrogram**. The spectrogram for the signal in figure 10.10 is shown in figure 8 (see Lab Manual).

10.6.3 *Fourier transforms of periodic signals*

In the previous two sections, we have seen the relationship between the DTFT of a finite signal and the DFT of the periodic signal that is constructed by repeating the finite signal. The periodic signal, however, also has a DTFT itself. In the interest of variety, we explore this concept in continuous time.

A periodic continuous-time signal has a Fourier transform. That Fourier transform, however, has Dirac delta functions in it, which are mathematically tricky. Recall that a Dirac delta function is an infinitely narrow and infinitely high pulse with unit area. The Fourier series enables us to examine the frequency domain representation of a periodic signal without dealing with Dirac delta functions.

It is certainly useful to simplify the mathematics. Using a Fourier series whenever possible allows us to do that. Moreover, in working with signals computationally, everything must be discrete and finite; computers cannot deal with infinitely narrow and infinitely high pulses numerically. Computation, therefore, must be done with the only transform that is completely discrete and finite: the DFS or its scaled cousin, the DFT.

To be concrete, suppose a continuous-time signal x has Fourier transform

$$\forall\, \omega \in Reals, \quad X(\omega) = 2\pi\,\delta(\omega - \omega_0) \qquad (10.15)$$

* MATLAB is a registered trademark of The MathWorks, Inc.

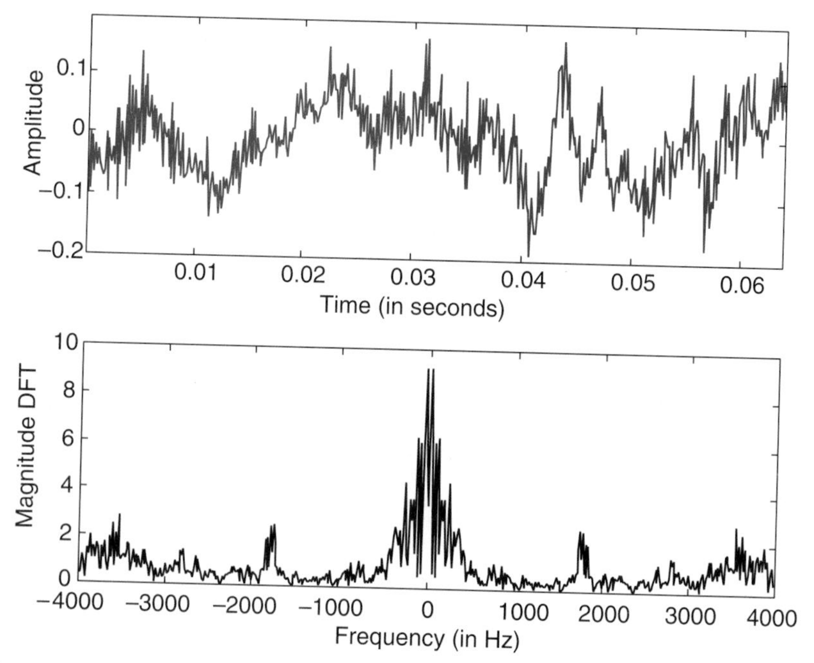

FIGURE 10.13: An unvoiced segment of speech (top) and one cycle of the magnitude of its DFT (bottom).

for some real value ω_0. We can find x by using the inverse CTFT (10.14):

$$\forall\, t \in Reals, \quad x(t) = \frac{1}{2\pi} \int_{-\infty}^{\infty} 2\pi\,\delta(\omega - \omega_0) e^{i\omega t} d\omega.$$

Using the sifting rule, this yields

$$x(t) = e^{i\omega_0 t}.$$

This is a periodic function with period $p = 2\pi/\omega_0$. From table 10.1, we see that the Fourier series for x is

$$\forall\, m \in Integers, \quad X_m = \begin{cases} 1 & \text{if } m = 1 \\ 0 & \text{otherwise.} \end{cases}$$

There is exactly one nonzero Fourier series coefficient, which corresponds to the exactly one Dirac delta pulse in the Fourier transform (10.15).

More generally, suppose x has multiple Dirac delta pulses in its Fourier transform, each with different weights:

$$\forall \omega \in Reals, \quad X(\omega) = 2\pi \sum_{m=-\infty}^{\infty} X_m \delta(\omega - m\omega_0). \tag{10.16}$$

The inverse CTFT (10.14) tells us that

$$\forall t \in Reals, \quad x(t) = \sum_{m=-\infty}^{\infty} X_m e^{im\omega_0 t}.$$

This is a periodic function with fundamental frequency ω_0 and harmonics in various weights. Its Fourier series coefficients are, by inspection, just X_m. Thus, for a periodic signal, (10.16) relates the CTFT and the Fourier series. The Fourier series gives the weights of a set of Dirac delta pulses in the Fourier transform, with a scaling factor of 2π.

Example 10.5: Consider x given by

$$\forall t \in Reals, \quad x(t) = \cos(\omega_0 t).$$

In table 10.1, we have seen that the Fourier series for x is

$$\forall m \in Integers, \quad X_m = \begin{cases} 1/2 & \text{if } |m| = 1 \\ 0 & \text{otherwise.} \end{cases}$$

There are only two nonzero Fourier series coefficients. We can use (10.16) to write down its Fourier transform:

$$\forall \omega \in Reals, \quad X(\omega) = \pi\delta(\omega + \omega_0) + \pi\delta(\omega - \omega_0).$$

We sketch this as shown in figure 10.14. See exercise 9 at the end of the chapter for a similar result for a sine function. ❐

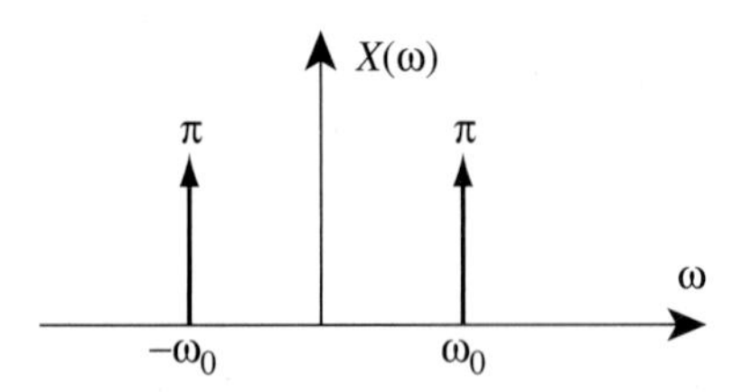

FIGURE 10.14: Fourier transform of a cosine.

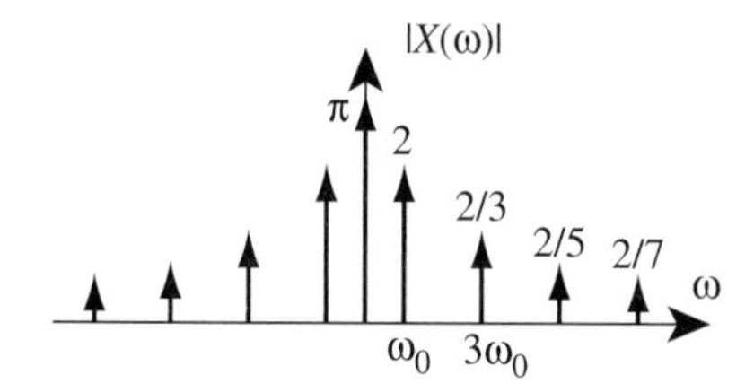

FIGURE 10.15: Fourier transform of a square wave.

Example 10.6: Consider the square wave of example 10.1. The Fourier series coefficients are

$$X_m = \begin{cases} 1/2 & \text{if } m = 0 \\ 0 & \text{if } m \text{ is even and } m \neq 0 \\ -i/m\pi & \text{if } m \text{ is odd.} \end{cases}$$

The Fourier transform therefore has Dirac delta pulses with these weights at multiples of ω_0, scaled by 2π, as shown in figure 10.15. ❐

The same concept applies in the discrete-time case, although some care is needed because both the DTFT and DFT are periodic, and not every function of the form $e^{i\omega n}$ is periodic (see section 7.6.1). In fact, in figure 10.11, the peaks in the spectrum hint at the Dirac delta functions in the DTFT. The signal appears to be approximately periodic with period 120 Hz, and so the DFT shows strong peaks at multiples of 120 Hz. If the signal were perfectly periodic, and if we were plotting the DTFT instead of the DFT, then these peaks would become infinitely narrow and infinitely high.

10.7 Properties of Fourier transforms

In this section, we describe a number of useful properties of the various Fourier transforms, together with a number of examples. These properties are summarized in tables 10.6 to 10.9. The properties and examples can often be used to avoid solving integrals or summations, to find the Fourier transform of some signal, or to find an inverse Fourier transform.

10.7.1 *Convolution*

Suppose a discrete-time LTI system has impulse response h and frequency response H. We have seen that if the input to this system is a complex exponential,

TABLE 10.6
Properties of the Fourier series.

Time domain	Frequency domain	Reference
$\forall\, t \in Reals$, $x(t)$ is real	$\forall\, m \in Integers$, $X_m = X^*_{-m}$	Section 10.7.2
$\forall\, t \in Reals$, $x(t) = x^*(-t)$	$\forall\, m \in Integers$, X_m is real	Section 10.7.2
$\forall\, t \in Reals$, $y(t) = x(t-\tau)$	$\forall\, m \in Integers$, $Y_m = e^{-im\omega_0\tau} X_m$	Section 13
$\forall\, t \in Reals$, $y(t) = e^{i\omega_1 t} x(t)$ where $\omega_1 = M\omega_0$, for some $M \in Integers$	$\forall\, m \in Integers$, $Y_m = X_{m-M}$	Section 13
$\forall\, t \in Reals$, $y(t) = \cos(\omega_1 t) x(t)$ where $\omega_1 = M\omega_0$, for some $M \in Integers$	$\forall\, m \in Integers$, $Y_m = (X_{m-M} + X_{m+M})/2$	Example 13
$\forall\, t \in Reals$, $y(t) = \sin(\omega_1 t) x(t)$ where $\omega_1 = M\omega_0$, for some $M \in Integers$	$\forall\, m \in Integers$, $Y_m = (X_{m-M} - X_{m+M})/2i$	Exercise 13
$\forall\, t \in Reals$, $y(t) = ax(t) + bw(t)$	$\forall\, m \in Integers$, $Y_m = aX_m + bW_m$	Section 10.7.4
$\forall\, t \in Reals$, $y(t) = x^*(t)$	$\forall\, m \in Integers$, $Y_m = X^*_{-m}$	—

* All time-domain signals are assumed to be periodic with period p, and fundamental frequency $\omega_0 = 2\pi/p$.

TABLE 10.7
Properties of the DFT.

Time domain	Frequency domain	Reference
$\forall n \in Integers$, $x(n)$ is real	$\forall k \in Integers$, $X'_k = X'^*_{-k}$	Section 10.7.2
$\forall n \in Integers$, $x(n) = x^*(-n)$	$\forall k \in Integers$, X'_k is real	Section 10.7.2
$\forall n \in Integers$, $y(n) = x(n - N)$	$\forall k \in Integers$, $Y'_k = e^{-ik\omega_0 N} X'_k$	—
$\forall n \in Integers$, $y(n) = e^{i\omega_1 n} x(n)$ where $\omega_1 = M\omega_0$, for some $M \in Integers$	$\forall k \in Integers$, $Y'_k = X'_{k-M}$	—
$\forall n \in Integers$, $y(n) = \cos(\omega_1 n) x(n)$ where $\omega_1 = M\omega_0$, for some $M \in Integers$	$\forall k \in Integers$, $Y'_k = (X'_{k-M} + X'_{k+M})/2$	—
$\forall n \in Integers$, $y(n) = \sin(\omega_1 n) x(n)$ where $\omega_1 = M\omega_0$, for some $M \in Integers$	$\forall k \in Integers$, $Y'_k = (X'_{k-M} - X'_{k+M})/2i$	—
$\forall n \in Integers$, $y(n) = ax(n) + bw(n)$	$\forall k \in Integers$, $Y'_k = aX'_k + bW'_k$	Section 10.7.4
$\forall n \in Integers$, $y(n) = x^*(n)$	$\forall k \in Integers$, $Y_k = X^*_{-k}$	—

* All time-domain signals are assumed to be periodic with period p, and fundamental frequency $\omega_0 = 2\pi/p$.

TABLE 10.8
Properties of the DTFT.

Time domain	Frequency domain	Reference
$\forall n \in Integers,$ $x(n)$ is real	$\forall \omega \in Reals,$ $X(\omega) = X^*(-\omega)$	Section 10.7.2
$\forall n \in Integers,$ $x(n) = x^*(-n)$	$\forall \omega \in Reals,$ $X(\omega)$ is real	Section 10.7.2
$\forall n \in Integers,$ $y(n) = x(n-N)$	$\forall \omega \in Reals,$ $Y(\omega) = e^{-i\omega N} X(\omega)$	Section 10.7.3
$\forall n \in Integers,$ $y(n) = e^{i\omega_1 n} x(n)$	$\forall \omega \in Reals,$ $Y(\omega) = X(\omega - \omega_1)$	Section 10.7.6
$\forall n \in Integers,$ $y(n) = \cos(\omega_1 n) x(n)$	$\forall \omega \in Reals,$ $Y(\omega) = (X(\omega - \omega_1) + X(\omega + \omega_1))/2$	Example 10.11
$\forall n \in Integers,$ $y(n) = \sin(\omega_1 n) x(n)$	$\forall \omega \in Reals,$ $Y(\omega) = (X(\omega - \omega_1) - X(\omega + \omega_1))/2i$	Exercise 12
$\forall n \in Integers,$ $x(n) = ax_1(n) + bx_2(n)$	$\forall \omega \in Reals,$ $X(\omega) = aX_1(\omega) + bX_2(\omega)$	Section 10.7.4
$\forall n \in Integers,$ $y(n) = (h * x)(n)$	$\forall \omega \in Reals,$ $Y(\omega) = H(\omega)X(\omega)$	Section 10.7.1
$\forall n \in Integers,$ $y(n) = x(n)p(n)$	$\forall \omega \in Reals,$ $Y(\omega) = \frac{1}{2\pi} \int_0^{2\pi} X(\Omega)P(\omega - \Omega) d\Omega$	Box on page 398
$\forall n \in Integers,$ $y(n) = \begin{cases} x(n/N) & n \text{ is a multiple of } N \\ 0 & \text{otherwise} \end{cases}$	$\forall \omega \in Reals,$ $Y(\omega) = X(N\Omega)$	Exercise 14

TABLE 10.9
Properties of the CTFT.

Time domain	Frequency domain	Reference
$\forall\, t \in Reals$, $x(t)$ is real	$\forall\, \omega \in Reals$, $X(\omega) = X^*(-\omega)$	Section 10.7.2
$\forall\, t \in Reals$, $x(t) = x^*(-t)$	$\forall\, \omega \in Reals$, $X(\omega)$ is real	Section 10.7.2
$\forall\, t \in Reals$, $y(t) = x(t - \tau)$	$\forall\, \omega \in Reals$, $Y(\omega) = e^{-i\omega\tau} X(\omega)$	Section 10.7.3
$\forall\, t \in Reals$, $y(t) = e^{i\omega_1 t} x(t)$	$\forall\, \omega \in Reals$, $Y(\omega) = X(\omega - \omega_1)$	Section 10.7.6
$\forall\, t \in Reals$, $y(t) = \cos(\omega_1 t) x(t)$	$\forall\, \omega \in Reals$, $Y(\omega) = (X(\omega - \omega_1) + X(\omega + \omega_1))/2$	Example 10.11
$\forall\, t \in Integers$, $y(t) = \sin(\omega_1 t) x(t)$	$\forall\, \omega \in Reals$, $Y(\omega) = (X(\omega - \omega_1) - X(\omega + \omega_1))/2i$	Exercise 12
$\forall\, t \in Reals$, $x(t) = a x_1(t) + b x_2(t)$	$\forall\, \omega \in Reals$, $X(\omega) = a X_1(\omega) + b X_2(\omega)$	Section 10.7.4
$\forall\, t \in Reals$, $y(t) = (h * x)(t)$	$\forall\, \omega \in Reals$, $Y(\omega) = H(\omega) X(\omega)$	Section 10.7.1
$\forall\, t \in Reals$, $y(t) = x(t) p(t)$	$\forall\, \omega \in Reals$, $Y(\omega) = \frac{1}{2\pi} \int_{-\infty}^{\infty} X(\Omega) P(\omega - \Omega) d\Omega$	—
$\forall\, t \in Reals$, $y(t) = x(at)$	$\forall\, \omega \in Reals$, $Y(\omega) = \frac{1}{\lvert a \rvert} X(\Omega/a)$	Exercise 11

$e^{i\omega n}$, then the output is $H(\omega)e^{i\omega n}$. Suppose the input is instead an arbitrary signal x with DTFT X. Using the inverse DTFT relation, we know that

$$\forall n \in Integers, \quad x(n) = \frac{1}{2\pi} \int_0^{2\pi} X(\omega)e^{i\omega n} d\omega.$$

View this as a summation of exponentials, each with weight $X(\omega)$. An integral, after all, is summation over a continuum. Each term in the summation is $X(\omega)e^{i\omega n}$. If this term were an input by itself, then the output would be $H(\omega)X(\omega)e^{i\omega n}$. Thus, according to linearity, if the input is x, the output should be

$$\forall n \in Integers, \quad y(n) = \frac{1}{2\pi} \int_0^{2\pi} H(\omega)X(\omega)e^{i\omega n} d\omega.$$

Comparing with the inverse DTFT relation for $y(n)$, we see that

$$\boxed{\forall \omega \in Reals, \quad Y(\omega) = H(\omega)X(\omega).} \tag{10.17}$$

This is the frequency-domain version of the convolution

$$\forall n \in Integers, \quad y(n) = (h * x)(n).$$

Thus, the frequency response of an LTI system multiplies the DTFT of the input. This is intuitive, inasmuch as the frequency response gives the weight imposed by the system on each frequency component of the input.

Equation (10.17) applies equally well to continuous-time systems, but in that case, H is the CTFT of the impulse response, and X is the CTFT of the input.

PROBING FURTHER

Multiplying signals

We have seen that convolution in the time domain corresponds to multiplication in the frequency domain. It turns out that this relationship is symmetric, in that multiplication in the time domain corresponds to a peculiar form of convolution in the frequency domain. In other words, given two discrete-time signals x and p with DTFTs X and P, if we multiply them in the time domain,

$$\forall n \in Integers, \quad y(n) = x(n)p(n),$$

PROBING FURTHER

then in the frequency domain, $Y(\omega) = (X \circledast P)(\omega)$, where the symbol "$\circledast$" indicates **circular convolution**, defined by

$$\forall\, \omega \in Reals, \quad (X \circledast P)(\omega) = \frac{1}{2\pi} \int_{0}^{2\pi} X(\Omega)P(\omega - \Omega)d\Omega.$$

To verify this, we can substitute into this integral the definitions for the DTFTs $X(\omega)$ and $P(\omega)$ to get

$$\begin{aligned}(X \circledast P)(\omega) &= \frac{1}{2\pi} \int_{0}^{2\pi} \left(\sum_{m=-\infty}^{\infty} x(m)e^{-i\Omega m} \right) \left(\sum_{k=-\infty}^{\infty} p(k)e^{-i(\omega-\Omega)k} \right) d\Omega \\ &= \sum_{k=-\infty}^{\infty} p(k)e^{-i\omega k} \sum_{m=-\infty}^{\infty} x(m) \frac{1}{2\pi} \int_{0}^{2\pi} e^{-i\Omega(m-k)} d\Omega \\ &= \sum_{k=-\infty}^{\infty} p(k)x(k)e^{-i\omega k},\end{aligned}$$

where the last equality follows from the observation that the integral in the middle expression is zero except when $m = k$, when it has value one. Thus, $(X \circledast P)(\omega)$ is the DTFT of $y = xp$, as we claimed.

The continuous-time case is somewhat simpler. If $\forall\, t \in Reals,\ y(t) = x(t)p(t)$, then in the frequency domain,

$$\forall\, \omega \in Reals, \quad Y(\omega) = \frac{1}{2\pi}(X * P)(\omega) = \frac{1}{2\pi} \int_{-\infty}^{\infty} X(\Omega)P(\omega - \Omega)d\Omega.$$

The "$*$" indicates ordinary convolution.

10.7.2 *Conjugate symmetry*

We have already shown (see (8.29)) that for real-valued signals, the Fourier series coefficients are conjugate symmetric:

$$X_k = X^*_{-k}.$$

In fact, all the Fourier transforms are conjugate symmetric if the time-domain function is real. We illustrate this with the CTFT. Suppose x: *Reals* $\rightarrow$ *Reals* is a real-valued signal and $X = CTFT(x)$. In general, $X(\omega)$ is complex valued. From (10.13), we have

$$
\begin{aligned}
[X(-\omega)]^* &= \int_{-\infty}^{\infty} [x(t)e^{i\omega t}]^* dt \\
&= \int_{-\infty}^{\infty} x(t)e^{-i\omega t} dt \\
&= X(\omega).
\end{aligned}
$$

Thus,

$$X(\omega) = X^*(-\omega);$$

that is, for real-valued signals, $X(\omega)$ and $X(-\omega)$ are complex conjugates of one another.

We can show that, conversely, if the Fourier transform is conjugate symmetric, then the time-domain function is real. To do this, write the inverse CTFT

$$x(t) = \frac{1}{2\pi} \int_{-\infty}^{\infty} X(\omega)e^{i\omega t} d\omega$$

and then conjugate both sides:

$$x^*(t) = \frac{1}{2\pi} \int_{-\infty}^{\infty} X^*(\omega)e^{-i\omega t} d\omega.$$

By changing variables (replacing ω with $-\omega$) and using the conjugate symmetry of X, we can show that

$$x^*(t) = x(t),$$

which implies that $x(t)$ is real for all t.

The inverse Fourier transforms can be used to show that if a time-domain function is conjugate symmetric,

$$x(t) = x^*(-t),$$

then its Fourier transform is real. The same property applies to all four Fourier transforms.

In summary, if a function in one domain (time or frequency) is conjugate symmetric, then the function in the other domain (frequency or time) is real.

10.7.3 *Time shifting*

Given a continuous-time function x and its Fourier transform $X = CTFT(x)$, let y be defined by

$$\forall t \in Reals, \quad y(t) = x(t - \tau)$$

for some real constant τ. This is called **time shifting**, or **delay**. We can find $Y = CTFT(y)$ in terms of X as follows:

$$\begin{aligned} Y(\omega) &= \int_{-\infty}^{\infty} y(t)e^{-i\omega t}dt \\ &= \int_{-\infty}^{\infty} x(t-\tau)e^{-i\omega t}dt \\ &= \int_{-\infty}^{\infty} x(t)e^{-i\omega(t+\tau)}dt \\ &= e^{-i\omega\tau} \int_{-\infty}^{\infty} x(t)e^{-i\omega t}dt \\ &= e^{-i\omega\tau}X(\omega). \end{aligned}$$

Thus, in summary,

$$\boxed{y(t) = x(t-\tau) \Leftrightarrow Y(\omega) = e^{-i\omega\tau}X(\omega).} \qquad (10.18)$$

The bidirectional arrow indicates that this relationship works both ways. If you know that $Y(\omega) = e^{-i\omega\tau}X(\omega)$, then you can conclude that $y(t) = x(t - \tau)$.

Example 10.7: One of the simplest Fourier transforms to compute is that of x when

$$\forall\, t \in Reals, \quad x(t) = \delta(t),$$

where δ is the Dirac delta function. Plugging into the formula, we find

$$\begin{aligned}\forall\, t \in Reals, \quad X(\omega) &= \int_{-\infty}^{\infty} x(t) e^{-i\omega t} dt \\ &= \int_{-\infty}^{\infty} \delta(t) e^{-i\omega t} dt \\ &= e^{-i\omega 0} \\ &= 1,\end{aligned}$$

where we have used the sifting property of the delta function. The Fourier transform is a constant, 1.* This indicates that the Dirac delta function, interestingly, contains all frequencies in equal amounts.

Moreover, this result is intuitive if we consider an LTI system with impulse response $h(t) = \delta(t)$. Such a system responds to an impulse by producing an impulse, which suggests that any input is simply passed through unchanged. Indeed, its frequency response is $H(\omega) = 1$ for all $\omega \in Reals$; therefore, given an input x with Fourier transform X, the output y has Fourier transform

$$Y(\omega) = H(\omega)X(\omega) = X(\omega).$$

Because the output has the same Fourier transform as the input, the output is the same as the input.

Using (10.18), we can now find the Fourier transform of another signal,

$$x(t) = \delta(t - \tau),$$

for some constant $\tau \in Reals$. It is

$$X(\omega) = e^{-i\omega\tau}.$$

Note that, as required, this is conjugate symmetric. Moreover, it has a magnitude of 1. Its phase is $-\omega\tau$, a linear function of ω.

* To see how the sifting property works, note that the integrand is 0 everywhere except where $t = 0$, at which point the complex exponential equals 1. Then the integral of the delta function itself has a value of 1.

Again, we can gain some intuition by considering an LTI system with impulse response

$$h(t) = \delta(t - \tau). \tag{10.19}$$

Such a system introduces a fixed time delay of τ. Its frequency response is

$$H(\omega) = e^{-i\omega\tau}. \tag{10.20}$$

Because this has a magnitude of 1 for all ω, it tells us that all frequencies get through the delay system unattenuated, as expected. However, each frequency ω undergoes a phase shift of $-\omega\tau$, corresponding to a time delay of τ. ❐

Discrete-time signals are similar. Consider a discrete-time signal x with $X = DTFT(x)$, and let y be defined by

$$\forall\, t \in Reals, \quad y(n) = x(n - N)$$

for some integer constant N. By similar methods, we can find that

$$\boxed{y(n) = x(n - N) \;\Leftrightarrow\; Y(\omega) = e^{-i\omega N} X(\omega).} \tag{10.21}$$

Example 10.8: Suppose we have a discrete-time signal x given by

$$\forall\, n \in Integers, \quad x(n) = \delta(n),$$

where δ is the Kronecker delta function. It is easy to show from the DTFT definition that

$$\forall\, w \in Reals, \quad X(\omega) = 1.$$

Using (10.21), we can now find the Fourier transform of another signal

$$\forall\, n \in Integers, \quad x(n) = \delta(n - N),$$

for some constant $N \in Integers$:

$$\forall\, w \in Reals, \quad X(\omega) = e^{-i\omega N}.$$

Notice that if $N = 0$, this reduces to $X(\omega) = 1$, as expected. ❐

10.7.4 *Linearity*

Consider three discrete-time signals x, x_1, and x_2, related by

$$\forall\, n \in Integers, \quad x(n) = ax_1(n) + bx_2(n).$$

Then it is easy to see from the definition of the DTFT that

$$\forall\, \omega \in Reals, \quad X(\omega) = aX_1(\omega) + bX_2(\omega),$$

where $X = DTFT(x)$, $X_1 = DTFT(x_1)$, and $X_2 = DTFT(x_2)$.

The same linearity property applies to the CTFT:

$$x(t) = ax_1(t) + bx_2(t) \Leftrightarrow X(\omega) = aX_1(\omega) + bX_2(\omega).$$

Linearity of the Fourier transform is one of the most useful properties for avoiding evaluation of integrals and summations.

Example 10.9: Consider the discrete-time signal x given by

$$x(n) = \delta(n+1) + \delta(n-1),$$

where δ is the Kronecker delta function. Using linearity, we know that the DTFT of x is the sum of the DTFT of $\delta(n+1)$ and the DTFT of $\delta(n-1)$. From the previous example, we know those two DTFTs, and so

$$X(\omega) = e^{i\omega} + e^{-i\omega}$$

because N is -1 and 1, respectively. Using Euler's relation, we can simplify this to get

$$X(\omega) = 2\cos(\omega).$$

Interestingly, the DTFT of this example turns out to be real. This is because the time-domain function is conjugate symmetric (the conjugate of something real is itself real). Moreover, because it is real in the time domain, the DTFT turns out to be conjugate symmetric. ❐

Linearity can also be used to find inverse Fourier transforms.

Example 10.10: Suppose you are told that a continuous-time signal has Fourier transform

$$X(\omega) = \cos(\omega).$$

How would you find the time-domain function? You could evaluate the inverse CTFT, but the integration that you would have to perform is quite difficult. Instead, use Euler's relation to write

$$X(\omega) = (e^{i\omega} + e^{-i\omega})/2.$$

Then use linearity. The inverse Fourier transform of this sum is the sum of the inverse Fourier transforms of the terms. These we can recognize from (10.19) and (10.20), and so

$$x(t) = (\delta(t+1) + \delta(t-1))/2,$$

where δ is the Dirac delta function. ❐

10.7.5 *Constant signals*

We have seen that the Fourier transform of a delta function is a constant. With the symmetries that we have observed between time and frequency, it should come as no surprise that the Fourier transform of a constant is a delta function.

Consider first a continuous-time signal x given by

$$\forall\, t \in Reals, \quad x(t) = K$$

for some real constant K. Its CTFT is

$$X(\omega) = K \int_{-\infty}^{\infty} e^{-i\omega t} dt,$$

which is not easy to evaluate. This integral is mathematically subtle. The answer is

$$\forall\, \omega \in Reals, \quad X(\omega) = 2\pi K \delta(\omega),$$

where δ is the Dirac delta function. What this says is that a constant in the time domain is concentrated at zero frequency in the frequency domain (which should not be surprising). Except for the multiplying constant of 2π, we probably could have guessed this answer. We can verify this answer by evaluating the inverse CTFT,

$$\begin{aligned} x(t) &= \frac{1}{2\pi} \int_{-\infty}^{\infty} X(\omega) e^{i\omega t} d\omega \\ &= K \int_{-\infty}^{\infty} \delta(\omega) e^{i\omega t} d\omega \\ &= K, \end{aligned}$$

where the final step follows from the sifting property of the Dirac delta function. Thus, in summary,

$$\boxed{x(t) = K \Leftrightarrow X(\omega) = 2\pi K \delta(\omega).} \tag{10.22}$$

The discrete-time case is similar, but there is one subtlety because the DTFT is periodic. Let x be a discrete-time signal where

$$\forall\, n \in Integers, \quad x(n) = K$$

for some real constant K. Its DTFT is

$$\forall\, \omega \in [-\pi, \pi], \quad X(\omega) = 2\pi K \delta(\omega),$$

where δ is the Dirac delta function. This is easy to verify by using the inverse DTFT. Again, what this says is that a constant in the time domain is concentrated at zero frequency in the frequency domain (which should not be surprising). However, recall that a DTFT is periodic with period 2π, meaning that for all integers N,

$$X(\omega) = X(\omega + N2\pi).$$

Thus, in addition to a delta function at $\omega = 0$, there must be one at $\omega = 2\pi$, $\omega = -2\pi$, $\omega = 4\pi$, and so forth. This can be written by using a shift-and-add summation:

$$\forall\, \omega \in Reals, \quad X(\omega) = 2\pi K \sum_{k=-\infty}^{\infty} \delta(\omega - k2\pi).$$

Thus, in summary,

$$\boxed{x(n) = K \Leftrightarrow X(\omega) = 2\pi K \sum_{k=-\infty}^{\infty} \delta(\omega - k2\pi).} \tag{10.23}$$

10.7.6 *Frequency shifting and modulation*

Suppose that x is a continuous-time signal with CTFT X. Let y be another continuous-time signal defined by

$$y(t) = x(t)e^{i\omega_0 t}$$

for some real constant ω_0. The CTFT of y is easy to compute:

$$\begin{aligned} Y(\omega) &= \int_{-\infty}^{\infty} y(t)e^{-i\omega t}dt \\ &= \int_{-\infty}^{\infty} x(t)e^{i\omega_0 t}e^{-i\omega t}dt \\ &= \int_{-\infty}^{\infty} x(t)e^{-i(\omega-\omega_0)t}dt \\ &= X(\omega - \omega_0). \end{aligned}$$

Thus, the Fourier transform of y is the same as that of x but shifted to the right by ω_0. In summary,

$$\boxed{y(t) = x(t)e^{i\omega_0 t} \Leftrightarrow Y(\omega) = X(\omega - \omega_0).} \tag{10.24}$$

This result can be used to determine the effect of multiplying a signal by a sinusoid, a process called **modulation** (see exercise 16 at the end of this chapter and lab 10 in Lab Manual.).

Example 10.11: Suppose

$$y(t) = x(t)\cos(\omega_0 t).$$

Use Euler's relation to rewrite this

$$y(t) = x(t)(e^{i\omega_0 t} + e^{-i\omega_0 t})/2.$$

Then use (10.24) to get the CTFT:

$$Y(\omega) = (X(\omega - \omega_0) + X(\omega + \omega_0))/2. \square$$

We can combine (10.24) with (10.22) to get the following fact:

$$\boxed{x(t) = e^{i\omega_0 t} \Leftrightarrow X(\omega) = 2\pi\,\delta(\omega - \omega_0).} \tag{10.25}$$

This says that a complex exponential signal with frequency ω_0 is concentrated in the frequency domain at frequency ω_0, which should not be surprising. Similarly,

$$\boxed{x(t) = \cos(\omega_0 t) \Leftrightarrow X(\omega) = \pi(\delta(\omega - \omega_0) + \delta(\omega + \omega_0)).} \tag{10.26}$$

We can get a similar set of results for discrete-time signals. We summarize the results here and leave their verification to the reader (see exercise 14 at the end of this chapter):

$$\boxed{y(n) = x(n)e^{i\omega_0 n} \Leftrightarrow Y(\omega) = X(\omega - \omega_0);} \tag{10.27}$$

$$\boxed{y(n) = x(n)\cos(\omega_0 n) \Leftrightarrow Y(\omega) = (X(\omega - \omega_0) + X(\omega + \omega_0))/2;} \tag{10.28}$$

$$\boxed{x(n) = e^{i\omega_0 n} \Leftrightarrow X(\omega) = 2\pi \sum_{k=-\infty}^{\infty} \delta(\omega - \omega_0 - k2\pi);} \tag{10.29}$$

$$\boxed{x(n) = \cos(\omega_0 n) \Leftrightarrow X(\omega) = \pi \sum_{k=-\infty}^{\infty} (\delta(\omega - \omega_0 - k2\pi) + \delta(\omega + \omega_0 - k2\pi)).} \tag{10.30}$$

Additional properties of Fourier transforms are explored in the exercises.

10.8 Summary

In previous chapters, we developed the result that signals can be represented as sums of sinusoids. We used this representation to analyze the effect that an LTI system has on signals. We variously considered periodic and nonperiodic signals and impulse and frequency responses of LTI systems.

In this chapter, we unified all frequency domain discussion by showing that there are four closely interrelated Fourier transforms. Two of these apply to discrete-time signals, and two apply to continuous-time signals. Two of them apply to periodic signals, and two of them apply to nonperiodic signals.

The four Fourier transforms can be viewed as generalizations of the Fourier series. Consequently, they share many properties that are rooted in the Fourier series. These properties can be leveraged to analyze more complicated signals by using an analysis of simpler signals.

EXERCISES

KEY: E = mechanical T = requires plan of attack C = more than 1 answer

E 1. Show that if two discrete-time systems with frequency responses $H_1(\omega)$ and $H_2(\omega)$ are connected in cascade, then the DTFT of the output is given by $Y(\omega) = H_1(\omega)H_2(\omega)X(\omega)$, where $X(\omega)$ is the DTFT of the input.

E 2. This exercise verifies some of the relations in table 10.1.

(a) Let x be defined by

$$\forall\, t \in Reals, \quad x(t) = e^{i\omega_0 t},$$

where $\omega_0 \neq 0$. Use (10.2) to verify that its Fourier series coefficients are

$$\forall\, m \in Integers, \quad X_m = \begin{cases} 1 & \text{if } m = 1 \\ 0 & \text{otherwise.} \end{cases}$$

(b) Let x be defined by

$$\forall\, t \in Reals, \quad x(t) = \cos(\omega_0 t),$$

where $\omega_0 \neq 0$. Use part (a) and properties of the Fourier series to verify that its Fourier series coefficients are

$$\forall\, m \in Integers, \quad X_m = \begin{cases} 1/2 & \text{if } |m| = 1 \\ 0 & \text{otherwise.} \end{cases}$$

(c) Let x be defined by

$$\forall\, t \in Reals, \quad x(t) = \sin(\omega_0 t),$$

where $\omega_0 \neq 0$. Use part (a) and properties of the Fourier series to verify that its Fourier series coefficients are

$$\forall\, m \in Integers, \quad X_m = \begin{cases} 1/2i & \text{if } m = 1 \\ -1/2i & \text{if } m = -1 \\ 0 & \text{otherwise.} \end{cases}$$

(d) Let x be defined by

$$\forall\, t \in Reals, \quad x(t) = 1.$$

Use (10.2) to verify that its Fourier series coefficients are

$$\forall\, m \in Integers, \quad X_m = \begin{cases} 1 & \text{if } m = 0 \\ 0 & \text{otherwise.} \end{cases}$$

Notice that the answer does not depend on your choice for p.

T 3. Consider a symmetric square wave $x \in ContPeriodic_p$, shown in figure 10.16. It is periodic with period p, and its fundamental frequency is $\omega_0 = 2\pi/p$ radians/second. Over one period, this is represented by

$$\forall\, t \in [0, p], \quad x(t) = \begin{cases} 1 & \text{if } t < T \text{ or } t > p - T \\ 0 & \text{otherwise.} \end{cases}$$

(a) Show that its Fourier series coefficients are

$$\forall\, m \in Integers, \quad X_m = \begin{cases} 2T/p & \text{if } m = 0 \\ (\sin(m\omega_0 T))/(m\pi) & \text{otherwise.} \end{cases}$$

You can use l'Hôpital's rule (see page 58) to verify that $(\sin(x))/x = 1$ when $x = 0$, and so this can be written more simply as

$$\forall\, m \in Integers, \quad X_m = \frac{\sin(m\omega_0 T)}{m\pi}. \tag{10.31}$$

Note that this is real, as expected, because x is symmetric. Hint: The integration formula (10.5) may be helpful.

(b) Let $T = 0.5$, and use MATLAB to plot the Fourier series coefficients as a stem plot (using the `stem` commmand) for m ranging from -20 to 20. Note that you will have to be careful to avoid a divide-by-zero error at $m = 0$. Construct plots for $p = 2$, $p = 4$, and $p = 8$.

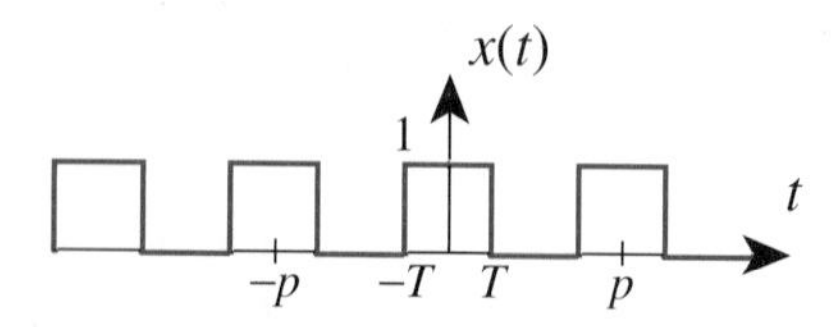

FIGURE 10.16: A symmetric square wave.

E 4. Let x be a periodic **impulse train**, expressed by

$$\forall\, t \in Reals, \quad x(t) = \sum_{n=-\infty}^{\infty} \delta(t - np),$$

where $p > 0$ is the period. This signal has a Dirac delta function at all multiples of p. Use (10.2) to verify that its Fourier series coefficients are

$$\forall\, m \in Integers, \quad X_m = 1/p.$$

Note that there is a subtlety in using (10.2) here: There are impulses at each end of the integration interval. This subtlety can be avoided by observing that (10.2) can, in fact, integrate over any interval that covers one period. Thus, it can equally well be written

$$\forall\, m \in Integers, \quad X_m = \frac{1}{p} \int_{-p/2}^{p/2} x(t) e^{-im\omega_0 t} dt.$$

This simplifies the problem considerably.

E 5. This exercise verifies some of the relations in table 10.2. In all cases, assume that the frequency f is rational and that it is related to the period p by $f = m/p$ for some integer m (see section 7.6.1).

(a) Let x be defined by

$$\forall\, n \in Integers, \quad x(n) = e^{i2\pi fn},$$

where $f \neq 0$. Use (10.9) to verify that its DFT coefficients are

$$\forall\, k \in Integers, \quad X'_k = \begin{cases} p & \text{if } k \in \{\ldots m - 2p,\ m - p,\ m,\ m + p, \\ & \quad m + 2p, \ldots\} \\ 0 & \text{otherwise.} \end{cases}$$

(b) Let x be defined by

$$\forall\, n \in Integers, \quad x(n) = \cos(2\pi fn),$$

where $f \neq 0$. Use part (a) and properties of the DFT to verify that its DFT coefficients are

$$X'_k = \begin{cases} p/2 & \text{if } k \in \{\ldots m-2p,\ m-p,\ m,\ m+p,\ m+2p, \ldots\} \\ p/2 & \text{if } k \in \{\ldots -m-2p,\ -m-p,\ -m,\ -m+p, \\ & \quad -m+2p, \ldots\} \\ 0 & \text{otherwise.} \end{cases}$$

(c) Let x be defined by

$$\forall\, n \in Integers, \quad x(n) = \sin(i2\pi fn),$$

where $f \neq 0$. Use part (a) and properties of the DFT to verify that its DFT coefficients are

$$X'_k = \begin{cases} p/2i & \text{if } k \in \{\ldots m-2p,\ m-p,\ m,\ m+p,\ m+2p, \ldots\} \\ -p/2i & \text{if } k \in \{\ldots -m-2p,\ -m-p,\ -m,\ -m+p, \\ & \quad -m+2p, \ldots\} \\ 0 & \text{otherwise.} \end{cases}$$

(d) Let x be defined by

$$\forall\, n \in Integers, \quad x(n) = 1.$$

Use (10.9) to verify that its DFT coefficients are

$$X_k = \begin{cases} p & \text{if } k \in \{\ldots -2p, -p, 0, p, 2p, \ldots\} \\ 0 & \text{otherwise.} \end{cases}$$

Notice that this result depends on your choice for p. This is a consequence of the unfortunate scaling that is used for the DFT, as opposed to the discrete Fourier series (DFS).

T 6. Consider a symmetric discrete square wave $x \in DiscPeriodic_p$, shown in figure 10.17. It is periodic with period p, and its fundamental

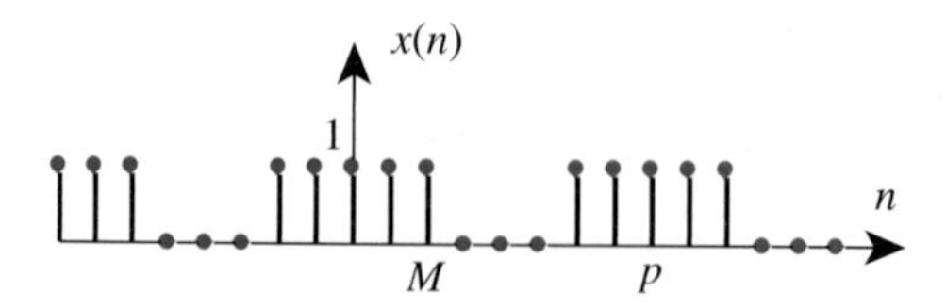

FIGURE 10.17: A symmetric discrete square wave.

frequency is $\omega_0 = 2\pi/p$ radians/sample. Over one period, this is represented by

$$\forall n \in \{0, 1, \ldots, p-1\}, \quad x(n) = \begin{cases} 1 & \text{if } n \leq M \text{ or } n \geq p - M \\ 0 & \text{otherwise,} \end{cases}$$

where, in the figure, $M = 2$ and $p = 8$. For this problem, however, assume that M and p can be any positive integers where $p > 2M$.

(a) Show that the DFT is represented by

$$\forall k \in \mathit{Integers},$$

$$X'_k = \begin{cases} 2M + 1 & \text{if } k \text{ is a multiple of } p \\ (\sin(k(M+0.5)\omega_0))/(\sin(k\omega_0/2)) & \text{otherwise.} \end{cases}$$

You can use l'Hôpital's rule (see page 58) to verify that $\sin(Kx)/\sin(x) = K$ when $x = 0$, and so this can be written more simply as

$$\forall k \in \mathit{Integers}, \quad X'_k = \frac{\sin(k(M+0.5)\omega_0)}{\sin(k\omega_0/2)}. \tag{10.32}$$

Note that this is real, as expected, because x is symmetric. Hint: The summation identity formula (10.10) may be helpful.

(b) Let $M = 2$, and use MATLAB to plot the Fourier series coefficients as a stem plot (using the **stem** commmand) for m ranging from $-p+1$ to $p-1$. Note that you will have to be careful to avoid a divide-by-zero error at $k = 0$. Construct plots for $p = 8$, $p = 16$, and $p = 32$.

E 7. Let x be a discrete periodic impulse train, represented by

$$\forall n \in \mathit{Integers}, \quad x(n) = \sum_{m=-\infty}^{\infty} \delta(n - mp),$$

where $p > 0$ is the integer period. This signal has a Kronecker delta function at all multiples of p. Use (10.9) to verify that its DFT coefficients are

$$\forall k \in \mathit{Integers}, \quad X'_k = 1.$$

T 8. Consider a symmetric discrete rectangle $x \in \mathit{DiscSignals}$, shown in figure 10.18. This is expressed by

$$\forall n \in \mathit{Integers}, \quad x(n) = \begin{cases} 1 & \text{if } |n| \leq M \\ 0 & \text{otherwise.} \end{cases}$$

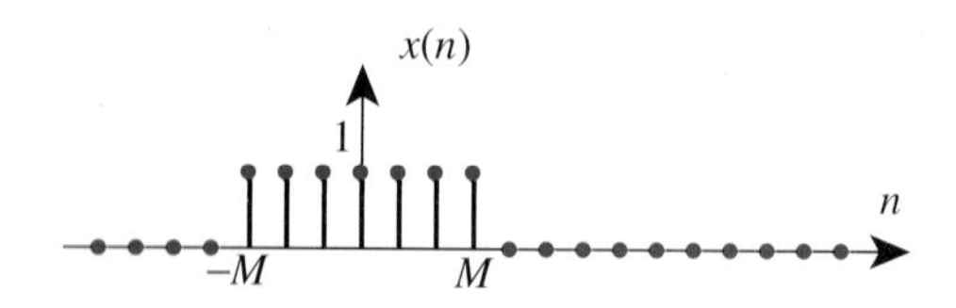

FIGURE 10.18: A symmetric discrete rectangle signal.

(a) Show that its DTFT is expressed by

$$\forall\, \omega \in Reals, \quad X(\omega) = \frac{\sin(\omega(M+0.5))}{\sin(\omega/2)}.$$

Hint: The summation identity formula (10.10) may be helpful.

(b) Let $M = 3$, and use MATLAB to plot the DTFT with ω ranging from $-\pi$ to π.

E 9. Consider x defined by

$$\forall\, t \in Reals, \quad x(t) = \sin(\omega_0 t).$$

Show that the CTFT is expressed by

$$\forall\, \omega \in Reals, \quad X(\omega) = (\pi/i)\delta(\omega - \omega_0) - (\pi/i)\delta(\omega + \omega_0).$$

C 10. In section 10.6, we explored the relationship between the CTFT of a periodic continuous-time signal and its Fourier series. In this problem, we do the same for discrete-time signals.

(a) Consider a discrete-time signal x with DTFT defined by

$$\forall\, \omega \in [-\pi, \pi], \quad X(\omega) = 2\pi\,\delta(\omega - \omega_0),$$

where $\omega_0 = 2\pi/p$ for some integer p. This DTFT is represented over only one cycle, but note that because it is a DTFT, it is periodic. Use the inverse DTFT to determine x.

(b) Is x in part (a) periodic? If so, what is the period, and what is its DFT?

(c) More generally, consider a discrete-time signal x with DTFT represented by

$$\forall\, \omega \in [-\pi, \pi], \quad X(\omega) = 2\pi \sum_{k=-\lfloor p/2 \rfloor}^{\lfloor p/2 \rfloor} X_k \delta(\omega - k\omega_0),$$

where $\omega_0 = 2\pi/p$ for some integer p. For simplicity, assume that p is odd, and let $\lfloor p/2 \rfloor$ be the largest integer less than $p/2$. Use the inverse DTFT to determine x.

(d) Is x in part (c) periodic? If so, what is the period, and what is its DFT? Express the DTFT in terms of the DFT coefficients.

T 11. Consider a continuous-time signal x with Fourier transform X. Let y be such that

$$\forall\, t \in Reals, \quad y(t) = x(at),$$

for some real number a. Show that its Fourier transform Y is

$$\forall\, \omega \in Reals, \quad Y(\omega) = \frac{1}{|a|} X(\omega/a).$$

E 12. Consider the discrete-time signal y defined by

$$\forall\, n \in Integers, \quad y(n) = \sin(\omega_1 n) x(n).$$

Show that the DTFT is

$$\forall\, \omega \in Reals, \quad Y(\omega) = (X(\omega - \omega_1) - X(\omega + \omega_1))/2i,$$

where $X = DTFT(x)$.

E 13. In this exercise, you verify various properties of the Fourier series in table 10.6. In all the following parts, x is a periodic continuous-time signal with period p and fundamental frequency $\omega_0 = 2\pi/p$.

(a) Let y be defined by

$$\forall\, t \in Reals, \quad y(t) = x(t - \tau),$$

for some real number τ. Show that the Fourier series coefficients of y are represented by

$$\forall\, m \in Reals, \quad Y_m = e^{-im\omega_0 \tau} X_m,$$

where X_m are the Fourier series coefficients of x.

(b) Let y be defined by

$$\forall\, t \in Reals, \quad y(t) = e^{i\omega_1 t} x(t),$$

where $\omega_1 = M\omega_0$, for some $M \in Integers$. Show that the Fourier series coefficients of y are represented by

$$\forall\, m \in Integers, \quad Y_m = X_{m-M},$$

where X_m are the Fourier series coefficients of x.

(c) Let y be defined by

$$\forall\, t \in Reals, \quad y(t) = \cos(\omega_1 t)x(t),$$

where $\omega_1 = M\omega_0$, for some $M \in Integers$. Show that the Fourier series coefficients of y are represented by

$$\forall\, m \in Integers, \quad Y_m = (X_{m-M} + X_{m+M})/2,$$

where X_m are the Fourier series coefficients of x.

(d) Let y be defined by

$$\forall\, t \in Reals, \quad y(t) = \sin(\omega_1 t)x(t),$$

where $\omega_1 = M\omega_0$, for some $M \in Integers$. Show that the Fourier series coefficients of y are represented by

$$\forall\, m \in Integers, \quad Y_m = (X_{m-M} - X_{m+M})/2i,$$

where X_m are the Fourier series coefficients of x.

T 14. Consider a discrete-time signal x with Fourier transform X. For each of the new signals defined as follows, show that its Fourier transform is as shown.

(a) If y is such that

$$\forall\, n \in Integers, \quad y(n) = \begin{cases} x(n/N) & \text{if } n \text{ is an integer multiple of } N \\ 0 & \text{otherwise.} \end{cases}$$

for some integer N, then its Fourier transform Y is

$$\forall\, \omega \in Reals, \quad Y(\omega) = X(\omega N).$$

(b) If w is such that

$$\forall\, n \in Integers, \quad w(n) = x(n)e^{i\alpha n},$$

for some real number α, then its Fourier transform W is

$$\forall\, \omega \in Reals, \quad W(\omega) = X(\omega - \alpha).$$

(c) If z is

$$\forall\, n \in Integers, \quad z(n) = x(n)\cos(\alpha n),$$

for some real number α, then its Fourier transform Z is such that

$$Z(\omega) = (X(\omega - \alpha) + X(\omega + \alpha))/2.$$

T 15. Consider the FIR system described by the following block diagram:

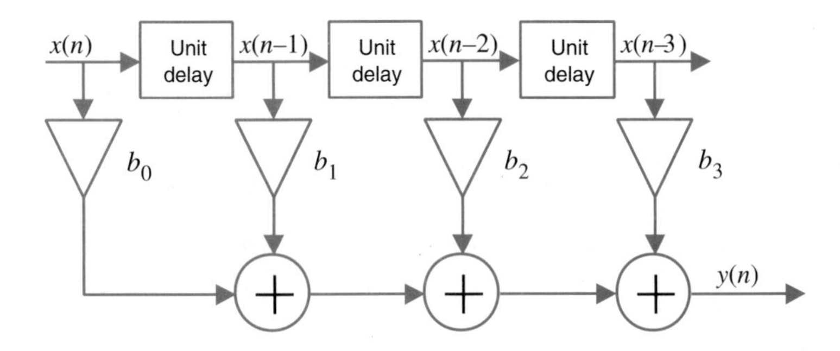

The notation here is the same as in figure 9.15. Suppose that this system has frequency response H. Define a new system with the identical structure as in this illustration, except that each unit delay is replaced by a double delay (two cascaded unit delays). Find the frequency response of that system in terms of H.

T 16. **Amplitude modulation** (**AM**) is a technique that is used to convert low-frequency signals into high-frequency signals for transmission over a radio channel. Conversion of the high-frequency signal back to a low-frequency signal is called **demodulation**. The system structure is depicted as follows:

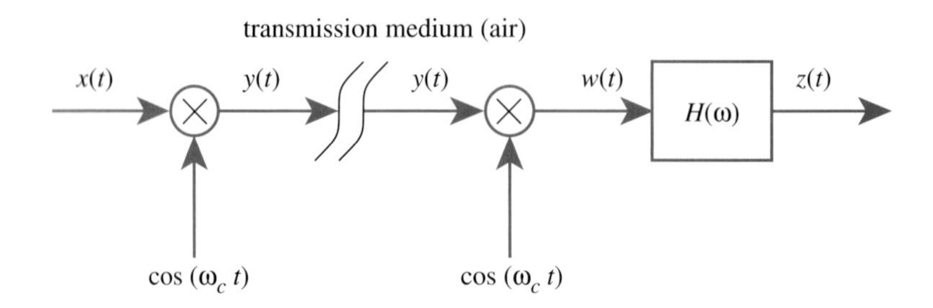

The circular components multiply their input signals. The transmission medium (air, for radio signals) is approximated here as a medium that passes the signal y unaltered.

Suppose your AM radio station is allowed to transmit signals at a carrier frequency of 740 kHz (this is the frequency of KCBS in San Francisco). Suppose you want to send the audio signal x:*Reals* $\to$ *Reals*. The AM signal that you would transmit is defined by, for all $t \in$ *Reals*,

$$y(t) = x(t)\cos(\omega_c t),$$

where $\omega_c = 2\pi \times 740{,}000$ is the **carrier frequency** (in radians per second). Suppose $X(\omega)$ is the Fourier transform of an audio signal with magnitude depicted as follows:

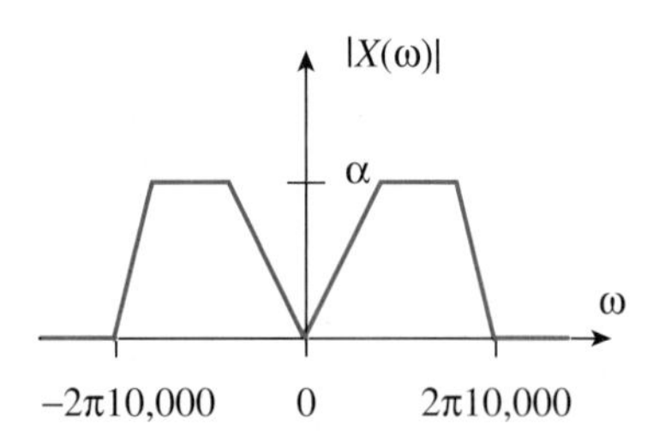

(a) Show that the Fourier transform Y of y in terms of X is

$$Y(\omega) = (X(\omega - \omega_c) + X(\omega + \omega_c))/2.$$

(b) Carefully sketch $|Y(\omega)|$, and note the important magnitudes and frequencies on your sketch.

Note that if $X(\omega) = 0$ for $|\omega| > 2\pi \times 10{,}000$, then $Y(\omega) = 0$ for $||\omega| - |\omega_c|| > 2\pi \times 10{,}000$. In words, if the signal x being modulated is bandlimited to less than 10 kHz, then the modulated signal is bandlimited to frequencies that are within 10 kHz of the carrier frequency. Thus, an AM radio station needs to occupy 20 kHz of the radio spectrum in order to transmit audio signals up to 10 kHz.

(c) At the receiver, the problem is to recover the audio signal x from y. One way is to demodulate by multiplying y by a sine wave at the carrier frequency to obtain the signal w, where

$$w(t) = y(t)\cos(\omega_c t).$$

What is the Fourier transform W of w in terms of X? Sketch $|W(\omega)|$, and note the important magnitudes and frequencies.

(d) After performing the demodulation of part (c), an AM receiver filters the received signal through a lowpass filter with frequency response $H(\omega)$ such that $H(\omega) = 1$ for $|\omega| \leq 2\pi \times 10{,}000$ and $|H(\omega)| = 0$ for $|\omega| > 2\pi \times 20{,}000$. Let z be the filtered signal, as shown in the illustration before part (a). What is the Fourier transform Z of z? What is the relationship between z and x?

T 17. In the following parts, assume that x is a discrete-time signal given by

$$\forall\, n \in Integers, \quad x(n) = \delta(n+1) + \delta(n) + \delta(n-1)$$

and that S is an LTI system with frequency response H given by

$$\forall\, \omega \in Reals, \quad H(\omega) = e^{-i\omega}.$$

(a) Find $X = DTFT(x)$, and make a well-labeled sketch for $\omega \in [-\pi, \pi]$ in radians/sample. Check that X is periodic with period 2π.

(b) Let $y = S(x)$. Find $Y = DTFT(y)$.

(c) Find $y = S(x)$.

(d) Sketch x and y, and comment on what the system S does.

T 18. Consider a causal discrete-time LTI system S with input x and output y such that

$$\forall\, n \in Integers, \quad y(n) = x(n) + ay(n-1),$$

where $a \in Reals$ is a given constant such that $|a| < 1$.

(a) Find the impulse response h of S.

(b) Find the frequency response of S by letting the input be $e^{i\omega n}$, letting the output be $H(\omega)e^{i\omega n}$, and solving for $H(\omega)$.

(c) Use your results in parts (a) and (b) and the fact that the DTFT of an impulse response is the frequency response to show that h represented by

$$\forall\, n \in Integers, \quad h(n) = a^n u(n)$$

has the discrete-time Fourier transform $H = DTFT(h)$ given by

$$H(\omega) = \frac{1}{1 - ae^{-i\omega}},$$

where $u(n)$ is the unit step.

(d) Use MATLAB to plot $h(n)$ and $|H(\omega)|$ for $a = -0.9$ and $a = 0.9$. You may choose the interval of n for your plot of h, but you should plot $|H(\omega)|$ in the interval $\omega \in [-\pi, \pi]$. Discuss the differences between these plots for the two different values of a.

T 19. Suppose a discrete-time signal x has DTFT represented by

$$X(\omega) = i \sin(K\omega)$$

for some positive integer K. Note that $X(\omega)$ is periodic with period 2π, as it must be to be a DTFT.

(a) Determine from the symmetry properties of X whether the time-domain signal x is real.

(b) Find x. Hint: Use Euler's relation and the linearity of the DTFT.

T 20. Consider a periodic continuous-time signal x with period p and Fourier series $X\colon Integers \to Complex$. Let y be another signal expressed by

$$y(t) = x(t - \tau)$$

for some real constant τ. Find the Fourier series coefficients of y in terms of those of X.

T 21. Consider the continuous-time signal expressed by

$$\forall\, t \in Reals, \quad x(t) = \frac{\sin(\pi t/T)}{(\pi t/T)}.$$

Show that its CTFT is represented by

$$\forall\, \omega \in Reals, \quad X(\omega) = \begin{cases} T, & \text{if } |\omega| \le \pi/T \\ 0, & \text{if } |\omega| > \pi/T. \end{cases}$$

The fact (10.5) from calculus may be useful.

T 22. If x is a continuous-time signal with CTFT X, then we can define a new time-domain function y such that

$$\forall t \in Reals, \quad y(t) = X(t);$$

that is, the new time-domain function has the same shape as the frequency-domain function X. Then the CTFT Y of y is expressed by

$$\forall \omega \in Reals, \quad Y(\omega) = 2\pi x(-\omega);$$

that is, the frequency domain of the new function has the shape of the time domain of the old function but is reversed and scaled by 2π. This property is called **duality** because it shows that time and frequency are interchangeable. Show that the property is true.

T 23. Use the results of exercises 21 and 22 to show that a continuous-time signal x defined by

$$\forall t \in Reals, \quad x(t) = \begin{cases} \pi/a & \text{if } |t| \leq a \\ 0 & \text{if } |t| > a, \end{cases}$$

where a is a positive real number, has CTFT X represented by

$$\forall \omega \in Reals \quad X(\omega) = 2\pi \frac{\sin(a\omega)}{(a\omega)}.$$

INTERVIEW

Jeff Bier

Jeff Bier is General Manager of Berkeley Design Technology, Inc. (BDTI), a firm he co-founded in 1991. BDTI provides independent analysis of DSP technology and develops optimized DSP application software. Jeff has overseen BDTI's growth from a two-person "garage" operation to one of the leading sources of practical DSP expertise for the electronics industry, with customers worldwide. Jeff has written and edited numerous articles, conference presentations, and industry reports; his monthly column, Impulse Response, appears in *EE Times.* Jeff also serves on the IEEE's technical committee on the design and implementation of signal processing systems.

How did you decide to study electrical engineering?

My father has a mechanical engineering background and worked in telecom and mainframe computers. At home he had a great model-train set-up, and he was often engaged in electrical or electronic projects. From a very young age I took an interest in electricity and electronics. I was a curious kid and asked a lot of questions, which fortunately my dad was always happy to answer. I gained a lot of basic knowledge through playing around at my dad's workbench. Later, I became interested in amateur radio and studied electronics in order to get my license.

Do you have any advice for students studying electrical and computer engineering?

Pursue breadth and depth in your education. By all means, dive into the details and really develop your technical skills—this is critical—but don't spend all of your time on technical topics. Learn how to communicate with people, how to work effectively with groups of people, and how to plan and organize projects. Think about how your work fits in to the world around you. After you've graduated, whatever career you pursue, these "soft" skills are essential for success.

Explore careers early. Just because you find circuits interesting doesn't necessarily mean you'll enjoy spending 50 hours a week designing them. Seek out people who are doing the kinds of jobs you think you might want. Spend time with them, seeing what their daily work is like. Seek out summer jobs, co-ops, and part-time work during your academic career so that you can direct your studies toward the kind of work that you'll most enjoy.

What is the most challenging part of your job?

I like to take an engineering approach to problems—to find an answer that can be proven to be correct. But when managing teams and projects, and making business decisions, things are seldom that simple. Every day I have to make dozens of decisions where I simply can't know what the optimal choice is. I have to do the best I can with the information and resources at hand. That can be frustrating when you've been trained to find the "right" answer.

What is your vision for the future of electrical and computer engineering?

The ideas, technology, and products produced by electrical and computer engineers are becoming more and more central to our economic, cultural, and social lives. Ten years ago, an engineer developing a product based on digital signal processing could have little hope that his friends and family would use that product in their everyday lives. But today we see DSP technology used in hearing aids, personal audio equipment, and cellular phones—an enormous variety of consumer goods. This trend will continue, and DSP technology will become pervasive in our lives, much the way simpler technology (like electric motors) has over the past century.

What would you say have been the greatest advances in the last decade in your area of expertise?

From the perspective of digital signal processing technology, it is the collective effect of countless advances in implementation technology—from integrated circuit manufacturing to processor architecture to software tools—that is making it possible to employ DSP techniques in a rapidly growing range of inexpensive, often battery-powered, products. As a result, in a brief time, digital signal processing has gone from being a specialty technology mostly used for cost-insensitive applications like military sonar and medical equipment to being an essential, mainstream enabler of the electronics industry.

CHAPTER 11
Sampling and reconstruction

Digital hardware, including that in computers, takes actions in discrete steps. Thus, it can deal with discrete-time signals, but it cannot directly handle the continuous-time signals that are prevalent in the physical world. This chapter is about the interface between these two worlds, the one continuous and the other discrete. A discrete-time signal is constructed by **sampling** a continuous-time signal, and a continuous-time signal is **reconstructed** by **interpolating** a discrete-time signal.

11.1 Sampling

A sampler for complex-valued signals is a system

$$Sampler_T\colon [Reals \to Complex] \to [Integers \to Complex], \qquad (11.1)$$

where T is the **sampling interval** (in units of seconds per sample). The system is depicted in figure 11.1. The **sampling frequency**, or **sample rate**, is $f_s = 1/T$, in units of samples per second (or, sometimes, Hertz) or $\omega_s = 2\pi/T$ in units of radians per second. If $y = Sampler_T(x)$, then y is defined by

$$\forall\, n \in Integers, \quad y(n) = x(nT). \qquad (11.2)$$

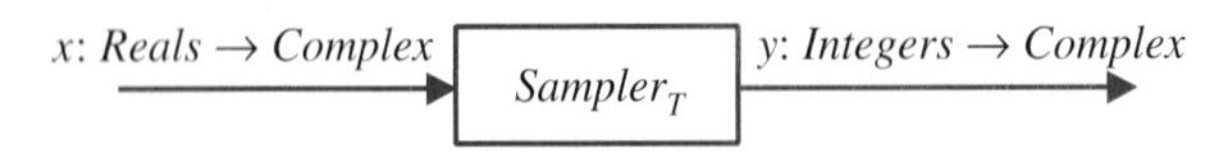

FIGURE 11.1: Sampler.

11.1.1 *Sampling a sinusoid*

Let $x: Reals \to Reals$ be the sinusoidal signal

$$\forall\, t \in Reals, \quad x(t) = \cos(2\pi f t), \tag{11.3}$$

where f is the frequency of the sine wave in Hertz. Let $y = Sampler_T(x)$. Then

$$\forall\, n \in Integers, \quad y(n) = \cos(2\pi f n T). \tag{11.4}$$

Although this looks similar to the continuous-time sinusoid, there is a fundamental difference. Because the index n is discrete, it turns out that the frequency f is indistinguishable from frequency $f + f_s$ when the discrete-time signal is observed. This phenomenon is called **aliasing**.

BASICS

Units

Recall that frequency can be given with any of various units. The units of f in (11.3) and (11.4) are Hertz, or cycles per second. In (11.3), it is sensible to give the frequency as $\omega = 2\pi f$, which is in units of radians per second. The constant 2π is in units of radians per cycle, and so the units are mathematically compatible. Moreover, the time argument t is in units of seconds, and so the argument to the cosine function, $2\pi f t$, is in units of radians, as expected.

In the discrete-time case (11.4), it is sensible to represent the frequency as $2\pi f T$, which is in units of radians per sample. The sampling interval T is in units of seconds per sample, and so, again, the units are compatible. Moreover, the integer n has units of samples, and so, again, the argument to the cosine function, $2\pi f n T$, is in units of radians, as expected.

In general, in discussions of continuous-time signals and their sampled discrete-time signals, it is important to be careful and consistent in the units used; otherwise, considerable confusion can result. Many texts mention **normalized frequency** when discussing discrete-time signals, by which they simply mean frequency in units of radians per sample. The frequency is **normalized** in the sense that it does not depend on the sampling interval.

11.1.2 *Aliasing*

Consider another sinusoidal signal, u, described by

$$\forall\, t \in Reals, \quad u(t) = \cos(2\pi (f + N f_s) t),$$

where N is some integer and $f_s = 1/T$. If $N \neq 0$, then this signal is clearly different from x in (11.3). Let

$$w = Sampler_T(u).$$

Then for all $n \in Integers$,

$$w(n) = \cos(2\pi(f + Nf_s)nT) = \cos(2\pi fnT + 2\pi Nn) = \cos(2\pi fnT) = y(n),$$

because Nn is an integer. Thus, even though $u \neq x$, $Sampler_T(u) = Sampler_T(x)$, and after being sampled, the signals x and u are indistinguishable. This phenomenon is called **aliasing**, presumably because it implies that any discrete-time sinusoidal signal has many continuous-time identities (its "identity" is presumably its frequency).

Example 11.1: A typical sample rate for voice signals is $f_s = 8{,}000$ samples/second, and so the sampling interval is $T = 0.125$ msec/sample. A continuous-time sinusoid with a frequency of 440 Hz, when sampled at this rate, is indistinguishable from a continuous-time sinusoid with a frequency of 8,440 Hz, when sampled at the same rate. ❐

Example 11.2: Compact discs are created by sampling audio signals at $f_s = 44{,}100$ Hz, and so the sampling interval is about $T = 22.7$ μsec/sample. A continuous-time sinusoid with frequency 440 Hz, when sampled at this rate, is indistinguishable from a continuous-time sinusoid with frequency 44,540 Hz, when sampled at the same rate. ❐

The frequency-domain analysis of the previous chapters relied heavily on complex exponential signals. Recall that a cosine can be expressed as a sum of two complex exponentials, by using Euler's relation:

$$\cos(2\pi ft) = 0.5(e^{i2\pi ft} + e^{-i2\pi ft}).$$

One of the complex exponentials is at frequency f, and the other is at frequency $-f$. Complex exponentials exhibit the same aliasing behavior that we have illustrated for sinusoids.

Let $x\colon Reals \to Complex$ be

$$\forall\, t \in Reals, \quad x(t) = e^{i2\pi ft},$$

where f is the frequency in Hertz. Let $y = Sampler_T(x)$. Then for all $n \in Integers$,

$$y(n) = e^{i2\pi fnT}.$$

Consider another complex exponential signal u,

$$u(t) = e^{i2\pi(f+Nf_s)t},$$

where N is some integer. Let

$$w = Sampler_T(u).$$

Then for all $n \in Integers$,

$$w(n) = e^{i2\pi(f+Nf_s)nT} = e^{i2\pi fnT}e^{i2\pi Nf_s nT} = e^{i2\pi fnT} = y(n),$$

because $e^{i2\pi Nf_s nT} = 1$. Thus, as with sinusoids, when we sample a complex exponential signal with frequency f at sample rate f_s, it is indistinguishable from one at frequency $f + f_s$ (or $f + Nf_s$ for any integer N).

There is considerably more to this story. Mathematically, aliasing relates to the periodicity of the frequency-domain representation (the discrete-time Fourier transform [DTFT]) of a discrete-time signal. Also, the effects of aliasing on real-valued signals (like the cosine, but unlike the complex exponential) depend strongly on the conjugate symmetry of the DTFT as well.

11.1.3 *Perceived pitch experiment*

Consider the following experiment.* Generate a discrete-time audio signal with an 8,000-samples/second sample rate according to the formula (11.4). Let the frequency f begin at 0 Hz and sweep upward through 4 kHz to at least 8 kHz. Use the audio output of a computer to listen to the resulting sound. The result is illustrated in figure 11.2. As the frequency of the continuous-time sinusoid rises, so does the perceived pitch, until the frequency reaches 4 kHz. At that point, the perceived pitch begins to fall rather than rise, even as the frequency of the continuous-time sinusoid continues to rise. It will fall until the frequency reaches 8 kHz, at which point no sound is heard at all (the perceived pitch is 0 Hz). Then the perceived pitch begins to rise again.

That the perceived pitch rises from 0 after the frequency f rises above 8,000 Hz is not surprising. We have already determined that in a discrete-time signal, a frequency of f is indistinguishable from a frequency $f + 8{,}000$, assuming the sample rate is 8,000 samples/second. But why does the perceived pitch drop when f rises above 4 kHz?

The frequency 4 kHz, $f_s/2$, is called the **Nyquist frequency**, after Harry Nyquist (1889–1976), an engineer at Bell Labs, who, in the 1920s and 1930s, laid much of the groundwork for digital transmission of information. The Nyquist frequency turns out to be a key threshold in the relationship between discrete-time

*Similar experiments are carried out in lab 11 (in Lab Manual). There is also an applet on the Web site demonstrating this phenomenon.

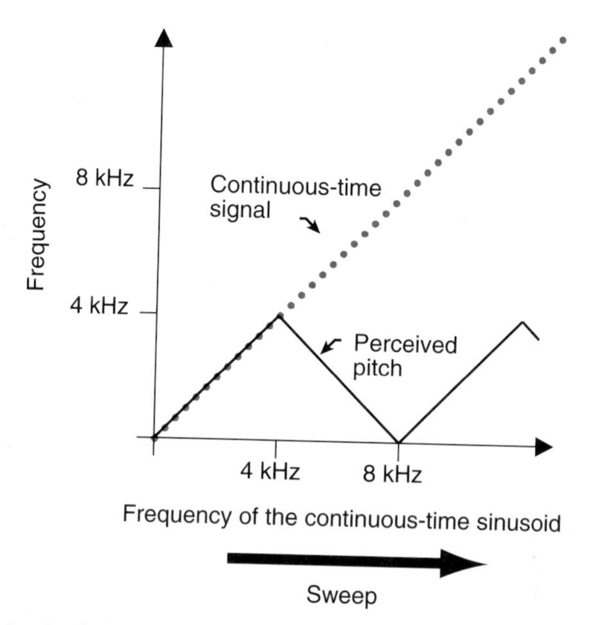

FIGURE 11.2: As the frequency of a continuous signal increases beyond the Nyquist frequency, the perceived pitch starts to drop.

and continuous-time signals, more important even than the sampling frequency. The intuitive reason is that if we sample a sinusoid with a frequency below the Nyquist frequency (below half the sampling frequency), then we take at least two samples per cycle of the sinusoid. It should be intuitively appealing that taking at least two samples per cycle of a sinusoid has some key significance. The two-sample minimum allows the samples to capture the oscillatory nature of the sinusoid. Fewer than two samples would not do this. However, what happens when fewer than two samples are taken per cycle is not necessarily intuitive. It turns out that the sinusoid masquerades as one of another frequency.

Consider the situation when the frequency f of a continuous-time sinusoid is 7,560 Hz. Figure 11.3 shows 4.5 msec of the continuous-time waveform, together with samples taken at 8 kHz. Notice that the samples trace out another sinusoid. We can determine the frequency of that sinusoid with the help of figure 11.2, which suggests that the perceived pitch is $8{,}000 - 7{,}560 = 440$ Hz (the slope of the perceived pitch line is -1 in this region). Indeed, if we listen to the sampled sinusoid, it will be the musical note A-440.

Recall that a cosine can be represented as a sum of complex exponentials with frequencies that are negatives of one another. Recall further that a complex exponential with frequency f is indistinguishable from one with frequency $f + Nf_s$, for any integer N. A variant of figure 11.2 that leverages this representation is depicted in figure 11.4.

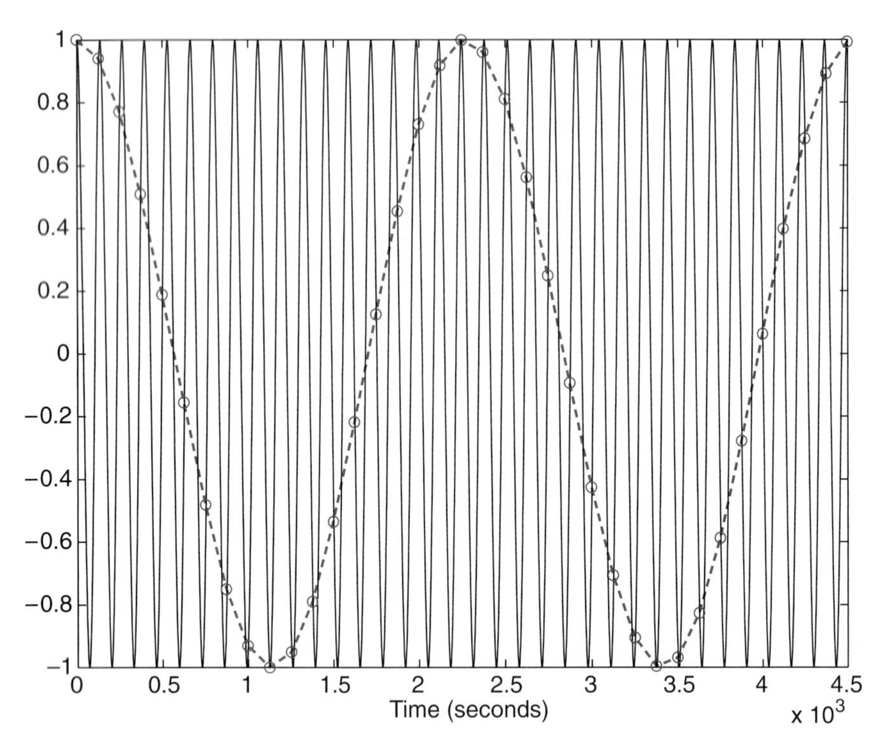

FIGURE 11.3: A sinusoid at 7.56 kHz and samples taken at 8 kHz.

In figure 11.4, as we follow the frequency of the continuous-time signal from 0 to 8 kHz, we move from left to right in the figure. The sinusoid consists not only of the rising frequency shown by the dotted line in figure 11.2 but also of a corresponding falling (negative) frequency, as shown in figure 11.4. Moreover, these two frequencies are indistinguishable, after sampling, from frequencies that are 8 kHz higher or lower, also shown by dotted lines in figure 11.4.

When the discrete-time signal is converted to a continuous-time audio signal, the hardware performing this conversion can choose any matching pair of positive and negative frequencies. By far the most common choice is to select the matching pair with lowest frequency, shown in figure 11.4 by the solid lines behind dotted lines. These result in a sinusoid with a frequency between 0 and the Nyquist frequency, $f_s/2$. This is why the perceived pitch falls after the frequency exceeds $f_s/2 = 4$ kHz.

Recall that the frequency-domain representation (i.e., the DTFT) of a discrete-time signal is periodic with period 2π radians/sample; that is, if X is a DTFT, then

$$\forall\, \omega \in Reals, \quad X(\omega) = X(\omega + 2\pi).$$

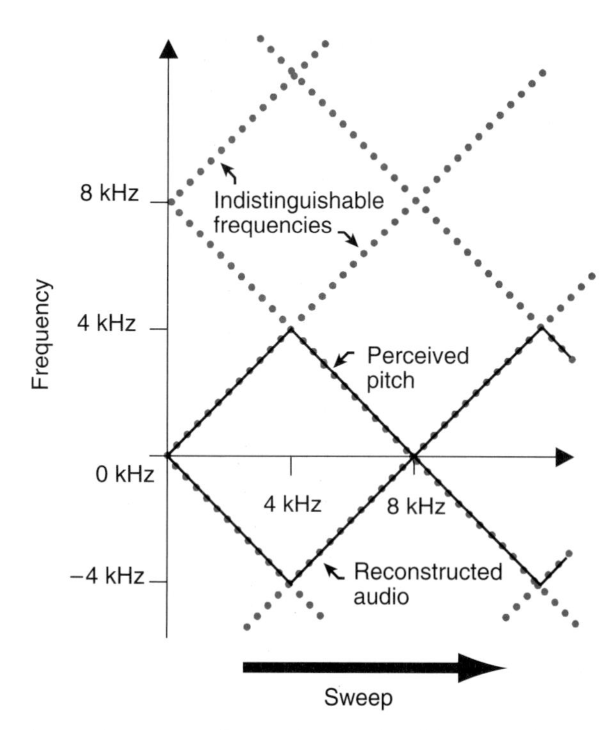

FIGURE 11.4: As the frequency of a continuous signal increases beyond the Nyquist frequency, the perceived pitch starts to drop because the frequency of the reconstructed continuous-time audio signal stays in the range $-f_s/2$ to $f_s/2$.

In radians per second, it is periodic with period $2\pi f_s$. In Hertz, it is periodic with period f_s, the sampling frequency. Thus, in figure 11.4, the dotted lines represent this periodicity. This periodicity is another way of stating that frequencies separated by f_s are indistinguishable.

11.1.4 *Avoiding aliasing ambiguities*

Figure 11.4 suggests that even though a discrete-time signal has ambiguous frequency content, it is possible to construct a uniquely defined continuous-time signal from the discrete-time waveform by choosing the one unique frequency for each component that is closest to 0. This always results in a reconstructed signal that contains only frequencies below the Nyquist frequency in magnitude.

Correspondingly, this suggests that if a continuous-time signal being sampled contains only frequencies below the Nyquist frequency, then this reconstruction strategy perfectly recovers the signal. This is an intuitive statement of the **Nyquist-Shannon sampling theorem**. This suggests that before sampling, it is reasonable to filter a signal to remove components with frequencies above $f_s/2$. A filter that realizes this is called an **antialiasing filter**.

Example 11.3: In the telephone network, speech is sampled at 8,000 samples per second before being digitized. Before this sampling, the speech signal goes through a lowpass filter to remove frequency components above 4,000 Hz. This lowpass-filtered speech can then be perfectly reconstructed at the far end of the telephone connection, which receives a stream of samples at 8,000 samples/second. ❒

Before probing this further, we examine in more detail what we mean by reconstruction.

PROBING FURTHER

Antialiasing for fonts

When rendering characters on a computer screen, it is common to use **antialiasing** to make the characters look better. Consider the following two images:

At the left is an image of the Greek letter omega. At the right is the result of sampling that rendition by taking only one pixel out of every 100 pixels in the original (every 10th pixel horizontally and vertically) and then rescaling the image so that it has the same size as the one on the left. The original image is discrete, and the resulting image is a smaller discrete image (this process is known as **subsampling**). Rendered with normal-sized pixels, the character on the right looks like this:

To the discerning eye, this can be improved considerably. The problem is that original image has hard edges and, hence, high (spatial) frequencies. Those high frequencies result in aliasing distortion during subsampling. To improve the result, we first subject the image to a lowpass filter (blurring it) and then subsample, as shown:

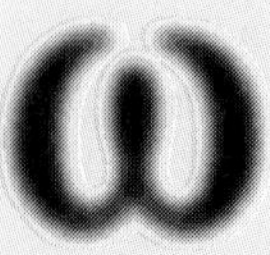

The result looks better to the discerning eye:

ω

11.2 Reconstruction

Consider a system that constructs a continuous-time signal x from a discrete-time signal y:

$$DiscToCont_T: DiscSignals \to ContSignals.$$

This is illustrated in figure 11.5. Systems that carry out such "discrete-to-continuous" conversion can be realized in any number of ways. Some common examples are illustrated in figure 11.6 and defined as follows:

- **Zero-order hold**: This means simply that the value of the each sample $y(n)$ is held constant for duration T, so that $x(t) = y(n)$ for the time interval from $t = nT$ to $t = (n+1)T$, as illustrated in figure 11.6(b). Let this system be denoted

  $$ZeroOrderHold_T: DiscSignals \to ContSignals.$$

- **Linear interpolation**: Intuitively, this means simply that we connect the dots with straight lines. Specifically, in the time interval from $t = nT$ to $t = (n+1)T$, $x(t)$ has values that vary along a straight line from $y(n)$ to $y(n+1)$, as illustrated in figure 11.6(c). Linear interpolation is sometimes called **first-order hold**. Let this system be denoted

  $$LinearInterpolator_T: DiscSignals \to ContSignals.$$

- **Ideal interpolation**: It is not yet clear what this should mean, but, intuitively, it should result in a smooth curve that passes through the samples, as illustrated in figure 11.6(d). We provide a precise meaning later. Let this system be denoted

  $$IdealInterpolator_T: DiscSignals \to ContSignals.$$

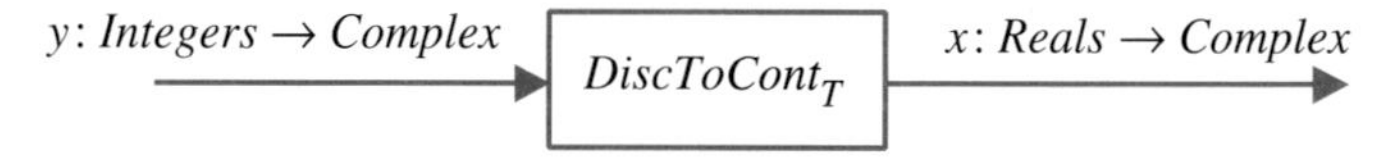

FIGURE 11.5: Discrete to continuous converter.

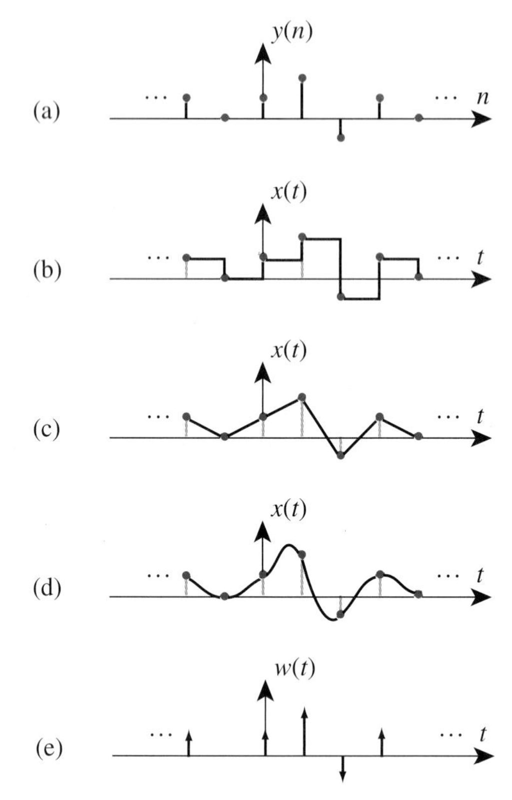

FIGURE 11.6: A discrete-time signal (a), a continuous-time reconstruction with zero-order hold (b), a reconstruction with linear interpolation (c), a reconstruction with ideal interpolation (d), and a reconstruction with weighted Dirac delta functions (e).

11.2.1 *A model for reconstruction*

A convenient mathematical model for reconstruction divides the reconstruction process into a cascade of two systems, as shown in figure 11.7. Thus,

$$x = S(ImpulseGen_T(y)),$$

where S is an LTI system to be determined. The first of these two subsystems,

$$ImpulseGen_T\colon DiscSignals \to ContSignals,$$

constructs a continuous-time signal, where for all $t \in Reals$,

$$w(t) = \sum_{k=-\infty}^{\infty} y(k)\delta(t - kT).$$

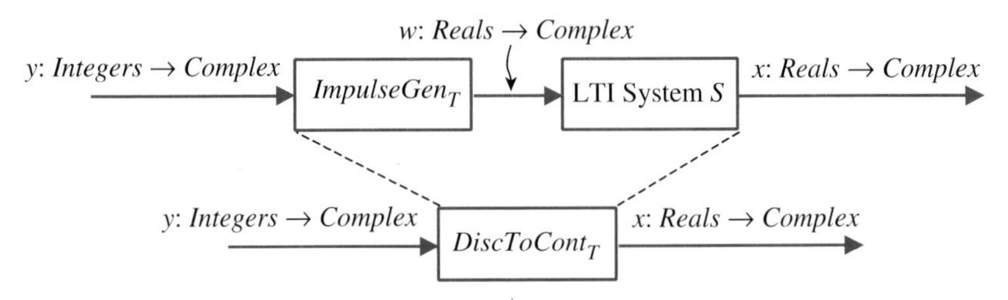

FIGURE 11.7: A model for reconstruction divides it into two stages.

This is a continuous-time signal that, at each sampling instant kT, produces a Dirac delta function with weight equal to sample value, $y(k)$. This signal is illustrated in figure 11.6(e). It is a mathematical abstraction, because everyday engineering systems do not exhibit the singularity-like behavior of the Dirac delta function. Nonetheless, it is a useful mathematical abstraction.

The second system in figure 11.7, S, is a continuous-time LTI filter with an impulse response that determines the interpolation method. The impulse responses that yield the interpolation methods in figure 11.6(b–e) are shown in figure 11.8(b–e). If

$$\forall\, t \in Reals, \quad h(t) = \begin{cases} 1 & 0 \le t < T \\ 0 & \text{otherwise,} \end{cases}$$

then the interpolation method is zero-order hold. If

$$\forall\, t \in Reals, \quad h(t) = \begin{cases} 1 + t/T & -T < t < 0 \\ 1 - t/T & 0 \le t < T \\ 0 & \text{otherwise,} \end{cases}$$

then the interpolation method is linear. If the impulse response is

$$\forall\, t \in Reals, \quad h(t) = \frac{\sin(\pi t/T)}{\pi t/T},$$

then the interpolation method is ideal. The impulse response just shown is called a **sinc function**, and its Fourier transform, from table 10.4, is represented by

$$\forall\, \omega \in Reals, \quad X(\omega) = \begin{cases} T & \text{if } |\omega| \le \pi/T \\ 0 & \text{otherwise.} \end{cases}$$

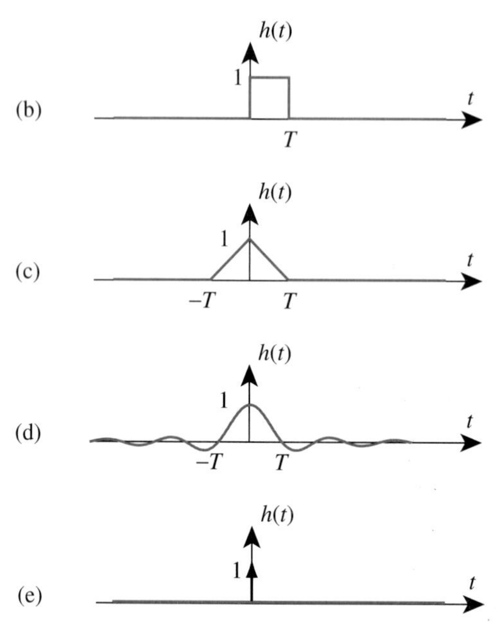

FIGURE 11.8: The impulse responses for the LTI system S in figure 11.7 that yield the interpolation methods in figure 11.6(b–e).

Notice that the Fourier transform is 0 at all frequencies above π/T radians/second, or $f_s/2$ Hz, the Nyquist frequency. It is this characteristic that makes it ideal. It precisely performs the strategy illustrated in figure 11.4, in which among all indistinguishable frequencies we select the ones between $-f_s/2$ and $f_s/2$.

If we let $Sinc_T$ denote the LTI system S when the impulse response is a sinc function, then

$$IdealInterpolator_T = Sinc_T \circ ImpulseGen_T.$$

In practice, ideal interpolation is difficult to accomplish. From the expression for the sinc function we can understand why. First, this impulse response is not causal. Second, it is infinite in extent. More important is that its magnitude decreases rather slowly as t increases or decreases (proportional to $1/t$ only). Thus, truncating it at a finite length leads to substantial errors.

If the impulse response of S is

$$h(t) = \delta(t),$$

where δ is the Dirac delta function, then the system S is a pass-through system, and the reconstruction consists of weighted delta functions.

PROBING FURTHER

Sampling

We can construct a mathematical model for sampling by using Dirac delta functions. Define a pulse stream by

$$\forall t \in Reals, \quad p(t) = \sum_{k=-\infty}^{\infty} \delta(t - kT).$$

Consider a continuous-time signal x that we wish to sample with sampling period T; that is, we define $y(n) = x(nT)$. Construct first an intermediate continuous-time signal $w(t) = x(t)p(t)$. We can show that the CTFT of w is equal to the DTFT of y. This gives us a way to relate the CTFT of x to the DTFT of its samples y. Recall that multiplication in the time domain results in convolution in the frequency domain (see table 10.9), and so

$$W(\omega) = \frac{1}{2\pi}(X * P)(\omega) = \frac{1}{2\pi}\int_{-\infty}^{\infty} X(\Omega)P(\omega - \Omega)d\Omega.$$

It can be shown (see the next box) that the CTFT of $p(t)$ is

$$P(\omega) = \frac{2\pi}{T}\sum_{k=-\infty}^{\infty} \delta(\omega - k\frac{2\pi}{T});$$

therefore,

$$\begin{aligned} W(\omega) &= \frac{1}{2\pi}\int_{-\infty}^{\infty} X(\Omega)\frac{2\pi}{T}\sum_{k=-\infty}^{\infty} \delta(\omega - \Omega - k\frac{2\pi}{T})d\Omega \\ &= \frac{1}{T}\sum_{k=-\infty}^{\infty}\int_{-\infty}^{\infty} X(\Omega)\delta(\omega - \Omega - k\frac{2\pi}{T})d\Omega \\ &= \frac{1}{T}\sum_{k=-\infty}^{\infty} X(\omega - k\frac{2\pi}{T}), \end{aligned}$$

where the last equality follows from the sifting property (9.11). The next step is to show that

$$Y(\omega) = W(\omega/T).$$

We leave this as an exercise. From this, the basic Nyquist-Shannon result follows:

$$Y(\omega) = \frac{1}{T}\sum_{k=-\infty}^{\infty} X\left(\frac{\omega - 2\pi k}{T}\right).$$

This relates the continuous-time Fourier transfrom (CTFT) X of the signal being sampled, x, to the DTFT Y of the discrete-time result, y.

PROBING FURTHER

Impulse trains

Consider a signal p consisting of periodically repeated Dirac delta functions with period T:

$$\forall\, t \in Reals, \quad p(t) = \sum_{k=-\infty}^{\infty} \delta(t - kT).$$

This signal has the Fourier series expansion

$$\forall\, t \in Reals, \quad p(t) = \sum_{m=-\infty}^{\infty} \frac{1}{T} e^{i\omega_0 mt},$$

where the fundamental frequency is $\omega_0 = 2\pi/T$. This can be verified by applying the formula from table 10.5. That formula, however, gives an integration range of 0 to the period, which in this case is T. This integral covers one period of the periodic signal but starts and ends on a delta function in p. To avoid the resultant mathematical subtleties, we can integrate from $-T/2$ to $T/2$, getting Fourier series coefficients

$$\forall\, m \in Integers, \quad P_m = \frac{1}{T} \int_{-T/2}^{T/2} \left[\sum_{k=-\infty}^{\infty} \delta(t - kT) \right] e^{i\omega_0 mt} dt.$$

The integral is now over a range that includes only one of the delta functions. The kernel of the integral is 0 except when $t = 0$; therefore, according to the sifting rule, the integral equals 1. Thus, all Fourier series coefficients are $P_m = 1/T$. Using the relationship between the Fourier series and the Fourier transform of a periodic signal (from section 10.6.3), we can write the CTFT of p as

$$\forall\, \omega \in Reals, \quad P(\omega) = \frac{2\pi}{T} \sum_{k=-\infty}^{\infty} \delta\left(\omega - \frac{2\pi}{T} k\right).$$

11.3 The Nyquist-Shannon sampling theorem

We can now give a precise statement of the **Nyquist-Shannon sampling theorem**:

> If x is a continuous-time signal with Fourier transform X and if $X(\omega)$ is zero outside the range $-\pi/T < \omega < \pi/T$ radians/second, then
>
> $$x = IdealInterpolator_T(Sampler_T(x)).$$

We can state this theorem slightly differently. Suppose x is a continuous-time signal with no frequency larger than some f_0 Hz. Then x can be recovered from its samples if $f_0 < f_s/2$, the Nyquist frequency.

A formal proof of this theorem involves some technical difficulties (it was first given by Claude Shannon [1916–2001] of Bell Labs in the late 1940s). But we can get the idea from the following three-step argument (see figure 11.9).

Step 1. Let x be a continuous-time signal with Fourier transform X. At this point we do not require that $X(\omega)$ be zero outside the range $-\pi/T < \omega < \pi/T$. We sample x with sampling interval T to get the discrete-time signal

$$y = Sampler_T(x).$$

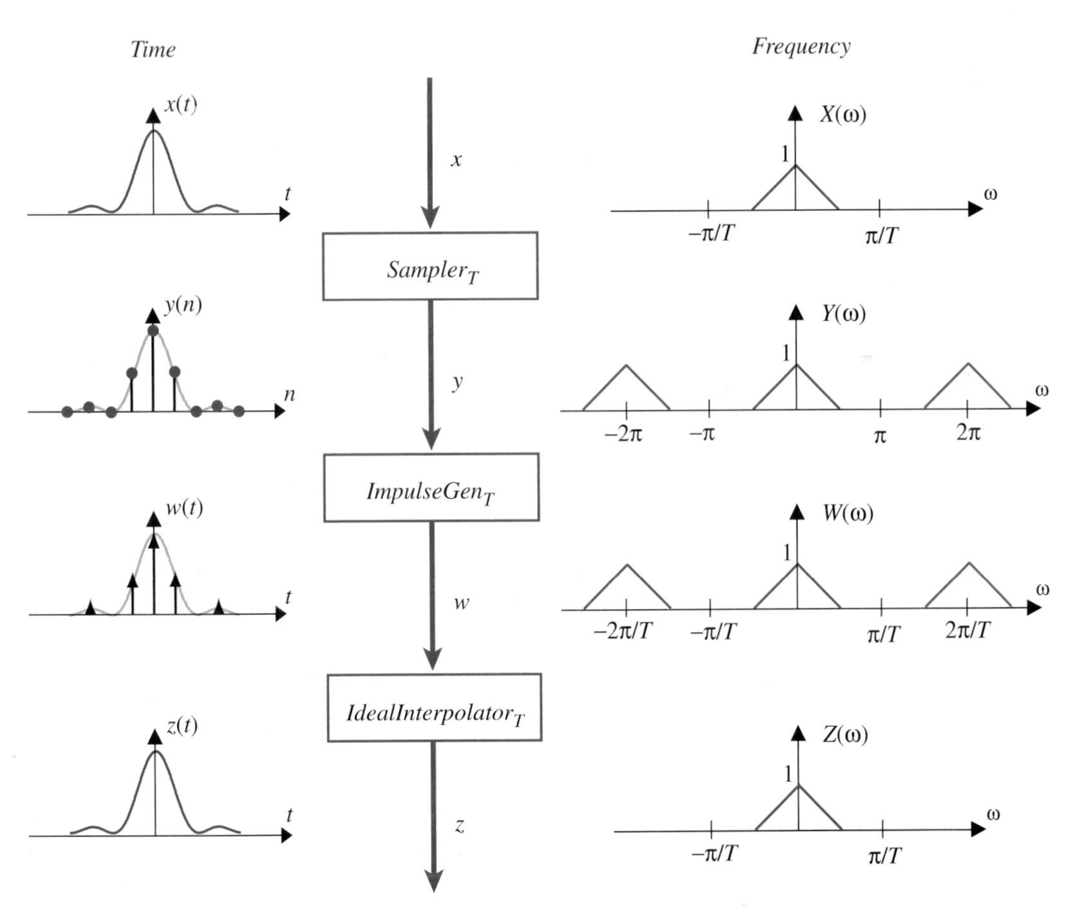

FIGURE 11.9: Steps in the justification of the Nyquist-Shannon sampling theorem.

It can be shown (see Probing Further: Sampling box on page 437) that the DTFT Y of y is related to the CTFT X of x by

$$Y(\omega) = \frac{1}{T} \sum_{k=-\infty}^{\infty} X\left(\frac{\omega}{T} - \frac{2\pi k}{T}\right).$$

This important relation says that the DTFT Y of y is the sum of the CTFT X with copies of it shifted by multiples of $2\pi/T$. Also, the frequency axis is normalized by dividing ω by T. There are two cases to consider, depending on whether the shifted copies overlap.

First, if $X(\omega) = 0$ outside the range $-\pi/T < \omega < \pi/T$, then the copies do not overlap, and in the range $-\pi < \omega < \pi$,

$$Y(\omega) = \frac{1}{T} X\left(\frac{\omega}{T}\right). \tag{11.5}$$

In this range of frequencies, Y has the same shape as X, scaled by $1/T$. This relationship between X and Y is illustrated in figure 11.10, where X is drawn with a triangular shape.

In the second case, illustrated in figure 11.11, X does have nonzero-frequency components higher than π/T. Notice that in the sampled signal, the frequencies in the vicinity of π are distorted by the overlapping of frequency components above and below π/T in the original signal. This distortion is called **aliasing distortion**.

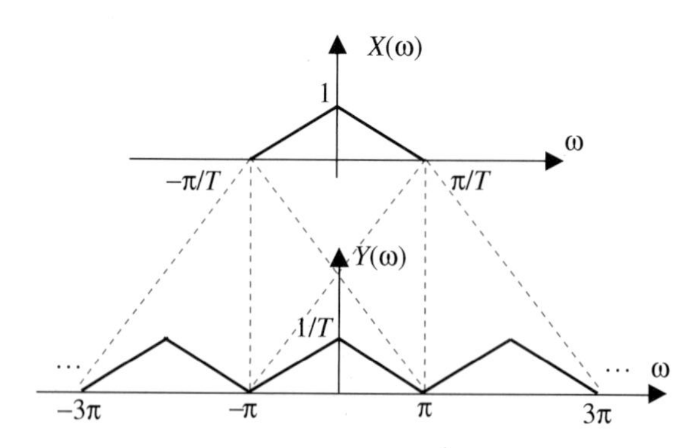

FIGURE 11.10: Relationship between the CTFT of a continuous-time signal and the DTFT of its discrete-time samples. The DTFT is the sum of the CTFT and its copies shifted by multiples of $2\pi/T$, the sampling frequency in radians per second. The frequency axis is also normalized by $1/T$.

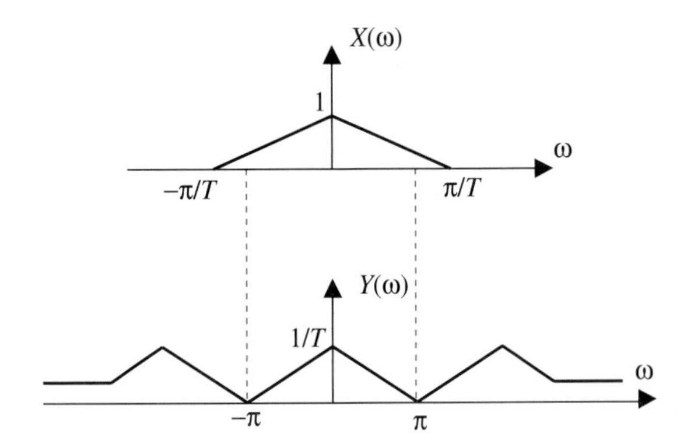

FIGURE 11.11: Relationship between the CTFT of a continuous-time signal and the DTFT of its discrete-time samples when the continuous-time signal has a broad enough bandwidth to introduce aliasing distortion.

We continue with the remaining steps, following the signals in figure 11.9.

Step 2. Let w be the signal produced by the impulse generator

$$\forall\, t \in Reals, \quad w(t) = \sum_{n=-\infty}^{\infty} y(n)\delta(t - nT).$$

The Fourier transform of w is $W(\omega) = Y(\omega T)$ (see Probing Further: Sampling box on page 437).

Step 3. Let z be the output of the *IdealInterpolator*$_T$. Its Fourier transform is simply

$$\begin{aligned} Z(\omega) &= W(\omega)S(\omega) \\ &= Y(\omega T)S(\omega), \end{aligned}$$

where $S(\omega)$ is the frequency response of the reconstruction filter *IdealInterpolator*$_T$. As seen in exercise 21 of chapter 10,

$$S(\omega) = \begin{cases} T & \text{if } |\omega| \le \pi/T \\ 0 & \text{if } |\omega| > \pi/T. \end{cases} \tag{11.6}$$

Substituting for S and Y, we get

$$Z(\omega) = \begin{cases} TY(\omega T) & \text{if } -\pi/T < \omega < \pi/T \\ 0 & \text{otherwise.} \end{cases}$$

$$= \begin{cases} \sum_{k=-\infty}^{\infty} X(\omega - 2\pi k/T) & \text{if } -\pi/T < \omega < \pi/T \\ 0 & \text{otherwise.} \end{cases}$$

If $X(\omega)$ is 0 for $|\omega|$ larger than the Nyquist frequency π/T, then we conclude that

$$\forall\, \omega \in Reals, \quad Z(\omega) = X(\omega);$$

that is, w is identical to x. This proves the Nyquist-Shannon result.

However, if $X(\omega)$ does have nonzero values for some $|\omega|$ larger than the Nyquist frequency, then z is different from x, as illustrated in figure 11.11.

11.4 Summary

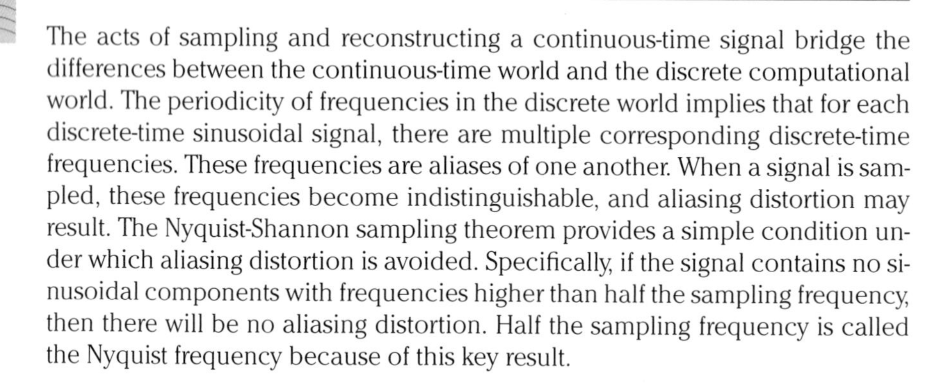

The acts of sampling and reconstructing a continuous-time signal bridge the differences between the continuous-time world and the discrete computational world. The periodicity of frequencies in the discrete world implies that for each discrete-time sinusoidal signal, there are multiple corresponding discrete-time frequencies. These frequencies are aliases of one another. When a signal is sampled, these frequencies become indistinguishable, and aliasing distortion may result. The Nyquist-Shannon sampling theorem provides a simple condition under which aliasing distortion is avoided. Specifically, if the signal contains no sinusoidal components with frequencies higher than half the sampling frequency, then there will be no aliasing distortion. Half the sampling frequency is called the Nyquist frequency because of this key result.

EXERCISES

KEY: **E** = mechanical **T** = requires plan of attack **C** = more than 1 answer

E 1. Consider the continuous-time signal

$$x(t) = \cos(10\pi t) + \cos(20\pi t) + \cos(30\pi t).$$

(a) Find the fundamental frequency. Give the units.

(b) Find the Fourier series coefficients $A_0, A_1, \ldots$ and $\phi_1, \phi_2, \ldots$ in the form of (7.4).

(c) Let y be the result of sampling this signal with sampling frequency 10 Hz. Find the fundamental frequency for y, and give the units.

(d) For the same y, find the discrete-time Fourier series coefficients, $A_0, A_1, \ldots$ and $\phi_1, \phi_2, \ldots$.

(e) Find

$$w = \mathit{IdealInterpolator}_T(\mathit{Sampler}_T(x))$$

for $T = 0.1$ seconds.

(f) Is there any aliasing distortion caused by sampling at 10 Hz? If there is, describe the aliasing distortion in words.

(g) Give the smallest sampling frequency that avoids aliasing distortion.

E 2. Verify that $\mathit{Sampler}_T$ defined by (11.1) and (11.2) is linear but not time invariant.

E 3. A real-valued sinusoidal signal with a negative frequency is always exactly equal to another sinusoid with positive frequency. Consider a real-valued sinusoid with a negative frequency -440 Hz:

$$y(n) = \cos(-2\pi 440 nT + \phi).$$

Find a positive frequency f and phase θ such that

$$y(n) = \cos(2\pi f nT + \theta).$$

T 4. Consider a continuous-time signal x where for all $t \in \mathit{Reals}$,

$$x(t) = \sum_{k=-\infty}^{\infty} r(t-k)$$

where

$$r(t) = \begin{cases} 1 & \text{if } 0 \le t < 0.5 \\ 0 & \text{otherwise.} \end{cases}$$

(a) Is $x(t)$ periodic? If so, what is the period?

(b) Suppose that $T = 1$. Give a simple expression for $y = \mathit{Sampler}_T(x)$.

(c) Suppose that $T = 0.5$. Give a simple expression for $y = \mathit{Sampler}_T(x)$ and $z = \mathit{IdealInterpolator}_T(\mathit{Sampler}_T(x))$.

(d) Find an upper bound for T (in seconds) such that $x = \mathit{IdealInterpolator}_T(\mathit{Sampler}_T(x))$, or argue that no value of T makes this assertion true.

T 5. Consider a continuous-time signal x with the following finite Fourier series expansion:

$$\forall\, t \in Reals, \quad x(t) = \sum_{k=0}^{4} \cos(k\omega_0 t),$$

where $\omega_0 = \pi/4$ radians/second.

(a) Give an upper bound on T (in seconds) such that $x = IdealInterpolator_T(Sampler_T(x))$.

(b) Suppose that $T = 4$ seconds. Give a simple expression for $y = Sampler_T(x)$.

(c) For the same $T = 4$ seconds, give a simple expression for

$$w = IdealInterpolator_T(Sampler_T(x)).$$

T 6. Consider a continuous-time audio signal x with CTFT shown in figure 11.12. Note that it contains no frequencies beyond 10 kHz. Suppose it is sampled at 40 kHz to yield a signal that we call x_{40}. Let X_{40} be the DTFT of x_{40}.

(a) Sketch $|X_{40}(\omega)|$, and carefully mark the magnitudes and frequencies.

(b) Suppose x is sampled at 20,000 samples/second. Let x_{20} be the resulting sampled signal and X_{20} its DTFT. Sketch and compare x_{20} and x_{40}.

(c) Now suppose x is sampled at 15,000 samples/second. Let x_{15} be the resulting sampled signal and X_{15} its DTFT. Sketch and compare X_{20} and X_{15}. Make sure that your sketch shows aliasing distortion.

C 7. Consider two continuous-time sinusoidal signals given by

$$x_1(t) = \cos(\omega_1 t)$$

and

$$x_2(t) = \cos(\omega_2 t),$$

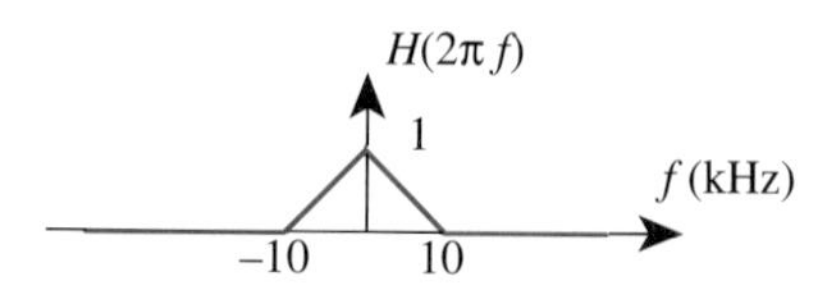

FIGURE 11.12: CTFT of an audio signal.

with frequencies of ω_1 and ω_2 radians/second such that

$$0 \le \omega_1 \le \pi/T \quad \text{and} \quad 0 \le \omega_2 \le \pi/T.$$

Show that if $\omega_1 \neq \omega_2$, then

$$Sampler_T(x_1) \neq Sampler_T(x_2);$$

that is, the two distinct sinusoids cannot be aliases of one another if both have frequencies below the Nyquist frequency. Hint: Try evaluating the sampled signals at $n = 1$.

CHAPTER 12

Stability

The four Fourier transforms prove to be useful tools for analyzing signals and systems. When a system is LTI, it is characterized by its frequency response H, and its input x and output y are related simply by

$$\forall\, \omega \in \mathit{Reals}, \quad Y(\omega) = H(\omega)X(\omega),$$

where Y is the Fourier transform of y, and X is the Fourier transform of x.

However, we ignored a lurking problem. Any of the three Fourier transforms, X, Y, or H, may not exist. Suppose, for example, that x is a discrete-time signal. Then its Fourier transform (the DTFT) is given by

$$\forall\, \omega \in \mathit{Reals}, \quad X(\omega) = \sum_{n=-\infty}^{\infty} x(n)e^{-i\omega n}. \tag{12.1}$$

This is an infinite sum, properly viewed as the limit

$$\forall\, \omega \in \mathit{Reals}, \quad X(\omega) = \lim_{N\to\infty} \sum_{n=-N}^{N} x(n)e^{-i\omega n}. \tag{12.2}$$

As with all such limits, there is a risk that it does not exist. If the limit does not exist for any $\omega \in \mathit{Reals}$, then the Fourier transform becomes mathematically treacherous at best (involving, for example, Dirac delta functions), and mathematical nonsense at worst.

Example 12.1: Consider the sequence

$$x(n) = \begin{cases} 0, & n \leq 0 \\ a^{n-1}, & n > 0 \end{cases},$$

where $a > 1$ is a constant. Plugging into (12.1), the Fourier transform should be

$$\forall\, \omega \in Reals, \quad X(\omega) = \sum_{n=1}^{\infty} a^{n-1} e^{-i\omega n}.$$

At $\omega = 0$, it is easy to see that this sum is infinite (every term in the sum is greater than or equal to one). At other values of ω, there are also problems. For example, at $\omega = \pi$, the terms of the sum alternate in sign and increase in magnitude as n gets larger. The limit (12.2) clearly will not exist. ❒

A similar problem arises with continuous-time signals. If x is a continuous-time signal, then its Fourier transform (the CTFT) is given by

$$\forall\, \omega \in Reals, \quad X(\omega) = \int_{-\infty}^{\infty} x(t) e^{-i\omega t} dt. \tag{12.3}$$

Again, there is a risk that this integral does not exist.

This chapter studies signals for which the Fourier transform does not exist. Such signals prove to be both common and useful. The signal in example 12.1 gives the bank balance of example 5.12 when an initial deposit of one dollar is made, and no further deposits or withdrawals are made (thus, it is the impulse response of the bank account). This signal grows without bound, and any signal that grows without bound will cause difficulties when using the Fourier transform.

The bank account is said to be an **unstable system**, because its output can grow without bound even when the input is always bounded. Such unstable systems are common, so it is unfortunate that the frequency domain methods we have studied so far do not appear to apply.

Example 12.2: A helicopter is intrinsically an unstable system, requiring an electronic or mechanical feedback control system to stabilize it. It has two rotors, one above, which provides lift, and one on the tail. Without the rotor on the tail, the body of the helicopter would start to spin. The rotor on the tail counteracts that spin. However, the force produced by the tail rotor must perfectly counter the friction with the main rotor, or the body will spin.

A highly simplified version of the helicopter is shown in figure 12.1. The body of the helicopter is modeled as a horizontal arm with moment of intertia M. The tail rotor goes on the end of this arm. The body of the helicopter rotates freely around the main rotor shaft. Friction with the main rotor will tend to cause it to rotate by applying a torque as suggested by the curved arrow. The tail rotor will have to counter that torque to keep the body of the helicopter from spinning.

Let the input x to the system be the net torque on the tail of the helicopter, as a function of time. That is, at time t, $x(t)$ is the difference between the frictional torque exerted by the main rotor shaft and the counteracting torque exerted by the tail rotor. Let the output y be the velocity of rotation of the body. From basic physics, torque equals moment of inertia times rotational acceleration. The rotational acceleration is $\dot{y}$, the derivative of y, so

$$\dot{y}(t) = x(t)/M.$$

Integrating both sides, assuming that the initial velocity of rotation is zero, we get the output as a function of the input,

$$\forall\, t \in Reals, \quad y(t) = \frac{1}{M} \int_{0}^{t} x(\tau) d\tau.$$

It is now easy to see that this system is unstable. Let the input be $x = u$, where u is the **unit step**, given by

$$\forall\, t \in Reals, \quad u(t) = \begin{cases} 0, & t < 0 \\ 1, & t \geq 0 \end{cases}. \tag{12.4}$$

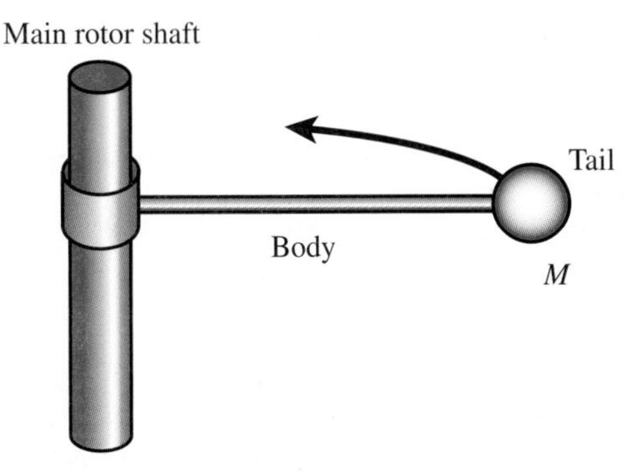

FIGURE 12.1: A highly simplified helicopter.

This input is clearly bounded. It never exceeds one in magnitude. However, the output grows without bound.

In practice, a helicopter uses a feedback system to determine how much torque to apply at the tail rotor to keep the body of the helicopter straight. We will see how to do this in chapter 14. ❐

In this chapter we develop the basics of modeling unstable systems in the frequency domain. We define two new transforms, called the **Z transform** and **Laplace transform**. The Z transform is a generalization of the DTFT and applies to discrete-time signals. The Laplace transform is a generalization of the CTFT and applies to continuous-time signals. These generalizations support frequency-domain analysis of signals that do not have a Fourier transform, and thus allow analysis of unstable systems.

In particular, let $\hat{X}$ denote the Laplace or Z transform of x, depending on whether it is a continuous or discrete-time signal. Then the Laplace or Z transform of the output of an LTI system is given by $\hat{Y} = \hat{H}\hat{X}$, where $\hat{H}$ is the Laplace or Z transform of the impulse response. This relation applies even when the system is unstable. Thus, these transforms take the place of the Fourier transform when the Fourier transform cannot be used. $\hat{H}$ is called the **transfer function** of the LTI system, and it is a generalization of the frequency response.

12.1 Boundedness and stability

Not all discrete-time signals have a DTFT. In this section, we identify a simple condition for the existence of the DTFT, which is that the signal be **absolutely summable**. We then define a **stable system** and show that an LTI system is stable if and only if its impulse response is absolutely summable. Continuous-time signals are slightly more complicated, requiring slightly more than that they be **absolutely integrable**. The conditions for the existence of the CTFT are called the **Dirichlet conditions**, and once again, if the impulse response of an LTI system satisfies these conditions, then it is stable.

12.1.1 *Absolutely summable and absolutely integrable*

A discrete-time signal x is said to be **absolutely summable** if

$$\sum_{n=-\infty}^{\infty} |x(n)|$$

exists and is finite. The "absolutely" in "absolutely summable" refers to the absolute value (or magnitude) in the summation. The sum is said to **converge**

absolutely. The following simple fact gives a condition for the existence of the DTFT:

> If a discrete-time signal x is absolutely summable, then its DTFT exists and is finite for all ω.

To see that this is true, note that the DTFT exists and is finite if

$$\forall\, \omega \in Reals, \quad |X(\omega)| = \left| \sum_{n=-\infty}^{\infty} x(n) e^{-i\omega n} \right|$$

exists and is finite. But

$$\left| \sum_{n=-\infty}^{\infty} x(n) e^{-i\omega n} \right| \leq \sum_{n=-\infty}^{\infty} |x(n) e^{-i\omega n}| \tag{12.5}$$

$$= \sum_{n=-\infty}^{\infty} |x(n)| \cdot |e^{-i\omega n}| \tag{12.6}$$

$$= \sum_{n=-\infty}^{\infty} |x(n)|. \tag{12.7}$$

This follows from the following facts about complex (or real) numbers:

$$|a + b| \leq |a| + |b|,$$

which is known as the **triangle inequality** (and generalizes to infinite sums),

$$|ab| = |a| \cdot |b|,$$

and

$$\forall\, \theta \in Reals, \quad |e^{i\theta}| = 1.$$

We can conclude from (12.5) through (12.7) that

$$\forall\, \omega \in Reals, \quad |X(\omega)| \leq \sum_{n=-\infty}^{\infty} |x(n)|.$$

This means that if x is absolutely summable, then the DTFT exists and is finite, because if a sum converges absolutely, then it also converges without the absolute value.

A continuous-time signal x is said to be **absolutely integrable** if

$$\int_{-\infty}^{\infty} |x(t)|dt$$

exists and is finite. A similar argument to that above (with summations replaced by integrals) suggests that if a continuous-time signal x is absolutely integrable, then its CTFT should exist and be finite for all ω. However, caution is in order. Integrals are more complicated than summations, and we need some additional conditions to ensure that the integral is well defined. We can use essentially the same conditions given on page 263 for the convergence of the continuous-time Fourier series. These are called the **Dirichlet conditions**, and require three things:

1. x is absolutely integrable;
2. in any finite interval, x is of **bounded variation**, meaning that there are no more than a finite number of maxima or minima; and
3. in any finite interval, x is continuous at all but a finite number of points.

Most any signal of practical engineering importance satisfies the last two conditions, so the important condition is that it be absolutely integrable. We will henceforth assume without comment that all continuous-time signals satisfy the last two conditions, so the only important condition becomes the first one. Under this assumption, the following simple fact gives a condition for the existence of the CTFT:

An absolutely integrable continuous-time signal x has a CTFT X, and its CTFT $X(\omega)$ is finite for all $\omega \in Reals$.

12.1.2 *Stability*

A system is said to be **bounded-input bounded-output stable** (**BIBO stable** or just **stable**) if the output signal is bounded for all input signals that are bounded. Consider a discrete-time system with input x and output y. An input is bounded if there is a real number $M < \infty$ such that $|x(k)| \le M$ for all $k \in Integers$. An output is bounded if there is a real number $N < \infty$ such that $|y(n)| \le N$ for all $n \in Integers$. The system is stable if for any input bounded by M, there is some bound N on the output.

Consider a discrete-time LTI system with impulse response h. The output y corresponding to the input x is given by the convolution sum,

$$\forall\, n \in Integers, \quad y(n) = (h * x)(n) = \sum_{m=-\infty}^{\infty} h(m)x(n-m). \qquad (12.8)$$

Suppose that the input x is bounded with bound M. Then, applying the triangle inequality, we see that

$$|y(n)| \leq \sum_{m=-\infty}^{\infty} |h(m)||x(n-m)| \leq M \sum_{m=-\infty}^{\infty} |h(m)|.$$

Thus, if the impulse response is absolutely summable, then the output is bounded with bound

$$N = M \sum_{m=-\infty}^{\infty} |h(m)|.$$

Thus, if the impulse response of an LTI system is absolutely summable, then the system is stable. The converse is also true, but more difficult to show. That is, if the system is stable, then the impulse response is absolutely summable (see following Probing Further box). The same argument applies for continuous-time signals.

> A discrete-time LTI system is stable if and only if its impulse response is absolutely summable. A continuous-time LTI system is stable if and only if its impulse response is absolutely integrable.

PROBING FURTHER

Stable systems and their impulse response

Consider a discrete-time LTI system with real-valued impulse response h. In this box, we show that if the system is stable, then its impulse response is absolutely summable. To show this, we show the contrapositive.* That is, we show that if the impulse response is not absolutely summable, then the system is not stable. To do this, suppose that the impulse response is not absolutely summable. That is, the sum

$$\sum_{n=-\infty}^{\infty} |h(n)|$$

continued on next page

PROBING FURTHER

is not bounded. To show that the system is not stable, we need only to find one bounded input for which the output either does not exist or is not bounded. Such an input is given by

$$\forall n \in Integers, \quad x(n) = \begin{cases} h(-n)/|h(-n)|, & h(n) \neq 0 \\ 0, & h(n) = 0. \end{cases}$$

This input is clearly bounded, with bound $M = 1$. Plugging this input into the convolution sum (12.8) and evaluating at $n = 0$ we get

$$\begin{aligned} y(0) &= \sum_{m=-\infty}^{\infty} h(m)x(-m) \\ &= \sum_{m=-\infty}^{\infty} (h(m))^2/|h(m)| \\ &= \sum_{m=-\infty}^{\infty} |h(m)|, \end{aligned}$$

where the last step follows from the fact that for real-valued $h(m)$, $(h(m))^2 = |h(m)|^2$. But since the impulse response is not absolutely summable, $y(0)$ does not exist or is not finite, so the system is not stable. A nearly identical argument works for continuous-time systems.*

* The contrapositive of a statement "if p then q" is "if not q then not p." The contrapositive is true if and only if the original statement is true.

The following example makes use of the **geometric series identity**, valid for any real or complex a where $|a| < 1$,

$$\sum_{m=0}^{\infty} a^m = \frac{1}{1-a}. \tag{12.9}$$

To verify this identity, just multiply both sides by $1 - a$ to get

$$\sum_{m=0}^{\infty} a^m - a\sum_{m=0}^{\infty} a^m = 1.$$

This can be written

$$a^0 + \sum_{m=1}^{\infty} a^m - \sum_{m=1}^{\infty} a^m = 1.$$

Now note that $a^0 = 1$ and that the two sums are identical. Since $|a| < 1$, the sums converge, and hence they cancel, so the identity is true.

Example 12.3: As in example 12.1, the impulse response of the bank account of example 5.12 is

$$h(n) = \begin{cases} 0, & n \le 0 \\ a^{n-1}, & n > 0 \end{cases},$$

where $a > 1$ is a constant that reflects the interest rate. This impulse response is not absolutely summable, so this system is not stable. A system with the same impulse response, but where $0 < a < 1$, however, would be stable, as is easily verified using (12.9). To use this identity, note that

$$\begin{aligned} \sum_{n=-\infty}^{\infty} |h(n)| &= \sum_{n=1}^{\infty} a^{n-1} \\ &= \sum_{m=0}^{\infty} a^{m} \\ &= \frac{1}{1-a}, \end{aligned}$$

where the second step results from a change of variables, letting $m = n - 1$. ❐

Example 12.4: Consider a continuous-time LTI system with impulse response

$$\forall\, t \in Reals, \quad h(t) = a^t u(t),$$

where $a > 0$ is a real number and u is the unit step, given by (12.4). To determine whether this system is stable, we need to determine whether the impulse response is absolutely integrable. That is, we need to determine whether the following integral exists and is finite,

$$\int_{-\infty}^{\infty} |a^t u(t)| dt.$$

Since $a > 0$ and u is the unit step, this simplifies to

$$\int_{0}^{\infty} a^t dt.$$

From calculus, we know that this integral is infinite if $a \geq 1$, so the system is unstable if $a \geq 1$. The integral is finite if $0 < a < 1$ and is equal to

$$\int_0^{\infty} a^t dt = -1/\ln(a).$$

Thus, the system is stable if $0 < a < 1$. ❐

When all pertinent signals are absolutely summable (or absolutely integrable), then we can use Fourier transform techniques with confidence. However, many useful signals do not fall in this category (the unit step and sinusoidal signals, for example). Moreover, many useful systems have impulse responses that are not absolutely summable (or absolutely integrable). Fortunately, we can generalize the DTFT and CTFT to get the Z transform and Laplace transform, which easily handle signals that are not absolutely summable.

12.2 The Z transform

Consider a discrete-time signal x that is not absolutely summable. The scaled signal x_r is given by

$$\forall\, n \in Integers, \quad x_r(n) = x(n)r^{-n}, \tag{12.10}$$

for some real number $r \geq 0$. Often, this signal is absolutely summable when r is chosen appropriately. This new signal, therefore, will have a DTFT, even if x does not.

Example 12.5: Continuing with example 12.3, the impulse response of the bank account is

$$h(n) = \begin{cases} 0, & n \leq 0 \\ a^{n-1}, & n > 0 \end{cases}$$

where $a > 1$. This system is not stable. However, the scaled signal

$$h_r(n) = h(n)r^{-n}$$

is absolutely summable if $r > a$. Its DTFT is

$$\forall\, r > a, \forall\, \omega \in Reals, \quad H_r(\omega) = \sum_{m=-\infty}^{\infty} h(m) r^{-m} e^{-i\omega m}$$

$$= \sum_{m=1}^{\infty} a^{m-1} (re^{i\omega})^{-m}$$

$$= \sum_{n=0}^{\infty} a^{n} (re^{i\omega})^{-n-1}$$

$$= (re^{i\omega})^{-1} \sum_{n=0}^{\infty} (a(re^{i\omega})^{-1})^{n}$$

$$= \frac{(re^{i\omega})^{-1}}{1 - a(re^{i\omega})^{-1}}.$$

The second step is by change of variables, $n = m - 1$, and the final step applies the geometric series identity (12.9). ❐

In general, the DTFT of the scaled signal x_r in (12.10) is

$$\forall\, \omega \in Reals, \quad X_r(\omega) = \sum_{m=-\infty}^{\infty} x(m) (re^{i\omega})^{-m}.$$

Notice that this is a function not just of ω, but also of r, and in fact, we are only sure it is valid for values of r that yield an absolutely summable signal h_r. If we define the complex number

$$z = re^{i\omega}$$

then we can write this DTFT as

$$\boxed{\forall\, z \in RoC(x), \quad \hat{X}(z) = \sum_{m=-\infty}^{\infty} x(m) z^{-m},} \tag{12.11}$$

where $\hat{X}$ is a function called the **Z transform** of x, and

$$\hat{X} : RoC(x) \to Complex$$

where $RoC(x) \subset Complex$ is defined by

$$\boxed{RoC(x) = \{z = re^{i\omega} \in Complex \mid x(n)r^{-n} \text{ is absolutely summable}\}.} \quad (12.12)$$

The term *RoC* is shorthand for **region of convergence**.

Example 12.6: Continuing example 12.5, we can recognize from the form of $H_r(\omega)$ that the Z transform of the impulse response h is

$$\forall\, z \in RoC(h), \quad \hat{H}(z) = \frac{z^{-1}}{1 - az^{-1}} = \frac{1}{z - a},$$

where the last step is the result of multiplying top and bottom by z. The *RoC* is

$$RoC(h) = \{z = re^{i\omega} \in Complex \mid r > a\}. \square$$

The Z tranform $\hat{H}$ of the impulse response h of an LTI system is called the **transfer function** of the system.

12.2.1 *Structure of the region of convergence*

When a signal has a Fourier transform, then knowing the Fourier transform is equivalent to knowing the signal. The signal can be obtained from its Fourier transform, and the Fourier transform can be obtained from the signal. The same is true of a Z transform, but there is a complication. The Z transform is a function $\hat{X}: RoC \to Complex$, and it is necessary to know the set *RoC* to know the function $\hat{X}$. The region of convergence is an essential part of the Z transform. We will see that very different signals can have very similar Z transforms that differ only in the region of convergence.

Given a discrete-time signal x, $RoC(x)$ is defined to be the set of all complex numbers $z = re^{i\omega}$ for which the following series converges:

$$\sum_{m=-\infty}^{\infty} |x(m)r^{-m}|.$$

Notice that if this series converges, then so does

$$\sum_{m=-\infty}^{\infty} |x(m)z^{-m}|$$

for any complex number z with magnitude r. This is because

$$|x(m)z^{-m}| = |x(m)(re^{i\omega})^{-m}| = |x(m)| \cdot |r^{-m}| \cdot |e^{-i\omega m}| = |x(m)| \cdot |r^{-m}|.$$

Thus, the set *RoC* could equally well be defined to be the set of all complex numbers z such that $x(n)z^{-n}$ is absolutely summable.

Whether this series converges depends only on r, the magnitude of the complex number $z = re^{i\omega}$, and not on ω, its angle. Thus, if any point $z = re^{i\omega}$ is in the set *RoC*, then all points z' with the same magnitude are also in *RoC*. This implies that the set *RoC*, a subset of *Complex*, will have circular symmetry.

The set *RoC* turns out to have even more structure. There are only three possible patterns, illustrated by the shaded areas in figure 12.2. Each figure illustrates the complex plane, and the shaded area is a region of convergence. Each possibility has circular symmetry, in that whether a point is in the *RoC* depends only on its magnitude.

Figure 12.2(a) shows the *RoC* of a causal signal. A discrete-time signal x is **causal** if $x(n) = 0$ for all $n < 0$. The *RoC* is the set of complex numbers $z = re^{i\omega}$ where the following series converges:

$$\sum_{m=-\infty}^{\infty} |x(m)r^{-m}|.$$

But if x is causal, then

$$\sum_{m=-\infty}^{\infty} |x(m)r^{-m}| = \sum_{m=0}^{\infty} |x(m)r^{-m}|.$$

If this series converges for some given r, then it must also converge for any $\tilde{r} > r$ (because for all $m \geq 0$, $\tilde{r}^{-m} < r^{-m}$). Thus, if $z \in RoC$, the *RoC* must include all points in the complex plane on the circle passing through z and every point outside that circle.

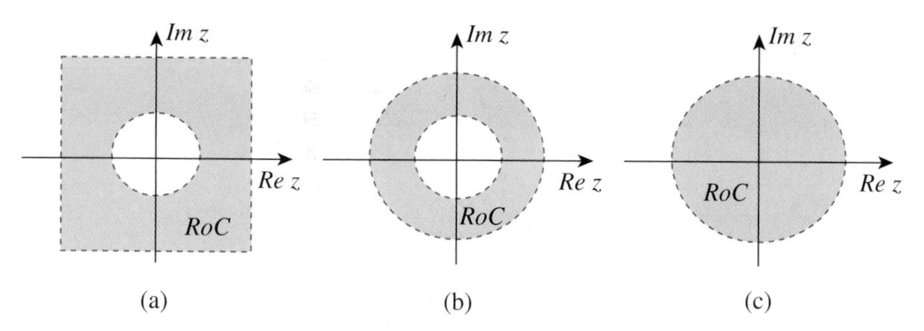

FIGURE 12.2: Three possible structures for the region of convergence of a Z transform: (a) causal or right-sided, (b) two-sided, and (c) anticausal.

Note further that not only must the *RoC* include every point outside the circle, but the series must also converge in the limit as $|z|$ goes to infinity. Thus, for example, $H(z) = z$ cannot be the Z transform of a causal signal because its *Roc* cannot possibly include infinity ($H(z)$ is not finite there).

Figure 12.2(c) shows the *RoC* of an anticausal signal. A discrete-time signal x is **anticausal** if $x(n) = 0$ for all $n > 0$. By a similar argument, if $z \in RoC$, then the *RoC* must include all points in the complex plane on the circle passing through z and every point inside that circle.

Figure 12.2(b) shows the *RoC* of a signal that is neither causal nor anticausal. Such a signal is called a **two-sided signal**. Such a signal can always be expressed as a sum of causal signal and an anticausal signal. The *RoC* is the intersection of the regions of convergence for these two components. To see this, just note that the *RoC* is the set of complex numbers $z = re^{i\omega}$ where the following series converges:

$$\sum_{m=-\infty}^{\infty} |x(m)r^{-m}| = \sum_{m=-\infty}^{-1} |x(m)r^{-m}| + \sum_{m=0}^{\infty} |x(m)r^{-m}|.$$

The first sum on the right corresponds to an anticausal signal, and the second sum on the right to a causal signal. For this series to converge, both sums must converge. Thus, for a two-sided signal, the *RoC* has a ring structure, or $RoC = \emptyset$.

Example 12.7: Consider the discrete-time unit step signal u, given by

$$u(n) = \begin{cases} 0, & n < 0 \\ 1, & n \geq 0. \end{cases} \tag{12.13}$$

The Z transform is, using the geometric series identity (12.9),

$$\hat{U}(z) = \sum_{m=-\infty}^{\infty} u(m)z^{-m} = \sum_{m=\infty}^{\infty} z^{-m} = \frac{1}{1-z^{-1}} \frac{z}{z-1},$$

with domain

$$RoC(u) = \{z \in Complex \mid \sum_{m=1}^{\infty} |z|^{-m} < \infty\} = \{z \mid |z| > 1\}.$$

This region of convergence has the structure of figure 12.2(a), where the dashed circle has radius one (that circle is called the **unit circle**). Indeed, this signal is causal, so this structure makes sense. ❒

Example 12.8: The signal v given by

$$v(n) = \begin{cases} -1, & n < 0 \\ 0, & n \geq 0 \end{cases},$$

has Z transform

$$\hat{V}(z) = \sum_{m=-\infty}^{\infty} v(m) z^{-m} = -\sum_{m=-\infty}^{1} z^{-m} = -z \sum_{k=0}^{\infty} z^{k} = \frac{z}{z-1},$$

with domain

$$RoC(v) = \{z \in Complex \mid \sum_{m=-\infty}^{1} |z|^{-m} < \infty\} = \{z \mid |z| < 1\}.$$

This region of convergence has the structure of figure 12.2(c), where the dashed circle is again the unit circle. Indeed, this signal is anticausal, so this structure makes sense. ❐

Notice that although the Z transform $\hat{U}$ of u and $\hat{V}$ of v have the *same* algebraic form, namely, $z/(z-1)$, they are *different* functions, because their domains are different. Thus, the Z transform of a signal comprises *both* the algebraic form of the Z transform as well as its *RoC*.

A **right-sided signal** x is where for some integer N,

$$x(n) = 0, \quad \forall n < N.$$

Of course, if $N \geq 0$, then this signal is also causal. However, if $N < 0$, then the signal is two sided. Suppose $N < 0$, then we can write the Z transform of x as

$$\sum_{m=-\infty}^{\infty} |x(m) r^{-m}| = \sum_{m=N}^{-1} |x(m) r^{-m}| + \sum_{m=0}^{\infty} |x(m) r^{-m}|.$$

The left summation on the right side is finite, and each term is finite for all $z \in Complex$, so therefore it converges for all $z \in Complex$. Thus, the region of convergence is determined entirely by the right summation, which is the Z transform of the causal part of x. Thus, the region of convergence of a right-sided signal has the same form as that of a causal sequence, as shown in figure 12.2(a). (However, if the signal is not causal, the Z transform does not converge at infinity.)

A **left-sided signal** x is where for some integer N,

$$x(n) = 0, \quad \forall n > N.$$

Of course, if $N \leq 0$, then this signal is also anticausal. However, if $N > 0$, the signal is two sided. Suppose $N > 0$, then we can write the Z transform of x as

$$\sum_{m=-\infty}^{\infty} |x(m)r^{-m}| = \sum_{m=-\infty}^{0} |x(m)r^{-m}| + \sum_{m=1}^{N} |x(m)r^{-m}|.$$

The right summation is finite, and therefore converges for all $z \in Complex$ except $z = 0$, where the individual terms of the sum are not finite. Thus, the region of convergence is that of the left summation, except for the point $z = 0$. Thus, the region of convergence of a left-sided signal has the same form as that of an anticausal sequence, as shown in figure 12.2(c), except that the origin $(z = 0)$ is excluded. This, of course, is simply the structure of 12.2(b) where the inner circle has zero radius.

Some signals have no meaningful Z transform.

Example 12.9: The signal x with $x(n) = 1$, for all n, does not have a Z transform. We can write $x = u - v$, where u and v are defined in the previous examples. Thus, the region of convergence of x must be the intersection of the regions of convergence of u and v. However, these two regions of convergence have an empty intersection, so $RoC(x) = \emptyset$.

Viewed another way, the set $RoC(x)$ is the set of complex numbers z such that

$$\sum_{m=-\infty}^{\infty} |x(m)z^{-m}| = \sum_{m=-\infty}^{\infty} |z^{-m}| < \infty.$$

But there is no such complex number z. ❐

Note that the signal x in example 12.9 is periodic with any integer period p (because $x(n+p) = x(n)$ for any $p \in Integers$). Thus, it has a Fourier series representation. In fact, as shown in section 10.6.3, a periodic signal also has a Fourier transform representation, as long as we are willing to allow Dirac delta functions in the Fourier transform. (Recall that this means that there are values of ω where $X(\omega)$ will not be finite.) With periodic signals, the Fourier series is by far the simplest frequency-domain tool to use. The Fourier transform can also be used if we allow Dirac delta functions. The Z transform, however, is more problematic, because the region of convergence is empty.

12.2.2 *Stability and the Z transform*

If a discrete-time signal x is absolutely summable, then it has a DTFT X that is finite for all $\omega \in Reals$. Moreover, the DTFT is equal to the Z transform evaluated on the unit circle,

$$\forall\, \omega \in Reals, \quad X(\omega) = \hat{X}(z)|_{z=e^{i\omega}} = \hat{X}(e^{i\omega}).$$

The complex number $z = e^{i\omega}$ has magnitude one, and therefore lies on the unit circle. Recall that an LTI system is stable if and only if its impulse response is absolutely summable. Thus,

> A discrete-time LTI system with impulse response h is stable if and only if the transfer function $\hat{H}$, which is the Z transform of h, has a region of convergence that includes the unit circle.

Example 12.10: Continuing example 12.6, the transfer function of the bank account system has the region of convergence

$$RoC(h) = \{z = re^{i\omega} \in Complex \mid r > a\},$$

where $a > 1$. Thus, the region of convergence includes only complex numbers with magnitude greater than one, and therefore does not include the unit circle. The bank account system is therefore not stable. ❐

12.2.3 *Rational Z tranforms and poles and zeros*

All of the Z transforms we have seen so far are **rational polynomials** in z. A rational polynomial is simply the ratio of two finite-order polynomials. For example, the bank account system has transfer function

$$\hat{H}(z) = \frac{1}{z-a}$$

(see example 12.6). The unit step of example 12.7 and its anticausal cousin of example 12.8 have Z transforms

$$\hat{U}(z) = \frac{z}{z-1}, \quad \hat{V}(z) = \frac{z}{z-1},$$

albeit with different regions of convergence.

In practice, most Z transforms of practical interest can be written as the ratio of two finite-order polynomials in z,

$$\hat{X}(z) = \frac{A(z)}{B(z)}.$$

The **order** of the polynomial A or B is the power of the highest power of z that appears in the polynomial. For the unit step, the numerator polynomial is $A(z) = z$, a first-order polynomial, and the denominator is $B(z) = z - 1$, also a first-order polynomial.

Recall from algebra that a polynomial of order N has N (possibly complex-valued) **roots**, which are values of z where the polynomial evaluates to zero. The roots of the numerator A are called the **zeroes** of the Z transform, and the roots of the denominator B are called the **poles** of the Z transform. The term "zero" refers to the fact that the Z transform evaluates to zero at a zero. The term "pole" suggests an infinitely high tent pole, where the Z transform evaluates to infinity. The locations in the complex plane of the poles and zeros turn out to yield considerable insight about a Z transform. A plot of these locations is called a **pole-zero plot**. The poles are shown as crosses and the zeros as circles.

Example 12.11: The unit step of example 12.7 and its anticausal cousin of example 12.8 have pole-zero plots shown in figure 12.3. In each case, the Z transform has the form

$$\frac{z}{z-1} = \frac{A(z)}{B(z)},$$

where $A(z) = z$ and $B(z) = z - 1$. $A(z)$ has only one root, at $z = 0$, so the Z transforms each have one zero, at the origin in the complex plane. $B(z)$ also has only one root, at $z = 1$, so the Z transform has one pole, at $z = 1$.

These plots also show the unit circle, with a dashed line, and the regions of convergence for the two examples, as shaded areas. Note that $RoC(u)$ has the form of a region of convergence of a causal signal, as it should, and $RoC(v)$ has the form of a region of convergence of an anticausal signal, as it should (see figure 12.2). Note that neither RoC includes the unit circle, so if these signals were impulse responses of LTI systems, these systems would be unstable. ❒

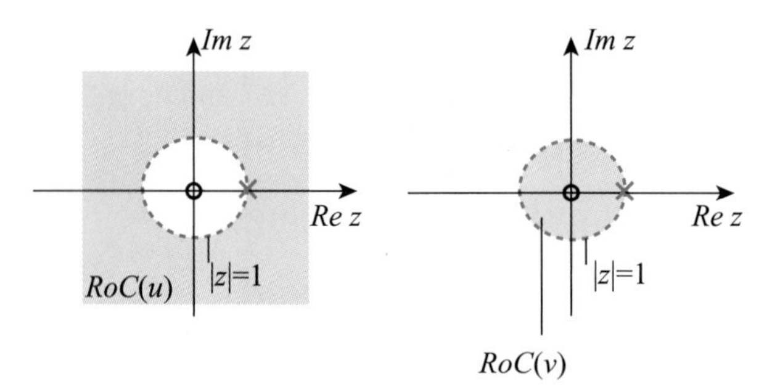

FIGURE 12.3: Pole-zero plots for the unit step u and its anticausal cousin v. The regions of convergence are the shaded area in the complex plane, not including the unit circle. Both Z tranforms, $\hat{U}$ and $\hat{V}$, have one pole at $z = 1$ and one zero at $z = 0$.

Consider a rational Z transform

$$\hat{X}(z) = \frac{A(z)}{B(z)}.$$

The denominator polynomial B evaluates to zero at a pole. That is, if there is a pole at location $z = p$ (a complex number), then $B(p) = 0$. Assuming that $A(p) \neq 0$, then $\hat{X}(p)$ is not finite. Thus, the region of convergence cannot include any pole p that is not canceled by a zero. This fact, combined with the fact that a causal signal always has a *RoC* of the form of the left one in figure 12.2, leads to the following simple **stability criterion for causal systems**:

> A discrete-time causal system is stable if and only if all the poles of its transfer function lie inside the unit circle.

A subtle fact about rational Z transforms is that the region of convergence is always bordered by the pole locations. This is evident in figure 12.3, where the single pole at $z = 1$ lies on the boundary of the two possible regions of convergence. In fact, the rational polynomial

$$\frac{z}{z-1}$$

can be associated with only three possible Z transforms, two of which have the two regions of convergence shown in figure 12.3, plus the one not shown where $RoC = \emptyset$.

Although a polynomial of order N has N roots, these roots are not necessarily distinct. Consider the (rather trivial) polynomial

$$A(z) = z^2.$$

This has order 2, and hence two roots, but both roots are at $z = 0$. Consider a Z transform given by

$$\forall z \in RoC(x), \quad \hat{X}(z) = \frac{z^2}{(z-1)^2}.$$

This has two zeros at $z = 0$, and two poles at $z = 1$. We say that the zero at $z = 0$ has **multiplicity** two. Similarly, the pole at $z = 1$ has multiplicity two. This multiplicity is indicated in a pole-zero plot by a number adjacent to the pole or zero, as shown in figure 12.4.

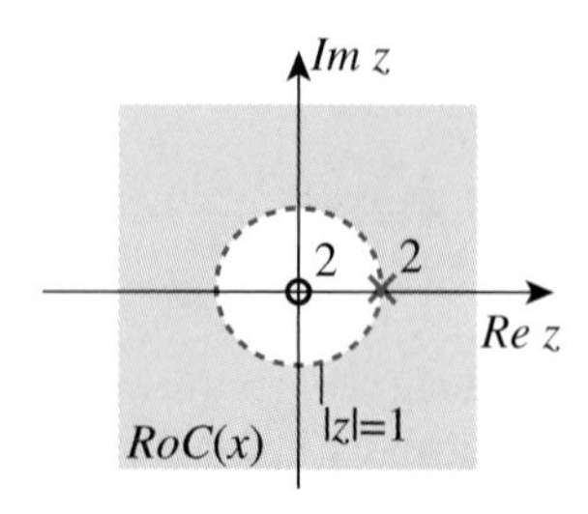

FIGURE 12.4: Poles and zeros with multiplicity greater than one are indicated by a number next to the cross or circle.

Example 12.12: Consider a signal x that is equal to the delayed Kronecker delta function,

$$\forall\, n \in Integers, \quad x(n) = \delta(n - M),$$

where $M \in Integers$ is a constant. Its Z transform is easy to find using the sifting rule,

$$\forall\, z \in RoC(x), \quad \hat{X}(z) = \sum_{m=-\infty}^{\infty} \delta(m - M) z^{-m} = z^{-M} = 1/z^{M}.$$

If $M > 0$, then this converges absolutely for any $z \neq 0$. Thus, if $M > 0$,

$$RoC(x) = \{z \in Complex \mid z \neq 0\}.$$

This Z transform has M poles at $z = 0$. Notice that this region of convergence, appropriately, has the form of that of a causal signal, figure 12.2(a), but where the circle has radius zero.

If $M < 0$, then the region of convergence is the entire set *Complex*, and the Z transform has M zeros at $z = 0$. This signal is anticausal, and its *RoC* matches the structure of 12.2(c), where the radius of the circle is infinite. Note that this Z transform does not converge at infinity, which it would have to do if the signal were causal.

If $M = 0$, then $\hat{X}(z) = 1$ for all $z \in Complex$, so $RoC = Complex$, and there are no poles or zeros. This is a particularly simple Z transform. ❐

Recall that for a causal signal, the Z transform must converge as $|z| \to \infty$. The region of convergence must include everything outside some circle, including infinity.* This implies that for a causal signal with a rational Z transform, the

* Some texts consider poles and zeros at infinity, in which case a causal signal cannot have a pole at infinity.

Z transform must be **proper**. A rational polynomial is proper when the order of the numerator is smaller than or equal to the order of the denominator. For example, if $M = -1$ in the previous example, then $x(n) = \delta(n+1)$ and $\hat{H}(z) = z$, which has numerator order one and denominator order zero. It is not proper, and indeed, it does not converge as $|z| \to \infty$. Any rational polynomial that has a numerator of higher order than the denominator will not converge as $z \to \infty$, and hence cannot be the Z transform of a causal signal.

In the following chapter, table 13.1 gives many common Z tranforms, all of which are rational polynomials. Together with the properties discussed in chapter 13, we can find the Z transforms of many signals.

12.3 The Laplace transform

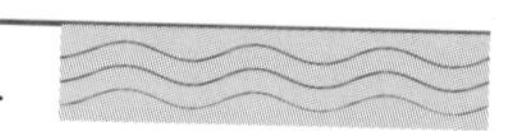

Consider a continuous-time signal x that is not absolutely integrable. Consider the scaled signal x_σ given by*

$$\forall t \in Reals, \quad x_\sigma(t) = x(t)e^{-\sigma t}, \tag{12.14}$$

for some real number σ. Often, this signal is absolutely integrable when σ is chosen appropriately. This new signal, therefore, will have a CTFT, even if x does not.

Example 12.13: Consider the impulse response of the simplified helicopter system described in example 12.2. The output as a function of the input is given by

$$\forall t \in Reals, \quad y(t) = \frac{1}{M} \int_0^t x(\tau) d\tau.$$

The impulse response is found by letting the input be a Dirac delta function and using the sifting rule to get

$$\forall t \in Reals, \quad h(t) = u(t)/M,$$

where u is the continuous-time unit step in (12.4). This is not absolutely integrable, so this system is not stable. However, the scaled signal

$$\forall t \in Reals, \quad h_\sigma(t) = h(t)e^{-\sigma t}$$

*The reason that this is different from the scaling by r^{-n} used to get the Z transform is somewhat subtle. The two methods are essentially equivalent, if we let $r = e^\sigma$. But scaling by $e^{-\sigma t}$ turns out to be more convenient for continuous-time systems, as we will see.

is absolutely integrable if $\sigma > 0$. Its CTFT is

$$\forall \sigma > 0, \forall \omega \in Reals, \quad H_\sigma(\omega) = \int_{-\infty}^{\infty} h(t) e^{-\sigma t} e^{-i\omega t} dt$$

$$= \frac{1}{M} \int_{0}^{\infty} e^{-\sigma t} e^{-i\omega t} dt$$

$$= \frac{1}{M} \int_{0}^{\infty} e^{-(\sigma + i\omega)t} dt$$

$$= \frac{1}{M(\sigma + i\omega)} . \square$$

The last step in example 12.13 uses the following useful fact from calculus,

$$\int_{a}^{b} e^{ct} dt = \frac{1}{c}(e^{cb} - e^{ca}), \tag{12.15}$$

for any $c \in Complex$ and $a, b \in Reals \cup \{-\infty, \infty\}$ where e^{cb} and e^{ca} are finite.

In general, the CTFT of the scaled signal x_σ in (12.14) is

$$\forall \omega \in Reals, \quad X_\sigma(\omega) = \int_{-\infty}^{\infty} x(t) e^{-(\sigma + i\omega)t} dt.$$

Notice that this is a function not just of ω, but also of σ. We are only sure it is valid for values of σ that yield an absolutely integrable signal X_σ.

Define the complex number

$$s = \sigma + i\omega.$$

Then we can write this CTFT as

$$\boxed{\forall s \in RoC(x), \quad \hat{X}(s) = \int_{-\infty}^{\infty} x(t) e^{-st} dt,} \tag{12.16}$$

where $\hat{X}$ is a function called the **Laplace transform** of x, and

$$\hat{X}: RoC(x) \to Complex$$

where $RoC(x) \subset Complex$ is given by

$$\boxed{RoC(x) = \{s = \sigma + i\omega \in Complex \mid x(t)e^{-\sigma t} \text{ is absolutely integrable}\}.} \quad (12.17)$$

The Laplace tranform $\hat{H}$ of the impulse response h of an LTI system is called the **transfer function** of the system, just as with discrete-time systems.

Example 12.14: Continuing example 12.13, we can recognize from the form of $H_\sigma(\omega)$ that the transfer function of the helicopter system is

$$\forall s \in RoC(h), \quad \hat{H}(s) = \frac{1}{Ms}.$$

The RoC is

$$RoC(h) = \{s = \sigma + i\omega \in Complex \mid \sigma > 0\}. \square$$

12.3.1 *Structure of the region of convergence*

As with the Z transform, the region of convergence is an essential part of a Laplace transform. It gives the domain of the function $\hat{X}$. Whether a complex number s is in the RoC depends only on σ, not on ω, as is evident in the definition (12.17). Since $s = \sigma + i\omega$, whether a complex number is in the region of convergence depends only on its real part. Once again, there are only three possible patterns for the region of convergence, shown in figure 12.5. Each figure illustrates the complex plane, and the shaded area is a region of convergence. Each possibility has vertical symmetry, in that whether a point is in the RoC depends only on its real part.

Figure 12.5(a) shows the RoC of a causal or right-sided signal. A continuous-time signal x is **right-sided** if $x(t) = 0$ for all $t < T$ for some $T \in Reals$. The

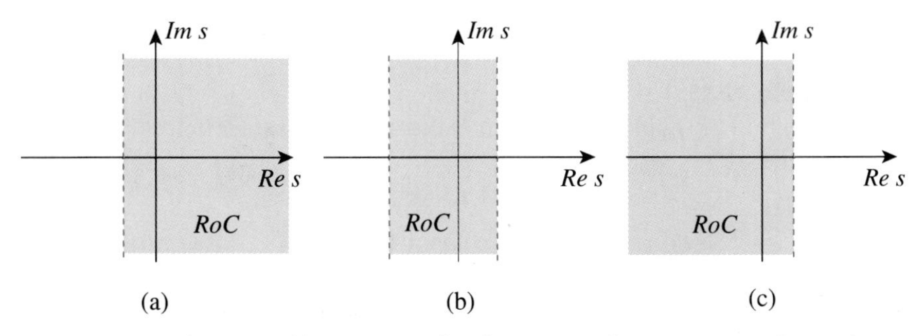

FIGURE 12.5: Three possible structures for the region of convergence of a Laplace transform: (a) causal or right-sided, (b) two-sided, and (c) anticausal or left-sided.

RoC is the set of complex numbers $s = \sigma + i\omega$ where following integral converges:

$$\int_{-\infty}^{\infty} |x(t)e^{-\sigma t}| dt.$$

But if x is right-sided, then

$$\int_{-\infty}^{\infty} |x(t)e^{-\sigma t}| dt = \int_{T}^{\infty} |x(t)e^{-\sigma t}| dt.$$

If $T \geq 0$ and this integral converges for some given σ, then it must also converge for any $\tilde{\sigma} > \sigma$ because for all $t \geq 0$, $e^{-\tilde{\sigma} t} < e^{-\sigma t}$. Thus, if $s = \sigma + i\omega \in RoC(x)$, then the $RoC(x)$ must include all points in the complex plane on the vertical line passing through s and every point to the right of that line.*

If $T < 0$, then

$$\int_{T}^{\infty} |x(t)e^{-\sigma t}| dt = \int_{T}^{0} |x(t)e^{-\sigma t}| dt + \int_{0}^{\infty} |x(t)e^{-\sigma t}| dt,$$

then the finite integral exists and is finite for all σ, so the same argument applies.

Figure 12.5(c) shows the *RoC* of a left-sided signal. A continuous-time signal x is **left-sided** if $x(t) = 0$ for all $t > T$ for some $T \in Reals$. By a similar argument, if $s = \sigma + i\omega \in RoC(x)$, then the $RoC(x)$ must include all points in the complex plane on the vertical line passing through s and every point to the left of that line.

Figure 12.5(b) shows the *RoC* of a signal that is a **two-sided signal**. Such a signal can always be expressed as a sum of a right-sided signal and left-sided signal. The *RoC* is the intersection of the regions of convergence for these two components.

Example 12.15: Using the same methods as in examples 12.13 and 12.14 we can find the Laplace transform of the continuous-time unit step signal u, given by

$$\forall t \in Reals, \quad u(t) = \begin{cases} 0, & t < 0 \\ 1, & t \geq 0. \end{cases} \tag{12.18}$$

* It is convenient but coincidental that the region of convergence is the right half of a plane when the sequence is right-sided.

The Laplace transform is

$$\forall s \in RoC(u), \quad \hat{U}(s) = \int_{-\infty}^{\infty} u(t)e^{-st}dt$$

$$= \int_{0}^{\infty} e^{-st}dt$$

$$= \frac{1}{s},$$

where again we have used (12.15). The domain of $\hat{U}$ is

$$RoC(u) = \{s \in Complex \mid Re\{s\} > 0\}.$$

This region of convergence has the structure of figure 12.5(a), where the dashed line sits exactly on the imaginary axis. The region of convergence, therefore, is simply the right half of the complex plane. ❐

Example 12.16: The signal v given by

$$\forall t \in Reals, \quad v(t) = -u(-t) = \begin{cases} -1, & t < 0 \\ 0, & t \geq 0 \end{cases}$$

has Laplace transform

$$\forall s \in RoC(v), \quad \hat{V}(s) = \int_{-\infty}^{\infty} v(t)e^{-st}dt$$

$$= -\int_{-\infty}^{0} e^{-st}dt$$

$$= \frac{1}{s}$$

with domain

$$RoC(v) = \{s \in Complex \mid Re\{s\} < 0\}.$$

This region of convergence has the structure of figure 12.5(c), where the dashed line coincides with the imaginary axis. ❐

Notice that although the Laplace transforms $\hat{U}$ and $\hat{V}$ have the same algebraic form, namely, $1/s$, they are in fact different functions, because their domains are different.

Some signals have no meaningful Laplace transform.

Example 12.17: The signal x with $x(t) = 1$, for all $t \in Reals$, does not have a Laplace transform. We can write $x = u - v$, where u and v are defined in the previous examples. Thus, the region of convergence of x must be the intersection of the regions of convergence of u and v. However, these two regions have an empty intersection, so $RoC(x) = \emptyset$.

Viewed another way, the set $RoC(x)$ is the set of complex numbers s such that

$$\int_{-\infty}^{\infty} |x(t)e^{-st}|dt = \int_{-\infty}^{\infty} |e^{-st}|dt < \infty.$$

But there is no such complex number s. ❒

Note that the signal x in example 12.17 is periodic with any period $p \in Reals$ (because $x(t+p) = x(t)$ for any $p \in Reals$). Thus, it has a Fourier series representation. In fact, as shown in section 10.6.3, a periodic signal also has a Fourier transform representation, as long as we are willing to allow Dirac delta functions in the Fourier transform. (Recall that this means that there are values of ω where $X(\omega)$ will not be finite.) In the continuous-time case as in the discrete-time case, with periodic signals, the Fourier series is by far the simplest frequency-domain tool to use. The Fourier transform can also be used if we allow Dirac delta functions. The Laplace transform, however, is more problematic, because the region of convergence is empty.

12.3.2 *Stability and the Laplace transform*

If a continuous-time signal x is absolutely integrable, then it has a CTFT X that is finite for all $\omega \in Reals$. Moreover, the CTFT is equal to the Laplace transform evaluated on the imaginary axis,

$$\boxed{\forall\, \omega \in Reals, \quad X(\omega) = \hat{X}(s)|_{s=i\omega} = \hat{X}(i\omega).}$$

The complex number $s = i\omega$ is pure imaginary, and therefore lies on the imaginary axis. Recall that an LTI system is stable if and only if its impulse response is absolutely integrable. Thus,

> A continuous-time LTI system with impulse response h is stable if and only if the transfer function $\hat{H}$, which is the Laplace transform of h, has a region of convergence that includes the imaginary axis.

Example 12.18: Consider the exponential signal h given by

$$\forall\, t \in Reals,\;\; h(t) = e^{-at}u(t),$$

for some real or complex number a, where, as usual, u is the unit step. The Laplace transform is

$$\begin{aligned}
\forall\, s \in RoC(h), \quad \hat{H}(s) &= \int_{-\infty}^{\infty} h(t)e^{-st}dt \\
&= \int_{0}^{\infty} e^{-at}e^{-st}dt \\
&= \int_{0}^{\infty} e^{-(s+a)t}dt \\
&= \frac{1}{s+a},
\end{aligned}$$

where again we have used (12.15). It is evident from (12.15) that for this integral to be valid, the domain of $\hat{H}$ must be

$$RoC(h) = \{s \in Complex \mid Re\{s\} > -Re\{a\}\}.$$

This region of convergence has the structure of figure 12.5(a), where the vertical dashed line passes through a.

Now suppose that h is the impulse response of an LTI system. That LTI system is stable if an only if $Re\{a\} > 0$. Indeed, if $Re\{a\} < 0$, then the impulse response grows without bound, because e^{-at} grows without bound as t gets large. If $Re\{a\} = 0$, we have the unit step. ❐

12.3.3 *Rational Laplace tranforms and poles and zeros*

All of the Laplace transforms we have seen so far are **rational polynomials** in s. In practice, most Laplace transforms of interest can be written as the ratio of two finite-order polynomials in s,

$$\hat{X}(s) = \frac{A(s)}{B(s)}.$$

An exception is illustrated in the following example.

Example 12.19: Consider a signal x that is equal to the delayed Dirac delta function,

$$\forall\, t \in Reals, \;\; x(t) = \delta(t - \tau),$$

where $\tau \in Reals$ is a constant. Its Laplace transform is easy to find using the sifting rule,

$$\forall\, s \in RoC(x), \quad \hat{X}(s) = \int_{-\infty}^{\infty} \delta(t - \tau) e^{-st} dt = e^{-s\tau}.$$

This has no finite-order rational polynomial representation. ❐

Unlike the discrete-time case, pure time delays turn out to be rather difficult to realize in many physical systems that are studied using Laplace transforms, so we need not be overly concerned with them. We focus henceforth on rational Laplace transforms.

For a rational Laplace transform, the **order** of the polynomial A or B is the power of the highest power of s that appears in the polynomial. For the exponential of example 12.18, the numerator polynomial is $A(s) = 1$, a zero-order polynomial, and the denominator is $B(s) = s + a$, a first-order polynomial. As with the Z transform, the roots of the numerator polynomial are called the **zeros** of the Laplace transform, and the roots of the denominator polynomial are called the **poles**.

Example 12.20: The exponential of example 12.18 has a single pole at $s = -a$, and no zeros.* A pole-zero plot is shown in figure 12.6, where we assume that a is a complex number with a positive real part. The region of convergence includes the imaginary axis, so this signal is absolutely integrable. ❐

* In some texts, it will be observed that as s approaches infinity, this Laplace transform approaches zero, and hence it will be said that there is a zero at infinity. So to avoid conflict with such texts, we might say that this Laplace transform has no finite zeros.

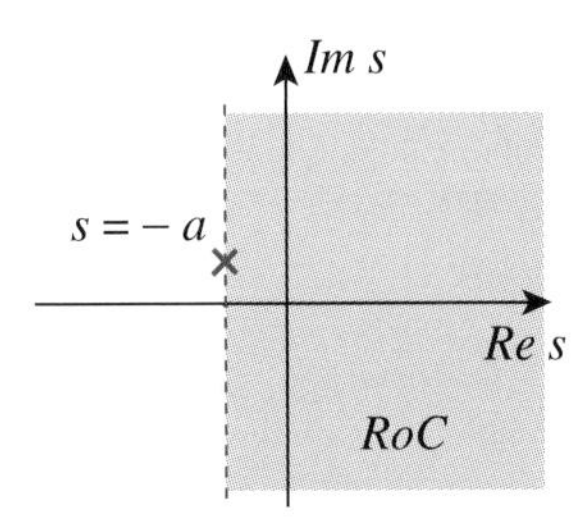

FIGURE 12.6: Pole-zero plot for the exponential signal of example 12.18, assuming a has a positive real part.

As with Z transforms, the region of convergence of a rational Laplace transform is bordered by the pole locations. Hence,

> A continuous-time causal system is stable if and only if all the poles of its transfer function lie in the left half of the complex plane. That is, all the poles must have negative real parts.

Table 13.3 in the following chapter gives many common Laplace tranforms.

12.4 Summary

Many useful signals have no Fourier transform. A sufficient condition for a signal to have a Fourier transform that is finite at all frequencies is that the signal be absolutely summable (if it is a discrete-time signal) or absolutely integrable (if it is a continuous-time system).

Many useful systems are not stable, which means that even with a bounded input, the output may be unbounded. An LTI system is stable if and only if its impulse response is absolutely summable (discrete time) or absolutely integrable (continuous time).

Many signals that are not absolutely summable (integrable) can be scaled by an exponential to get a new signal that is absolutely summable (integrable). The DTFT (CTFT) of the scaled signal is called the Z transform (Laplace transform) of the signal.

The Z transform (Laplace transform) is defined over a region of convergence, where the structure of the region of convergence depends on whether the signal is causal, anticausal, or two-sided. The Z transform (Laplace transform) of the impulse response is called the transfer function of an LTI system. The region of convergence includes the unit circle (imaginary axis), if and only if the system is stable.

A rational Z transform (Laplace transform) has poles and zeros, and the poles bound the region of convergence. The locations of the poles and zeros yield considerable information about the system, including whether it is stable.

EXERCISES

KEY: E = mechanical T = requires plan of attack C = more than 1 answer

E 1. Consider the signal x given by

$$\forall\, n \in \mathit{Integers},\ \ x(n) = a^n u(-n),$$

where a is a complex constant.

(a) Find the Z transform of x. Be sure to give the region of convergence.

(b) Where are the poles and zeros?

(c) Under what conditions on a is x absolutely summable?

(d) Assuming that x is absolutely summable, find its DTFT.

T 2. Consider the signal x given by

$$\forall\, n \in \mathit{Integers},\ \ x(n) = \begin{cases} 1, & |n| \leq M \\ 0, & \text{otherwise} \end{cases}$$

for some integer $M > 0$.

(a) Find the Z transform of x. Simplify so that there remain no summations. Be sure to give the region of convergence.

(b) Where are the poles and zeros? Do not give poles and zeros that cancel each other out.

(c) Under what conditions is x absolutely summable?

(d) Assuming that x is absolutely summable, find its DTFT.

T 3. Consider the **unit ramp** signal w given by

$$\forall\, n \in \mathit{Integers},\ \ w(n) = nu(n),$$

where u is the unit step. The following identity will be useful,

$$\sum_{m=0}^{\infty}(m+1)a^m = (\sum_{m=0}^{\infty} a^m)^2 = \frac{1}{(1-a)^2}. \tag{12.19}$$

This is a generalization of the geometric series identity, given by (12.9). This series converges for any complex number a with $|a| < 1$, because

$$\begin{aligned}\sum_{m=0}^{\infty}(m+1)|a|^m &= 1 + 2|a| + 3|a|^2 + \cdots \\ &= (1 + |a| + |a|^2 + \cdots)(1 + |a| + |a|^2 + \cdots) \\ &= (\sum_{m=0}^{\infty} |a|^m)^2 \\ &< \infty.\end{aligned}$$

(a) Use the given identity to find the Z transform of the unit ramp. Be sure to give the region of convergence. Check your answer against that given in the Probing Further box on pages 491–492.

(b) Sketch the pole-zero plot of the Z transform.

(c) Is the unit ramp absolutely summable?

E 4. Sketch the pole-zero plots and regions of convergence for the Z transforms of the following impulse responses, and indicate whether a discrete-time LTI system with these impulse responses is stable:

(a) $h_1(n) = \delta(n) + 0.5\delta(n-1)$.

(b) $h_2(n) = (0.5)^n u(n)$.

(c) $h_3(n) = 2^n u(n)$.

E 5. Consider the anticausal continuous-time exponential signal x given by

$$\forall\, t \in Reals, \;\; x(t) = -e^{-at}u(-t),$$

for some real or complex number a, where, as usual, u is the unit step.

(a) Show that the Laplace transform of x is

$$\hat{X}(s) = \frac{1}{s+a}$$

with region of convergence

$$RoC(x) = \{s \in Complex \mid Re\{s\} < -Re\{a\}\}.$$

(b) Where are the poles and zeros?

(c) Under what conditions on a is x absolutely integrable?

(d) Assuming that x is absolutely integrable, find its CTFT.

E 6. This exercise demonstrates that the Laplace transform is linear. Show that if x and y are continuous-time signals, a and b are complex (or real) constants, and w is given by

$$\forall\, t \in Reals, \quad w(t) = ax(t) + by(t),$$

then the Laplace transform is

$$\forall\, s \in RoC(w), \quad \hat{W}(s) = a\hat{X}(s) + b\hat{Y}(s),$$

where

$$RoC(w) \supset RoC(x) \cap RoC(y).$$

T 7. Let the causal sinusoidal signal y be given by

$$\forall\, t \in Reals, \quad y(t) = \cos(\omega_0 t)u(t),$$

where ω_0 is a real number and u is the unit step.

(a) Show that the Laplace transform is

$$\forall\, s \in \{s \mid Re\{s\} > 0\}, \quad \hat{Y}(s) = \frac{s}{s^2 + \omega_0^2}.$$

Hint: Use linearity, demonstrated in exercise 6, and Euler's relation.

(b) Sketch the pole-zero plot and show the region of convergence.

E 8. Consider a discrete-time LTI system with impulse response

$$\forall\, n, \quad h(n) = a^n \cos(\omega_0 n)u(n),$$

for some $\omega_0 \in Reals$. Determine for what values of a this system is stable.

T 9. The continuous-time unit ramp signal w is given by

$$\forall\, t \in Reals, \quad x(t) = tu(t),$$

where u is the unit step.

(a) Find the Laplace transform of the unit ramp, and give the region of convergence.
Hint: Use integration by parts in (12.16) and the fact that $\int_0^\infty te^{-\sigma t}dt < \infty$ for $\sigma > 0$.

(b) Sketch the pole-zero plot of the Laplace transform.

E 10. Let h and g be the impulse response of two stable systems. They may be discrete time or continuous time. Let a and b be two complex numbers. Show that the system with impulse response $ah + bg$ is stable.

T 11. Consider a series composition of two (continuous- or discrete-time) systems with impulse response h and g. The output v of the first system is related to its input x by $v = h * x$. The output y of the second system (and of the series composition) is $y = g * v$. Suppose both systems are stable. Show that the series composition is stable. Hint: Use the definition of stability.

T 12. Let h be the impulse response of a stable discrete-time system, so it is absolutely summable, and denote

$$\|h\| = \sum_{n=-\infty}^{\infty} |h(n)|.$$

($\|h\|$ is called the **norm** of the impulse response.)

(a) Suppose the input signal x is bounded by M, that is, $\forall\, n,\ |x(n)| \leq M$. Show that the output $y = h * x$ is bounded by $\|h\|M$.

(b) Consider the input signal x where

$$\forall\, n \in Integers, \quad x(n) = \begin{cases} h(-n)/|h(-n)|, & h(n) \neq 0 \\ 0, & h(n) = 0. \end{cases}$$

Show that $\|h\|$ is the smallest bound of the output $y = h * x$.

(c) Let g be the impulse response of another stable system with norm $\|g\|$. Show that the norm satisfies the triangle inequality,

$$\|h + g\| \leq \|h\| + \|g\|.$$

(d) Suppose the two systems are placed in series. The composition has the impulse response $h * g$. Show that

$$\|h * g\| \leq \|h\| \times \|g\|.$$

E 13. Show that the series-parallel composition of figure 12.7 is stable if the four component systems are stable. Let h be the impulse response of the composition. Express h in terms of the component impulse responses and then estimate $\|h\|$ in terms of the norms of the components.

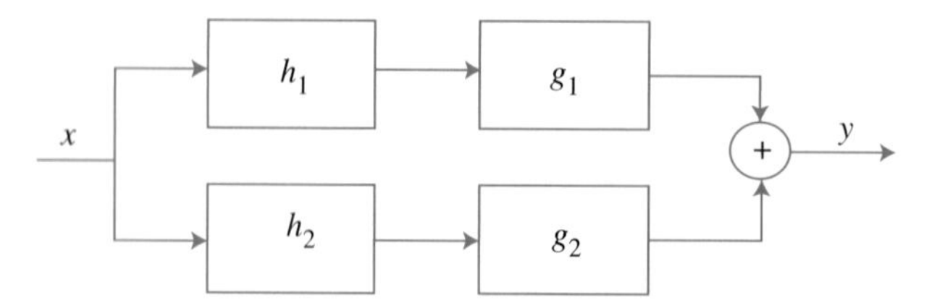

FIGURE 12.7: System composition.

E 14. Let x be a discrete-time signal of finite duration,that is, $x(n) = 0$ for $n < M$ and $n > N$ where M and N are finite integers (positive or negative). Let $\hat{X}$ be its Z transform.

(a) Show that all its poles (if any) are at $z = 0$.

(b) Show that if x is causal it has N poles at $z = 0$.

T 15. This problem relates the Z and Laplace transforms. Let x be a discrete-time signal with Z transform $\hat{X} : RoC(x) \to Complex$. Consider the continuous-time signal y related to x by

$$\forall t \in Reals, \quad y(t) = \sum_{n=-\infty}^{\infty} x(n)\delta(t - nT).$$

Here $T > 0$ is a fixed period. So y comprises delta functions located at $t = nT$ of magnitude $x(n)$.

(a) Use the sifting property and the definition (12.16) to find the Laplace transform $\hat{Y}$ of y. What is $RoC(y)$?

(b) Show that $\hat{Y}(s) = \hat{X}(e^{sT})$, where $\hat{X}(e^{sT})$ is $\hat{X}(z)$ evaluated at $z = e^{sT}$.

(c) Suppose $\hat{X}(z) = \frac{1}{z-1}$ with $RoC(x) = \{z \mid |z| > 1\}$. What are $\hat{Y}$ and $RoC(y)$?

INTERVIEW
Xavier Rodet

Xavier Rodet's research interests are in the areas of signal and pattern analysis, recognition, and synthesis. He has been working particularly on digital signal processing for speech, speech and singing voice synthesis, and automatic speech recognition. Computer music is his other main domain of interest. He has been working on understanding spectro-temporal patterns of musical sounds and on synthesis-by-rules. He has been developing new methods, programs, and patents for musical sound signal analysis, synthesis, and control. He is also working on physical models of musical instruments and nonlinear dynamic systems applied to sound signal synthesis.

How did you decide to study engineering?

When I learned that sound and music could be performed by computers calculating the sound samples, this was the start. It meant that any sound that you are able to hear could be built by a computer program provided you specify the sound properties adequately. That is, all the sounds that I was imagining in my head, I would be able to construct them.

What was your first job in the industry?

My PhD research was on speech synthesis from text. There was already a lot of interest in the industry on text to speech synthesis, but it took a long time before commercial applications appeared.

Which person in the field has inspired you most?

James A. Moorer

Do you have any advice for students studying electrical and computer engineering?

Study mathematics as well, as much as possible!

What is your vision for the future of electrical and computer engineering?

In my domain, interdisciplinary research is more and more necessary. If we take information extraction from audio signals as an example, several other fields, such as utilization of large Data Bases and Artificial Intelligence methods in addition to just signal processing, seem to contribute largely to progress.

CHAPTER 13

Laplace and Z transforms

In the previous chapter, we defined Laplace and Z transforms to deal with signals that are not absolutely summable and systems that are not stable. The Z transform of the discrete-time signal x is given by

$$\forall z \in RoC(x), \quad \hat{X}(z) = \sum_{m=-\infty}^{\infty} x(m) z^{-m},$$

where $RoC(x)$ is the **region of convergence**, the region in which the sum above converges absolutely.

The Laplace transform of the continuous-time signal x is given by

$$\forall s \in RoC(x), \quad \hat{X}(s) = \int_{-\infty}^{\infty} x(t) e^{-st} dt,$$

where $RoC(x)$ is again the region of convergence, the region in which the preceding integral converges absolutely.

In this chapter, we explore key properties of the Z and Laplace transforms and give examples of transforms. We also explain how, given a rational polynomial in z or s, plus a region of convergence, one can find the corresponding time-domain function. This **inverse transform** proves particularly useful, because compositions of LTI systems, studied in the next chapter, often lead to rather complicated rational polynomial descriptions of a transfer function.

Z transforms of common signals are given in table 13.1. Properties of the Z transform are summarized in table 13.2 and elaborated in the following section.

TABLE 13.1
Z transforms of key signals.*

Discrete-time signal $\forall n \in Integers$	**Z transform** $\forall z \in RoC(x)$	$Roc(x) \subset Complex$	**Reference**
$x(n) = \delta(n - M)$	$\hat{X}(z) = z^{-M}$	$Complex$	Example 12.12
$x(n) = u(n)$	$\hat{X}(z) = \frac{z}{z-1}$	$\{z \mid \lvert z\rvert > 1\}$	Example 12.7
$x(n) = a^n u(n)$	$\hat{X}(z) = \frac{z}{z-a}$	$\{z \mid \lvert z\rvert > \lvert a\rvert\}$	Example 13.3
$x(n) = a^n u(-n)$	$\hat{X}(z) = \frac{1}{1-a^{-1}z}$	$\{z \mid \lvert z\rvert < \lvert a\rvert\}$	Exercise 1 in chapter 12
$x(n) = \cos(\omega_0 n)u(n)$	$\hat{X}(z) = \frac{z^2 - z\cos(\omega_0)}{z^2 - 2z\cos(\omega_0) + 1}$	$\{z \mid \lvert z\rvert > 1\}$	Example 13.3
$x(n) = \sin(\omega_0 n)u(n)$	$\hat{X}(z) = \frac{z\sin(\omega_0)}{z^2 - 2z\cos(\omega_0) + 1}$	$\{z \mid \lvert z\rvert > 1\}$	Exercise 1
$x(n) = \frac{1}{(N-1)!}(n-1)\cdots(n-N+1)a^{n-N}u(n-N)$	$\hat{X}(z) = \frac{1}{(z-a)^N}$	$\{z \mid \lvert z\rvert > \lvert a\rvert\}$	(13.13)
$x(n) = \frac{(-1)^N}{(N-1)!}(N-1-n)\cdots(1-n)a^{n-N}u(-n)$	$\hat{X}(z) = \frac{1}{(z-a)^N}$	$\{z \mid \lvert z\rvert < \lvert a\rvert\}$	(13.14)

* The signal u is the unit step (12.13); δ is the Kronecker delta; a is any complex constant; ω_0 is any real constant; M is any integer constant; and $N > 0$ is any integer constant.

TABLE 13.2
Properties of the Z transform.*

Time domain $\forall n \in Integers$	**Frequency domain** $\forall z \in RoC$	RoC	**Name** (Reference)
$w(n) = ax(n) + by(n)$	$\hat{W}(z) = a\hat{X}(z) + b\hat{Y}(z)$	$RoC(w) \supset RoC(x) \cap RoC(y)$	**Linearity** (section 13.1.1)
$y(n) = x(n - N)$	$\hat{Y}(z) = z^{-N}\hat{X}(z)$	$RoC(y) = RoC(x)$	**Delay** (section 13.1.2)
$y(n) = (x * h)(n)$	$\hat{Y}(z) = \hat{X}(z)\hat{H}(z)$	$RoC(y) \supset RoC(x) \cap RoC(h)$	**Convolution** (section 13.1.3)
$y(n) = x^*(n)$	$\hat{Y}(z) = [\hat{X}(z^*)]^*$	$RoC(y) = RoC(x)$	**Conjugation** (section 13.1.4)
$y(n) = x(-n)$	$\hat{Y}(z) = \hat{X}(z^{-1})$	$RoC(y) = \{z \mid z^{-1} \in RoC(x)\}$	**Time reversal** (section 13.1.5)
$y(n) = nx(n)$	$\hat{Y}(z) = -z\frac{d}{dz}\hat{X}(z)$	$RoC(y) = RoC(x)$	**Scaling by *n*** (page 491)
$y(n) = a^{-n}x(n)$	$\hat{Y}(z) = \hat{X}(az)$	$RoC(y) = \{z \mid az \in RoC(x)\}$	**Exponential scaling** (section 13.1.6)

* a and b are complex constants, and N is any integer constant.

13.1 Properties of the Z tranform

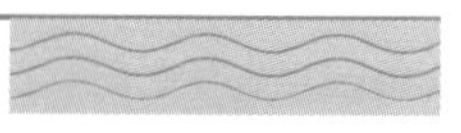

The Z transform has useful properties that are similar to those of the four Fourier transforms.

13.1.1 *Linearity*

Suppose x and y have Z transforms $\hat{X}$ and $\hat{Y}$, a, b are two complex constants, and

$$w = ax + by.$$

Then the Z transform of w is

$$\forall\, z \in RoC(w), \quad \hat{W}(z) = a\hat{X}(z) + b\hat{Y}(z).$$

This follows immediately from the definition of the Z transform,

$$\begin{aligned}\hat{W}(z) &= \sum_{m=-\infty}^{\infty} w(m)z^{-m} \\ &= \sum_{m=-\infty}^{\infty} (ax(m) + by(m))z^{-m} \\ &= a\hat{X}(z) + b\hat{Y}(z).\end{aligned}$$

The region of convergence of w must include at least the regions of convergence of x and y, because if $x(n)r^{-n}$ and $y(n)r^{-n}$ are absolutely summable, then certainly $(ax(n) + by(n))r^{-n}$ is absolutely summable. Conceivably, however, the region of convergence may be larger. Thus, all we can assert in general is

$$RoC(w) \supset RoC(x) \cap RoC(y). \tag{13.1}$$

Linearity is extremely useful because it makes it easy to find the Z transform of complicated signals that can be expressed as a linear combination of signals with known Z transforms.

> **Example 13.1:** We can use the results of example 12.12 plus linearity to find, for example, the Z transform of the signal x given by
>
> $$\forall\, n \in Integers, \quad x(n) = \delta(n) + 0.9\delta(n-4) + 0.8\delta(n-5).$$

This is simply

$$\hat{X}(z) = 1 + 0.9z^{-4} + 0.8z^{-5}.$$

We can identify the poles by writing this as a rational polynomial in z (multiply top and bottom by z^5),

$$\hat{X}(z) = \frac{z^5 + 0.9z + 0.8}{z^5},$$

from which we see that there are five poles at $z = 0$. The signal is causal, so the region of convergence is the region outside the circle passing through the pole with the largest magnitude, or in this case,

$$RoC(x) = \{z \in Complex \mid z \neq 0\}. \square$$

Example 13.1 illustrates how to find the transfer function of any finite impulse response (FIR) filter. It also suggests that the transfer function of an FIR filter always has a region of convergence that includes the entire complex plane, except possibly $z = 0$. The region of convergence will also not include $z = \infty$ if the FIR filter is not causal.

Linearity can also be used to invert a Z transform. That is, given a rational polynomial and a region of convergence, we can find the time-domain function that has this Z transform. The general method for doing this will be considered in section 13.5, but for certain simple cases, we just have to recognize familiar Z transforms.

Example 13.2: Suppose we are given the Z transform

$$\forall z \in \{z \in Complex \mid z \neq 0\}, \quad \hat{X}(z) = \frac{z^5 + 0.9z + 0.8}{z^5}.$$

We can immediately recognize this as the Z transform of a causal signal, because it is a proper rational polynomial and the region of convergence includes the entire complex plane except $z = 0$ (thus, it has the form of figure 12.2(a)).

If we divide through by z^5, this becomes

$$\forall z \in \{z \in Complex \mid z \neq 0\}, \quad \hat{X}(z) = 1 + 0.9z^{-4} + 0.8z^{-5}.$$

By linearity, we can see that

$$\forall n \in Integers, \quad x(n) = x_1(n) + 0.9x_2(n) + 0.8x_3(n),$$

where x_1 has Z transform 1, x_2 has Z transform z^{-4}, and x_3 has Z transform z^{-5}.

The regions of convergence for each Z transform must be at least that of x, or at least $\{z \in Complex \mid z \neq 0\}$. From example 12.12, we recognize these Z transforms, and hence obtain

$$\forall\, n \in Integers, \quad x(n) = \delta(n) + 0.9\delta(n-4) + 0.8\delta(n-5). \; \square$$

Another application of linearity uses Euler's relation to deal with sinusoidal signals.

Example 13.3: Consider the exponential signal x given by

$$\forall\, n \in Integers, \quad x(n) = a^n u(n),$$

where a is a complex constant. Its Z transform is

$$\hat{X}(z) = \sum_{m=-\infty}^{\infty} x(m) z^{-m} = \sum_{m=0}^{\infty} a^m z^{-m} = \frac{1}{1 - az^{-1}} = \frac{z}{z-a}, \tag{13.2}$$

where we have used the geometric series identity (12.9). This has a zero at $z = 0$ and a pole at $z = a$. The region of convergence is

$$RoC(x) = \{z \in Complex \mid \sum_{m=0}^{\infty} |a|^m |z|^{-m} < \infty\} = \{z \mid |z| > |a|\}, \tag{13.3}$$

the region of the complex plane outside the circle that passes through the pole. A pole-zero plot is shown in figure 13.1(a).

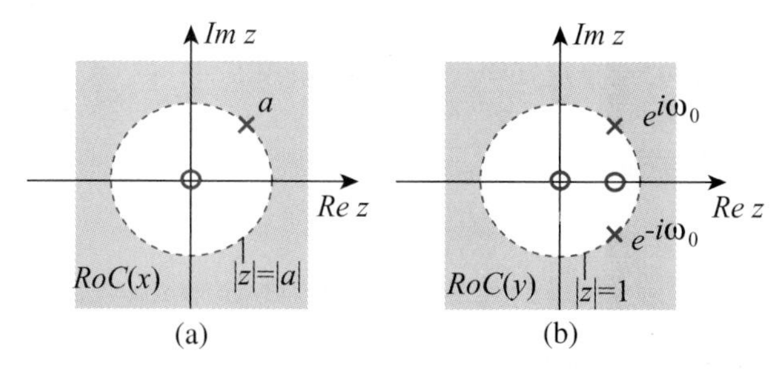

FIGURE 13.1: Pole-zero plots for (a) the exponential signal x and (b) the sinusoidal signal y.

We can use this result plus linearity of the Z transform to determine the Z transform of the causal sinusoidal signal y given by

$$\forall n \in Integers, \quad y(n) = \cos(\omega_0 n)u(n).$$

Euler's relation implies that

$$y(n) = \frac{1}{2}\{e^{i\omega_0 n}u(n) + e^{-i\omega_0 n}u(n)\}.$$

Using (13.2) and linearity,

$$\begin{aligned}\hat{Y}(z) &= \frac{1}{2}\left\{\frac{z}{z - e^{i\omega_0}} + \frac{z}{z - e^{-i\omega_0}}\right\} \\ &= \frac{1}{2}\frac{2z^2 - z(e^{i\omega_0} + e^{-i\omega_0})}{(z - e^{i\omega_0})(z - e^{-i\omega_0})} \\ &= \frac{z(z - \cos(\omega_0))}{z^2 - 2z\cos(\omega_0) + 1}.\end{aligned}$$

This has a zero at $z = 0$, another zero at $z = \cos(\omega_0)$, and two poles, one at $z = e^{i\omega_0}$ and the other at $z = e^{-i\omega_0}$. Both of these poles lie on the unit circle. A pole-zero plot is shown in figure 13.1(b), in which we assume that $\omega_0 = \pi/4$. We know from (13.1) and (13.3) that the region of convergence is at least the area outside the unit circle. In this case, we can conclude that it is exactly the area outside the unit circle, because it must be bordered by the poles, and it must have the form of a region of convergence of a causal signal. ❐

13.1.2 *Delay*

For any integer N (positive or negative) and signal x, let $y = D_N(x)$ be the signal given by

$$\forall n \in Integers, \quad y(n) = x(n - N).$$

Suppose x has Z transform $\hat{X}$ with domain $RoC(x)$. Then $RoC(y) = RoC(x)$ and

$$\begin{aligned}\forall z \in RoC(y), \quad \hat{Y}(z) &= \sum_{m=-\infty}^{\infty} y(m)z^{-m} \\ &= \sum_{m=-\infty}^{\infty} x(m - N)z^{-m} = z^{-N}\hat{X}(z). \qquad (13.4)\end{aligned}$$

Thus,

> If a signal is delayed by N samples, its Z transform is multiplied by z^{-N}.

13.1.3 *Convolution*

Suppose x and h have Z transforms $\hat{X}$ and $\hat{H}$. Let

$$y = x * h.$$

Then

$$\forall z \in RoC(y), \quad \hat{Y}(z) = \hat{X}(z)\hat{H}(z). \tag{13.5}$$

This follows from using the definition of convolution,

$$\forall n \in Integers, \quad y(n) = \sum_{m=-\infty}^{\infty} x(m)h(n-m),$$

in the definition of the Z transform,

$$\hat{Y}(z) = \sum_{n=-\infty}^{\infty} y(n)z^{-n} = \sum_{n=-\infty}^{\infty} \sum_{m=-\infty}^{\infty} x(m)z^{-m}h(n-m)z^{-(n-m)}$$

$$= \sum_{l=-\infty}^{\infty} \sum_{m=-\infty}^{\infty} x(m)z^{-m}h(l)z^{-l} = \hat{X}(z)\hat{H}(z).$$

The Z transform of y converges absolutely at least at values of z where both $\hat{X}$ and $\hat{H}$ converge absolutely. Thus,

$$RoC(y) \supset RoC(x) \cap RoC(h).$$

This is true because the double sum above can be written as

$$\sum_{n=-\infty}^{\infty} y(n)z^{-n} = \left(\sum_{m=-\infty}^{\infty} x(m)z^{-m} \right) \left(\sum_{l=-\infty}^{\infty} h(l)z^{-l} \right).$$

This obviously converges absolutely if each of the two factors converges absolutely. Note that the region of convergence may actually be larger than $RoC(x) \cap RoC(h)$. This can occur, for example, if the product (13.5) results in zeros of $\hat{X}(z)$ canceling poles of $\hat{H}(z)$ (see exercise 3 at the end of this chapter).

If h is the impulse response of an LTI system, then its Z transform is called the **transfer function** of the system. The result (13.5) tells us that the Z transform of the output is the product of the Z transform of the input and the transfer function. The transfer function, therefore, serves the same role as the frequency response. It converts convolution into simple multiplication.

13.1.4 Conjugation

Suppose x is a complex-valued signal. Let y be defined by

$$\forall\, n \in Integers, \quad y(n) = [x(n)]^*.$$

Then

$$\forall\, z \in RoC(y), \quad \hat{Y}(z) = [\hat{X}(z^*)]^*,$$

where

$$RoC(y) = RoC(x).$$

This follows because

$$\begin{aligned}\forall\, z \in RoC(x), \quad \hat{Y}(z) &= \sum_{n=-\infty}^{\infty} y(n) z^{-n} \\ &= \sum_{n=-\infty}^{\infty} x^*(n) z^{-n} \\ &= \left[\sum_{n=-\infty}^{\infty} x(n)(z^*)^{-n}\right]^* \\ &= [\hat{X}(z^*)]^*.\end{aligned}$$

If x happens to be a real signal, then $y = x$, so $\hat{Y} = \hat{X}$, so

$$\hat{X}(z) = [\hat{X}(z^*)]^*.$$

The key consequence is:

> For the Z transform of a real-valued signal, poles and zeros occur in complex-conjugate pairs. That is, if there is a zero at $z = q$, then there must be a zero at $z = q^*$, and if there is a pole at $z = p$, then there must be a pole at $z = p^*$.

This is because

$$0 = \hat{X}(q) = (\hat{X}(q^*))^*.$$

Similarly, if there is a pole at $z = p$, then there must also be a pole at $z = p^*$.

Example 13.4: Example 13.3 gave the Z transform of a signal of the form $x(n) = a^n u(n)$, where a is allowed to be complex, and the Z tranform of a signal of the form $y(n) = \cos(\omega_0 n)u(n)$, which is real-valued. The pole-zero plots are shown in figure 13.1. In that figure, the complex signal has a pole at $z = a$, and none at $z = a^*$. But the real signal has a pole at $z = e^{i\omega_0}$ and a matching pole at the complex conjugate, $z = e^{-i\omega_0}$. ❐

13.1.5 *Time reversal*

Suppose x has Z transform $\hat{X}$ and y is obtained from x by reversing time, so that

$$\forall\, n \in Integers, \quad y(n) = x(-n).$$

Then

$$\forall\, z \in \{z \in Complex \mid z^{-1} \in Roc(x)\}, \quad \hat{Y}(z) = \hat{X}(z^{-1}).$$

This is evident from the definition of the Z transform, which implies that

$$\hat{Y}(z) = \sum_{m=-\infty}^{\infty} x(-m)z^{-m} = \sum_{n=-\infty}^{\infty} x(n)(z^{-1})^{-n} = \hat{X}(z^{-1}),$$

where $\hat{X}(z^{-1})$ is $\hat{X}$ evaluated at z^{-1}.

Derivatives of Z transforms

PROBING FURTHER

Calculus on complex-valued functions of complex variables can be somewhat intricate. Suppose $\hat{X}$ is a function of a complex variable. The derivative can be defined as a limit,

$$\frac{d}{dz}\hat{X}(z) = \lim_{\epsilon \to 0} \frac{\hat{X}(z+\epsilon) - \hat{X}(z)}{\epsilon},$$

where ϵ is a complex variable that can approach zero from any direction in the complex plane. The derivative exists if the limit does not depend on the direction. If the derivative exists at all points within a distance $\epsilon > 0$ of a point z in the complex plane, then $\hat{X}$ is said to be **analytic** at z. A Z transform is a series of the form

$$\forall\, z \in RoC(x), \quad \hat{X}(z) = \sum_{n=-\infty}^{\infty} x(n)z^{-n}.$$

continued on next page

PROBING FURTHER

This is called a **Laurent series** in the theory of complex variables. It can be shown that a Laurent series is analytic at all points $z \in RoC(x)$, and that the derivative is

$$\forall z \in RoC(x), \quad \frac{d}{dz}\hat{X}(z) = \sum_{m=-\infty}^{\infty} -mx(m)z^{-m-1}.$$

We can use this fact to show that the Z transform of y given by $y(n) = nx(n)$ is

$$\forall z \in Roc(x), \quad \hat{Y}(z) = -z\frac{d}{dz}\hat{X}(z).$$

This is because

$$\begin{aligned}\hat{Y}(z) &= \sum_{n=-\infty}^{\infty} nx(n)z^{-n} \\ &= \sum_{n=-\infty}^{\infty} (-z)\frac{d}{dz}x(n)z^{-n} \\ &= -z\frac{d}{dz}\hat{X}(z).\end{aligned}$$

It is not difficult to show that $Roc(y) = Roc(x)$ (see exercise 5 at the end of this chapter).

This property can be used to find other Z transforms. For example, the Z transform of the unit step, $x = u$, is $\hat{X}(z) = z/(z-1)$, with $RoC(x) = \{z \in Complex \mid |z| > 1\}$. So the Z transform of the unit ramp y, given by $y(n) = nu(n)$, is

$$\hat{Y}(z) = -z\frac{d}{dz}\frac{z}{z-1} = \frac{z}{(z-1)^2},$$

with $RoC(y) = \{z \in Complex \mid |z| > 1\}$. Another method for finding the Z transform of the unit ramp is given in exercise 3 in chapter 12.

13.1.6 *Multiplication by an exponential*

Suppose x has Z transform $\hat{X}$, a is a complex constant, and $y(n) = a^{-n}x(n)$ for all n. Then

$$\forall z \in \{z \in Complex \mid az \in RoC(x)\}, \quad \hat{Y}(z) = \hat{X}(az),$$

where $\hat{X}(az)$ is $\hat{X}$ evaluated at az. This is because

$$\hat{Y}(z) = \sum_{m=-\infty}^{\infty} y(m)z^{-m} = \sum_{m=-\infty}^{\infty} x(m)(az)^{-m} = \hat{X}(az).$$

Note that if $\hat{X}$ has a pole at p (or a zero at q), then $\hat{Y}$ has a pole at p/a (or a zero at q/a).

Example 13.5: Suppose x is given by

$$\forall\, n \in Integers, \quad x(n) = a^n u(n).$$

Then we know from example 13.3 that

$$\forall\, z \in \{z \mid |z| > |a|\}, \quad \hat{X}(z) = \frac{z}{z-a}.$$

This has a pole at $z = a$. Now let $y(n) = a^{-n}x(n) = u(n)$. The Z transform is

$$\hat{Y}(z) = \hat{X}(az) = \frac{az}{az - a} = \frac{z}{z-1},$$

as expected. Moreover, this has a pole at $z = a/a = 1$, as expected, and the region of convergence is indeed given by

$$\{z \in Complex \mid az \in RoC(x)\} = \{z \in Complex \mid |z| > 1\}. \square$$

13.1.7 *Causal signals and the initial value theorem*

Consider a causal discrete-time signal x. Its Z transform is

$$\forall\, z \in \{z \in Complex \mid |z| > r\}, \quad \hat{X}(z) = \sum_{m=0}^{\infty} x(m)z^{-m},$$

for some r (the largest magnitude of a pole). Then

$$\lim_{z\to\infty} \sum_{m=0}^{\infty} x(m)z^{-m} = x(0) + \lim_{z\to\infty} \sum_{m=1}^{\infty} x(m)z^{-m} = x(0).$$

This is because as z goes to ∞, each term $x(m)z^{-m}$ goes to zero. Thus, if x is causal, $x(0) = \lim_{z\to\infty} \hat{X}(z)$. This is called the **initial value theorem**.

Example 13.6: The Z transform of the unit step $x(n) = u(n)$ is $\hat{X}(z) = z/(z-1)$, so, as expected,

$$x(0) = \lim_{z\to\infty} \hat{X}(z) = \lim_{z\to\infty} \frac{z}{z-1} = \lim_{z\to\infty} \frac{1}{1-z^{-1}} = 1,$$

because

$$\lim_{z\to\infty} z^{-1} = 0. \square$$

Suppose a Z transform $\hat{X}$ is the rational polynomial

$$\hat{X}(z) = \frac{a_M z^M + a_{M-1} z^{M-1} + \cdots + a_0}{z^N + b_{N-1} z^{N-1} + \cdots + b_0}.$$

If x is causal, then this rational polynomial must be **proper**. Were this not the case, if $M > N$, then by the initial value theorem, we would have

$$x(0) = \lim_{z \to \infty} \hat{X}(z) = \infty,$$

which is certainly incorrect.

Example 13.7: Consider the Z transform

$$\forall z \in Complex, \quad \hat{X}(z) = z.$$

This is not a proper rational polynomial (the numerator has order 1 and the denominator, which is 1, has order 0). From example 12.12, we know that this corresponds to

$$\forall n \in Integers, \quad x(n) = \delta(n+1).$$

This is not a causal signal. ❐

13.2 Frequency response and pole-zero plots

A pole-zero plot can be used to get a quick estimate of key properties of an LTI system. We have already seen that it reveals whether the system is stable. It also reveals key features of the frequency response, such as whether the system is highpass or lowpass.

Consider a stable discrete-time LTI system with impulse response h, frequency response H, and rational transfer function $\hat{H}$. We know that the frequency response and transfer function are related by

$$\forall \omega \in Reals, \quad H(\omega) = \hat{H}(e^{i\omega}).$$

That is, the frequency response is equal to the Z transform evaluated on the unit circle. The unit circle is in the region of convergence because the system is stable.

Assume that $\hat{H}$ is a rational polynomial, in which case we can express it in terms of the first-order factors of the numerator and denominator polynomials,

$$\hat{H}(z) = \frac{(z - q_1) \cdots (z - q_M)}{(z - p_1) \cdots (z - p_N)},$$

with zeros at $q_1, \ldots, q_M$ and poles at $p_1, \ldots, p_N$. The zeros and poles may be repeated (i.e., they may have multiplicity greater than one). The frequency response is

$$\forall \omega \in Reals, \quad H(\omega) = \frac{(e^{i\omega} - q_1) \cdots (e^{i\omega} - q_M)}{(e^{i\omega} - p_1) \cdots (e^{i\omega} - p_N)}.$$

The magnitude response is

$$\forall \omega \in Reals, \quad |H(\omega)| = \frac{|e^{i\omega} - q_1| \cdots |e^{i\omega} - q_M|}{|e^{i\omega} - p_1| \cdots |e^{i\omega} - p_N|}.$$

Each of these factors has the form

$$|e^{i\omega} - b|$$

where b is the location of either a pole or a zero. The factor $|e^{i\omega} - b|$ is just the distance from $e^{i\omega}$ to b in the complex plane.

Of course, $e^{i\omega}$ is a point on the unit circle. If that point is close to a zero at location q, then the factor $|e^{i\omega} - q|$ is small, so the magnitude response will be small. If that point is close to a pole at p, then the factor $|e^{i\omega} - p|$ is small, but since this factor is in the denominator, the magnitude response will be large. Thus:

> The magnitude response of a stable LTI system may be estimated from the pole-zero plot of its transfer function. Starting at $\omega = 0$, trace counter-clockwise around the unit circle as ω increases. If you pass near a zero, then the magnitude response should dip. If you pass near a pole, then the magnitude response should rise.

Example 13.8: Consider the causal LTI system of example 9.16, which is defined by the difference equation

$$\forall n \in Integers, \quad y(n) = x(n) + 0.9y(n-1).$$

We can find the transfer function by taking Z transforms on both sides, using linearity, to get

$$\hat{Y}(z) = \hat{X}(z) + 0.9z^{-1}\hat{Y}(z).$$

The transfer function is

$$\hat{H}(z) = \frac{\hat{Y}(z)}{\hat{X}(z)} = \frac{1}{1 - 0.9z^{-1}} = \frac{z}{z - 0.9}.$$

This has a pole at $z = 0.9$, which is closest to $z = 1$ on the unit circle, and a zero at $z = 0$, which is equidistant from all points on the unit circle. The zero, therefore, has no effect on the magnitude response. The pole is closest to $z = 1$, which corresponds to $\omega = 0$, so the magnitude response peaks at $\omega = 0$, as shown in figure 9.12. ❒

Example 13.9: Consider a length-4 moving average. Using methods like those in example 9.12, we can show that the transfer function is

$$\forall\, z \in \{z \in Complex \mid z \neq 0\}, \quad \hat{H}(z) = \frac{1}{4} \cdot \frac{1 - z^{-4}}{1 - z^{-1}} = \frac{1}{4} \frac{z^4 - 1}{z^3(z-1)}.$$

The numerator polynomial has roots at the four roots of unity, which are $z = 1$, $z = e^{i\pi/2}$, $z = -1$, and $z = e^{i3\pi/2}$. Thus, we can write this transfer function as

$$\forall\, z \in \{z \in Complex \mid z \neq 0\},$$

$$\hat{H}(z) = \frac{1}{4} \frac{(z-1)(z - e^{i\pi/2})(z+1)(z - e^{i3\pi/2})}{z^3(z-1)}$$

$$= \frac{1}{4} \frac{(z - e^{i\pi/2})(z+1)(z - e^{i3\pi/2})}{z^3}.$$

The $(z - 1)$ factors in the numerator and denominator cancel (fortunately, or we would have a pole at $z = 1$, on the unit circle and would have to conclude that the system was unstable). A pole-zero plot is shown in figure 13.2.

The magnitude response is shown in figure 9.8. Relating that figure to the pole-zero plot, we see that the frequency response peaks at $z = 1$, and as we move around the unit circle, we pass through zero at $\omega = \pi/2$, or $z = e^{i\pi/2}$, and again through zero at $\omega = \pi$. The magnitude response is periodic with period 2π, so the zero at $z = e^{3i\pi/2}$ is also a zero at $z = e^{-i\pi/2}$, corresponding to a frequency of $\omega = -\pi/2$. ❒

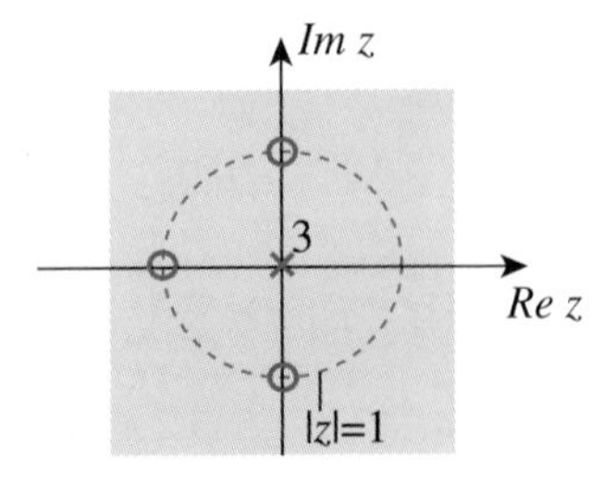

FIGURE 13.2: Pole-zero plot for a length-4 moving average system.

13.3 Properties of the Laplace transform

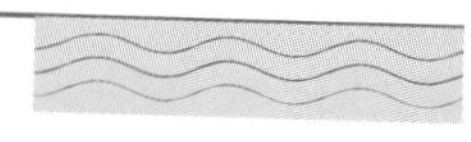

Key Laplace transforms are given in table 13.3. The Laplace transform has useful properties that are similar to those of the Z transform. They are summarized in table 13.4 and elaborated mostly in the exercises at the end of this chapter. In this section, we elaborate on one of the properties that is not shared by the Z transform, integration, and then use the properties to develop some examples.

13.3.1 *Integration*

Let y be defined by

$$\forall t \in Reals, \quad y(t) = \int_{-\infty}^{t} x(\tau)d\tau.$$

The Laplace transform is

$$\forall s \in RoC(y), \quad \hat{Y}(s) = \hat{X}(s)/s,$$

where

$$RoC(y) \supset RoC(x) \cap \{s \mid Re\{s\} > 0\}.$$

TABLE 13.3
Laplace transforms of key signals.*

Continuous-time signal $\forall t \in Reals$	**Laplace transform** $\forall s \in RoC(x)$	$Roc(x)$	**Reference**
$x(t) = \delta(t-\tau)$	$\hat{X}(s) = e^{-s\tau}$	$Complex$	Exercise 12.19
$x(t) = u(t)$	$\hat{X}(s) = 1/s$	$\{s \in Complex \mid Re\{s\} > 0\}$	Example 12.15
$x(t) = e^{-at}u(t)$	$\hat{X}(s) = \frac{1}{s+a}$	$\{s \in Complex \mid Re\{s\} > -Re\{a\}\}$	Example 12.18
$x(t) = -e^{-at}u(-t)$	$\hat{X}(s) = \frac{1}{s+a}$	$\{s \in Complex \mid Re\{s\} < -Re\{a\}\}$	Exercise 5
$x(t) = \cos(\omega_0 t)u(t)$	$\hat{X}(s) = \frac{s}{s^2+\omega_0^2}$	$\{s \mid Re\{s\} > 0\}$	Exercise 7
$x(t) = \sin(\omega_0 t)u(t)$	$\hat{X}(s) = \frac{\omega_0}{s^2+\omega_0^2}$	$\{s \mid Re\{s\} > 0\}$	Example 13.10
$x(t) = \frac{t^{N-1}}{(N-1)!}e^{-at}u(t)$	$\hat{X}(z) = \frac{1}{(s+a)^N}$	$\{s \in Complex \mid Re\{s\} > -Re\{a\}\}$	—
$x(t) = -\frac{t^{N-1}}{(N-1)!}e^{-at}u(-t)$	$\hat{X}(z) = \frac{1}{(s+a)^N}$	$\{s \in Complex \mid Re\{s\} < -Re\{a\}\}$	—

* The signal u is the unit step (12.18); δ is the Dirac delta; a is any complex constant; ω_0 is any real constant; τ is any real constant; and N is a positive integer.

TABLE 13.4
Properties of the Laplace transform.*

Time domain $\forall t \in Reals$	***s* domain** $\forall s \in RoC$	RoC	**Name** (Reference)
$w(t) = ax(t) + by(t)$	$\hat{W}(s) = a\hat{X}(s) + b\hat{Y}(s)$	$RoC(w) \supset RoC(x) \cap RoC(y)$	**Linearity** (exercise 6)
$y(t) = x(t - \tau)$	$\hat{Y}(s) = e^{-s\tau}\hat{X}(s)$	$RoC(y) = RoC(x)$	**Delay** (exercise 7)
$y(t) = (x * h)(r)$	$\hat{Y}(s) = \hat{X}(s)\hat{H}(s)$	$RoC(y) \supset RoC(x) \cap RoC(h)$	**Convolution** (exercise 8)
$y(t) = x^*(t)$	$\hat{Y}(s) = [\hat{X}(s^*)]^*$	$RoC(y) = RoC(x)$	**Conjugation** (exercise 9)
$y(t) = x(ct)$	$\hat{Y}(s) = \hat{X}(s/c)/\lvert c\rvert$	$RoC(y) = \{s \mid s/c \in RoC(x)\}$	**Time scaling** (exercise 10)
$y(t) = tx(t)$	$\hat{Y}(s) = -\frac{d}{ds}\hat{X}(s)$	$RoC(y) = RoC(x)$	**Scaling by *t*** —
$y(t) = e^{at}x(t)$	$\hat{Y}(s) = \hat{X}(s - a)$	$RoC(y) = \{s \mid s - a \in RoC(x)\}$	**Exponential scaling** (exercise 11)
$y(t) = \int_{-\infty}^{t} x(\tau)d\tau$	$\hat{Y}(s) = \hat{X}(s)/s$	$RoC(y) \supset RoC(x) \cap \{s \mid Re\{s\} > 0\}$	**Integration** (section 13.3.1)
$y(t) = \frac{d}{dt}x(t)$	$\hat{Y}(s) = s\hat{X}(s)$	$RoC(y) \supset RoC(x)$	**Differentiation** (page 515)

* a and b are complex constants; c and τ are real constants.

This follows from the convolution property in table 13.4. We recognize that

$$y(t) = (x * u)(t),$$

where u is the unit step. Hence, from the convolution property,

$$\hat{Y}(s) = \hat{X}(s)\hat{U}(s)$$

and

$$RoC(y) \supset RoC(x) \cap RoC(u).$$

$\hat{U}$ and $RoC(u)$ are given in example 12.15, from which the property follows.

13.3.2 *Sinusoidal signals*

Sinusoidal signals have Laplace transforms with poles on the imaginary axis, as illustrated in example 13.10.

Example 13.10: Let the causal sinusoidal signal y be given by

$$\forall\, t \in Reals,\quad y(t) = \sin(\omega_0 t)u(t),$$

where ω_0 is a real number and u is the unit step. Euler's relation implies that

$$y(t) = \frac{1}{2i}[e^{i\omega_0 t}u(t) - e^{-i\omega_0 t}u(t)].$$

Using (12.18) and linearity,

$$\begin{aligned}\hat{Y}(s) &= \frac{1}{2i}\left\{\frac{1}{s + i\omega_0} - \frac{1}{s - i\omega_0}\right\} \\ &= \frac{\omega_0}{s^2 + \omega_0^2}\,.\end{aligned}$$

This has no finite zeros and two poles, one at $s = i\omega_0$ and the other at $s = -i\omega_0$. Both of these poles lie on the imaginary axis, as shown in figure 13.3. The region of convergence is the right half of the complex plane. Note that if this were the impulse response of an LTI system, that system would not be stable. The region of convergence does not include the imaginary axis. ❐

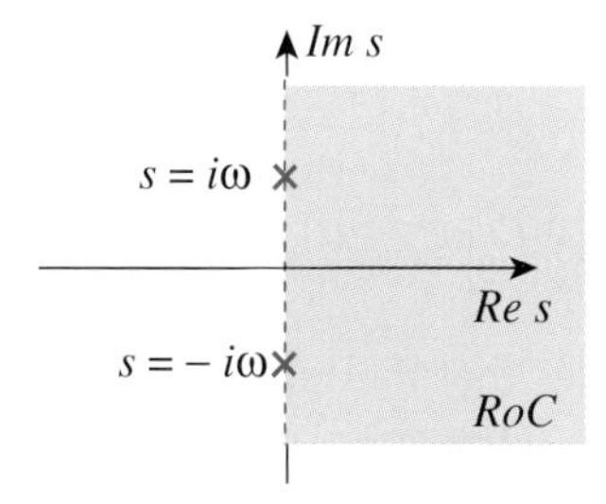

FIGURE 13.3: Pole-zero plot for the sinusoidal signal y.

13.3.3 *Differential equations*

We can use the differentiation property in table 13.4 to solve differential equations with constant coefficients.

Example 13.11: In the tuning fork example of example 2.17, the displacement y of a tine is related to the acceleration of the tine by

$$\ddot{y}(t) = -\omega_0^2 y(t),$$

where ω_0 is a real constant. Let us assume that the tuning fork is initially at rest, and an external input x (representing, say, a hammer strike) adds to the acceleration as follows,

$$\ddot{y}(t) = -\omega_0^2 y(t) + x(t).$$

We can use Laplace transforms to find the impulse response of this LTI system. Taking Laplace transforms on both sides, using linearity and the differentiation property,

$$\forall\, s \in RoC(y) \cap RoC(x), \quad s^2 \hat{Y}(s) = -\omega_0^2 \hat{Y}(s) + \hat{X}(s).$$

From this, we can find the transfer function of the system,

$$\hat{H}(s) = \frac{\hat{Y}(s)}{\hat{X}(s)} = \frac{1}{s^2 + \omega_0^2}.$$

Comparing this with example 13.10, we see that this differs only by a scaling by ω_0 from the Laplace transform in that example. Thus, the pole-zero plot of the tuning fork is shown in figure 13.3, and the impulse response is given by

$$\forall\, t \in Reals, \quad h(t) = \sin(\omega_0 t) u(t)/\omega_0.$$

Interestingly, this implies that the tuning fork is unstable. This impulse response is not absolutely integrable. However, this model of the tuning fork is idealized. It fails to account for loss of energy due to friction. A more accurate model would be stable. ❐

Example 13.11 can be easily generalized to find the transfer function of any LTI system described by a differential equation. In fact, Laplace transforms offer a powerful and effective way to solve differential equations.

In example 13.10, we inverted the Laplace transform by recognizing that it matched the example before that. In section 13.5, we will give a more general method for inverting a Laplace transform.

13.4 Frequency response and pole-zero plots, continuous time

Just as with Z transforms, the pole-zero plot of a Laplace transform can be used to get a quick estimate of key properties of an LTI system. Consider a stable continuous-time LTI system with impulse response h, frequency response H, and rational transfer function $\hat{H}$. We know that the frequency response and transfer function are related by

$$\forall\, \omega \in Reals, \quad H(\omega) = \hat{H}(i\omega).$$

That is, the frequency response is equal to the Laplace transform evaluated on the imaginary axis. The imaginary axis is in the region of convergence because the system is stable.

Assume that $\hat{H}$ is a rational polynomial, in which case we can express it in terms of the first-order factors of the numerator and denominator polynomials,

$$\hat{H}(s) = \frac{(s - q_1) \cdots (s - q_M)}{(s - p_1) \cdots (s - p_N)},$$

with zeros at $q_1, \ldots, q_M$ and poles at $p_1, \ldots, p_N$. The zeros and poles may be repeated (i.e., they may have multiplicity greater than one). The frequency response is

$$\forall\, \omega \in Reals, \quad H(\omega) = \frac{(i\omega - q_1) \cdots (i\omega - q_M)}{(i\omega - p_1) \cdots (i\omega - p_N)}.$$

The magnitude response is

$$\forall\, \omega \in Reals, \quad |H(\omega)| = \frac{|i\omega - q_1| \cdots |i\omega - q_M|}{|i\omega - p_1| \cdots |i\omega - p_N|}.$$

Each of these factors has the form

$$|i\omega - b|$$

where b is the location of either a pole or a zero. The factor $|i\omega - b|$ is just the distance from $i\omega$ to b in the complex plane.

Of course, $i\omega$ is a point on the imaginary axis. If that point is close to a zero at location q, then the factor $|i\omega - q|$ is small, so the magnitude response will be small. If that point is close to a pole at p, then the factor $|i\omega - p|$ is small, but since this factor is in the denominator, the magnitude response will be large. Thus:

> The magnitude response of a stable LTI system may be estimated from the pole-zero plot of its transfer function. Starting at $i\omega = 0$, trace upwards and downwards along the imaginary axis to increase or decrease ω. If you pass near a zero, then the magnitude response should dip. If you pass near a pole, then the magnitude response should rise.

Example 13.12: Consider an LTI system with transfer function given by

$$\forall\, s \in \{s \mid Re\{s\} > Re\{a\}\}, \quad H(s) = \frac{s}{(s-a)(s-a^*)}.$$

Suppose that $a = c + i\omega_0$. Figure 13.4 shows three pole-zero plots for $\omega_0 = 1$ and three values of c, namely $c = -1$, $c = -0.5$, and $c = -0.1$. The magnitude frequency responses can be calculated and plotted using the following MATLAB®* code:

```
omega = [-10:0.05:10];
a1 = -1.0 + i;
H1 = i*omega./((i*omega - a1).*(i*omega-conj(a1)));
a2 = -0.5 + i;
H2 = i*omega./((i*omega - a2).*(i*omega-conj(a2)));
a3 = -0.1 + i;
H3 = i*omega./((i*omega - a3).*(i*omega-conj(a3)));
plot(omega, abs(H1), omega, abs(H2), omega, abs(H3))
```

The plots are shown together at the bottom of figure 13.4. The plot with the higher peaks corresponds to the pole-zero plot with the poles closer to the imaginary axis. ❐

* MATLAB is a registered trademark of The MathWorks, Inc.

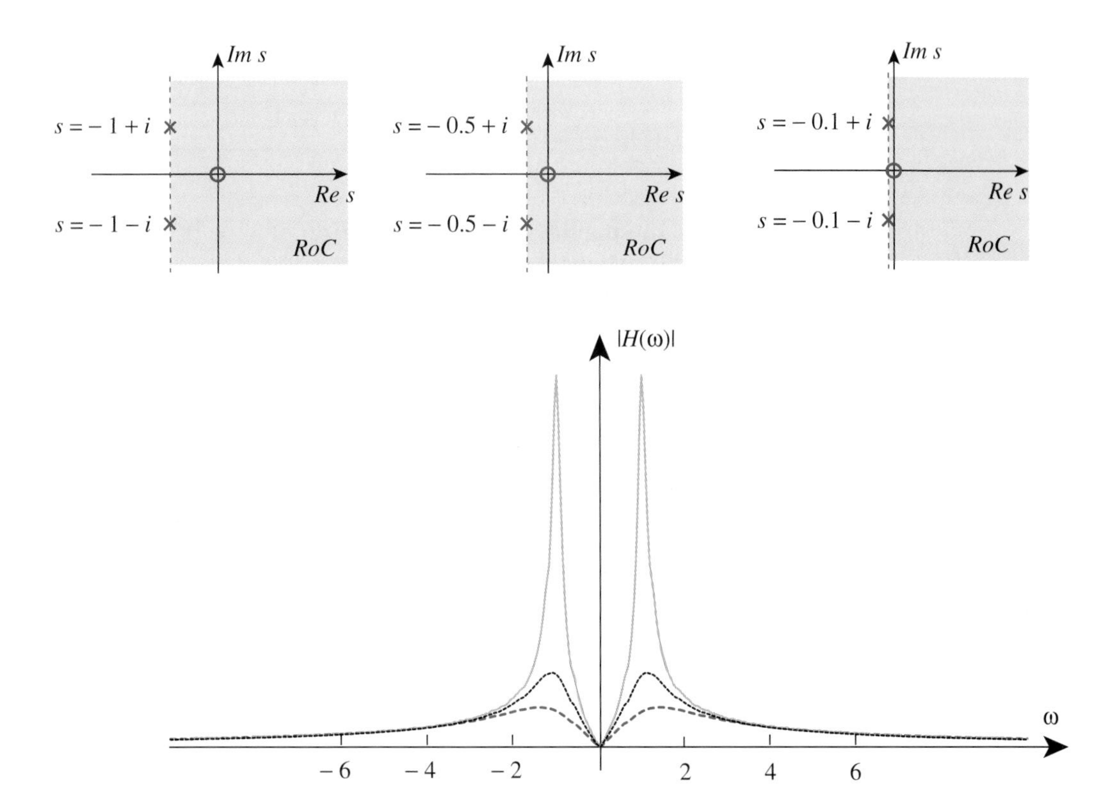

FIGURE 13.4: Pole-zero plots for the three transfer functions, and the three corresponding magnitude frequency responses.

13.5 The inverse transforms

There are two inverse transforms. The inverse Z transform recovers the discrete-time signal x from its Z transform $\hat{X}$. The inverse Laplace transform recovers the continuous-time signal x from its Laplace transform $\hat{X}$. We study the inverse Z transform in detail. The inverse Laplace transform is almost identical. The general approach is to break down a complicated rational polynomial into a sum of simple rational polynomials whose inverse transforms we recognize. We consider only the case where $\hat{X}$ can be expressed as a rational polynomial.

13.5.1 *Inverse Z transform*

The procedure is to construct the **partial fraction expansion** of $\hat{X}$, which breaks it down into a sum of simpler rational polynomials.

Example 13.13: Consider a Z transform given by

$$\forall z \in RoC(x), \quad \hat{X}(z) = \frac{1}{(z-1)(z-2)} = \frac{-1}{z-1} + \frac{1}{z-2}. \qquad (13.6)$$

This sum is called the **partial fraction expansion** of $\hat{X}$, and we will see below how to find it systematically. We can write this as

$$\forall z \in RoC(x), \quad \hat{X}(z) = \hat{X}_1(z) + \hat{X}_2(z),$$

where $\hat{X}_1(z) = -1/(z-1)$ and $\hat{X}_2(z) = 1/(z-2)$ are the two terms.

To determine the inverse Z transforms of the two terms, we need to know their regions of convergence. Recall from the linearity property that $RoC(x)$ includes the intersection of the regions of convergence of the two terms,

$$RoC(x) \supset RoC(x_1) \cap RoC(x_2). \qquad (13.7)$$

Once we know these two regions of convergence, we can use table 13.1 to obtain the inverse Z transform of each term. By the linearity property the sum of these inverses is the inverse Z transform of $\hat{X}$.

$\hat{X}$ given by (13.6) has one pole at $z = 1$ and one pole at $z = 2$. From section 12.2.3 we know that $RoC(x)$ is bordered by these poles, so it has one of three forms:

1. $RoC(x) = \{z \in Complex \mid |z| < 1\}$,
2. $RoC(x) = \{z \in Complex \mid 1 < |z| < 2\}$, or
3. $RoC(x) = \{z \in Complex \mid |z| > 2\}$.

Suppose we have case (1), which implies that x is anticausal. From (13.7), the region of convergence of the term $-1/(z-1)$ must be $\{z \in Complex \mid |z| < 1\}$. The only other possibility is $\{z \in Complex \mid |z| > 1\}$, which would violate (13.7) unless the intersection is empty (which would not be an interesting case). Thus, from table 13.1, the inverse Z transform of the first term must be the anticausal signal $x_1(n) = u(-n)$, for all $n \in Integers$.

For the second term, $1/(z-2)$, its region of convergence could be either $\{z \in Complex \mid |z| < 2\}$ or $\{z \in Complex \mid |z| > 2\}$. Again, the second possibility would violate (13.7), so we must have the first possibility. This results in $x_2(n) = -2^{n-1}u(-n)$, from the last entry in table 13.1. Hence, the inverse Z transform is

$$\forall n \in Integers, \quad x(n) = u(-n) - 2^{n-1}u(-n).$$

If $RoC(x)$ is given by case (2), we rewrite (13.6) slightly as

$$\hat{X}(z) = -z^{-1}\frac{z}{z-1} + \frac{1}{z-2}.$$

The inverse Z transform of the first term is obtained from table 13.1, together with the delay property in table 13.2. The inverse Z transform of the second term is the same as in case (1). We conclude that in case (2) the inverse Z transform is the two-sided signal

$$\forall\, n, \quad x(n) = -u(n-1) - 2^{n-1}u(-n).$$

In case (3), we write (13.6) as

$$\hat{X}(z) = -z^{-1}\frac{z}{z-1} + z^{-1}\frac{z}{z-2},$$

and conclude that the inverse Z transform is the causal signal

$$\forall\, n, \quad x(n) = -u(n-1) + 2^{n-1}u(n-1). \;\square$$

We can generalize this example. Consider any **strictly proper** rational polynomial

$$\hat{X}(z) = \frac{A(z)}{B(z)} = \frac{a_M z^M + \cdots + a_1 z + a_0}{z^N + b_{N-1}z^{N-1} + \cdots + b_1 z + b_0}.$$

The numerator is of order M, the denominator is of order N. *Strictly proper* means that $M < N$. We can factor the denominator,

$$\hat{X}(z) = \frac{a_M z^M + \cdots + a_1 z + a_0}{(z-p_1)^{m_1}(z-p_2)^{m_2}\cdots(z-p_k)^{m_k}}. \tag{13.8}$$

Thus, $\hat{X}$ has k distinct poles at k locations p_i, each with multiplicy m_i. Since the order of the denominator is N, it must be true that

$$N = \sum_{i=1}^{k} m_i\,. \tag{13.9}$$

The partial fraction expansion of (13.8) is

$$\boxed{\hat{X}(z) = \sum_{i=1}^{k}\left[\frac{R_{i1}}{(z-p_i)} + \frac{R_{i2}}{(z-p_i)^2} + \cdots + \frac{R_{im_i}}{(z-p_i)^{m_i}}\right].} \tag{13.10}$$

A pole with multiplicity m_i contributes m_i terms to the partial fraction expansion, so the total number of terms is N, the order of the denominator, from (13.9). The coefficients R_{ij} are complex numbers called the **residues** of the pole p_i.

We assume that the poles $p_1, \ldots, p_N$ are indexed so that $|p_1| \le \cdots \le |p_N|$. The $RoC(x)$ must have one of the following three forms:

1. $RoC = \{z \in Complex \mid |z| < |p_1|\}$,
2. $RoC = \{z \in Complex \mid |p_{j-1}| < |z| < |p_j|\}$, for $j \in \{2, \ldots, k\}$, or
3. $RoC = \{z \in Complex \mid |z| > |p_k|\}$.

As in example 13.13, each term in the partial fraction expansion has two possible regions of convergence, only one of which overlaps with $RoC(x)$. Thus, if we know $RoC(x)$, we can determine the region of convergence of each term of the partial fraction expansion, and then use table 13.1 to find its inverse.

The following example illustrates how to find the residues.

Example 13.14: We will find the inverse Z transform of

$$\hat{X}(z) = \frac{2z+3}{(z-1)(z+2)} = \frac{R_1}{z-1} + \frac{R_2}{z+2}.$$

The residues R_1, R_2 can be found by matching coefficients on both sides. Rewrite the right side as

$$\frac{(R_1 + R_2)z + (2R_1 - R_2)}{(z-1)(z+2)}.$$

Matching the coefficients of the numerator polynomials on both sides we conclude that $R_1 + R_2 = 2$ and $2R_1 - R_2 = 3$. We can solve these simultaneous equations to determine that $R_1 = 5/3$ and $R_2 = 1/3$.

Alternatively, we can find residue R_1 by multiplying both sides by $(z-1)$ and evaluating at $z = 1$. That is,

$$R_1 = \left. \frac{2z+3}{z+2} \right|_{z=1} = \frac{5}{3}.$$

Similarly, we can find R_2 by multiplying both sides by $z + 2$ and evaluating at $z = -2$, to get

$$\left. \frac{2z+3}{z-1} \right|_{z=-2} = R_2,$$

so $R_2 = 1/3$. Thus, the partial fraction expansion is

$$\hat{X}(z) = \frac{5/3}{z-1} + \frac{1/3}{z+2}.$$

$RoC(x)$ is either

1. $\{z \in Complex \mid |z| < 1\}$,
2. $\{z \in Complex \mid 1 < |z| < 2\}$, or
3. $\{z \in Complex \mid |z| > 2\}$.

Knowing which case holds, we can find the inverse Z transform of each term from table 13.1. In the first case, x is the anticausal signal

$$\forall n, \quad x(n) = -\frac{5}{3}u(-n) - \frac{1}{3}(-2)^{n-1}u(-n).$$

In the second case, x is the two-sided signal

$$\forall n, \quad x(n) = \frac{5}{3}u(n-1) - \frac{1}{3}(-2)^{n-1}u(-n).$$

In the third case, x is the causal signal

$$\forall n, \quad x(n) = \frac{5}{3}u(n-1) + \frac{1}{3}(-2)^{n-1}u(n-1). \square$$

If some pole of $\hat{X}$ has multiplicity greater than one, it is slightly more difficult to carry out the partial fraction expansion. The following example illustrates the method.

Example 13.15: Consider the expansion

$$\hat{X}(z) = \frac{2z+3}{(z-1)(z+2)^2} = \frac{R_1}{z-1} + \frac{R_{21}}{z+2} + \frac{R_{22}}{(z+2)^2}.$$

Again we can match coefficients and determine the residues. Alternatively, to obtain R_1 we multiply both sides by $(z-1)$ and evaluate the result at $z = 1$, to get $R_1 = 5/9$. To obtain R_{22} we multiply both sides by $(z+2)^2$ and evaluate the result at $z = -2$, to get $R_{22} = 1/3$.

To obtain R_{21} we multiply both sides by $(z+2)^2$,

$$\frac{2z+3}{z-1} = \frac{(z+2)^2 R_1}{z-1} + R_{21}(z+2) + R_{22},$$

and then differentiate both sides with respect to z. We evaluate the result at $z = -2$, to get

$$\frac{d}{dz}\frac{2z+3}{z-1}\bigg|_{z=-2} = R_{21}.$$

Hence, $R_{21} = -5/9$. So the partial fraction expansion is

$$\frac{2z+3}{(z-1)(z+2)^2} = \frac{5/9}{z-1} - \frac{5/9}{z+2} + \frac{1/3}{(z+2)^2}.$$

Knowing the *RoC*, we can now obtain the inverse Z transform of $\hat{X}$. For instance, in the case where $RoC = \{z \in Complex \mid |z| > 2\}$, the inverse Z transform is the causal signal

$$\forall n, \quad x(n) = \frac{5}{9}u(n-1) - \frac{5}{9}(-2)^{n-1}u(n-1) + \frac{1}{3}(n-1)(-2)^{n-2}u(n-2). \ \square$$

In example 13.15, we used the next to the last entry in table 13.1 to find the inverse transform of the term $(1/3)/(z+2)^2$. That entry in the table is based on a generalization of the geometric series identity, given by (12.9). The first generalization is

$$\sum_{n=0}^{\infty}(n+1)a^n = (\sum_{n=0}^{\infty} a^n)^2 = \frac{1}{(1-a)^2}. \tag{13.11}$$

The series above converges for any complex number a with $|a| < 1$ (see exercise 3 in chapter 12). The broader generalization, for any integer $k \geq 1$, is

$$\frac{1}{k!}\sum_{n=0}^{\infty}(n+k)(n+k-1)\cdots(n+1)a^n = \frac{1}{(1-a)^{k+1}}, \tag{13.12}$$

for any complex number a with $|a| < 1$.

Consider then a Z transform $\hat{X}$ that has a pole at p of multiplicity m and no zeros. Since the pole p cannot belong to *RoC*, the *RoC* is either

$$\{z \in Complex \mid |z| > |p|\} \text{ or } \{z \in Complex \mid |z| < |p|\}.$$

In the first case, we expand $\hat{X}$ in a series involving only the terms $z^{-n}, n \geq 0$,

$$\begin{aligned}
\hat{X}(z) &= \frac{1}{(z-p)^m} \\
&= \frac{z^{-m}}{(1-pz^{-1})^m} \\
&= z^{-m}\frac{1}{(m-1)!}\sum_{n=0}^{\infty}(m+n-1)\cdots(n+1)(pz^{-1})^n, \text{ using (13.12)} \\
&= \frac{1}{(m-1)!}\sum_{k=m}^{\infty}(k-1)\cdots(k-m+1)p^{k-m}z^{-k}, \text{ defining } k=n+m,
\end{aligned}$$

and the series converges for any z with $|z| > |p|$. We can match the coefficients of the powers of z in the Z transform definition,

$$\hat{X}(z) = \sum_{k=-\infty}^{\infty} x(k)z^{-k},$$

from which we can recognize that

$$\begin{aligned}
\forall\, k \in \mathit{Integers}, \quad x(k) &= \begin{cases} 0, & k < m \\ \frac{1}{(m-1)!}(k-1)\cdots(k-m+1)p^{k-m}, & k \geq m \end{cases} \\
&= \frac{1}{(m-1)!}(k-1)\cdots(k-m+1)p^{k-m}u(k-m).
\end{aligned} \tag{13.13}$$

In the second case, $RoC = \{z \in \mathit{Complex} \mid |z| < |p|\}$, we expand $\hat{X}$ in a series involving only the terms $z^{-n}, n \leq 0$,

$$\begin{aligned}
\hat{X}(z) &= \frac{1}{(z-p)^m} \\
&= \frac{1}{(-p)^m}\frac{1}{(1-p^{-1}z)^m} \\
&= \frac{1}{(-p)^m}\frac{1}{(m-1)!}\sum_{n=0}^{\infty}(m+n-1)\cdots(n+1)(p^{-1}z)^n, \text{ using (13.12)} \\
&= \frac{(-1)^m}{(m-1)!}\sum_{k=-\infty}^{0}(m-k-1)\cdots(1-k)p^{k-m}z^{-k}, \text{ defining } k=-n,
\end{aligned}$$

and the series converges for any z with $|z| < |p|$. Again, we match powers of z in the Z transform definition to get

$$\forall\, k \in Integers, \quad x(k) = \begin{cases} \frac{(-1)^m}{(m-1)!}(m-1-k)\cdots(1-k)p^{k-m}, & k \le 0 \\ 0, & k > 0 \end{cases}$$
$$= \frac{(-1)^m}{(m-1)!}(m-1-k)\cdots(1-k)p^{k-m}u(-k). \tag{13.14}$$

Example 13.16: Suppose

$$\hat{X}(z) = \frac{1}{(z-2)^2}$$

with $RoC = \{z \in Complex \mid |z| > 2\}$. Then, by (13.13), $\hat{X}$ is the Z transform of the signal

$$\forall\, k \in Integers, \quad x(k) = \begin{cases} 0, & k < 2 \\ (k-1)2^{k-2}, & k \ge 2. \end{cases}$$

Suppose

$$\hat{Y}(z) = \frac{1}{(z-2)^2}$$

with $RoC = \{z \in Complex \mid |z| < 2\}$. Then, by (13.14), $\hat{Y}$ is the Z transform of the signal

$$\forall\, k \in Integers, \quad y(k) = \begin{cases} (1-k)2^{k-2}, & k \le 0 \\ 0, & k > 0. \end{cases}$$

Since the unit circle $\{z \in Complex \mid |z| = 1\} \subset RoC$, the DTFT of y is defined and given by

$$\forall\, \omega \in Reals, \quad Y(\omega) = \hat{Y}(e^{i\omega}) = \frac{1}{(e^{i\omega}-2)^2}. \quad \square$$

Now that we know how to inverse transform all the terms of the partial fraction expansion, we can generalize the method used in example 13.15 to calculate the inverse Z transform of any $\hat{X}$ of the form

$$\hat{X}(z) = \frac{a_M z^M + \cdots + a_0}{z^N + b_{N-1}z^{N-1} + \cdots + b_0}.$$

Step 1. If $M \geq N$, divide through to obtain

$$\hat{X}(z) = c_{M-N} z^{M-N} + \cdots + c_0 + \hat{W}(z),$$

where $\hat{W}$ is strictly proper.

Step 2. Carry out the partial fraction expansion of $\hat{W}$ and, knowing the *RoC*, obtain the inverse Z transform w. Then from table 13.1,

$$\forall n, \quad x(n) = c_{m+l-N}\delta(n + m + l - N) + \cdots + c_0\delta(n) + w(n).$$

Example 13.17: We follow the procedure for

$$\hat{X}(z) = \frac{z^2 + z + 1 + z^{-1}}{(z+2)^2}.$$

First, to get this into the proper form, as a rational polynomial in z, notice that

$$\hat{X}(z) = z^{-1}\hat{Y}(z),$$

where

$$\hat{Y}(z) = \frac{z^3 + z^2 + z + 1}{(z+2)^2}.$$

Since z^{-1} corresponds to a one-step delay,

$$x(n) = y(n-1),$$

so if we find the inverse Z transform of $\hat{Y}$, then we have found the inverse Z transform of $\hat{X}$.

Working now with $\hat{Y}$, step 1 yields

$$\hat{Y}(z) = z - 3 + \frac{9z + 13}{z^2 + 4z + 4}.$$

Step 2 gives

$$\hat{W}(z) = \frac{9z + 13}{(z+2)^2} = \frac{-5}{(z+2)^2} + \frac{9}{z+2}.$$

Suppose $RoC = \{z \in Complex \mid |z| > 2\}$. Then from table 13.1,

$$\forall\, n, \quad w(n) = -5(n-1)(-2)^{n-2}u(n-2) + 9(-2)^{n-1}u(n-1),$$

$$\forall\, n, \quad y(n) = \delta(n+1) - 3\delta(n) + w(n),$$

$$\forall\, n, \quad x(n) = y(n-1).$$

Hence, for all $n \in Integers$,

$$x(n) = \delta(n) - 3\delta(n-1) - 5(n-2)(-2)^{n-3}u(n-3) + 9(-2)^{n-2}u(n-2). \ \square$$

13.5.2 *Inverse Laplace transform*

The procedure to calculate the inverse Laplace transform is virtually identical. Suppose the Laplace transform $\hat{X}$ is a rational polynomial

$$\hat{X}(s) = \frac{a_M s^M + \cdots + a_0}{s^N + b_{N-1}s^{N-1} + \cdots + b_0}.$$

We follow Steps 1 and 2 above. We divide through in case $M \geq N$ to obtain

$$\hat{X}(s) = c_{M-N}s^{M-N} + \cdots + c_0 + \hat{W}(s),$$

where $\hat{W}$ is strictly proper. We carry out the partial fraction expansion of $\hat{W}$. Knowing $RoC(x)$, we can again infer the region of convergence of each term. We then obtain the inverse Laplace transform term by term using table 13.3,

$$\forall\, t \in Reals, \quad x(t) = c_{m-n}\delta^{(M-N)}(t) + \cdots + c_0\delta(t) + w(t).$$

Here w is the inverse Laplace transform of $\hat{W}$, δ is the Dirac delta function, and $\delta^{(i)}$ is the ith derivative of the Dirac delta function.*

* The derivative of δ is a function only in a formal sense, and we obtain its Laplace transform using the differentiation property in table 13.4.

Example 13.18: We follow the procedure and obtain the partial fraction expansion of

$$\begin{aligned}\hat{X}(s) &= \frac{s^3+s^2+s+1}{s(s+2)^2} \\ &= 1+\frac{-3s^2-3s+1}{s(s+2)^2} \\ &= 1+\frac{1/4}{s}+\frac{-13/4}{s+2}+\frac{5/2}{(s+2)^2}.\end{aligned}$$

$\hat{X}$ has one pole at $s=0$ and a pole at $s=-2$ of multiplicity two. So its *RoC* has one of three forms:

1. $RoC = \{s \in Complex \mid Re\{s\} < -2\}$,
2. $RoC = \{s \in Complex \mid -2 < Re\{s\} < 0\}$, or
3. $RoC = \{s \in Complex \mid Re\{s\} > 0\}$.

We now use table 13.3 to obtain the inverse Laplace transform of each term. In case (1), the continuous-time signal is the anticausal signal

$$\forall t, \quad x(t) = \delta(t) - \frac{1}{4}u(-t) + \frac{13}{4}e^{-2t}u(-t) - \frac{5}{2}te^{-2t}u(-t).$$

In case (2), it is the two-sided signal,

$$\forall t, \quad x(t) = \delta(t) - \frac{1}{4}u(-t) - \frac{13}{4}e^{-2t}u(t) + \frac{5}{2}te^{-2t}u(t).$$

In case (3), it is the causal signal,

$$\forall t, \quad x(t) = \delta(t) + \frac{1}{4}u(t) - \frac{13}{4}e^{-2t}u(t) + \frac{5}{2}te^{-2t}u(t). \square$$

PROBING FURTHER

Inverse transform as an integral

Even if the Z transform is not a rational polynomial, we can recover the signal x from its Z transform, $\hat{X}: RoC(x) \to Complex$, using the DTFT. A nonempty $RoC(x)$ contains the circle of radius r for some $r > 0$. So the series in the equation

$$\hat{X}(re^{i\omega}) = \sum_{m=-\infty}^{\infty} x(m)(re^{i\omega})^{-m} = \sum_{m=-\infty}^{\infty} (x(m)r^{-m})e^{-i\omega m}$$

is absolutely summable. Hence the signal $x_r: \forall\, n, x_r(n) = x(n)r^{-n}$, has DTFT $X_r: \forall\, \omega, X_r(\omega) = \hat{X}(re^{i\omega})$. We can, therefore, obtain x_r as the inverse DTFT of X_r

$$\forall\, n, \quad x_r(n) = r^{-n}x(n) = \frac{1}{2\pi}\int_0^{2\pi} \hat{X}(re^{i\omega})e^{i\omega n}d\omega.$$

Multiplying both sides by r^n, we can recover x as

$$\forall\, n \in Integers, \quad x(n) = \frac{1}{2\pi}\int_0^{2\pi} \hat{X}(re^{i\omega})(re^{i\omega})^n d\omega. \tag{13.15}$$

This formula defines the inverse Z transform as an integral of the real variable ω. It is conventional to write the inverse Z transform differently. Express z as $z = re^{i\omega}$. Then as ω varies from 0 to 2π, z varies as

$$dz = re^{i\omega}id\omega = zid\omega, \text{ or } d\omega = \frac{dz}{iz}.$$

Substituting this in (13.15) gives,

$$\forall\, n, \quad x(n) = \frac{1}{2\pi}\oint \hat{X}(z)z^n\frac{dz}{iz} = \frac{1}{2\pi i}\oint \hat{X}(z)z^{n-1}dz.$$

Here the "circle" in the integral sign, $\oint$, means that the integral in the complex z-plane is along any closed counterclockwise circle contained in $RoC(x)$. (An integral along a closed contour is called a **contour integral**.) In summary,

$$\boxed{\forall\, n \in Integers, \quad x(n) = \frac{1}{2\pi i}\oint \hat{X}(z)z^{n-1}dz,} \tag{13.16}$$

where the integral is along any closed counterclockwise circle inside $RoC(x)$.

We can similarly use the CTFT to recover any continuous-time signal x from its Laplace transform by

$$\boxed{\forall\, t \in Reals, \quad x(t) = \frac{1}{2\pi i}\int_{\sigma - i\infty}^{\sigma + i\infty} \hat{X}(s)e^{st}ds}$$

where the integral is along any vertical line $(\sigma - i\infty, \sigma + i\infty)$ contained in $RoC(x)$.

PROBING FURTHER

Differentiation property of the Laplace transform

We can use the inverse Laplace transform as given in the box on page 514 to demonstrate the differentiation property in table 13.4. Let y be defined by

$$\forall t \in Reals, \quad y(t) = \frac{d}{dt}x(t).$$

We can write x in terms of its Laplace transform as

$$\forall t \in Reals, \quad x(t) = \frac{1}{2\pi i}\int_{\sigma - i\infty}^{\sigma + i\infty} \hat{X}(s)e^{st}ds.$$

Differentiating this with respect to t is easy,

$$\forall t \in Reals, \quad \frac{d}{dt}x(t) = \frac{1}{2\pi i}\int_{\sigma - i\infty}^{\sigma + i\infty} s\hat{X}(s)e^{st}ds.$$

Consequently, $y(t) = dx(t)/dt$ is the inverse transform of $s\hat{X}(s)$, so

$$\forall s \in RoC(y), \quad \hat{Y}(s) = s\hat{X}(s),$$

where $RoC(y) \supset RoC(x)$.

13.6 Steady-state response

Although it has been a fair amount of work, being able to compute an inverse transform for an arbitrary rational polynomial proves useful. Our first use will be to study the **steady-state response** of a causal and stable LTI system that has a sinusoidal input that starts at time zero.

If the input to an LTI system is a complex exponential,

$$\forall t \in Reals, \quad x(t) = e^{i\omega t},$$

then the output y is an exponential of the same frequency but with amplitude and phase given by $H(\omega)$,

$$\forall t \in Reals, \quad y(t) = H(\omega)e^{i\omega t},$$

where H is the frequency response. However, this result requires the exponential input to start at $t = -\infty$. In practice, of course, an input may start at some finite time, say at $t = 0$, but this result does not describe the output if the input is

$$\forall t \in Reals, \quad x(t) = e^{i\omega t}u(t). \tag{13.17}$$

We will see that if the system is stable and causal,* then the output y decomposes into two parts, a **transient output** and a **steady-state output**,

$$y = y_{tr} + y_{ss},$$

where the transient becomes vanishingly small for large t. That is,

$$\lim_{t\to\infty} y_{tr}(t) = 0.$$

Moreover, the steady-state signal is the exponential,

$$\forall t, \quad y_{ss}(t) = H(\omega)e^{i\omega t}u(t). \tag{13.18}$$

Thus for stable systems, we can use the frequency response to describe the eventual output to sinusoidal signals that start at some finite time.

For the special case $\omega = 0$, the input (13.17) is the unit step, $x = u$, and $y_{ss} = H(0)u$. So for stable systems, the steady-state response to a unit step input is a step of size $H(0)$. ($H(0)$ is called the **dc gain**.) This case is important in the design of feedback control, considered in chapter 14.

Let h be the impulse response and $\hat{H}$ be the Laplace transform of a stable and causal LTI system. We assume for simplicity that $\hat{H}$ is a strictly proper rational polynomial all of whose poles have multiplicity one,

$$\hat{H}(s) = \frac{A(s)}{(s-p_1)\cdots(s-p_N)}.$$

Because the system is causal, $RoC(h)$ has the form

$$RoC(h) = \{s \mid Re\{s\} > q\},$$

where q is the largest real part of any pole. Because the system is stable, $q < 0$, so that the region of convergence includes the imaginary axis.

*This result can be generalized to noncausal systems, but causal systems will be sufficient for our purposes.

From table 13.3 the Laplace transform $\hat{X}$ of the input signal (13.17) is

$$\hat{X}(s) = \frac{1}{s - i\omega},$$

with $RoC(x) = \{s \in Complex \mid Re\{s\} > 0\}$. The Laplace transform of the output $y = h * x$ is

$$\hat{Y} = \hat{H}\hat{X},$$

with

$$RoC(y) \supset RoC(h) \cap RoC(x) = \{s \in Complex \mid Re\{s\} > 0\}.$$

The partial fraction expansion of $\hat{Y}$ is

$$\hat{Y}(s) = \hat{H}(s)\hat{X}(s) = \frac{A(s)}{(s - p_1) \cdots (s - p_N)} \cdot \frac{1}{s - i\omega} \tag{13.19}$$

$$= \frac{R_1}{s - p_1} + \cdots + \frac{R_N}{s - p_N} + \frac{R_\omega}{s - i\omega}. \tag{13.20}$$

Because everything is causal, each term must be causal, so from table 13.3 we obtain

$$\forall t, \quad y(t) = \sum_{k=1}^{N} R_k e^{p_k t} u(t) + R_\omega e^{i\omega t} u(t).$$

We decompose $y = y_{tr} + y_{ss}$, with

$$\forall t, \quad y_{tr}(t) = \sum_{k=1}^{N} R_k e^{p_k t} u(t),$$

$$\forall t, \quad y_{ss}(t) = R_\omega e^{i\omega t} u(t).$$

Since $Re\{p_k\} < 0$ for $k = 1, \ldots, N$,

$$\lim_{t \to \infty} y_{tr}(t) = 0.$$

Thus, the steady-state response y_{ss} is eventually all that is left.

Finally, the residue R_ω is obtained by multiplying both sides of (13.19) by $s - i\omega$ and evaluating at $s = i\omega$ to get $R_\omega = \hat{H}(i\omega) = H(\omega)$, so that

$$\forall t, \quad y_{ss}(t) = H(\omega) e^{i\omega t} u(t).$$

This analysis reveals several interesting features of the total response y. First, from (13.20) we see the poles $p_1, \ldots, p_N$ of the transfer function contribute to the transient response y_{tr}, and the pole of the input $\hat{X}$ at $i\omega$ contributes to the steady-state response. Second, we can determine how quickly the transient response dies down. The transient response is

$$\forall\, t, \quad y_{tr}(t) = R_1 e^{p_1 t} u(t) + \cdots + R_N e^{p_N t} u(t).$$

The magnitude of the terms is

$$|R_1| e^{Re\{p_1\}t}, \ldots, |R_N| e^{Re\{p_N\}t}.$$

Each term decreases exponentially with t, since the real parts of the poles are negative. The slowest decrease is due to the pole with the least negative part. Thus, the pole of the stable, causal transfer function with the least negative part determines how fast the transient response goes to zero. Indeed for large t, we can approximate the response y as

$$y(t) \approx R_i e^{p_i t} + H(\omega) e^{i\omega t},$$

where p_i is the pole with the largest (least negative) real part.

There is a similar result for discrete-time systems, and it is obtained in the same way. Suppose an exponential input

$$\forall\, n \in Integers, \quad x(n) = e^{i\omega n} u(n),$$

is applied to a stable and causal system with impulse response h, transfer function $\hat{H}$, and frequency response H. Then the output $y = h * x$ can again be decomposed as

$$\forall\, n, \quad y(n) = y_{tr}(n) + y_{ss}(n),$$

where the transient $y_{tr}(n) \to 0$ as $n \to \infty$, and the steady-state response is

$$\forall\, n, \quad y_{ss}(n) = \hat{H}(e^{i\omega}) e^{i\omega n} u(n) = H(\omega) e^{i\omega n} u(n).$$

For large n, the transient response decays exponentially as p_i^n, namely,

$$y(n) \approx R_i p_i^n + y_{ss}(n),$$

where p_i is the pole with the largest magnitude (which must be less than one, since the system is stable).

13.7 Linear difference and differential equations*

Many natural and artificial systems can be modeled as linear differential equations or difference equations. We have seen that when such systems are initially at rest, they are LTI systems. Hence, we can use their transfer functions (Z transforms or Laplace transforms) to analyze the response of these systems to external inputs.

However, physical systems often are not initially at rest. Dealing with nonzero initial conditions introduces some complexity in the analysis. Mathematicians call such systems with nonzero initial conditions **initial value problems**. We can adapt our methods to deal with initial conditions. The rest of this chapter is devoted to these methods.

Example 13.19: In example 13.8 we considered the LTI system described by the difference equation

$$y(n) - 0.9y(n-1) = x(n).$$

The transfer function of this system is $\hat{H}(z) = z/(z - 0.9)$. If the system is initially at rest, we can calculate its response y from its Z transform $\hat{Y} = \hat{H}\hat{X}$. For instance, if the input is the unit step, $\hat{X}(z) = z/(z-1)$,

$$\hat{Y}(z) = \frac{z^2}{(z-0.9)(z-1)} = \frac{-9z}{z-0.9} + \frac{10z}{z-1},$$

and so $y(n) = -9(0.9)^n + 10, n \geq 0$.

We cannot use the transfer function, however, to determine the response if the initial condition at time $n = 0$ is $y(-1) = \bar{y}(-1)$, a number different than zero, and the input is $x(n) = 0, n \geq 0$. The response to this initial condition is

$$y(n) = \bar{y}(-1)(0.9)^{n+1}, \ n \geq -1.$$

We can check that this expression is correct by verifying that it satisfies both the initial condition and the difference equation.

If the initial condition is $y(-1) = \bar{y}(-1)$, and the input is a unit step, the response turns out to be the sum of the response due to the input (with zero initial condition) and the response due to the initial condition (with zero input),

$$y(n) = [-9(0.9)^n + 10] + [\bar{y}(-1)(0.9)^{n+1}], \ n \geq 0.$$

* This section may be skipped on a first reading.

For small values of n the response depends heavily on the initial condition, especially if $\bar{y}(-1)$ is large. Because this system is stable, the effect of the initial condition becomes vanishingly small for large n. ❐

An **LTI difference equation** has the form

$$\begin{aligned} y(n) + a_1 y(n-1) + \cdots + a_m y(n-m) \\ = b_0 x(n) + \cdots + b_k x(n-k),\ n \geq 0. \end{aligned} \tag{13.21}$$

We interpret this equation as describing a causal discrete-time LTI system, in which $x(n)$ is the input and $y(n)$ is the output at time n. The a_i and b_j are constant coefficients that specify the system.

We have used difference equations before. In section 8.2.1 we used this form and the discrete time Fourier transform to find the frequency response of this system. In section 9.5 we showed how to realize such systems as IIR filters. In example 13.19 we used the transfer function to find the response. But in all these cases, we had to assume that the system was initially at rest. We now develop a method to find the response for arbitrary initial conditions.

We assume the input signal x starts at some finite time, which we take to be zero, $x(n) = 0, n < 0$. We wish to calculate $y(n), n \geq 0$. From (13.21) we can see that we need to be given m initial conditions, which are m numbers,

$$y(-1) = \bar{y}(-1), \ldots, y(-m) = \bar{y}(-m).$$

Given the input signal and these initial conditions, there is a straightforward procedure to calculate the output response $y(n), n \geq 0$: Rewrite (13.21) as

$$y(n) = -a_1 y(n-1) - \cdots - a_m y(n-m) + b_0 x(n) + \cdots + b_k x(n-k), \tag{13.22}$$

and recursively use (13.22) to obtain $y(0), y(1), y(2), \ldots$. For $n = 0$, (13.22) yields

$$\begin{aligned} y(0) &= -a_1 y(-1) - \cdots - a_m y(-m) + b_0 x(0) + \cdots + b_k x(-k) \\ &= -a_1 \bar{y}(-1) - \cdots - a_m \bar{y}(-m) + b_0 x(0). \end{aligned}$$

All the terms on the right are known from the initial conditions and the input $x(0)$, so we can calculate $y(0)$. Next, taking $n = 1$ in (13.22),

$$y(1) = -a_1 y(0) + \cdots + a_m y(1-m) + b_0 x(1) + \cdots + b_k x(1-k).$$

All the terms on the right are known either from the given data or from precalculated values—$y(0)$ in this case. Proceeding in this way we can calculate the remaining values of the output sequence $y(2), y(3), \ldots$, one at a time.

We now use the Z transform to calculate the *entire* output sequence. Multiplying both sides of (13.21) by $u(n)$, the unit step, gives us a relation that holds among signals whose domain is *Integers*:

$$y(n)u(n) + a_1 y(n-1)u(n) + \cdots + a_m y(n-m)u(n)$$
$$= b_0 x(n)u(n) + \cdots + b_k x(n-k)u(n), \; n \in \textit{Integers}.$$

We can now take the Z transforms of both sides. We multiply both sides by z^{-n} and sum,

$$\sum_{n=0}^{\infty} y(n)z^{-n} + a_1 \sum_{n=0}^{\infty} y(n-1)z^{-n} + \cdots + a_m \sum_{n=0}^{\infty} y(n-m)z^{-n}$$
$$= b_0 \sum_{n=0}^{\infty} x(n)z^{-n} + \cdots + b_k \sum_{n=0}^{\infty} x(n-k)z^{-n}. \quad (13.23)$$

Define

$$\hat{X}(z) = \sum_{n=0}^{\infty} x(n)z^{-n}, \quad \hat{Y}(z) = \sum_{n=0}^{\infty} y(n)z^{-n}.$$

Each sum in (13.23) can be expressed in terms of $\hat{Y}$ or $\hat{X}$. In evaluting the Z transforms of the signals $y(n-1)u(n), y(n-2)u(n), \ldots$, we need to include the initial conditions:

$$\sum_{n=0}^{\infty} y(n-1)z^{-n} = \bar{y}(-1)z^0 + z^{-1} \sum_{n=1}^{\infty} y(n-1)z^{-(n-1)} = \bar{y}(-1)z^0 + z^{-1}\hat{Y}(z),$$

$$\sum_{n=0}^{\infty} y(n-2)z^{-n} = \bar{y}(-2)z^0 + \bar{y}(-1)z^{-1} + z^{-2} \sum_{n=2}^{\infty} y(n-2)z^{-(n-2)}$$
$$= \bar{y}(-2)z^0 + \bar{y}(-1)z^{-1} + z^{-2}\hat{Y}(z),$$
$$\cdots$$

$$\sum_{n=0}^{\infty} y(n-m)z^{-n} = \bar{y}(-m)z^0 + \cdots + \bar{y}(-1)z^{-(m-1)} + z^{-m} \sum_{n=m}^{\infty} y(n-m)z^{-(n-m)}$$
$$= \bar{y}(-m)z^0 + \cdots + \bar{y}(-1)z^{-(m-1)} + z^{-m}\hat{Y}(z).$$

Because $x(n) = 0, n < 0$, by assumption, the sums on the right in (13.23) are simpler:

$$\sum_{n=0}^{\infty} x(n-1)z^{-n} = x(-1)z^0 + z^{-1}\hat{X}(z) = z^{-1}\hat{X}(z)$$

$$\sum_{n=0}^{\infty} x(n-2)z^{-n} = x(-2)z^0 + x(-1)z^{-1} + z^{-2}\hat{X}(z) = z^{-2}\hat{X}(z)$$

$$\cdots$$

$$\sum_{n=0}^{\infty} x(n-k)z^{-n} = x(-k)z^0 + \cdots + x(-1)z^{-(k-1)} + z^{-k}\hat{X}(z) = z^{-k}\hat{X}(z).$$

(If there were nonzero initial conditions for $x(-1), \ldots, x(-k)$, we could include them in the Z transforms of $x(n-1)u(n), \ldots, x(n-k)u(n)$.) Substituting these relations in (13.23) yields

$$\hat{Y}(z) + a_1[z^{-1}\hat{Y}(z) + \bar{y}(-1)z^0] + \cdots + a_m[z^{-m}\hat{Y}(z) + \bar{y}(-m)z^0$$
$$+ \cdots + \bar{y}(-1)z^{-(m-1)}] = b_0\hat{X}(z) + b_1z^{-1}\hat{X}(z) + \cdots + b_k\hat{X}z^{-k}, \quad (13.24)$$

from which, by rearranging terms, we obtain

$$[1 + a_1z^{-1} + \cdots + a_mz^{-m}]\hat{Y}(z) = [b_0 + b_1z^{-1} + \cdots + b_kz^{-k}]\hat{X}(z) + \hat{C}(z),$$

where $\hat{C}(z)$ is an expression involving only the initial conditions $\bar{y}(-1), \ldots, \bar{y}(-m)$. Therefore,

$$\hat{Y}(z) = \frac{b_0 + b_1z^{-1} + \cdots b_kz^{-k}}{1 + a_1z^{-1} + \cdots + a_mz^{-m}}\hat{X}(z) + \frac{\hat{C}(z)}{1 + a_1z^{-1} + \cdots + a_mz^{-m}}.$$

We rewrite this relation as

$$\hat{Y}(z) = \hat{H}(z)\hat{X}(z) + \frac{\hat{C}(z)}{1 + a_1z^{-1} + \cdots + a_mz^{-m}}. \quad (13.25)$$

where

$$\boxed{\hat{H}(z) = \frac{b_0 + b_1z^{-1} + \cdots + b_kz^{-k}}{1 + a_1z^{-1} + \cdots + a_mz^{-m}}}. \quad (13.26)$$

Observe that if the initial conditions are all zero, $\hat{C}(z) = 0$, and we only have the first term on the right in (13.25); and if the input is zero—that is, $x(n) = 0$ for all n, then $\hat{X}(z) = 0$, and we only have the second term.

By definition, $\hat{Y}(z)$ is the Z transform of the causal signal $y(n)u(n), n \in$ *Integers*. So its $RoC = \{z \in Complex \mid |z| > |p|\}$, in which p is the pole of the right side of (13.25) with the largest magnitude. The inverse Z transform of $\hat{Y}$ can be expressed as

$$\forall n \geq 0, \quad y(n) = y_{zs}(n) + y_{zi}(n), \tag{13.27}$$

where $y_{zs}(n)$, the inverse Z transform of $\hat{H}\hat{X}$, is the **zero-state response**, and $y_{zi}(n)$, the inverse Z transform of $\hat{C}(z)/[1 + a_1 z^{-1} + \cdots + a_m z^{-m}]$, is the **zero-input response**. The zero-state response, also called the **forced response**, is the output when all initial conditions are zero. The zero-input response, also called the **natural response**, is the output when the input is zero.

Thus, the (total) response is the sum of the zero-state and zero-input response. We first encountered this property of linearity in chapter 5.

By definition, the **transfer function** is the Z transform of the zero-state impulse response. Taking $\hat{C} = 0$ and $\hat{X} = 1$ in (13.25) shows that the transfer function is $\hat{H}(z)$. From (13.26) we see that $\hat{H}$ can be written down by inspection of the difference equation (13.21). If the system is stable—all poles of $\hat{H}$ are inside the unit circle—the frequency response is

$$\forall \omega, \quad H(\omega) = \hat{H}(e^{i\omega}) = \frac{b_0 + b_1 e^{-i\omega} + \cdots + b_k e^{-ik\omega}}{1 + a_1 e^{-i\omega} + \cdots + a_m e^{-im\omega}}.$$

We saw this relation in (8.21).

Example 13.20: Consider the difference equation

$$y(n) - \frac{5}{6}y(n-1) + \frac{1}{6}y(n-2) = x(n), \; n \geq 0.$$

Taking Z transforms, as in (13.24), yields

$$\hat{Y}(z) - \frac{5}{6}[z^{-1}\hat{Y}(z) + \bar{y}(-1)] + \frac{1}{6}[z^{-2}\hat{Y}(z) + \bar{y}(-2) + \bar{y}(-1)z^{-1}] = \hat{X}(z).$$

Therefore,

$$\hat{Y}(z) = \frac{1}{1 - \frac{5}{6}z^{-1} + \frac{1}{6}z^{-2}}\hat{X}(z) + \frac{\frac{5}{6}\bar{y}(-1) + \frac{1}{6}\bar{y}(-2) + \frac{1}{6}\bar{y}(-1)z^{-1}}{1 - \frac{5}{6}z^{-1} + \frac{1}{6}z^{-2}}$$

$$= \frac{z^2}{z^2 - \frac{5}{6}z + \frac{1}{6}}\hat{X}(z) + \frac{[\frac{5}{6}\bar{y}(-1) + \frac{1}{6}\bar{y}(-2)]z^2 + \frac{1}{6}\bar{y}(-1)z}{z^2 - \frac{5}{6}z + \frac{1}{6}},$$

from which we can obtain $\hat{Y}$ for a specified $\hat{X}$ and initial conditions $\bar{y}(-1)$, $\bar{y}(-2)$.

The transfer function is

$$\hat{H}(z) = \frac{z^2}{z^2 - \frac{5}{6}z + \frac{1}{6}} = \frac{z^2}{(z - \frac{1}{3})(z - \frac{1}{2})},$$

which has poles at $z = 1/3$ and $z = 1/2$ (and two zeros at $z = 0$). The system is stable. The zero-state impulse response h is the inverse Z transform of $\hat{H}(z)$, which we obtain using partial fraction expansion,

$$\hat{H}(z) = z\left[\frac{-2}{z - \frac{1}{3}} + \frac{3}{z - \frac{1}{2}}\right]$$

so that

$$\forall\, n \in Integers, \quad h(n) = -2\left(\frac{1}{3}\right)^n u(n) + 3\left(\frac{1}{2}\right)^n u(n).$$

We can recognize that the impulse response consists of two terms, each contributed by one pole of the transfer function.

Suppose the initial conditions are $\bar{y}(-1) = 1, \bar{y}(-2) = 1$ and the input x is the unit step, so that $\hat{X}(z) = z/(z - 1)$. Then the zero-input response, y_{zi}, has Z transform

$$\hat{Y}_{zi}(z) = \frac{[\frac{5}{6}\bar{y}(-1) + \frac{1}{6}\bar{y}(-2)]z^2 + \frac{1}{6}\bar{y}(-1)z}{(z - \frac{1}{3})(z - \frac{1}{2})}$$

$$= \frac{z^2 + \frac{1}{6}z}{(z - \frac{1}{3})(z - \frac{1}{2})} = z\left[\frac{-3}{z - \frac{1}{3}} + \frac{4}{z - \frac{1}{2}}\right],$$

so

$$\forall\, n, \quad y_{zi}(n) = -3\left(\frac{1}{3}\right)^n u(n) + 4\left(\frac{1}{2}\right)^n u(n).$$

The zero-state response, y_{zs}, has Z transform

$$\hat{Y}_{zs}(z) = \hat{H}(z)\hat{X}(z) = \frac{z^3}{(z-\frac{1}{3})(z-\frac{1}{2})(z-1)}$$

$$= z\left[\frac{1}{z-\frac{1}{3}} + \frac{-3}{z-\frac{1}{2}} + \frac{3}{z-1}\right],$$

so

$$\forall n, \quad y_{zs}(n) = \left(\frac{1}{3}\right)^n u(n) - 3\left(\frac{1}{2}\right)^n u(n) + 3u(n).$$

The (total) response

$$\forall n \in Integers, \quad y(n) = y_{zs}(n) + y_{zi}(n) = 3u(n) + [-2(1/3)^n + (1/2)^n]u(n),$$

can also be expressed as the sum of the steady-state and the transient response with $y_{ss}(n) = 3u(n)$ and $y_{tr}(n) = -2(1/3)^n u(n) + (1/2)^n u(n)$. Note that the decomposition of the response into the sum of the zero-state and zero-input responses is different from its decomposition into the steady-state and transient responses. ❐

13.7.1 *LTI differential equations*

The analogous development for continuous time concerns systems described by an **LTI differential equation** of the form

$$\frac{d^m y}{dt^m}(t) + a_{m-1}\frac{d^{m-1}y}{dt^{m-1}}(t) + \cdots + a_1\frac{dy}{dt}(t) + a_0 y(t)$$

$$= b_k\frac{d^k x}{dt^k}(t) + \cdots + b_1\frac{dx}{dt}(t) + b_0 x(t), \; t \geq 0. \quad (13.28)$$

We interpret this equation as describing a causal continuous-time LTI system in which $x(t)$ is the input and $y(t)$ is the output at time t. The constant coefficients a_i and b_j specify the system.

In section 8.2.1 we used this form to find the frequency response. In example 13.11, we used the Laplace transform to find the transfer function of a tuning fork. In both cases we assumed that the system was initially at rest. We now develop a method to find the response to arbitrary initial conditions. We begin with a simple circuit example.

Example 13.21: A series connection of a resistor R, a capacitor C, and a voltage source x, is described by the differential equation

$$\frac{dy}{dt}(t) + \frac{1}{RC}y(t) = x(t),$$

in which y is the voltage across the capacitor. The differential equation is obtained from Kirchhoff's voltage law. The transfer function of this system is $\hat{H}(s) = 1/(s + 1/RC)$. So if the system is initially at rest, we can calculate the response y from its Laplace transform $\hat{Y} = \hat{H}\hat{X}$. For instance, if the input is a unit step, $\hat{X}(s) = 1/s$,

$$\hat{Y}(s) = \frac{1}{(s + 1/RC)s} = \frac{-RC}{s + 1/RC} + \frac{RC}{s},$$

therefore, $y(t) = -RCe^{-t/RC} + RC,\ t \geq 0$.

We cannot use this transfer function, however, to determine the response if the initial capacitor voltage is $y(0) = \bar{y}(0)$, some number different from zero, and $x(t) = 0, t \geq 0$. In this case, the response is

$$y(t) = \bar{y}(0)e^{-t/RC},\ t \geq 0.$$

We can check that this expression is correct by verifying that it satisfies the given initial condition and the differential equation.

If the initial condition is $y(0) = \bar{y}(0)$, and the input is a unit step, the response turns out to be the sum of the response due to the input (with zero initial condition) and the response due to the initial condition (with zero input),

$$y(t) = [-RCe^{-t/RC} + RC] + [\bar{y}(0)e^{-t/RC}],\ t \geq 0. \ \square$$

For the general case (13.28) we assume that the input x starts at some finite time that we take to be zero, so $x(t) = 0, t < 0$. We wish to calculate $y(t), t \geq 0$. From the theory of differential equations we know that we need to be given m initial conditions,

$$y(0) = \bar{y}(0),\ \frac{dy}{dt}(0) = \bar{y}^{(1)}(0), \ldots, \frac{d^{m-1}y}{dt^{m-1}}(0) = \bar{y}^{(m-1)}(0),$$

in order to calculate $y(t), t \geq 0$.

Because time is continuous, there is no recursive procedure for calculating the output from the given data as we did in (13.22). Instead, we calculate the output signal using the Laplace transform. We define the Laplace transforms of the signals $y(t)u(t), y^{(1)}(t)u(t), \ldots, y^{(m)}(t)u(t), x(t)u(t)$:

$$\hat{Y}(s) = \int_{-\infty}^{\infty} y(t)u(t)e^{-st}dt = \int_{0}^{\infty} y(t)e^{-st}dt$$

$$\hat{Y}^{(i)}(s) = \int_{-\infty}^{\infty} y^{(i)}(t)u(t) = \int_{0}^{\infty} y^{(i)}(t)e^{-st}dt, i = 1, \ldots, m$$

$$\hat{X}(s) = \int_{-\infty}^{\infty} x(t)u(t)e^{-st}dt = \int_{0}^{\infty} x(t)e^{-st}dt.$$

Here we use the notation $y^{(i)}(t) = \frac{d^i y}{dt^i}y(t), t \geq 0$. We now derive the relations between these Laplace transforms.

The derivative $y^{(1)}(t) = \frac{dy}{dt}(t)$ and y are related by

$$y(t)u(t) = y(0)u(t) + \int_{0}^{t} y^{(1)}(\tau)u(\tau)d\tau = \bar{y}(0)u(t) + \int_{0}^{t} y^{(1)}(\tau)u(\tau)d\tau,\ t \in Reals.$$

Using integration by parts,

$$\begin{aligned}
\hat{Y}(s) &= \int_{0}^{\infty} y(t)e^{-st}dt = \int_{0}^{\infty} \bar{y}(0)e^{-st}dt + \int_{0}^{\infty} \left(\int_{0}^{t} y^{(1)}(\tau)d\tau \right) e^{-st}dt \\
&= \frac{1}{s}\bar{y}(0) - \frac{1}{s}\int_{0}^{t} y^{(1)}(\tau)d\tau e^{-st}\Big|_{t=0}^{\infty} + \frac{1}{s}\int_{0}^{\infty} y^{(1)}(t)e^{-st}dt \\
&= \frac{1}{s}[\hat{Y}^{(1)}(s) + \bar{y}(0)].
\end{aligned}$$

Therefore,

$$\boxed{\hat{Y}^{(1)}(s) = s\hat{Y}(s) - \bar{y}(0).} \tag{13.29}$$

Repeating this procedure, we get the Laplace transforms of the higher-order derivatives,

$$\begin{aligned}
\hat{Y}^{(2)}(s) &= s\hat{Y}^{(1)}(s) - \bar{y}^{(1)}(0) \\
&= s^2\hat{Y}(s) - s\bar{y}(0) - \bar{y}^{(1)}(0) \\
&\cdots \\
\hat{Y}^{(m)}(s) &= s^m\hat{Y}(s) - s^{m-1}\bar{y}(0) - s^{m-2}\bar{y}^{(1)}(0) - \cdots - \bar{y}^{(m-1)}(0).
\end{aligned}$$

On the other hand, because $x^{(i)}(t) = \frac{d^i x}{dt^i}(t)$ for all $t \in Reals$, using the differentiation property in table 13.4, we obtain

$$\hat{X}^{(1)}(s) = s\hat{X}(s)$$
$$\dots$$
$$\hat{X}^{(k)}(s) = s^k\hat{X}(s).$$

By substituting from the relations just derived, we obtain the Laplace transforms of all the terms in (13.28),

$$\begin{aligned}[s^m\hat{Y}(s) - s^{m-1}\bar{y}(0) - \cdots - \bar{y}^{m-1}(0)] + a_{m-1}[s^{m-1}\hat{Y}(s) - s^{m-2}\bar{y}(0) \\ - \cdots - \bar{y}^{m-2}(0)] + \cdots + a_1[s\hat{Y}(s) - \bar{y}(0)] + a_0\hat{Y}(s) \\ = b_k s^k\hat{X}(s) + \cdots + b_1 s\hat{X}(s) + b_0\hat{X}(s).\end{aligned} \tag{13.30}$$

Rearranging terms yields

$$[s^m + a_{m-1}s^{m-1} + \cdots + a_1 s + a_0]\hat{Y}(s) = [b_k s^k + \cdots + b_1 s + b_0]\hat{X}(s) + \hat{C}(s),$$

in which $\hat{C}$ is an expression involving only the intial conditions $\bar{y}(0), \dots, \bar{y}^{(m-1)}(0)$. Therefore,

$$\begin{aligned}\hat{Y}(s) = \frac{b_k s^k + b_{k-1}s^{k-1} + \cdots + b_1 s + b_0}{s^m + a_{m-1}s^{m-1} + \cdots + a_1 s + a_0}\hat{X}(s) \\ + \frac{\hat{C}(s)}{s^m + a_{m-1}s^{m-1} + \cdots + a_1 s + a_0},\end{aligned} \tag{13.31}$$

which we also write as

$$\hat{Y}(s) = \hat{H}(s)\hat{X}(s) + \frac{\hat{C}(s)}{s^m + a_{m-1}s^{m-1} + \cdots + a_1 s + a_0}, \tag{13.32}$$

in which

$$\boxed{\hat{H}(s) = \frac{b_k s^k + \cdots + b_1 s + b_0}{s^m + \cdots + a_1 s + a_0}.} \tag{13.33}$$

If the initial conditions are all zero, $\hat{C}(s) = 0$, and we only have the first term on the right in (13.32); if the input is zero—that is, $x(t) = 0$ for all t, then $\hat{X}(s) = 0$ and we only get the second term in (13.32).

By definition, $\hat{Y}(s)$ is the Laplace transform of the causal signal $y(t)u(t), t \in Reals$. So its $RoC = \{s \in Complex \mid Re\{s\} > Re\{p\}\}$, where p is a pole of the right side of (13.32) with the largest real part.

Taking the inverse Laplace transform of $\hat{Y}$, we can decompose the output signal y as

$$\forall\, t, \quad y(t) = y_{zs}(t) + y_{zi}(t),$$

where y_{zs}, the inverse Laplace transform of $\hat{H}\hat{X}$, is the **zero-state** or **forced response** and y_{zi}, the inverse Laplace transform of $\hat{C}(s)/[s^m + \cdots + a_0]$, is the **zero-input** or **natural response**. The (total) response is the sum of the zero-state and zero-input response, which is a general property of linear systems.

By definition, the **transfer function** is the Laplace transform of the zero-state impulse response. Taking $\hat{C} = 0$ and $\hat{X} = 1$ (the Laplace transform of the unit impulse) in (13.32) shows that the transfer function is $\hat{H}(s)$ which, as we see from (13.33), can be written down by inspection of the differential equation (13.28). If the system is stable—all poles of $\hat{H}(s)$ have real parts strictly less than zero—the frequency response is

$$\forall\, \omega, \quad H(\omega) = \hat{H}(i\omega) = \frac{b_k(i\omega)^k + \cdots + b_1 i\omega + b_0}{(i\omega)^m + \cdots + a_1 i\omega + a_0}.$$

We saw this relation in (8.23).

Example 13.22: We find the response $y(t), t \geq 0$, for the differential equation

$$\frac{d^2y}{dt^2} + 3\frac{dy}{dt} + 2y = 3x(t) + \frac{dx}{dt},$$

when the input is a unit step $x(t) = u(t)$ and the initial conditions are $y(0) = 1, y^{(1)}(0) = 2$. Taking Laplace transforms of both sides as in (13.30),

$$[s^2\hat{Y}(s) - s\bar{y}(0) - \bar{y}^{(1)}(0)] + 3[s\hat{Y}(s) - \bar{y}(0)] + 2\hat{Y}(s) = 3\hat{X}(s) + s\hat{X}(s).$$

Therefore,

$$\hat{Y}(s) = \frac{s+3}{s^2+3s+2}\hat{X}(s) + \frac{s\bar{y}(0) + \bar{y}^{(1)}(0) + 3\bar{y}(0)}{s^2+3s+2}.$$

Substituting $\hat{X}(s) = 1/s$, $\bar{y}(0) = 1$, $\bar{y}^{(1)} = 2$, yields

$$\hat{Y}(s) = \frac{s+3}{s(s^2+3s+2)} + \frac{s+5}{s^2+3s+2}$$

$$= [\frac{3/2}{s} - \frac{2}{s+1} + \frac{1/2}{s+2}] + [\frac{4}{s+1} - \frac{3}{s+2}].$$

Taking inverse Laplace transforms gives

$$\forall\, t, \quad y(t) = y_{zs}(t) + y_{zi}(t)$$

$$= [\frac{3}{2}u(t) - 2e^{-t}u(t) + \frac{1}{2}e^{-2t}u(t)] + [4e^{-t}u(t) - 3e^{-2t}u(t)]$$

$$= \frac{3}{2}u(t) + [2e^{-t} - \frac{5}{2}e^{-2t}]u(t)$$

$$= y_{ss}(t) + y_{tr}(t).$$

As in the case of difference equations, the decomposition of the response into zero-state and zero-input responses is different from the decomposition into transient and steady-state responses. (Indeed, the steady-state response does not exist if the system is unstable, whereas the former decomposition always exists.) ❒

13.8 State-space models*

In section 5.3 we introduced single-input, single-output (SISO) multidimensional state-space models of discrete-time and continuous-time LTI systems. For LTI systems, state-space models provide an alternative description to difference or differential equation representations. The advantage of state-space models is that by using matrix notation we have a very compact representation of the response, independent of the order of the system. We develop a method that combines this matrix notation with transform techniques to calculate the response.

The discrete-time SISO state-space model is

$$\forall\, n \geq 0, \quad s(n+1) = As(n) + bx(n), \tag{13.34}$$

$$y(n) = c^T s(n) + dx(n), \tag{13.35}$$

in which $s(n) \in Reals^N$ is the state, $x(n) \in Reals$ is the input, and $y(n) \in Reals$ is the output at time n. In this $[A, b, c, d]$ representation, A is an $N \times N$ (square)

* This section is mathematically more advanced in that it uses the operation of matrix inverse. It may be skipped on a first reading.

matrix, b, c are N-dimensional column vectors, and d is a scalar. If the initial state is $s(0)$, and the input sequence is $x(0), x(1), \ldots,$ by recursively using (13.34) and (13.35) we obtain the state and output responses:

$$s(n) = A^n s(0) + \sum_{m=0}^{n-1} A^{n-1-m} b x(m), \tag{13.36}$$

$$y(n) = c^T A^n s(0) + \{\sum_{m=0}^{n-1} c^T A^{n-1-m} b x(m) + d x(n)\}, \tag{13.37}$$

for all $n \geq 0$. Notice that these closed-form formulas for the response are independent of the order N. Difference equation representations do not have such a closed-form formula.

Example 13.23: Consider the system described by the difference equation

$$y(n) - 2y(n-1) - 3y(n-2) = x(n).$$

As in section 5.3, we can construct a state-space model for this system by noting that the state at time n should remember the previous two inputs $y(n-1), y(n-2)$. Define the two-dimensional state vector $s(n) = [s_1(n)\ s_2(n)]^T$ by $s_1(n) = y(n-1), s_2(n) = ay(n-2)$, in which $a \neq 0$ is a constant. Exercise 23 at the end of this chapter asks you to show that the $[A, b, c, d]$ representation for this choice of state is given by

$$A = \begin{bmatrix} 2 & 3/a \\ a & 0 \end{bmatrix}, \quad b = \begin{bmatrix} 1 \\ 0 \end{bmatrix}, \quad c^T = [2 \quad 3/a], \text{ and } d = 1.$$

Different choices of a give a different state-space model. However, they all have the same input–output relation because they all have the same transfer function. ❐

We obtain the Z transforms of the response sequences (13.36), (13.37) by computing the Z transform of the entire $N \times N$ matrix sequence, $A^n u(n), n \in Integers$. This Z transform is

$$\boxed{\sum_{n=0}^{\infty} z^{-n} A^n = [I - z^{-1}A]^{-1} = z[zI - A]^{-1}.} \tag{13.38}$$

Here z is a complex number and I is the $N \times N$ identity matrix. The series on the left is an infinite sum of $N \times N$ matrices that converges to the $N \times N$ matrix on the right, for $z \in RoC$. The RoC is determined later.

Assuming the series converges, it is easy to check the equality (13.38): Just multiply both sides by $[I - z^{-1}A]$ and verify that

$$[I - z^{-1}A]\sum_{n=0}^{\infty} z^{-n}A^{n} = \sum_{n=0}^{\infty} z^{-n}A^{-n} - \sum_{n=0}^{\infty} z^{-(n+1)}A^{n+1} = z^{0}A^{0} = I.$$

Next, denote by F the matrix inverse,

$$F(z) = [I - z^{-1}A]^{-1} = z[zI - A]^{-1}, \tag{13.39}$$

and the coefficients of A^n and $F(z)$ by

$$A^{n} = [a_{ij}(n) \mid 1 \le i,j \le N], \quad F(z) = [f_{ij}(z) \mid 1 \le i,j \le N].$$

Then $f_{ij}(z) = \sum_{n=0}^{\infty} z^{-n}a_{ij}(n)$ is the Z transform of the sequence $a_{ij}(n), n \ge 0$, $1 \le i,j \le N$. So we can obtain $A^n, n \ge 0$, by taking the inverse Z transform of each of the N^2 coefficients of $F(z)$.

Example 13.24: Let

$$A = \begin{bmatrix} 2 & 1 \\ 3 & 4 \end{bmatrix},$$

then

$$[zI - A]^{-1} = \begin{bmatrix} z-2 & -1 \\ -3 & z-4 \end{bmatrix}^{-1} = \frac{1}{\det[zI - A]}\begin{bmatrix} z-4 & 1 \\ 3 & z-2 \end{bmatrix},$$

in which $\det[zI - A]$ denotes the determinant of $[zI - A]$,

$$\det[zI - A] = (z-2)(z-4) - 3 = z^2 - 6z + 5 = (z-1)(z-5).$$

Hence,

$$F(z) = z[zI - A]^{-1} = \frac{z}{(z-1)(z-5)}\begin{bmatrix} z-4 & 1 \\ 3 & z-2 \end{bmatrix}$$

$$= \begin{bmatrix} \frac{z(z-4)}{(z-1)(z-5)} & \frac{z}{(z-1)(z-5)} \\ \frac{3z}{(z-1)(z-5)} & \frac{z(z-2)}{(z-1)(z-5)} \end{bmatrix}.$$

The partial fraction expansion of the coefficients of F is

$$F(z) = \begin{bmatrix} \frac{(3/4)z}{z-1} + \frac{(1/4)z}{z-5} & \frac{(-1/4)z}{z-1} + \frac{(1/4)z}{z-5} \\ \frac{(-3/4)z}{z-1} + \frac{(3/4)z}{z-5} & \frac{(1/4)z}{z-1} + \frac{(3/4)z}{z-5} \end{bmatrix}.$$

Using table 13.1 we find the inverse Z transform of every coefficient of $F(z)$: For all $n \in Integers$,

$$A^n u(n) = \begin{bmatrix} \frac{3}{4}u(n) + \frac{1}{4}5^n u(n) & -\frac{1}{4}u(n) + \frac{1}{4}5^n u(n) \\ -\frac{3}{4}u(n) + \frac{3}{4}5^n u(n) & \frac{1}{4}u(n) + \frac{3}{4}5^n u(n) \end{bmatrix}.$$

This is more revealingly expressed as

$$A^n = \begin{bmatrix} 3/4 & -1/4 \\ -3/4 & 1/4 \end{bmatrix} + 5^n \begin{bmatrix} 1/4 & 1/4 \\ 3/4 & 3/4 \end{bmatrix}, \quad n \geq 0,$$

because it shows that the variation in n of A^n is determined by the two poles, at $z = 1$ and $z = 5$, in the coefficients of $F(z)$. Moreover, these two poles are the zeros of

$$\det[zI - A] = (z - 1)(z - 5).$$

This determinant is called the **characterstic polynomial** of the matrix A and its zeros are called the **eigenvalues** of A. The domain of convergence is $RoC = \{z \in Complex \mid |z| > 5\}$. ❒

We return to the general case in (13.39). Denote the matrix inverse of $[zI - A]$ as

$$[zI - A]^{-1} = \frac{1}{\det[zI - A]} G(z),$$

in which $G(z)$ is the $N \times N$ matrix of cofactors of $[zI - A]$. It follows that each coefficient $f_{ij}(z)$ of $F(z) = z[zI - A]^{-1}$ is a rational polynomial whose denominator is the characteristic polynomial of A, $\det[zI - A]$. Therefore, if there are no pole-zero cancellations, all coefficients of $F(z)$ have the same poles, which are the zeros of $\det[zI - A]$. These zeros are called the eigenvalues of A. The polynomial $\det[zI - A]$ is of order N, and so A has N eigenvalues.

Because $A^n u(n), n \in Integers$, is a causal sequence, the region of convergence is $RoC = \{z \in Complex \mid |z| > |p|\}$, in which p is the pole of F (or eigenvalue of A) with the largest magnitude. For the system (13.34), (13.35) to be stable, the poles of F must have magnitudes strictly smaller than 1.

Suppose A has N distinct eigenvalues $p_1, \ldots, p_N$,

$$\det[zI - A] = (z - p_1) \cdots (z - p_N).$$

Then the partial fraction expansion of $F(z)$ has the form

$$F(z) = \frac{z}{z - p_1} R_1 + \cdots + \frac{z}{z - p_N} R_N,$$

in which R_i is the matrix of residues of the coefficients of F at the pole p_i. R_i is a constant matrix, possibly with complex coefficients if p_i is complex. Recalling that $\frac{z}{z-p_i}$ is the inverse Z transform of $p_i^n u(n)$, we can take the inverse Z transform of $F(z)$ to conclude that

$$\boxed{A^n = p_1^n R_1 + \cdots p_N^n R_N, \quad n \geq 0.} \tag{13.40}$$

Thus, A^n is a linear combination of $p_1^n, \ldots, p_N^n$.

We can decompose the response (13.37) into the zero-input and zero-state responses, expressing the latter as a convolution sum,

$$y(n) = c^T A^n s(0) + \sum_{m=0}^{n} h(n-m)x(m), \; n \geq 0,$$

where the (zero-state) impulse response is

$$h(n) = \begin{cases} 0, & n < 0 \\ d, & n = 0 \\ c^T A^{n-1} b, & n \geq 1. \end{cases}$$

Let $\hat{X}, \hat{Y}, \hat{H}, \hat{Y}_{zi}$ be the Z transforms:

$$\hat{X}(z) = \sum_{n=0}^{\infty} x(n) z^{-n}, \quad \hat{Y}(z) = \sum_{n=0}^{\infty} y(n) z^{-n},$$

$$\hat{H}(z) = \sum_{n=0}^{\infty} h(n) z^{-n}, \quad \hat{Y}_{zi}(z) = \sum_{n=0}^{\infty} c^T z^{-n} A^n s(0).$$

Then

$$\hat{Y} = \hat{H}\hat{X} + \hat{Y}_{zi}.$$

Because $\sum_{n=0}^{\infty} z^{-n} A^n = z[zI - A]^{-1}$, we obtain

$$\boxed{\hat{H}(z) = c^T [zI - A]^{-1} b + d,}$$

and

$$\boxed{\hat{Y}_{zi}(z) = z c^T [zI - A]^{-1} s(0).}$$

Example 13.25: Suppose A is as in example 13.24, $b^T = [1\ 1], c^T = [2\ 0]$, $d = 3$, and $(s(0))^T = [0\ 4]$. Then the transfer function is

$$\hat{H}(z) = [2\ 0]\begin{bmatrix} \frac{(z-4)}{(z-1)(z-5)} & \frac{1}{(z-1)(z-5)} \\ \frac{3}{(z-1)(z-5)} & \frac{(z-2)}{(z-1)(z-5)} \end{bmatrix}\begin{bmatrix} 1 \\ 1 \end{bmatrix} + 3 = \frac{2(z-4)+2}{(z-1)(z-5)} + 3,$$

and the Z transform of the zero-input response is

$$\hat{Y}_{zi}(z) = [2\ 0]\begin{bmatrix} \frac{z(z-4)}{(z-1)(z-5)} & \frac{z}{(z-1)(z-5)} \\ \frac{3z}{(z-1)(z-5)} & \frac{z(z-2)}{(z-1)(z-5)} \end{bmatrix}\begin{bmatrix} 0 \\ 4 \end{bmatrix} = \frac{8z}{(z-1)(z-5)}.$$

The transfer function is

$$\hat{H}(z) = \frac{2(z-4)+2}{(z-1)(z-5)} + 3 = \frac{3z^2 - 16z + 9}{z^2 - 6z + 5} = \frac{3 - 16z^{-1} + 9z^{-2}}{1 - 6z^{-1} + 5z^{-2}}.$$

From (13.26) we recognize that $\hat{H}$ is also the transfer function of the difference equation

$$y(n) - 6y(n-1) + 5y(n-2) = 3x(n) - 16x(n-1) + 9x(n-2).$$

This difference equation describes the same input–output relation as the state-space model of this example. ❒

13.8.1 *Continuous-time state-space models*

The continuous-time SISO state-space model introduced in section 5.4 has the $[A, b, c, d]$ representation

$$\dot{v}(t) = Av(t) + bx(t), \tag{13.41}$$

$$y(t) = c^T v(t) + dx(t), \tag{13.42}$$

in which $v(t) \in Reals^N$ is the state, $x(t) \in Reals$ is the input, and $y(t) \in Reals$ is the output at time $t \in Reals$. A is an $N \times N$ matrix, and b, c are N-dimensional column vectors, and d is a scalar. (We use v instead of s to denote the state, because s is reserved for the Laplace transform variable.)

Given the initial state $v(0)$ and the input signal $x(t), t \geq 0$, we show that the state response and the output response obey the formulas

$$v(t) = e^{tA}v(0) + \int_0^t e^{(t-\tau)A}bx(\tau)d\tau, \tag{13.43}$$

$$y(t) = c^T e^{tA}v(0) + [\int_0^t c^T e^{(t-\tau)A}bx(\tau)d\tau] + dx(t). \tag{13.44}$$

In these formulas, e^{tA} or $\exp(tA)$ is the name of the $N \times N$ matrix

$$e^{tA} = \sum_{k=0}^{\infty} \frac{(tA)^k}{k!} = I + tA + \frac{(tA)^2}{2!} + \frac{(tA)^3}{3!} + \cdots, \tag{13.45}$$

where $(tA)^k$ is the matrix tA multiplied by itself k times, and $(tA)^0 = I$, the $N \times N$ identity matrix. Definition (13.45) of the **matrix exponential** is the natural generalization of the exponential of a real or complex number. (The series in (13.45) is absolutely summable because of the factor $k!$ in the denominator.)

Unlike in the discrete-time case, there is no recursive procedure to compute the responses (13.43), (13.44). This is because time is continuous, and the difficulty has to do with the integrals in these formulas. For numerical calculation, we resort to a finite sum approximation of the integrals, as indicated in section 5.4. The Laplace transform provides an alternative approach that is exact.

The key to showing (13.43) is the fact that $e^{tA}, t \geq 0$, is the solution to the differential equation

$$\frac{d}{dt}e^{tA} = Ae^{tA}, \quad t \geq 0, \tag{13.46}$$

with initial condition $e^{0A} = I$. Note that (13.44) follows immediately from (13.43) and (13.42).

To verify (13.46) we substitute for e^{tA} from (13.45) and differentiate the sum term by term,

$$\frac{d}{dt}e^{tA} = \sum_{k=0}^{\infty} \frac{d}{dt}\frac{(tA)^k}{k!} = \sum_{k=1}^{\infty} \frac{kA}{k!}(tA)^{k-1} = A\sum_{k=1}^{\infty} \frac{(tA)^{k-1}}{(k-1)!} = Ae^{tA}.$$

We can now check that (13.43) is indeed the solution of (13.41) by taking derivatives of both sides and using (13.46):

$$\begin{aligned}\dot{v}(t) &= Ae^{tA}v(0) + e^{0A}bx(t) + \int_0^t Ae^{(t-\tau)A}bx(\tau)d\tau \\ &= A[e^{tA}v(0) + \int_0^t Ae^{(t-\tau)A}bx(\tau)d\tau] + bx(t) \\ &= Av(t) + bx(t).\end{aligned}$$

We turn to the main difficulty in calculating the terms on the right in the responses (13.43), (13.44)—the calculation of the $N \times N$ matrix $e^{tA}, t \geq 0$. We determine the Laplace transform of $e^{tA}u(t), t \in Reals$, denoting it by

$$G(s) = \int_0^\infty e^{tA}e^{-st}dt.$$

This means that $g_{ij}(s)$ is the Laplace transform of $a_{ij}(t), t \geq 0$, denoting by $a_{ij}(t)$ and $g_{ij}(s)$ the coefficients of the $N \times N$ matrices e^{tA} and $G(s)$, respectively. The region of convergence of G, RoC is determined later.

Using the derivative formula (13.29) in (13.46) we see that

$$sG(s) - I = AG(s),$$

which gives $G(s) = [sI - A]^{-1}$, so that the Laplace transform of $e^{tA}u(t)$ is

$$\boxed{G(s) = \int_0^\infty e^{tA}e^{-st}dt = [sI - A]^{-1}.} \qquad (13.47)$$

Example 13.26: Let

$$A = \begin{bmatrix} 1 & 2 \\ -2 & 1 \end{bmatrix},$$

then

$$[sI - A]^{-1} = \begin{bmatrix} s-1 & -2 \\ 2 & s-1 \end{bmatrix}^{-1} = \frac{1}{\det[sI - A]}\begin{bmatrix} s-1 & 2 \\ -2 & s-1 \end{bmatrix}.$$

The determinant is

$$\det[sI - A] = (s-1)^2 + 4 = (s-1+2i)(s-1-2i),$$

so that

$$[sI - A]^{-1} = \begin{bmatrix} \frac{s-1}{(s-1+2i)(s-1-2i)} & \frac{2}{(s-1+2i)(s-1-2i)} \\ \frac{-2}{(s-1+2i)(s-1-2i)} & \frac{s-1}{(s-1+2i)(s-1-2i)} \end{bmatrix}$$
$$= \begin{bmatrix} \frac{1/2}{s-1+2i} + \frac{1/2}{s-1-2i} & \frac{i/2}{s-1+2i} + \frac{-i/2}{s-1-2i} \\ \frac{-i/2}{s-1+2i} + \frac{i/2}{s-1-2i} & \frac{1/2}{s-1+2i} + \frac{1/2}{s-1-2i} \end{bmatrix}.$$

The region of convergence $RoC = \{s \in Complex \mid Re\{s\} > 1\}$. We find the inverse Laplace transform using table 13.3 and express it in two ways: For all $t \geq 0$,

$$e^{tA} = e^{(1-2i)t} \begin{bmatrix} 1/2 & i/2 \\ -i/2 & 1/2 \end{bmatrix} + e^{(1+2i)t} \begin{bmatrix} 1/2 & -i/2 \\ i/2 & 1/2 \end{bmatrix}$$
$$= e^{t} \begin{bmatrix} \cos 2t & \sin 2t \\ -\sin 2t & \cos 2t \end{bmatrix}.$$

The first expression shows e^{tA} as a linear combination of the exponentials $e^{(1-2i)t}$ and $e^{(1+2i)t}$, in which the exponents $1-2i$ and $1+2i$ are the two eigenvalues of A—that is, the zeros of its characteristic polynomial, $\det[sI - A]$. The second expression shows that e^{tA} is sinusoidal with frequency 2 radians/sec equal to the imaginary part of the eigenvalues whose amplitude grows exponentially corresponding to the real part of the eigenvalues. ❐

Returning to the general case (13.47), denote the matrix inverse of $[sI - A]$ as

$$G(s) = [sI - A]^{-1} = \frac{1}{\det[sI - A]} K(s),$$

in which $K(s)$ is the $N \times N$ matrix of cofactors of $[sI - A]$. Each coefficient $g_{ij}(s)$ of $G(s)$ is a rational polynomial of A whose denominator is the characterstic polynomial of A, $\det[sI - A]$. Therefore, if there are no pole-zero cancellations, all coefficients of $G(s)$ have the same poles—the eigenvalues of A. Because $e^{tA}u(t), t \in Reals$ is a causal signal, the region of convergence of its Laplace transform $G(s)$ is $\{s \in Complex \mid Re\{s\} > Re\{p\}\}$, in which p is the pole of G with the largest real part.

Because $\det[sI - A]$ is a polynomial of order N, G has N poles. For the system (13.41), (13.42) to be stable, the poles of $G(s)$ must have strictly negative real parts. The system of example (13.26) is unstable because the real part of the eigenvalues is $+1$.

Suppose the characteristic polynomial has N distinct zeros $p_1, \ldots, p_N$,

$$\det[sI - A] = (s - p_1) \cdots (s - p_N).$$

Then the partial fraction expansion of $G(s)$ has the form

$$G(s) = [sI - A]^{-1} = \frac{1}{s - p_1} R_1 + \cdots + \frac{1}{s - p_N} R_N,$$

in which R_i is the matrix of residues at the pole p_i of the coefficients of $G(s)$. R_i is a constant matrix, possibly with complex coefficients, if p_i is complex. Because the inverse Laplace transform of $\frac{1}{s-p_i}$ is $e^{p_i t}u(t)$, the inverse Laplace transform of $[sI - A]^{-1}$ is

$$\boxed{e^{tA}u(t) = [e^{p_1 t}R_1 + \cdots + e^{p_N t}R_N]u(t).} \tag{13.48}$$

Thus, the matrix e^{tA} as a function of t is a linear combination of $e^{p_1 t}, \ldots, e^{p_N t}$, where the p_i are the eigenvalues of A—that is, the zeros of $\det[sI - A]$.

We decompose the response (13.44) into the sum of the zero-input and zero-state responses, expressing the latter as a convolution integral,

$$y(t) = c^T e^{tA} v(0) + \int_0^t h(t - \tau)x(\tau)d\tau, \; t \geq 0,$$

in which the (zero-state) impulse response is: For all $t \in Reals$,

$$h(t) = c^T e^{tA} bu(t) + d\delta(t)$$

in which δ is the Dirac delta function.

Let $\hat{X}, \hat{Y}, \hat{H}, \hat{Y}_{zi}$ be the Laplace transforms

$$\hat{X}(s) = \int_0^\infty x(t)e^{-st}dt, \; \hat{Y}(s) = \int_0^\infty y(t)e^{-st}dt,$$

$$\hat{H}(s) = \int_{-\infty}^\infty h(t)e^{-st}dt, \; \hat{Y}_{zi}(s) = \int_0^\infty c^T e^{tA} v(0)e^{-st}dt.$$

Then

$$\hat{Y} = \hat{H}\hat{X} + \hat{Y}_{zi},$$

in which

$$\boxed{\hat{H}(s) = c^T[sI - A]^{-1}b + d,}$$

and

$$\boxed{\hat{Y}_{zi}(s) = c^T[sI - A]^{-1}v(0).}$$

Example 13.27 continues example 13.26.

Example 13.27: Suppose A is as in example 13.26, $b^T = [1\ 1]^T, c^T = [2\ 0]^T$, $d = 3$, and $v(0)^T = [0\ 4]^T$. Then the transfer function is

$$\hat{H}(s) = [2\ 0]\begin{bmatrix} \frac{s-1}{(s-1)^2-4} & \frac{-2}{(s-1)^2-4} \\ \frac{2}{(s-1)^2-4} & \frac{s-1}{(s-1)^2-4} \end{bmatrix}\begin{bmatrix} 1 \\ 1 \end{bmatrix} + 3 = \frac{2s-6}{(s-1)^2-4} + 3,$$

and the Laplace transform of the zero-input response is

$$\hat{Y}_{zi}(s) = [2\ 0]\begin{bmatrix} \frac{s-1}{(s-1)^2-4} & \frac{-2}{(s-1)^2-4} \\ \frac{2}{(s-1)^2-4} & \frac{s-1}{(s-1)^2-4} \end{bmatrix}\begin{bmatrix} 0 \\ 4 \end{bmatrix} = \frac{-16}{(s-1)^2-4}.$$

The transfer function is

$$\hat{H}(s) = \frac{2s-6}{(s-1)^{-}4} + 3 = \frac{3s^2-4s-15}{s^2-2s-3}.$$

From (13.33) we know that $\hat{H}$ is also the transfer function of the differential equation

$$\frac{d^2y}{dt^2}(t) - 2\frac{dy}{dt}(t) - 3y(t) = 3\frac{d^2x}{dt^2}(t) - 4\frac{dx}{dt}(t) - 15x(t).$$

Thus, this differential equation describes the same system as the state-space model of example 13.24. ❐

This example illustrates a general way of obtaining a differential equation description of a continuous-time state-space model by means of its transfer function. It is also easy to obtain a state-space model with a specified proper transfer function,

$$\hat{H}(s) = \frac{b_{N-1}s^{N-1} + \cdots + b_1 s + b_0}{s^N + \cdots + a_1 s + a_0} + b_N.$$

The first term in $\hat{H}$ is strictly proper. Some of the coefficients b_i, a_j may be zero. Then the N-dimensional $[A, b, c, d]$ representation

$$A = \begin{bmatrix} 0 & 1 & 0 & \cdots & 0 \\ 0 & 0 & 1 & \cdots & 0 \\ \cdots & \cdots & & \cdots & \cdots \\ 0 & 0 & 0 & \cdots & 1 \\ -a_0 & -a_1 & -a_2 & \cdots & -a_{N-1} \end{bmatrix}, \quad b = \begin{bmatrix} 0 \\ 0 \\ \cdots \\ 0 \\ 1 \end{bmatrix}, \tag{13.49}$$

$$c^T = [b_0 b_1 \cdots b_{N-1}], \quad d = b_N.$$

has the same transfer function as $\hat{H}$, that is,

$$c^T[sI - A]^{-1}b + d = \hat{H}(s). \tag{13.50}$$

Exercise 30 at the end of this chapter asks you to verify (13.50).

Simply by interchanging the variables s and z we see that the proper rational polynomial

$$\hat{H}(z) = \frac{b_{N-1}z^{N-1} + \cdots + b_1 z + b_0}{z^N + \cdots + a_1 z + a_0} + b_N = d + c^T[zI - A]^{-1}b$$

is the transfer function of the discrete-time $[A, b, c, d]$ representation.

Thus, we can use any of three equivalent representations of LTI systems:

1. difference or differential equations, used to describe many physical systems,
2. transfer functions used for frequency-domain analysis, and in feedback design considered in the next chapter,
3. state-space models, used in modern control theory.

13.9 Summary

The Z transform and Laplace transform have many of the same properties as the Fourier transforms. They are linear, which greatly facilitates computation of the transforms and their inverses. Moreover, the Z transform (Laplace transform) of the output of an LTI system is the product of the Z transforms (Laplace transforms) of the input and the transfer function. Thus, the Z transform (Laplace transform) plays the same role as the frequency response, describing the relationship between the input and the output as a product rather than a convolution.

Linear difference and differential equations, and state-space models of LTI systems were introduced in chapters 5 and 8. However, we lacked a method to calculate the response of these models for nonzero initial conditions. The Z transform and the Laplace transform provide such a method.

EXERCISES

KEY: E = mechanical T = requires plan of attack C = more than 1 answer

E 1. Consider the signal x defined by

$$\forall n, \quad x(n) = \sin(\omega_0 n)u(n).$$

(a) Show that the Z transform is

$$\forall z \in RoC(x), \quad \hat{X}(z) = \frac{z\sin(\omega_0)}{z^2 - 2z\cos(\omega_0) + 1},$$

where

$$RoC(x) = \{z \in Complex \mid |z| > 1\}.$$

(b) Where are the poles and zeros?

(c) Is x absolutely summable?

T 2. Consider the signal x defined by

$$\forall n \in Integers, \quad x(n) = a^{|n|},$$

where $a \in Complex$.

(a) Find the Z transform of x. Be sure to give the region of convergence.

(b) Where are the poles?

(c) Under what conditions is x absolutely summable?

E 3. Consider a discrete-time LTI system with transfer function defined by

$$\forall z \in \{z \mid |z| > 0.9\}, \quad \hat{H}(z) = \frac{z}{z - 0.9}.$$

Suppose that the input x is given by

$$\forall n \in Integers, \quad x(n) = \delta(n) - 0.9\delta(n-1).$$

Find the Z transform of the output y, including its region of convergence.

E 4. Consider the exponentially modulated sinusoid y given by

$$\forall n \in Integers, \quad y(n) = a^{-n}\cos(\omega_0 n)u(n),$$

where a is a real number, ω_0 is a real number, and u is the unit step signal.

(a) Find the Z transform. Be sure to give the region of convergence. Hint: Use example 13.3 and section 13.1.6.

(b) Where are the poles?

(c) For what values of a is this signal absolutely summable?

T 5. Suppose $x \in DiscSignals$ satisfies

$$\sum_{n=-\infty}^{\infty} |x(n)r^{-n}| < \infty, \quad 0 < r_1 < r < r_2,$$

for some real numbers r_1 and r_2 such that $r_1 < r_2$. Show that

$$\sum_{n=-\infty}^{\infty} |nx(n)r^{-n}| < \infty, \quad 0 < r_1 < r < r_2.$$

Hint: Use the fact that for any $\epsilon > 0$ there exists $N < \infty$ such that $n(1+\epsilon)^{-n} < 1$ for all $n > N$.

T 6. Consider a causal discrete-time LTI system where the input x and output y are related by the difference equation

$$\forall n \in Integers, \ y(n) + b_1 y(n-1) + b_2 y(n-2)$$
$$= a_0 x(n) + a_1 x(n-1) + a_2 x(n-2),$$

where b_1, b_2, a_0, a_1, and a_2 are real-valued constants.

(a) Find the transfer function.

(b) Say as much as you can about the region of convergence.

(c) Under what conditions is the system stable?

E 7. This exercise verifies the time delay property of the Laplace transform. Show that if x is a continuous-time signal, τ is a real constant, and y is defined by

$$\forall t \in Reals, \ y(t) = x(t - \tau),$$

then its Laplace transform is

$$\forall s \in RoC(y), \quad \hat{Y}(s) = e^{-s\tau}\hat{X}(s),$$

with region of convergence

$$RoC(y) = RoC(x).$$

E 8. This exercise verifies the convolution property of the Laplace transform. Suppose x and h have Laplace transforms $\hat{X}$ and $\hat{H}$. Let y be defined by

$$\forall t \in Reals, \quad y(t) = (x * h)(t) = \int_{-\infty}^{\infty} x(\tau)h(t-\tau)d\tau.$$

Then show that the Laplace transform is

$$\forall s \in RoC(y), \quad \hat{Y}(s) = \hat{X}(s)\hat{H}(s),$$

with

$$RoC(y) \supset RoC(x) \cap RoC(h).$$

T 9. This exercise verifies the conjugation property of the Laplace transform, and then uses this property to demonstrate that for real-valued signals, poles and zeros come in complex-conjugate pairs.

(a) Let x be a complex-valued continuous-time signal and y be defined by

$$\forall t \in Reals, \quad y(t) = [x(t)]^*.$$

Show that

$$\forall s \in RoC(y), \quad \hat{Y}(s) = [\hat{X}(s^*)]^*,$$

where

$$RoC(y) = RoC(x).$$

(b) Use this property to show that if x is real, then complex poles and zeros occur in complex-conjugate pairs. That is, if there is a zero at $s = q$, then there must be a zero at $s = q^*$, and if there is a pole at $s = p$, then there must also be a pole at $s = p^*$.

T 10. This exercise verifies the time scaling property of the Laplace transform. Let y be defined by

$$\forall t \in Reals, \quad y(t) = x(ct),$$

for some real number c. Show that

$$\forall s \in RoC(y), \quad \hat{Y}(s) = \hat{X}(s/c)/|c|,$$

where

$$RoC(y) = \{s \mid s/c \in RoC(x)\}.$$

E 11. This exercise verifies the exponential scaling property of the Laplace transform. Let y be defined by

$$\forall\, t \in Reals, \quad y(t) = e^{at}x(t),$$

for some complex number a. Show that

$$\forall\, s \in RoC(y), \quad \hat{Y}(s) = \hat{X}(s - a),$$

with

$$RoC(y) = \{s \mid s - a \in RoC(x)\}.$$

T 12. Consider a discrete-time LTI system with the impulse response

$$\forall\, n, \quad h(n) = a^n \cos(\omega_0 n)u(n),$$

for some $\omega_0 \in Reals$. Show that if the input is

$$\forall\, n \in Integers, \quad x(n) = e^{i\omega_0 n}u(n),$$

then the output y is unbounded.

E 13. Find and plot the inverse Z transform of

$$\hat{X}(z) = \frac{1}{(z-3)^3},$$

with

(a) $Roc(x) = \{z \in Complex \mid |z| > 3\}$
(b) $Roc(x) = \{z \in Complex \mid |z| < 3\}$.

E 14. Obtain the partial fraction expansions of the following rational polynomials. Divide through first if necessary to get a strictly proper rational polynomial.

(a)
$$\frac{z+2}{(z+1)(z+3)},$$

(b)
$$\frac{(z+2)^2}{(z+1)(z+3)},$$

(c)
$$\frac{z+2}{z^2+4}.$$

E 15. Find the inverse Z transform x for each of the three possible regions of convergence associated with

$$\hat{X}(z) = \frac{(z+2)^2}{(z+1)(z+3)}.$$

For which region of convergence is x causal? For which is x strictly anti-causal? For which is x two-sided?

E 16. Find the inverse Z transform x for each of the two possible regions of convergence associated with

$$\hat{X}(z) = \frac{z+2}{z^2+4}.$$

E 17. Consider a stable system with the impulse response

$$h(n) = (0.5)^n x(n).$$

Find the steady-state response to a unit step input.

E 18. Let $h(n) = 2^n u(-n)$, all n, and $g(n) = 0.5^n u(n)$, for all n. Find $h * u$ and $g * u$, where u is the unit step.

19. This exercise shows how we can determine the transfer function and frequency response of an LTI system from its step response. Suppose a causal system with step input $x = u$ produces the output

$$\forall\, n \in Integers, \quad y(n) = (1 - 0.5^n)u(n).$$

(a) Find the transfer function (including its region of convergence).
(b) If the system is stable, find its frequency response.
(c) Find the impulse response of the system.

20. Consider an LTI system with impulse response h defined by

$$\forall\, n \in Integers, \quad h(n) = 2^n u(n).$$

(a) Find the transfer function, including its region of convergence.

(b) Use the transfer function to find the Z transform of the step response.

(c) Find the inverse transform of the result of part (b) to obtain the step response in the time domain.

E 21. Determine the zero-input and zero-state responses and the transfer function for the following. In both cases, take $y(-1) = y(-2) = 0$ and $x(n) = u(n)$.

(a) $y(n) + y(n-2) = x(n), n \geq 0.$

(b) $y(n) + 2y(n-1) + y(n-2) = x(n), n \geq 0.$

E 22. Determine the zero-input and the zero-state responses for the following:

(a) $5\dot{y} + 10y = 2x, y(0) = 2, x(t) = u(t).$

(b) $\ddot{y} + 5\dot{y} + 6y = -4x - 3\dot{x}, y(0) = -1, \dot{y}(0) = 5, x(t) = e^{-t}u(t).$

(c) $\ddot{y} + 4y = 8x, y(0) = 1, \dot{y}(0) = 2, x(t) = u(t).$

(d) $\ddot{y} + 2\dot{y} + 5y = \dot{x}, y(0) = 2, \dot{y}(0) = 0, x(t) = e^{-t}u(t).$

E 23. Show that the $[A, b, c, d]$ representation in example 13.23 is correct. Then show that the transfer function of the state-space model is the same as that of the difference equation.

T 24. Consider the circuit of figure 13.5. The input is the voltage x, the output is the capacitor voltage v. The inductor current is called i.

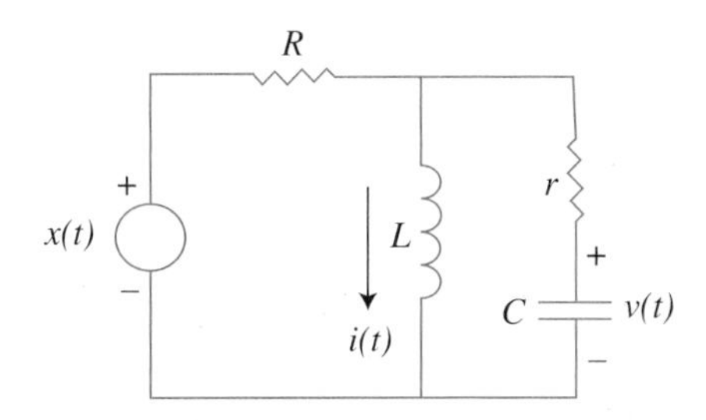

FIGURE 13.5: Circuit.

(a) Derive the $[A, b, c, d]$ representation for this system using $s(t) = [i(t), v(t)]^T$ as the state.

(b) Obtain an $[F, g, h, k]$ representation for a discrete-time model of the same circuit by sampling at times $kT, k = 0, 1, \ldots$, and using the approximation $\dot{s}(kT) = 1/T(s((k+1)T) - s(kT))$. (This is called a forward-Euler approximation.)

E 25. For the matrix A in example 13.24, determine $e^{tA}, t \geq 0$.

E 26. For the matrix A in example 13.26, determine $A^n, n \geq 0$.

T 27. A continuous-time SISO system has $[A, b, c, d]$ representation with

$$A = \begin{bmatrix} a & b \\ -b & a \end{bmatrix},$$

in which a, b are real constants.

(a) Find the eigenvalues of A.

(b) For what values of a, b is the SISO system stable?

(c) Calculate $e^{tA}, t \geq 0$.

(d) Suppose $b = c = [1\ 0]^T$ and $d = 0$. Find the transfer function.

T 28. Let A be an $N \times N$ matrix. Let p be an eigenvalue of A. An N-dimensional (column) vector e, possibly complex-valued, is said to be an **eigenvector** of A corresponding to p, if $e \neq 0$ and $Ae = pe$. Note that an eigenvector always exists since $\det[pI - A] = 0$. Find eigenvectors for each of the two eigenvalues of the matrices in examples 13.24 and 13.26.

E 29. Let A be a square matrix with eigenvalue p and corresponding eigenvector e. Determine the response of the following:

(a) $s(k+1) = As(k), k \geq 0;\ s(0) = e$.

(b) $\dot{s}(t) = As(t), t \geq 0;\ s(0) = e$.

Hint: Show that $A^n e = p^n e$ and $e^{tA} e = e^{pt} e$.

T 30. Verify (13.50). Hint: First show that

$$[sI - A]^{-1} b = \frac{1}{s^N + a_{N-1} s^{N-1} + \cdots + a_0} \begin{bmatrix} 1 \\ s \\ \cdots \\ s^{N-1} \end{bmatrix}$$

by multiplying both sides by $[sI - A]$. Then check (13.50).

CHAPTER 14
Composition and feedback control

A major theme of this book is that interesting systems are often compositions of simpler systems. Systems are functions, so their composition is function composition, as discussed in section 2.1.5. However, systems are often not directly described as functions, so function composition is not the easiest tool to use to understand the composition. We have seen systems described as state machines, frequency responses, and transfer functions. In chapter 4, we obtained the state machine of the composite system from its component state machines. In section 8.5, we obtained the frequency response of the composite system from the frequency response of its component linear time-invariant (LTI) systems. We extend the latter study in this chapter to the composition of LTI systems described by their transfer functions. This important extension allows us to consider unstable systems whose impulse response has a Z or Laplace transform, but not a Fourier transform.

As before, feedback systems prove challenging. A particularly interesting issue is how to maintain stability, and how to construct stable systems out of unstable ones. We will find that some feedback compositions of stable systems result in unstable systems, and conversely, some compositions of unstable systems result in stable systems. For example, we can stabilize the helicopter in example 12.2 using feedback; in fact we can precisely control its orientation, despite the intrinsic instability. The family of techniques for doing this is known as **feedback control**. This chapter serves as an introduction to that topic. Feedback control can also be used to drive stable systems, in which case it serves to improve their response. For example, feedback can result in faster or more precise responses, and can also prevent overshoot, where a system overreacts to a command.

We consider three styles of composition, **cascade composition**, **parallel composition**, and **feedback composition**. In each case, two LTI systems with transfer functions $\hat{H}_1$ and $\hat{H}_2$ are combined to get a new system. The transfer functions $\hat{H}_1$ and $\hat{H}_2$ are the (Z or Laplace) transforms of the respective impulse responses h_1 and h_2. Much of our discussion applies equally well whether the system is a continuous-time system or a discrete-time system, so in many cases we leave this unspecified.

14.1 Cascade composition

Consider the **cascade composition** shown in figure 14.1. The composition is the shaded box, which has transfer function

$$\hat{H} = \hat{H}_1 \hat{H}_2.$$

Notice that because of this simple form, if we know the pole and zero locations of the component systems, then it is easy to determine the pole and zero locations of the composition. Unless a pole of one is canceled by a zero of the other, the poles and zeros of the composition are simply the aggregate of the poles and zeros of the components. Moreover, any pole of $\hat{H}$ must be a pole of either $\hat{H}_1$ or $\hat{H}_2$, so if $\hat{H}_1$ and $\hat{H}_2$ are both stable, then so is $\hat{H}$.

14.1.1 Stabilization

The possibility for pole-zero cancelation suggests that cascade composition might be used to stabilize an unstable system.

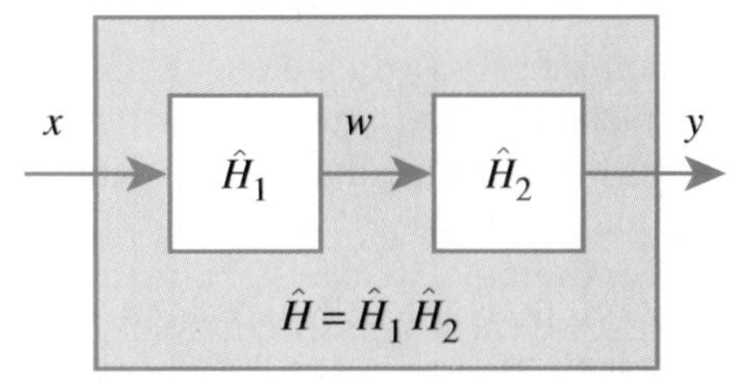

FIGURE 14.1: Cascade composition of two LTI systems with transfer functions H_1 and H_2.

Example 14.1: Consider a discrete-time system with transfer function

$$\forall z \in \{z \mid |z| > 1.1\}, \quad \hat{H}_1(z) = \frac{z}{z - 1.1}.$$

This is a proper rational polynomial with a region of convergence of the form for a causal signal, so it must be a causal system. It is unstable, however, because the region of convergence does not include the unit circle.

To stabilize this system, we might consider putting it in cascade with

$$\forall z \in Complex, \quad \hat{H}_2(z) = \frac{z - 1.1}{z}.$$

This is a causal and stable system. The transfer function of the cascade composition is

$$\hat{H}(z) = \frac{z}{z - 1.1} \frac{z - 1.1}{z} = 1.$$

The pole at $z = 1.1$ has been canceled, and the resulting region of convergence is the entire complex plane. Thus, the cascade composition is a causal and stable system, and we can recognize from table 13.1 that the impulse response is $h(n) = \delta(n)$. ❐

Stabilizing systems by canceling their poles in a cascade composition, however, is rarely a good idea. If the pole is not precisely canceled, then, no matter how small the error, the resulting system is still unstable.

Example 14.2: Suppose that in the previous example the pole location is not known precisely and turns out to be at $z = 1.1001$ instead of $z = 1.1$. Then the cascade composition has transfer function

$$\hat{H}(z) = \frac{z}{z - 1.1001} \frac{z - 1.1}{z} = \frac{z - 1.1}{z - 1.1001},$$

which is unstable. ❐

14.1.2 *Equalization*

While cascade compositions do not usually work well for stabilization, they do often work well for **equalization**. An **equalizer** is a **compensator** that reverses **distortion**. The source of the distortion, which is often called a **channel**, must be an LTI system, and the equalizer is composed in cascade with it. At first sight this is easy to do. If the channel has transfer function $\hat{H}_1$, then the equalizer could have transfer function

$$\hat{H}_2 = \hat{H}_1^{-1},$$

in which case the cascade composition will have transfer function

$$\hat{H} = \hat{H}_1 \hat{H}_2 = 1,$$

which is certainly distortion free.

Example 14.3: Some acoustic environments for audio have **resonances**, where certain frequencies are enhanced as the sound propagates through the environment (see lab 8 in Lab Manual). This typically occurs if the physics of the acoustic environment results in a transfer function with poles near the unit circle (for a discrete-time model) or near the imaginary axis (for a continuous-time model). Suppose, for example, that the acoustic environment is well modeled by a discrete-time LTI system with transfer function

$$\forall z \in \{z \mid |z| > 0.95\}, \quad \hat{H}_1(z) = \frac{z^2}{(z-a)(z-a^*)},$$

where $a = 0.95e^{i\omega_1}$ for some frequency ω_1. Using the methods of section 13.2, we can infer that the magnitude response will have a strong peak at frequencies ω_1 and $-\omega_1$, because the positions on the unit circle $e^{i\omega_1}$ and $e^{-i\omega_1}$ are very close to the poles. This will result in distortion of the audio signal, because frequencies near ω_1 will be amplified.

An equalizer that compensates for this distortion has transfer function

$$\hat{H}_2(z) = [\hat{H}_1(z)]^{-1} = \frac{(z-a)(z-a^*)}{z^2} = \frac{z^2 - 2Re\{a\}z + |a|^2}{z^2}.$$

As in example 13.2, we can recognize this as the Z transform of an FIR filter with impulse response

$$\forall n \in Integers, \quad h_2(n) = \delta(n) - 2Re\{a\}\delta(n-1) + |a|^2\delta(n-2).$$

This filter is causal and stable, and hence can serve as an effective equalizer. ❐

There are a number of potential problems with this approach, however. First, the transfer function of the channel is probably not known, or at least not known precisely. Second, the channel may not have a stable and causal inverse.

Let us examine the first difficulty, that the channel may not be known (precisely). If the channel model $\hat{H}_1$ and its inverse $\hat{H}_2$ are both stable, then the cascade composition is at least assured of being stable, even if the channel has been misconstrued. Moreover, if the equalizer is close to the inverse of the true channel, then often the distortion is significantly reduced despite the errors (see exercise 1 at the end of this chapter).

This difficulty can sometimes be dealt with by adaptively varying the equalizer based on measurements of the distortion. One way to measure the distortion is to send a known sequence called a **training sequence** through the channel and observe the output. Suppose that the training sequence is a signal x with Z transform $\hat{X}$, and that the channel $\hat{H}_1$ is unknown. If we can observe the output y of the channel, and calculate its Z transform $\hat{Y}$, then the channel transfer function is simply

$$\forall\, z \in RoC(h_1), \quad \hat{H}_1(z) = \frac{\hat{Y}(z)}{\hat{X}(z)},$$

where $RoC(h_1)$ is determined by identifying the poles and zeros of the rational polynomial $\hat{Y}(z)/\hat{X}(z)$ and finding the one ring-shaped region that includes the unit circle and is bordered by poles. This results in a stable channel model.

Training sequences are commonly used in digital communication systems, where, for example, a radio channel introduces distortion. However, it is also common for such channels to change over time. Radio channels, for example, change if either the transmitter or receiver moves, if the weather changes, or if obstacles appear or disappear. Repeatedly transmitting training sequences is an expensive waste of radio bandwidth, and, fortunately, is usually unnecessary, as illustrated in the following example.

Example 14.4: Consider a digital communication system where the channel is modeled as a discrete-time LTI system with transfer function $\hat{H}_1$, representing, for example, a radio transmission subsystem. Suppose that this digital communication system transmits a bit sequence represented as a discrete-time signal x of form

$$x\colon Integers \to \{0, 1\}.$$

Suppose further that we use a training sequence to obtain an initial estimate $\hat{H}_2$ of the inverse of the channel. But, over time, the channel drifts so that $\hat{H}_2$ is no longer the inverse of $\hat{H}_1$. Assuming that the drift is relatively slow, after a short time, $\hat{H}_2$ is still close to the inverse of $\hat{H}_1$, in that the cascade $\hat{H}_1\hat{H}_2$ yields only mild distortion. That is, if $x(n) = 0$ for some n, then $y(n) \approx 0$. Similarly, if $x(n) = 1$ for some n, then $y(n) \approx 1$. Thus, we can quantize y, getting an accurate estimate x, without it having to be a known training sequence. That is, when $y(n) \approx 0$, we assume that $x(n) = 0$, and when $y(n) \approx 1$, we assume that $x(n) = 1$. These assumptions are called **decisions**, and, in fact, such decisions must be made anyway for digital communication to occur. We have to decide whether a 1 or a 0 was transmitted, and closeness to 1 or 0 seems like an eminently reasonable basis on which to make such a decision.

Assuming there are no errors in these decisions, we can infer that

$$\hat{H}_1 \hat{H}_2 \hat{X}_d = \hat{Y},$$

where $\hat{X}_d$ is the Z transform of the decision sequence. So, without using another training sequence, we can revise our estimate of the channel transfer function as follows,

$$\hat{H}_1 = \frac{\hat{Y}}{\hat{H}_2 \hat{X}_d}.$$

We replace our equalizer $\hat{H}_2$ with

$$\hat{H}_2' = [\hat{H}_1]^{-1} = \frac{\hat{H}_2 \hat{X}_d}{\hat{Y}}.$$

Of course, we now start using $\hat{H}_2'$, which comes closer to correcting the channel distortion, making our decisions more reliable for the next update. ❒

Example 14.4 outlines a widely used technique called **decision-directed adaptive equalization**. It is so widely used, in fact, that it may be found in every digital cellular telephone and almost every modem, including voiceband data modems, radio modems, cable modems, and DSL modems. The algorithms used in practice to update the transfer function of the equalizer are not exactly as shown in the example, and their details are beyond the scope of this text, but they follow the general principle in the example.

Let us now turn our attention to the second difficulty with equalization, that the channel may not have a stable and causal inverse. We begin with an example.

Example 14.5: Suppose that, similar to example 14.3, a channel has transfer function

$$\forall z \in \{z \mid |z| > 0.95\}, \quad \hat{H}_1(z) = \frac{z}{(z-a)(z-a^*)},$$

where $a = 0.95e^{i\omega_1}$ for some frequency ω_1. The inverse is

$$[\hat{H}_1(z)]^{-1} = \frac{(z-a)(z-a^*)}{z} = \frac{z^2 - 2Re\{a\}z + |a|^2}{z},$$

which is not a proper rational polynomial. Thus, this cannot be the Z transform of a causal signal. Implementing a noncausal equalizer is usually impossible, since it requires knowing future inputs. However, suppose we simply force the equalizer to have a proper rational polynomial transfer function by dividing by a high enough power M of z to make $[\hat{H}_1(z)]^{-1}/z^M$ proper. In this example, $M = 1$ is sufficient, so we define the equalizer to be

$$\hat{H}_2(z) = \frac{[\hat{H}_1(z)]^{-1}}{z} = \frac{z^2 - 2Re\{a\}z + |a|^2}{z^2},$$

which we again recognize as the Z transform of an FIR filter with impulse response

$$\forall\, n \in Integers, \quad h_2(n) = \delta(n) - 2Re\{a\}\delta(n-1) + |a|^2\delta(n-2).$$

This filter is causal and stable, but does it serve as an effective equalizer? Consider now the cascade,

$$\hat{H}(z) = \hat{H}_1(z)\hat{H}_2(z) = \frac{1}{z}.$$

From section 13.1.2, we recognize this as the transfer function of the unit delay system—the equalizer completely compensates for the distortion, but at the expense of introducing a one sample delay. This is usually a perfectly acceptable cost. ❒

Example 14.5 demonstrates that when the channel inverse is not a proper rational polynomial, then introducing a delay may enable construction of a stable and causal equalizer. Not all equalization stories have such a happy ending, however. Consider the following example.

Example 14.6: Consider a channel with the following transfer function,

$$\forall\, z \in \{z \in Complex \mid z \neq 0\}, \quad \hat{H}_1(z) = \frac{z-2}{z}.$$

This is a stable and causal channel. Its inverse is

$$[\hat{H}_1(z)]^{-1} = \frac{z}{z-2}.$$

This has a pole at $z = 2$, so, in order to be stable, it would have to be anticausal (so that the region of convergence includes the unit circle). Implementing an anticausal equalizer is usually impossible. ❒

Example 14.6 shows that not all channels can be inverted by an equalizer. All is not lost, however. Given a channel $\hat{H}_1(z)$ that has a rational Z transform, we can usually find a transfer function $\hat{H}_2(z)$ that compensates for the magnitude response part of the distortion. That is, we can find a transfer function $\hat{H}_2(z)$ that is stable and causal such that the magnitude response of the composite satisfies

$$|H_1(\omega)H_2(\omega)| = |\hat{H}_1(e^{i\omega})\hat{H}_2(e^{i\omega})| = 1.$$

For some applications, audio equalization, for example, this is almost always sufficient because the human ear is not very sensitive to the phase of audio signals. It hears only the magnitude of the frequency components.

Example 14.7: Continuing example 14.6, let $\hat{H}_2$ be given by

$$\hat{H}_2 = \frac{z}{1-2z} = \frac{-0.5z}{z-0.5}.$$

This has a pole at $z = 0.5$, and is a proper rational polynomial, so it can be the transfer function of a causal and stable filter. Consider the cascade composition,

$$\hat{H}(z) = \hat{H}_1(z)\hat{H}_2(z) = \frac{z-2}{z} \cdot \frac{-0.5z}{z-0.5} = \frac{1-0.5z}{z-0.5}.$$

This hardly looks like what we want, but if we rewrite it slightly, it is easy to show that the magnitude frequency response has value one for all ω,

$$\hat{H}(z) = \frac{1-0.5z}{z-0.5} = z\frac{z^{-1}-0.5}{z-0.5}.$$

The magnitude frequency response is

$$|H(\omega)| = |\hat{H}(e^{i\omega})| = |e^{i\omega}| \cdot \frac{|e^{-i\omega}-0.5|}{|e^{i\omega}-0.5|} = 1.$$

This magnitude is equal to 1 because the numerator, $e^{-i\omega} - 0.5$, is the complex conjugate of the denominator, $e^{i\omega} - 0.5$, so they have the same magnitude. ❒

The method in example 14.7 can be generalized so that it is possible to cancel any magnitude distortion for most channels. The key is that if the channel transfer function has a zero outside the unit circle, say at $z = a$, then its inverse has a pole at the same location, $z = a$. A pole outside the unit circle makes it impossible to have a stable and causal filter. So the trick is to place a pole instead at $z = 1/a^*$. This pole cancels the effect on the magnitude (but not the phase) of the zero at $z = a$.

There are still channels for which this method will not work.

Example 14.8: Consider a channel given by

$$\forall z \in \{z \in Complex \mid z \neq 0\}, \quad \hat{H}_1(z) = \frac{z-1}{z}.$$

This has a pole at $z = 0$ and a zero at $z = 1$. Its inverse cannot be stable because it will have a pole at $z = 1$. In fact, no equalization is possible. This is intuitive because the frequency response is zero at $\omega = 0$, and no stable equalizer in cascade with this channel can reconstruct the original component at $\omega = 0$. It would have to have infinite gain at $\omega = 0$, which would make it unstable. ❐

14.2 Parallel composition

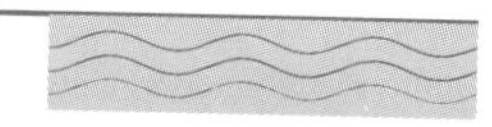

Consider the **parallel composition** shown in figure 14.2. The transfer function of the composition system is

$$\hat{H} = \hat{H}_1 + \hat{H}_2.$$

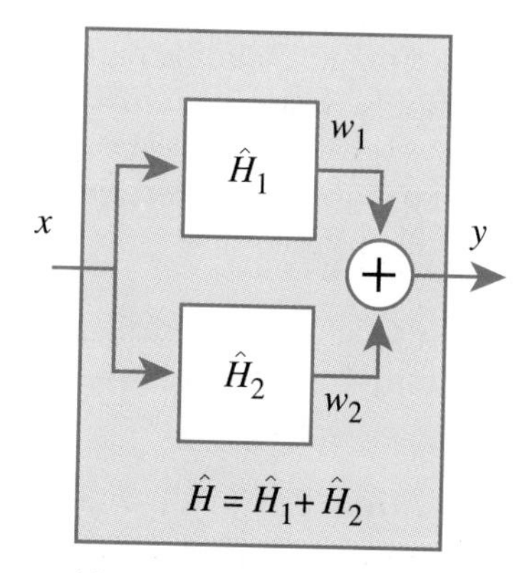

FIGURE 14.2: Parallel composition of two LTI systems with transfer functions $\hat{H}_1$ and $\hat{H}_2$.

This is valid whether these are Laplace transforms or Z transforms. Once again, notice that a pole of $\hat{H}$ must be a pole of either $\hat{H}_1$ or $\hat{H}_2$, so if $\hat{H}_1$ and $\hat{H}_2$ are stable, then so is $\hat{H}$. At the poles of $\hat{H}_1$, $\hat{H}_1(z)$ is infinite, so very likely a pole of $\hat{H}_1$ is also a pole of $\hat{H}$. However, just as in the cascade composition, this pole may be canceled by a zero.

Determining the location of the zeros of the composition, however, is slightly more complicated than for cascade composition. The sum has to be put into rational polynomial form and the polynomials then need to be factored.

14.2.1 *Stabilization*

Just as with cascade composition, stabilizing systems by canceling their poles in a parallel composition is possible but is rarely a good idea.

Example 14.9: Consider a discrete-time system with transfer function

$$\forall\, z \in \{z \mid |z| > 1.1\}, \quad \hat{H}_1(z) = \frac{z}{z - 1.1}.$$

This describes a causal but unstable system. If we combine this in parallel with a system with transfer function

$$\forall\, z \in \{z \mid |z| > 1.1\}, \quad \hat{H}_2(z) = \frac{-1.1}{z - 1.1},$$

this is again causal and unstable. The transfer function of the parallel composition is

$$\hat{H}(z) = \frac{z}{z - 1.1} + \frac{-1.1}{z - 1.1} = \frac{z - 1.1}{z - 1.1} = 1.$$

The pole at $z = 1.1$ has been canceled, and the resulting region of convergence is the entire complex plane. Thus, the parallel composition is a causal and stable system with impulse response $h(n) = \delta(n)$. ❐

However, if the pole is not precisely canceled, then, no matter how small the error, the resulting system is still unstable.

Example 14.10: Assume that in the previous example the pole location is not known precisely and turns out to be at $z = 1.1001$ instead of $z = 1.1$. Then the parallel composition has transfer function

$$\hat{H}(z) = \frac{z}{z - 1.1001} + \frac{-1.1}{z - 1.1} = \frac{z^2 - 2.2z + 1.21001}{(z - 1.1001)(z - 1.1)},$$

which is unstable. ❐

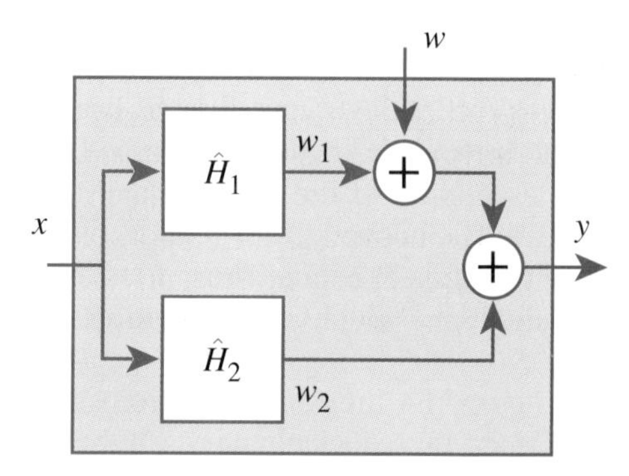

FIGURE 14.3: Structure of a noise canceler.

14.2.2 *Noise cancelation*

While parallel compositions do not usually work well for stabilization, with a small modification they do often work well for **noise cancelation**. A **noise canceler** is a compensator that removes an unwanted component from a signal. The unwanted component is called **noise**.

The pattern of a noise cancelation problem is shown in figure 14.3. The signal x is a noise source. This signal is filtered by $\hat{H}_1$ and added to the **desired signal** w. The result is a noisy signal. To cancel the noise, the signal from the noise source is filtered by a noise-canceling filter $\hat{H}_2$ and the result is added to the noisy signal. If x has (Laplace or Z) transform $\hat{X}$, w has transform $\hat{W}$, and y has transform $\hat{Y}$, then

$$\hat{Y} = \hat{W} + (\hat{H}_1 + \hat{H}_2)\hat{X}.$$

From this it is evident that if we choose

$$\hat{H}_2 = -\hat{H}_1,$$

then y will be a clean (noise-free) signal, equal to w. The following examples describe real-world applications of this pattern.

Example 14.11: A connection to the telephone network uses two wires (called a **twisted pair**, consisting of **tip** and **ring**) to connect a telephone to a central office. The central office may be, perhaps, four kilometers away. The two wires carry voice signals to and from the customer premises, representing the voice signals as a voltage difference across the two wires. Since two wires can only have one voltage difference across them, the incoming voice signal and the outgoing voice signal share the same twisted pair.

The central office needs to separate the voice signal from the local customer premises (called the **near-end signal**) from the voice signal that comes from the other end of the connection (called the **far-end signal**). The near-end signal is typically digitized (sampled and quantized), and a discrete-time representation of the voice signal is transmitted over the network to the far end. The network itself consists of circuits that can carry voice signals in one direction at a time. Thus, in the network, four wires (or equivalent) are required for a telephone connection, one wire pair for each direction.

As indicated in figure 14.4, the conversion from a two-wire to a four-wire connection is performed by a device called a **hybrid**.* A connection between subscribers A and B involves two hybrids, one in each subscriber's central office. The hybrid in B's central office ideally passes all of the incoming signal x to B's two-wire circuit, and none back into the network. However, the hybrid is not perfect, and some of the incoming signal x leaks through the hybrid into the return path back to A. The signal y in the figure is the sum of the signal from B and the leaked signal from A. A hears the leaked signal as an echo, because it is A's own signal, delayed by propagation through the telephone network.

If the telephone connection includes a satellite link, then the delay from one end of the connection to the other is about 300 ms. This is the time it takes for a radio signal to propagate to a geosynchronous satellite and back. The echo traverses this link twice, once going from A to B, and the second time coming back. Thus, the echo is A's own signal delayed by about 600 ms. For voice signals, 600 ms of delay is enough to create a very annoying echo that can make it difficult to speak. Humans have difficulty speaking when they hear their own voices 600 ms later. Consequently, the designers of the telephone network have installed echo cancelers to prevent the echo.

Let $\hat{H}_1$ be the transfer function of the hybrid leakage path. The echo canceler is the filter $\hat{H}_2$ placed in parallel composition with the hybrid, as shown in the figure. The output w_2 of this filter is added to the output $w_1 + w$ of the hybrid, so the signal that actually goes back is $y = w_2 + w_1 + w$. If

$$\hat{H}_2 = -\hat{H}_1,$$

then $y = w$ and the echo is canceled perfectly. Moreover, note that as long as $\hat{H}_1$ is stable and causal, so is the echo canceler $\hat{H}_2$.

However, $\hat{H}_1$ is not usually known in advance and also changes over time. So either a fixed $\hat{H}_2$ is designed to match a "typical" $\hat{H}_1$, or an adaptive

*A hybrid is a Wheatstone bridge, a circuit that can separate two signals based on the electrical impedance looking into the local twisted pair and the electrical impedance looking into the network. The design of this circuit is a suitable topic for a text on electrical circuits.

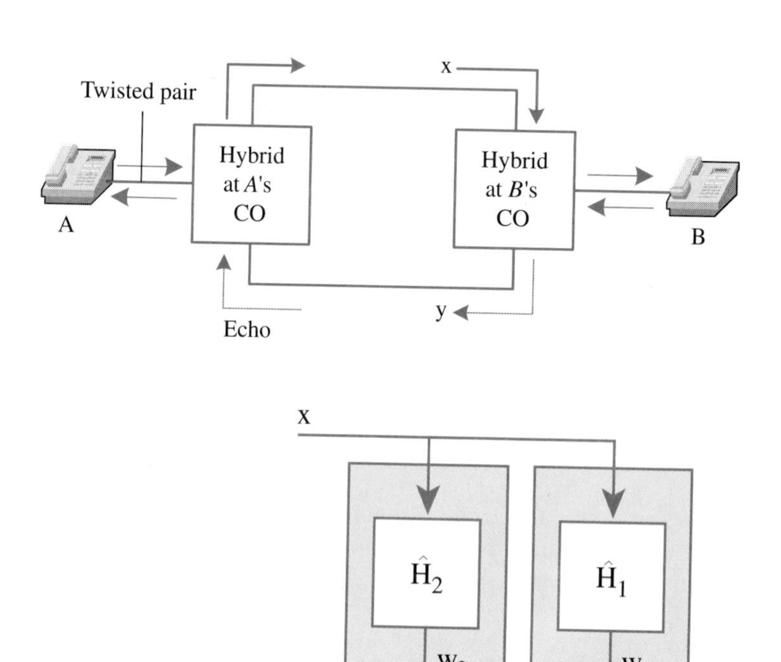

FIGURE 14.4: A telephone central office converts the two-wire connection with a customer telephone into a four-wire connection with the telephone network using a device called a hybrid. An imperfect hybrid leaks, causing echo. An echo canceler removes the leaked signal. x, signal from A; y, leakage of A's signal plus signal from B.

echo canceler is designed that estimates the characteristics of the echo path ($\hat{H}_1$) and changes $\hat{H}_2$ accordingly. Adaptive echo cancelers are common in the telephone network today. ❒

The following example combines cascade and parallel composition to achieve noise cancelation.

Example 14.12: Consider a microphone in a noisy environment. For example, a traffic helicopter might be used to deliver live traffic reports over the radio, but the (considerable) background noise of the helicopter would be highly undesirable on the radio. Fortunately, the background noise can be canceled. Referring to figure 14.5, suppose that w is the announcer's voice,

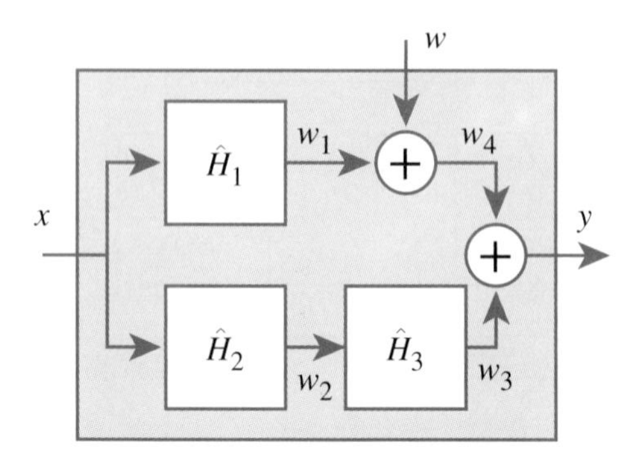

FIGURE 14.5: Traffic helicopter noise cancelation/equalization problem.

x is the engine noise, and $\hat{H}_1$ represents the acoustic path from the engine noise to the microphone. The microphone picks up both the engine noise and the announcer's voice, producing the noisy signal w_4. We can place a second microphone somewhere far enough from the announcer so as to not pick up much of his or her voice. Since this microphone is in a different place, say on the back of the announcer's helmet, the acoustic path is different, so we model that path with another transfer function $\hat{H}_2$. To cancel the noise, we design a filter $\hat{H}_3$. This filter needs to equalize (invert) $\hat{H}_2$ and cancel $\hat{H}_1$. That is, its ideal value is

$$\hat{H}_3 = -\hat{H}_1/\hat{H}_2.$$

Of course, as with the equalization scenario, we have to ensure that this filter remains stable. Once again, in practice, it is necessary to make the filter adaptive. ❒

14.3 Feedback composition

Consider the **feedback composition** in figure 14.6. It is a composition of two systems with transfer functions $\hat{H}_1$ and $\hat{H}_2$. We assume that these systems are causal and that $\hat{H}_1$ and $\hat{H}_2$ are proper rational polynomials in z or s. The regions of convergence of these two transfer functions are those suitable for causal systems (the region outside the largest circle passing through a pole for discrete time, and the region to the right of the pole with the largest real part for continuous time).

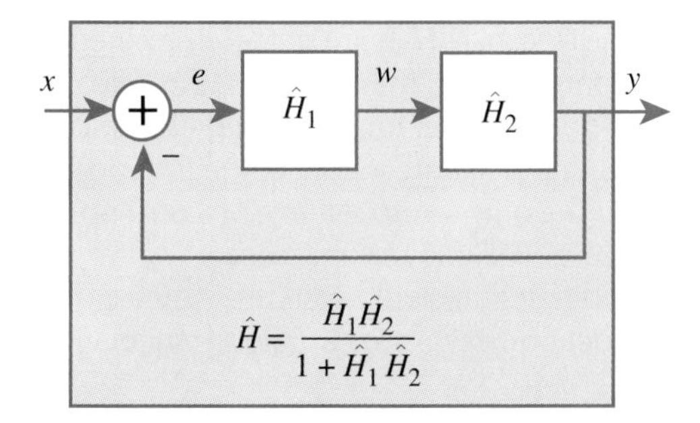

FIGURE 14.6: Negative feedback composition of two LTI systems with transfer functions H_1 and H_2.

In terms of Laplace or Z transforms, the signals in the figure are related by

$$\hat{Y} = \hat{H}_2\hat{H}_1\hat{E},$$

and

$$\hat{E} = \hat{X} - \hat{Y}.$$

Notice that, by convention, the feedback term is subtracted, as indicated by the minus sign adjacent to the adder (for this reason, this composition is called **negative feedback**). Combining these two equations to elimintate $\hat{E}$, we get

$$\hat{Y} = \hat{H}_1\hat{H}_2(\hat{X} - \hat{Y}),$$

which we can solve for the transfer function of the composition,

$$\hat{H} = \frac{\hat{Y}}{\hat{X}} = \frac{\hat{H}_1\hat{H}_2}{1+\hat{H}_1\hat{H}_2}. \quad (14.1)$$

This is often called the **closed-loop transfer function**, to contrast it with the **open-loop transfer function**, which is simply $\hat{H}_1\hat{H}_2$. We assume that this resulting system is causal, and that the region of convergence of this transfer function is therefore determined by the roots of the denominator polynomial $1+\hat{H}_1\hat{H}_2$.

The closed-loop transfer function is valid as long as the denominator $1+\hat{H}_1\hat{H}_2$ is not identically zero (that is, it is not zero for *all* s or z in *Complex*; it may be zero *some* s or z in *Complex*). This is sufficient for the feedback loop to be well-formed, although, in general, this fact is not trivial to demonstrate (exercise 8 at the end of this chapter considers the easier case where $\hat{H}_1\hat{H}_2$ is causal and strictly proper, in which case the system $\hat{H}_1\hat{H}_2$ has a state-determined output). We assume henceforth, without comment, that the denominator is not identically zero.

Feedback composition is useful for stabilizing unstable systems. In the case of cascade and parallel composition, a pole of the composite must be a pole of one of the components. The only way to remove or alter a pole of the components is to cancel it with a zero. For this reason, cascade and parallel composition are *not* effective for stabilizing unstable systems. Any error in the specification of the unstable pole location results in a failed cancelation and an unstable composition.

In contrast, the poles of the feedback composition are the roots of the denominator $1 + \hat{H}_1\hat{H}_2$, which are generally quite different from the poles of $\hat{H}_1$ and $\hat{H}_2$. This leads to the following important conclusion:

> The poles of a feedback composition can be different from the poles of its component subsystems. Consequently, unstable system can be effectively and robustly stabilized by feedback.

The stabilization is **robust** in that small changes in the pole or zero locations do not result in the composition going unstable. We will be able to quantify this robustness.

14.3.1 *Proportional controllers*

In control applications, one of the two systems being composed, say $\hat{H}_2$, is called the **plant**. This is a physical system that is given to us to control. Its transfer function is determined by its physics. The second system being composed, say $\hat{H}_1$, is the **controller**. We design this system to get the plant to do what we want. The following example illustrates a simple strategy called a **proportional controller** or **P controller**.

Example 14.13: For this example, we take as the plant the simplified continuous-time helicopter model of example 12.2,

$$\dot{y}(t) = \frac{1}{M} w(t).$$

Here $y(t)$ is the angular velocity at time t and $w(t)$ is the torque. M is the moment of inertia.

We have renamed the input w (instead of x) because we wish to control the helicopter, and the control input signal will not be the torque. Instead, we define the input x to be the desired angular velocity. So, to get the helicopter not to rotate, we provide input $x(t) = 0$.

We call the impulse response of the plant h_2, to conform with the notation in figure 14.6; it is given by

$$\forall\, t \in Reals, \quad h_2(t) = u(t)/M,$$

where u is the unit step. The transfer function is $\hat{H}_2(s) = 1/(Ms)$, with $RoC(h) = \{s \in Complex \mid Re(s) > 0\}$. $\hat{H}_2$ has a pole at $s = 0$, so this is an unstable system.

As a compensator we can simply place gain K in a negative feedback composition, as shown in figure 14.7. The intuition is as follows. Assume we wish to keep the helicopter from rotating; that is, we would like the output angular velocity to be zero, $y(t) = 0$. Then we should apply an input of zero, $x(t) = 0$. However, the plant is unstable, so even with a zero input, the output diverges (even the smallest nonzero initial condition or the smallest input disturbance will cause it to diverge). With the feedback arrangement in figure 14.7, if the output angular velocity rises above zero, then the input is modified downward (the feedback is negative), which results in a negative torque being applied to the plant, which counters the rising velocity. If the output angular velocity drops below zero, then the torque is modified upward, which again tends to counter the dropping velocity. The output velocity stabilizes at zero.

To get the helicopter to rotate, for example, to execute a turn, we simply apply a nonzero input. The feedback system again compensates so that the helicopter rotates at the angular velocity specified by the input.

The signal e is the difference between the input x, which is the desired angular velocity, and the output y, which is the actual angular velocity. It is called the **error signal**. Intuitively, this signal is zero when everything is as desired, when the output angular velocity matches the input. ❐

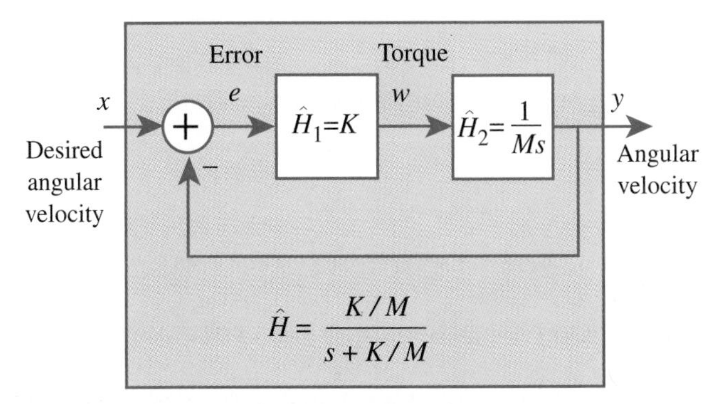

FIGURE 14.7: A negative feedback proportional controller with gain K.

A compensator like that in example 14.13 and figure 14.7 is called a **proportional controller** or **P controller**. The input w to the plant is proportional to the error e. The objective of the control system is to have the output y of the plant track the input x as closely as possible—that is, the error e needs to be small. We can use (14.1) to find the transfer function of the closed-loop system.

Example 14.14: Continuing with the helicopter of example 14.13, the closed-loop system transfer function is

$$\hat{G}(s) = \frac{K\hat{H}(s)}{1+K\hat{H}(s)} = \frac{K/M}{s+K/M}, \tag{14.2}$$

which has a pole at $s = -K/M$. If $K > 0$, the closed-loop system is stable, and if $K < 0$, it is unstable. Thus, we have considerable freedom to choose K. How should we choose its value?

As K increases from 0 to ∞, the pole at $s = -K/M$ moves left from 0 to $-\infty$. As K decreases from 0 to $-\infty$, the pole moves to the right from 0 to ∞. The locus of the pole as K varies is called the **root locus**, since the pole is a root of the denominator polynomial.

Figure 14.8 shows the root locus as a thick color line, on which are marked the locations of the pole for $K = 0, \pm 2, \pm\infty$. Since there is only one pole, the root locus comprises only one "branch." In general the root locus has as many branches as the number of poles, with each branch showing the movement of one pole as K varies. ❒

Note that, in principle, the same transfer function as the closed-loop transfer function can be achieved by a cascade composition. But as in example 14.1, the resulting system is not robust, in that even the smallest change in the pole location of the plant can cause the system to go unstable (see exercise 6 at the end of this chapter). The feedback system, however, is robust, as shown in the following example.

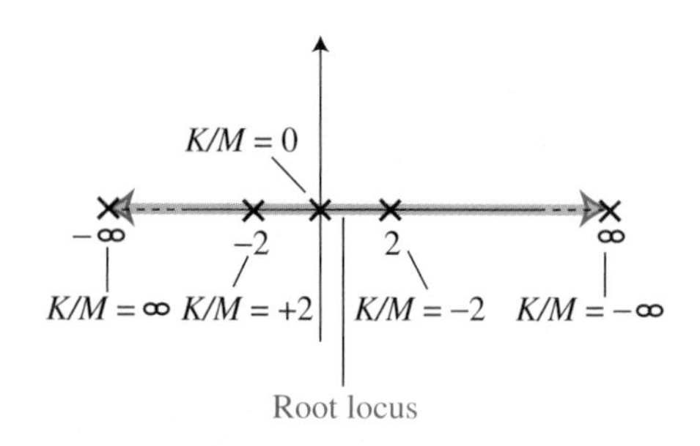

FIGURE 14.8: Root locus of the helicopter P controller.

Example 14.15: Continuing with the P controller for the helicopter, suppose that our model of the plant is not perfect, and its actual transfer function is

$$\hat{H}_2(s) = \frac{1}{M(s - \epsilon)},$$

for some small value of $\epsilon > 0$. In that case, the closed-loop transfer function is

$$\hat{H}(s) = \frac{K/M}{s - \epsilon + K/M},$$

which has a pole at $s = \epsilon - K/M$. The feedback system remains stable as long as

$$\epsilon < K/M. \square$$

In practice, when designing feedback controllers, we first quantify our uncertainty about the plant and then determine the controller parameters so that, under all possible plant transfer functions, the closed-loop system is stable.

Example 14.16: Continuing the helicopter example, we might say that $\epsilon < 0.5$. In that case, if we choose K so that $K/M > 0.5$, we would guarantee stability for all values of $\epsilon < 0.5$. We then say that the proportional feedback controller is **robust** for all plants with $\epsilon < 0.5$. ❐

We still have a large number of choices for K, the controller gain. How do we select one? To understand the implications of different choices of K we need to study the behavior of the output y (or the error signal e) for different choices of K. In the following examples, we use the closed-loop transfer function to analyze the response of a proportional controller system to various inputs. The first example studies the response to a step function input.

Example 14.17: Continuing the helicopter example, suppose that the input is a step function, $\forall\, t, x(t) = au(t)$, where a is a constant and u is the unit step. This input declares that at time $t = 0$, we wish for the helicopter to begin rotating with angular velocity a. The closed-loop transfer function is given by (14.2), and the Laplace transform of x is $\hat{X}(s) = a/s$, from table 13.3, so the Laplace transform of the output is

$$\hat{Y}(s) = \hat{G}(s)\hat{X}(s) = \frac{K/M}{s + K/M} \cdot \frac{a}{s}.$$

Carrying out the partial fraction expansion, this becomes

$$\hat{Y}(s) = \frac{-a}{s + K/M} + \frac{a}{s}.$$

We can use this to find the inverse Laplace transform,

$$\forall\, t, \quad y(t) = -ae^{-Kt/M}u(t) + au(t).$$

The second term is the **steady-state response** y_{ss}, which in this case equals the input. So the first term is the **tracking error** y_{tr}, which goes to zero faster for a larger K. Hence, for step inputs, the larger the gain K, the smaller the tracking error. ❐

In the previous example, we find that the error goes to zero when the input is a step function. Moreover, the error goes to zero faster if the gain K is larger than if it is smaller. In the following example, we find that if the input is sinusoidal, then larger gain K results in an ability to track higher frequency inputs.

Example 14.18: Suppose the input to the P controller helicopter system is a sinusoid of amplitude A and frequency ω_0,

$$\forall\, t \in Reals, \quad x(t) = A(\cos \omega_0 t)u(t).$$

We know that the response can be decomposed as $y = y_{tr} + y_{ss}$. The transient response y_{tr} is due to the pole at $s = -K/M$, and so it is of the form

$$\forall\, t \in Reals, \quad y_{tr}(t) = Re^{-Kt/M},$$

for some constant R. The steady-state response is determined by the frequency response at ω_0. The frequency response is

$$\forall\, \omega \in Reals, \quad H(\omega) = \hat{H}(i\omega) = \frac{K/M}{i\omega + K/M},$$

with magnitude and phase defined by

$$|H(\omega)| = \frac{K/M}{[\omega^2 + (K/M)^2]^{1/2}}, \quad \angle H(\omega) = -\tan^{-1}\frac{\omega M}{K}.$$

So the steady-state response is

$$\forall\, t, \quad y_{ss}(t) = |H(\omega_0)| A \cos(\omega_0 t + \angle H(\omega_0)).$$

Thus, the steady-state output is a sinusoid of the same frequency as the input but with a smaller amplitude (unless $\omega_0 = 0$). The larger ω_0 is, the smaller the output amplitude. Hence, the ability of the closed-loop system to track a sinusoidal input decreases as the frequency of the sinusoidal input increases. However, increasing K reduces this effect. Thus, larger gain in the feedback loop improves its ability to track higher frequency sinusoidal inputs.

In addition to the reduction in amplitude, the output has a phase difference. Again, if $\omega_0 = 0$, there is no phase error, because $\tan^{-1}(0) = 0$. As ω_0 increases, the phase lag increases (the phase angle decreases). Once again, however, increasing the gain K reduces the effect. ❐

Examples 14.17 and 14.18 suggest that a large gain in the feedback loop is always better. For a step function input, it causes the transient error to die out faster. For a sinusoidal input, it improves the ability to track higher frequency inputs and it reduces the phase error in the tracking. A large gain is not always a good idea, however, as seen in the next example, a DC motor.

Example 14.19: The angular position y of a DC motor is determined by the input voltage w according to the differential equation

$$M\ddot{y}(t) + D\dot{y}(t) = Lw(t),$$

where M is the moment of inertia of the rotor and attached load, $D\dot{y}$ is the damping force, and the torqe Lw is proportional to the voltage. The transfer function is

$$\hat{H}_2(s) = \frac{L}{Ms^2 + Ds} = \frac{L/M}{s(s + D/M)},$$

which has one pole at $s = 0$ and one pole at $s = -D/M$. By itself, the DC motor is unstable because of the pole at $s = 0$. The transfer function of the feedback composition with proportional gain K is

$$\hat{H}(s) = \frac{K\hat{H}_2(s)}{1 + K\hat{H}_2(s)} = \frac{KL}{Ms^2 + Ds + KL}.$$

There are two closed-loop poles—the roots of $Ms^2 + Ds + KL$—located at

$$s = -\frac{D}{2M} \pm \sqrt{\frac{D^2}{4M^2} - \frac{KL}{M}}.$$

The closed-loop system is stable if both poles have negative real parts, which is the case if $K > 0$. If $K < D^2/(4ML)$ both poles are real. But if $K > D^2/(4ML)$, the two poles form a complex conjugate pair located at

$$s = -\frac{D}{2M} \pm i\sqrt{\frac{KL}{M} - \frac{D^2}{4M^2}}.$$

The real part is fixed at $D/2M$, but the imaginary part increases with K. We investigate performance for the parameter values $L/M = 10, D/M = 0.1$. The transfer function is

$$\hat{H}(s) = \frac{10K}{s^2 + 0.1s + 10K}.$$

Because there are two poles, the root locus has two branches, as shown in figure 14.9. For $K = 0$, the two poles are located at 0 and -0.1, as illustrated by crosses in the figure. As K increases the two poles move toward each other, coinciding at -0.05 when $K = 0.00025$. For larger values of K, the two branches split into a pair of complex-conjugate poles.

To appreciate what values of $K > 0$ to select for good tracking, we consider the response to a step input $x = u(t)$ for two different values of K. For $K = 0.00025$, the Laplace transform of the output y is

$$\hat{Y}(s) = \frac{10K}{s^2 + 0.1s + 10K}\frac{1}{s} = \frac{0.0025}{(s+0.05)^2}\frac{1}{s} = -\frac{1}{s+0.05} - \frac{0.05}{(s+0.05)^2} + \frac{1}{s}.$$

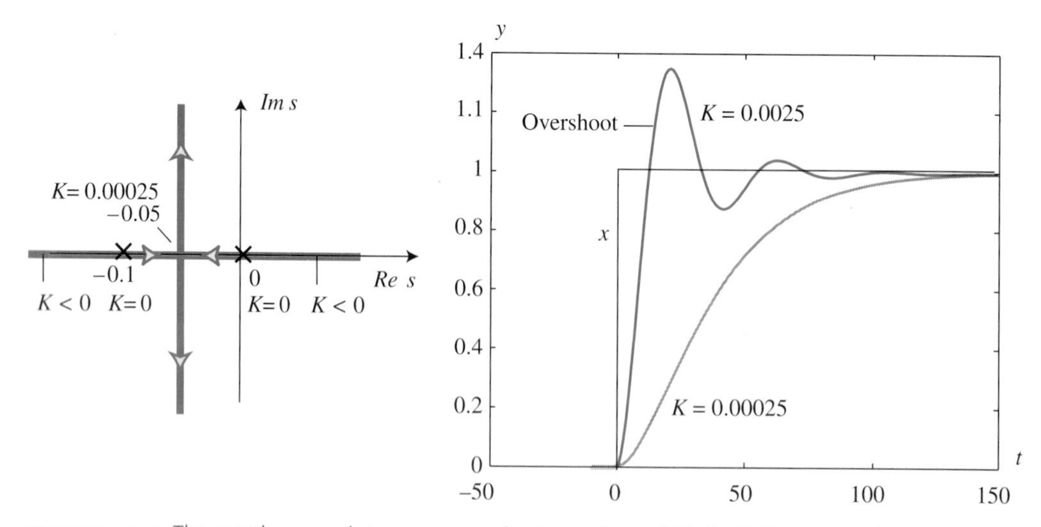

FIGURE 14.9: The root locus and step response for two values of K of a DC motor with proportional feedback.

So the time domain response is

$$\forall t, \quad y(t) = \{-e^{-0.05t} - 0.05te^{-0.05t}\}u(t) + u(t). \tag{14.3}$$

For $K = 0.0025$, the Laplace transform of the output y is

$$\hat{Y}(s) = \frac{0.025}{s^2 + 0.1s + 0.025}\frac{1}{s} \approx \frac{-0.5 + i0.17}{s + 0.05 - i0.15} + \frac{-0.5 - 0.17i}{s + 0.05 + i0.15} + \frac{1}{s}.$$

So,

$$\begin{aligned}\forall t, \quad y(t) &= e^{-0.5t}[0.527e^{i(0.15t+2.82)} + 0.527e^{-i(0.15t+2.82)}]u(t) + u(t) \\ &= 0.527e^{-0.5t} \times 2\cos(0.15t + 2.82)u(t) + u(t). \end{aligned} \tag{14.4}$$

The right-hand panel in figure 14.9 shows plots of the responses (14.3) and (14.4) that illustrate the design tradeoffs. In both cases, the output approaches the input as $t \to \infty$, so the asymptotic tracking error is zero. The response for the higher gain is faster but it overshoots the asymptotic value. The response for the lower gain is slower but there is no overshoot. In this example, K must be selected to balance speed of response versus the magnitude of the overshoot. In some applications, overshoot may be completely unacceptable. ❐

We can now investigate the proportional feedback control in a general setting. Suppose the plant transfer function is a proper rational polynomial

$$\hat{H}_2(s) = \frac{A(s)}{B(s)},$$

where A has degree M, B has degree N, and $M \leq N$ ($\hat{H}_2$ is proper). The closed-loop transfer function is

$$\hat{H}(s) = \frac{K\hat{H}_2(s)}{1 + K\hat{H}_2(s)} = \frac{KA(s)}{B(s) + KA(s)}. \tag{14.5}$$

The closed-loop poles are the N roots of the equation $B(s) + KA(s) = 0$. These roots will depend on K, so we denote them $p_1(K), \ldots, p_n(K)$. As K varies, these roots will trace out the N branches of the root locus. At $K = 0$, the poles are the roots of $B(s) = 0$, which are the poles of the plant transfer function $A(s)/B(s)$. The stability of the closed-loop plant requires that K must be such that

$$Re\{p_1(K)\} < 0, \ldots, Re\{p_n(K)\} < 0. \tag{14.6}$$

Within those values of K that satisfy (14.6) we must select K to get a good response.

The following example shows that a proportional compensator may be unable to guarantee closed-loop stability.

Example 14.20: Consider a plant transfer function given by

$$\hat{H}_2(s) = \frac{A(s)}{B(s)} = \frac{s+1}{(s-1)(s^2+s+1.25)} .$$

There is one zero at $s = -1$, one pole at $s = 1$ and a pair of complex conjugate poles at $s = -0.5 \pm i$. The plant is unstable because of the pole at $s = 1$. The closed-loop poles are the roots of the polynomial

$$P(K, s) = KA(s) + B(s) = K(s+1) + (s-1)((s+0.5)^2 + 1).$$

Figure 14.10 shows the three branches of the root locus plot for $K > 0$. As K increases, the unstable pole moves toward the zero, while the complex-conjugate poles move into the right-half plane. We need to find the values of K that satisfy the stability condition (14.6). The value of K for which the pole at $s = 1$ moves to $s = 0$ is obtained from the condition $P(K, 0) = 0$, which gives $-K - (0.5^2 + 1) = 0$ or $K = 1.25$. So the first condition for stability is $K > 1.25$. The complex-conjugate poles cross the imaginary axis at $s = \pm i1.23$. This occurs at K given by $P(K, i1.23) = 0$, which gives $K = 1.1$. So the second condition for stability is $K < 1.1$. The two conditions $K > 1.25$ and $K < 1.1$ are inconsistent, so no proportional compensator can stabilize this system. ❐

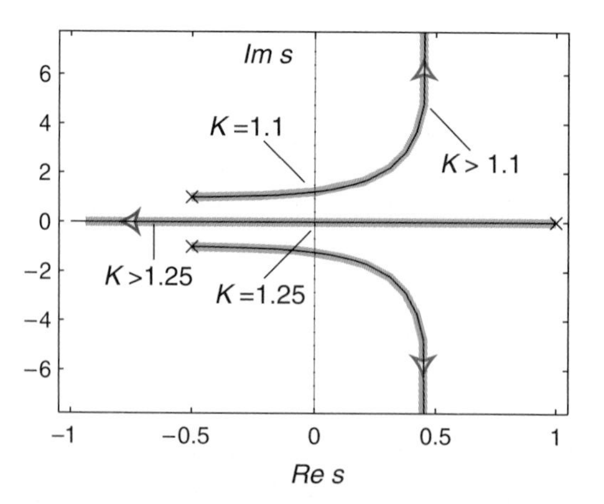

FIGURE 14.10: Root locus. Stability requires $K > 1.25$ and $K < 1.1$.

We return to the general discussion. Assume the stability condition (14.6) can be met. Among the values of K that achieve stability, we select that value for which the output y closely tracks a step input, $x = u$. In this case, the Laplace transform of the input is $1/s$, so the Laplace transform of y from (14.5) is

$$\hat{Y}(s) = \hat{H}(s)\frac{1}{s} = \frac{KA(s)}{B(s) + KA(s)}\frac{1}{s}. \tag{14.7}$$

Assuming for simplicity that all the poles $p_1(K), \ldots, p_n(K)$ have multiplicity 1, $\hat{Y}$ has the partial fraction expansion

$$\hat{Y}(s) = \sum_{i=1}^{N} \frac{R_i}{s - p_i(K)} + \frac{R_0}{s},$$

and, hence, the time-domain behavior of

$$\forall t, \quad y(t) = \left[\sum_{i=1}^{N} R_i e^{p_i(K)t}\right] u(t) + R_0 u(t).$$

The first term is the transient response, y_{tr}, and the second term is the steady-state response $y_{ss} = R_0 u$. The transient response goes to zero because, from (14.6), $Re\{p_i(K)\} < 0$ for all i. The input is the unit step, $x = u$. So the steady-state tracking error is $|R_0 - 1|$, which depends on R_0. It is easy to find the residue R_0. We simply multiply both sides of (14.7) by s and evaluate both sides at $s = 0$, to get

$$R_0 = \hat{G}(0) = \frac{K\hat{H}_2(0)}{1 + K\hat{H}_2(0)}.$$

To have zero steady-state error, we want $R_0 = 1$, which can only happen if $\hat{H}_2(0) = \infty$. But this means $s = 0$ must be a pole of the plant transfer function $\hat{H}_2$. (This is the case in the examples of the helicopter and the DC motor.) If the plant does not have a pole at $s = 0$, the steady-state error is

$$\left|1 - \frac{K\hat{H}_2(0)}{1 + K\hat{H}_2(0)}\right|.$$

This error is smaller the larger the gain K. So to minimize the steady-state error, we should choose as large a gain as possible, subject to the stability requirement (14.6).

However, a large value of K may lead to poor transient behavior by causing overshoots, as happened in the DC motor example in figure 14.9 for the larger gain, $K = 0.0025$. Deciding on the appropriate K is a matter of trial and error, one studies the transient response for different (stabilizing) values of K (as we did for the DC motor) and selects K that gives a satisfactory transient behavior.

14.4 PID controllers

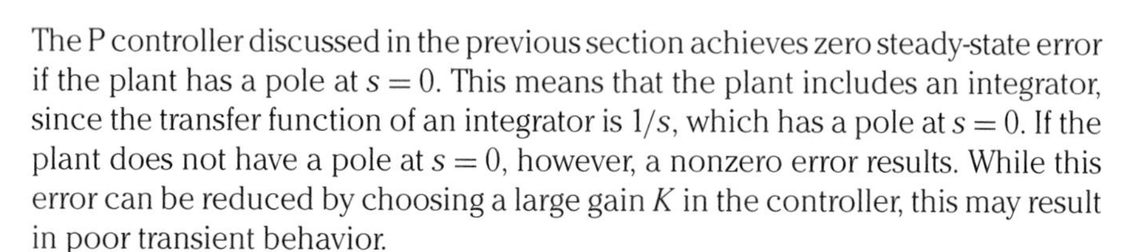

The P controller discussed in the previous section achieves zero steady-state error if the plant has a pole at $s = 0$. This means that the plant includes an integrator, since the transfer function of an integrator is $1/s$, which has a pole at $s = 0$. If the plant does not have a pole at $s = 0$, however, a nonzero error results. While this error can be reduced by choosing a large gain K in the controller, this may result in poor transient behavior.

In this section, we develop the well-known **PID controller**, which includes an integrator in the controller. It can achieve zero steady-state error even if the plant does not have a pole at $s = 0$, and still achieve reasonable transient behavior. The PID controller is a generalization of the P controller, in that with certain choices of parameters, it becomes a P controller.

We begin with an example that has rich enough dynamics to demonstrate the strengths of the PID controller. This example describes a mechanical system, but almost any physical system that is modeled by a linear second-order differential equation is subject to similar analysis. This includes, for example, electrical circuits having resistors, capacitors, and inductors.

Example 14.21: A basic **mass-spring-damper system** is illustrated in figure 14.11. This system has a mass M that slides on a frictionless surface, a spring that attaches the mass to a fixed physical object, and a damper, which absorbes mechanical energy. A damper might be, for example, a dashpot, which is a cylinder filled with oil plus a piston, such as a shock absorber in the suspension system of a car.

Suppose that an external force w is applied to the mass, where w is a continuous-time signal. The differential equation governing the system is obtained by setting the sum of all forces to zero,

$$M\ddot{y}(t) + D\dot{y}(t) + Cy(t) = w(t).$$

The output $y(t)$ is the position of the mass at time t, $M\ddot{y}(t)$ is the inertial force, $D\dot{y}(t)$ is the damping force due to the damper, $Cy(t)$ is the restoring force of the spring, and $w(t)$ is the externally applied force. We assume that $y(t) = 0$ when the spring is in its equilibrium position (neither extended nor

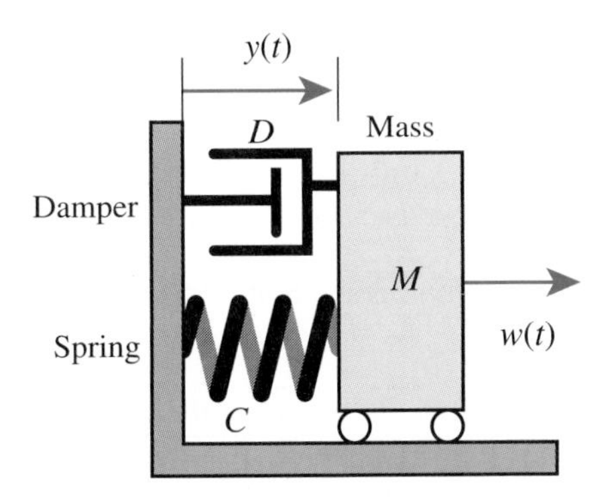

$w \longrightarrow \boxed{\hat{H}_2(s) = \dfrac{1}{Ms^2 + Ds + C}} \longrightarrow y$

FIGURE 14.11: A mass-spring-damper system.

compressed). M, D, and C are constants. Taking the Laplace transform, using the differentiation property from table 13.4, gives

$$s^2\hat{Y}(s) + Ds\hat{Y}(s) + C\hat{Y}(s) = \hat{W}(s),$$

so the plant or open-loop transfer function is

$$\hat{H}_2(s) = \frac{\hat{Y}(s)}{\hat{W}(s)} = \frac{1}{Ms^2 + Ds + C}.$$

Suppose, for example, that the constants have values $M = 1$, $D = 1$, and $C = 1.25$. Then,

$$\hat{H}_2(s) = \frac{1}{s^2 + s + 1.25}. \tag{14.8}$$

In this case, the transfer function has a pair of complex poles at $s = -0.5 \pm i$. Since their real part is strictly negative, the system is stable.

Suppose we wish to drive the system to move the mass to the right one unit of distance at time $t = 0$. We can apply an input force that is a unit step, scaled so that the steady-state response places the mass at position

$y(t) = 1$. The final steady-state output is determined by the DC gain, which is $\hat{H}_2(0) = 1/1.25 = 0.8$, so we can apply an input given by

$$\forall t, \quad w(t) = \frac{1}{0.8} u(t) = 1.25 u(t),$$

where u is the unit step signal. The resulting response y_o has Laplace transform

$$\hat{Y}_o(s) = \frac{1}{s^2 + s + 1.25} \cdot \frac{1.25}{s} = \frac{-0.5 + 0.25i}{s + 0.5 - i} + \frac{-0.5 - 0.25i}{s + 0.5 + i} + \frac{1}{s}.$$

We call this the open-loop step response because there is no control loop (yet).

Taking the inverse transform gives the open-loop step response

$$\forall t, \quad y_o(t) = e^{-0.5t}[(-0.5 + 0.25i)e^{it} + (-0.5 - 0.25i)e^{-it}]u(t) + u(t).$$

By combining the complex-conjugate terms, this can be expressed as

$$\forall t, \quad y_o(t) = Re^{-0.5t} \cos(t + \theta) u(t) + u(t),$$

where $R = 1.12$ and $\theta = 2.68$. Figure 14.12 displays a plot of this open-loop step response y_o. Notice that the mass settles to position $y(t) = 1$ for large t. ❐

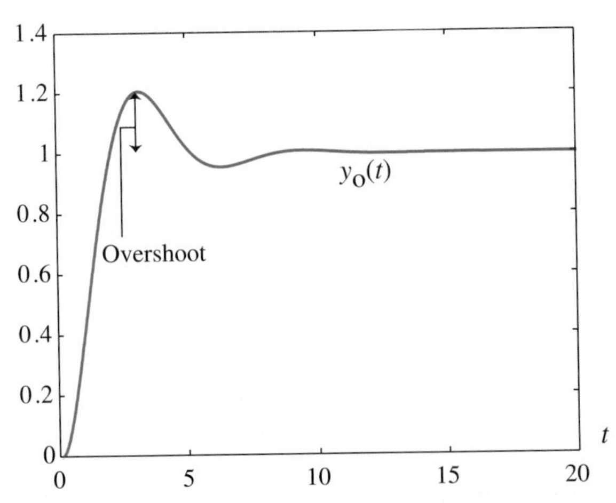

FIGURE 14.12: The open-loop step response y_o of the mass-spring-damper system.

The system in the previous example is stable, and therefore does not need a feedback control loop to stabilize it. However, there are two difficulties with its open-loop response, shown in figure 14.12, that can be corrected using a controller. First, it takes approximately 10 units of time for the transient to disappear, which may be too slow for some applications. Second, there is an overshoot of 20 percent beyond the final steady-state value, which may be too much.

We can correct for the slow response and the large overshoot using a **PID controller**. The term *PID* stands for proportional plus integral plus derivative. A PID controller generalizes the P controller of the previous section by adding an integral and derivative term.

The general form of the transfer function of a PID controller is

$$\hat{H}_1(s) = K_1 + \frac{K_2}{s} + K_3 s = \frac{K_3 s^2 + K_1 s + K_2}{s}. \tag{14.9}$$

We compose this with the plant in a feedback loop, as shown in figure 14.13. Here K_1, K_2, K_3 are constants to be selected by the designer. If $K_2 = K_3 = 0$, then we have a P controller. If $K_1 = K_3 = 0$, $\hat{H}_1(s) = K_2/s$, then we have an **integral controller**, so called because $1/s$ is the transfer function of an integrator. That is, if the input to the integral controller is e, and the output is w, then

$$\forall\, t, \quad w(t) = K_2 \int_{-\infty}^{t} e(\tau) d\tau.$$

If $K_1 = K_2 = 0$, $\hat{H}_2(s) = K_3 s$, then we have a **derivative controller**, so called because s is the transfer function of a differentiator. That is, if the input to the derivative controller is e, and the output is w, then

$$\forall\, t, \quad w(t) = K_3 \dot{e}(t).$$

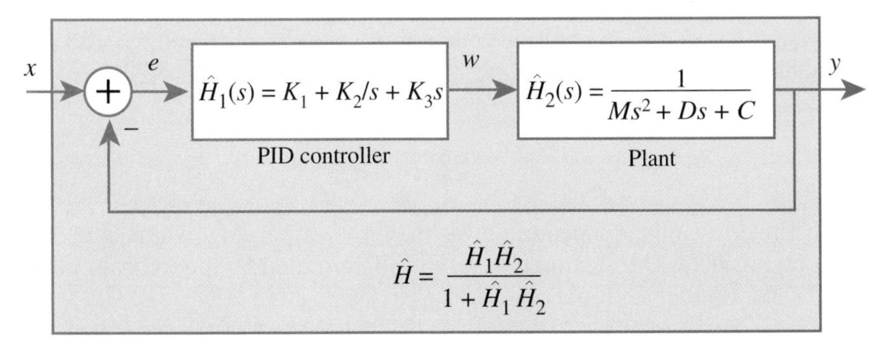

FIGURE 14.13: The mass-spring-damper system composed with a PID controller in a feedback composition.

The following table offers guidelines for selecting the parameters of a PID controller. Of course, these are guidelines only—the actual performance of the closed-loop system depends on the plant transfer function and must be checked in detail.

Parameter	Response speed	Overshoot	Steady-state error
K_1	Faster	Larger	Decreases
K_2	Faster	Larger	Zero
K_3	Minor change	Smaller	Minor change

Example 14.22: We now evaluate a PID controller for the mass-spring-damper system of figure 14.11, using the feedback composition of figure 14.6. We assume the parameters values $M = 1$, $D = 1$, and $C = 1.25$, as in example 14.21. The closed-loop transfer function with the PID controller is

$$\hat{H}(s) = \frac{\hat{H}_1(s)\hat{H}_2(s)}{1 + \hat{H}_1(s)\hat{H}_2(s)} = \frac{K_3 s^2 + K_1 s + K_2}{s^3 + (1 + K_3)s^2 + (1.25 + K_1)s + K_2}.$$

Assume we provide a unit step as input. This means that we wish to move the mass to the right one unit of distance, starting at time $t = 0$. The controller attempts to track this input. The response to a unit step input has Laplace transform

$$\hat{Y}_{pid}(s) = \hat{H}(s) \cdot \frac{1}{s} = \frac{K_3 s^2 + K_1 s + K_2}{s^3 + (1 + K_3)s^2 + (1.25 + K_1)s + K_2} \cdot \frac{1}{s}. \quad (14.10)$$

We now need to select the values for the parameters of the PID controller, K_1, K_2, and K_3. We first try proportional control with $K_1 = 10$, and $K_2 = K_3 = 0$. In this case, the step response has the Laplace transform

$$\hat{Y}_p(s) = \frac{10}{s^2 + s + 11.25} \cdot \frac{1}{s}.$$

The inverse Laplace transform gives the time response y_p, which is plotted in figure 14.14. The steady-state value is determined by the DC gain of the closed-loop transfer function,

$$\left.\frac{10}{s^2 + s + 11.25}\right|_{s=0} = \frac{10}{11.25} \approx 0.89.$$

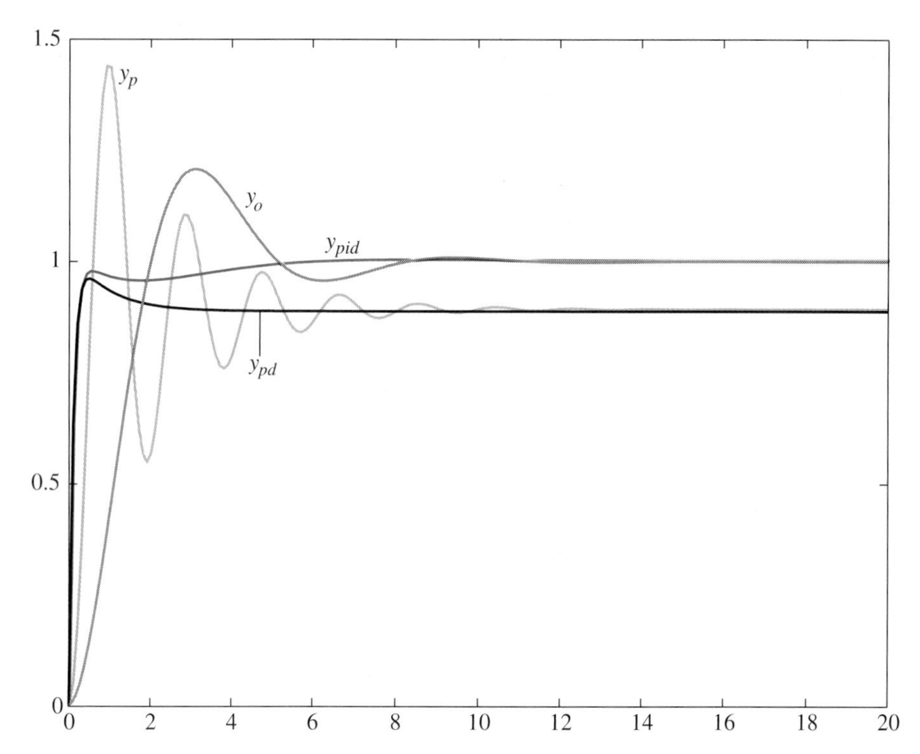

FIGURE 14.14: The step response for open loop, y_o, with P control, y_p, PD control, y_{pd}, and PID control, y_{pid}.

This yields an error of 11 percent, and the overshoot of 50 percent is much worse than that of the open-loop response y_o, also shown in the figure. Thus, a P controller with gain $K = 10$ is useless for this application.

Following the guidelines in the table on page 578, we add derivative control to reduce the overshoot. The result is a so-called **PD controller**, because it adds a proportional and a derivative term. For the PD controller we choose $K_1 = 10$ and $K_3 = 10$. Substitution in (14.10) gives the Laplace transform of the step response,

$$\hat{Y}_{pd}(s) = \frac{10s + 10}{s^2 + 11s + 11.25} \cdot \frac{1}{s}.$$

The steady-state value is given by the DC gain of the closed-loop transfer function,

$$\left. \frac{10s + 10}{s^2 + 11s + 11.25} \right|_{s=0} \approx 0.89,$$

which is the same as the steady-state value for the P controller. The inverse Laplace transform gives the time response y_{pd}, which is plotted in figure 14.14. The overshoot is reduced to 10 percent—a large improvement. Also, the response is quicker—the transient disappears in about 4 time units.

Finally, to eliminate the steady-state error, we add integral control. For the PID controller we choose $K_1 = 10, K_2 = 5, K_3 = 10$. Substitution in (14.10) gives the Laplace transform of the step response

$$\hat{Y}_{pid}(s) = \frac{10s^2 + 10s + 5}{s^3 + 11s^2 + 11.25s + 5} \cdot \frac{1}{s}.$$

The steady-state value is again given by the DC gain of the closed-loop transfer function,

$$\left. \frac{10s^2 + 10s + 5}{s^3 + 11s^2 + 11.25s + 5} \right|_{s=0} = 1.$$

So the steady-state error is eliminated, as expected. The time response y_{pid} is plotted in figure 14.14. It shows significant improvement over the other responses. There is no overshoot, and the transient disappears in about four time units. Further tuning of the parameters K_1, K_2, K_3 could yield small improvements. ❐

14.5 Summary

This chapter considers cascade, parallel, and feedback compositions of LTI systems described by Z or Laplace transforms. Cascade composition is applied to equalization; parallel composition is applied to noise cancelation; and feedback composition is applied to control.

Because we are using Z and Laplace transforms rather than Fourier transforms, we are able to consider unstable systems. In particular, we find that while, in principle, cascade and parallel compositions can be used to stabilize unstable systems, the result is not robust. Small changes in parameter values can result in the system again becoming unstable. Feedback composition, on the other hand, can be used to robustly stabilize unstable systems. We illustrate this first with a simple helicopter example. The second example, a DC motor, benefits from more sophisticated controllers. The third example, a mass-spring-damper system, motivates the development of the well-known PID controller structure. PID controllers can be used to stabilize unstable systems and to improve the response time, precision, and overshoot of stable systems.

KEY: E = mechanical T = requires plan of attack C = more than 1 answer

EXERCISES

E 1. This exercise studies equalization when the channel is only known approximately. Consider the cascade composition of figure 14.1, where $\hat{H}_1$ is the channel to be equalized, and $\hat{H}_2$ is the equalizer. If the equalizer is working perfectly, then $x = y$. For example, if

$$\hat{H}_1(z) = \frac{z}{z - 0.5} \quad \text{and} \quad \hat{H}_2(z) = \frac{z - 0.5}{z},$$

then $x = y$ because $\hat{H}_1(z)\hat{H}_2(z) = 1$.

(a) Suppose that $\hat{H}_2(z)$ remains the same, but the plant is a bit different,

$$\hat{H}_1(z) = \frac{z}{z - 0.5 - \epsilon}.$$

Suppose that $x = \delta$, the Kronecker delta function. Plot $y - x$ for $\epsilon = 0.1$ and $\epsilon = -0.1$.

(b) Now suppose that the equalizer is

$$\hat{H}_2(z) = \frac{z - 2}{z},$$

and the channel is

$$\hat{H}_1(z) = \frac{z}{z - 2 - \epsilon}.$$

Again suppose that $x = \delta$, the Kronecker delta function. Plot $y - x$ for $\epsilon = 0.1, -0.1$.

(c) For part (b), show that equalization error $y - x$ grows without bound whenever $\epsilon \neq 0$, $|\epsilon| < 1$.

E 2. This exercise studies equalization for continuous-time channels. Consider the cascade composition of figure 14.1, where $\hat{H}_1$ is the channel to be equalized, and $\hat{H}_2$ is the equalizer. Both are causal. If

$$\hat{H}_1(s) = \frac{s + 1}{s + 2} \quad \text{and} \quad \hat{H}_2(s) = \frac{s + 2}{s + 1},$$

then $x = y$ because $\hat{H}_1(s)\hat{H}_2(s) = 1$.

(a) Suppose $\hat{H}_2$ remains the same but

$$\hat{H}_1(s) = \frac{s+1}{s+2+\epsilon}.$$

Assume $x = u$ for the unit step. Plot $y - x$ for $\epsilon = 0.1$ and $\epsilon = -0.1$, and calculate the steady-state error.

(b) Now suppose the equalizer is

$$\hat{H}_2(s) = \frac{s-1}{s+2},$$

and the channel is

$$\hat{H}_1(s) = \frac{s+2}{s-1-\epsilon}.$$

Again suppose that $x = u$. Plot $y - x$ for $\epsilon = 0.1, -0.1$.

(c) For part (b) show that the error $y - x$ grows without bound for any $\epsilon \neq 0$, $|\epsilon| < 1$.

T 3. This exercise explores decision-directed equalization. The arrangement is shown in figure 14.15. The transmitted signal is a binary sequence x: *Integers* $\to \{0, 1\}$. The causal channel transfer function is $\hat{H}_1$ and the equalizer transfer function is $\hat{H}_2$. The channel output is the real-valued signal v: *Integers* $\to$ *Reals*. The equalizer output is the real-valued signal y: *Integers* $\to$ *Reals*. This signal is fed to a decision unit whose binary output at time n, $w(n) = 0$ if $y(n) < 0.5$ and $w(n) = 1$ if $y(n) \geq 0.5$. Thus, the decision unit is a (nonlinear) memoryless system,

$$\textit{Decision}: [\textit{Integers} \to \textit{Reals}] \to [\textit{Integers} \to \{0, 1\}],$$

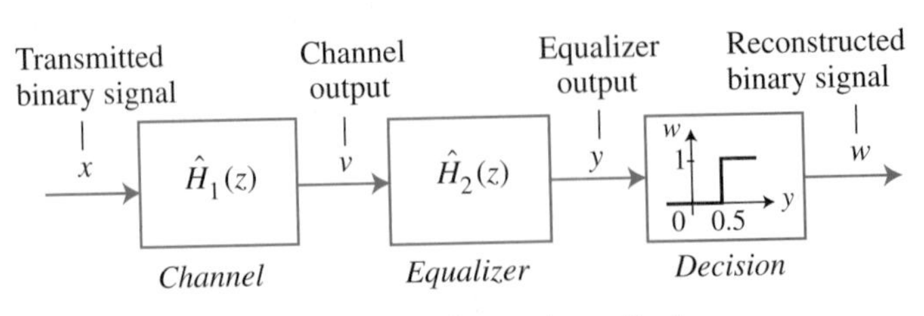

FIGURE 14.15: Arrangement of decision-directed equalization.

defined by a threshold rule

$$\forall n, \quad (Decision(y))(n) = \begin{cases} 0, & y(n) < 0.5 \\ 1, & y(n) \geq 0.5. \end{cases}$$

At each point in time, the receiver has an estimate $\hat{H}_1^e$ of the true channel transfer function, $\hat{H}_1$. The equalizer is set at

$$\hat{H}_2(z) = [\hat{H}_1^e(z)]^{-1}. \tag{14.11}$$

(a) Suppose that initially $\hat{H}_1(z) = \frac{z}{z-0.2}$, and the estimate is perfect, $\hat{H}_1^e = \hat{H}_1$. (This perfect estimate is achieved using a known training sequence for x.) Determine the respective impulse responses h_1 and h_2.

Now suppose the signal x is

$$\forall n, \quad x(n) = \begin{cases} 0, & n < 0 \\ 1, & n \geq 0, n \text{ even} \\ 0, & n \geq 0, n \text{ odd}. \end{cases} \tag{14.12}$$

Calculate the channel output $v(n) = (h_1 * x)(n), n \leq 3$, then calculate the equalizer output $y(n) = (h_2 * v)(n), n \leq 3$, and check that $y(n) = x(n), n \leq 3$. Also check that $w(n) = x(n), n \leq 3$.

(b) Now suppose the channel transfer function has changed to

$$\hat{H}_1(z) = \frac{z}{z - 0.3},$$

but the receiver's estimate hasn't changed—that is,

$$\hat{H}_1^e(z) = \frac{z}{z - 0.2},$$

so the equalizer (14.11) hasn't changed either. For the same input signal, again calculate the channel and equalizer outputs $v(n), y(n), n \leq 3$. Check that $y(n) \neq x(n), n > 0$. But show that the decision $w(n) = x(n), n \leq 3$, so the equalizer correctly determines x.

(c) Since the receiver's decision $w = x$, we can make a new estimate of the channel using the fact that $\hat{Y} = \hat{H}_1\hat{H}_2\hat{X} = \hat{H}_1\hat{H}_2\hat{W}$. The new estimate is

$$\hat{H}_1^e = \frac{\hat{Y}}{\hat{H}_2\hat{W}}. \tag{14.13}$$

Suppose the time is 3, and the receiver has observed $y(n), w(n), n \leq 3$. Since the Z transforms $\hat{Y}$ and $\hat{W}$ also depend on values of $y(n), w(n)$ for $n > 3$, these Z transforms *cannot* be calculated at time $n = 3$, and so the estimator (14.13) cannot be used. The following approach will work, however.

Suppose the receiver knows that the unknown channel transfer function is of the form

$$\hat{H}_1(z) = \frac{z}{z - a},$$

so that only the parameter a has to be estimated. Using this information, we have

$$\hat{Y}(z) = \frac{z}{z - a} \frac{z - 0.2}{z} \hat{W}(z) = \frac{z - 0.2}{z - a} W(z).$$

Now take the inverse Z transform and express the time-domain relation between y and w. Show that you can estimate a knowing $y(0), y(1), w(0), w(1)$.

E 4. This continues exercise 3. It shows that if the channel estimate $\hat{H}_1^e$ is not sufficiently close to the true channel $\hat{H}_1$, the decision may become incorrect. Suppose the true channel is $\hat{H}_1(z) = \frac{z}{z-a}$, the estimate is $\hat{H}_1^e(z) = \frac{z}{z-0.2}$, the equalizer is $\hat{H}_2(z) = [\hat{H}_1^e(z)]^{-1} = \frac{z-0.2}{z}$, and the decision is as in figure 14.15. Assume the input signal x to be the same as in (14.12). Show that if $a = 0.6$ then $w(0) = x(0), w(1) = x(1), w(2) = x(2)$, but $w(3) \neq x(3)$.

E 5. This exercise continues the discussion in examples 14.5 and 14.6 for the continuous-time, causal and stable channel with impulse response h_1 and transfer function

$$\forall\, s \in RoC(h_1) = \{s \mid Re\{s\} > -1\}, \quad \hat{H}_1(s) = \frac{s - 2}{s + 1}.$$

(a) Calculate h_1 and sketch it. (Observe how the zero in the right-half plane at $s = 2$ accounts for the negative values.)

(b) The inverse of $\hat{H}_1$,

$$\hat{H}_2(s) = \frac{s + 1}{s - 2},$$

has a pole at $s = 2$. So as a causal system, the inverse is unstable. But as a noncausal system it is stable with $RoC = \{s \mid Re\{s\} < 2\}$, which includes the imaginary axis. Evaluate the impulse response h_2 of $\hat{H}_2$ as an anticausal system, and give a sketch.

(c) The impulse response h_2 calculated in (a) is nonzero for $t \leq 0$. Consider the finite-duration, anticausal impulse response h_3 obtained by truncating h_2 before time -5,

$$\forall t \in Reals, \quad h_3(t) = \begin{cases} h_2(t), & t \geq -5 \\ 0, & t < -5 \end{cases}$$

and sketch h_3. Calculate the transfer function $\hat{H}_3$, including its RoC, by using the definition of the Laplace transform.

(d) Obtain the causal impulse response h_4 by delaying h_3 by time $T = 5$—that is,

$$\forall t \in Reals, \quad h_4(t) = h_3(t + T).$$

Sketch h_4 and find its transfer function, $\hat{H}_4$. Then $\hat{H}_4$ is an approximate inverse of $\hat{H}_1$ with a delay of 5 time units. (Note: h_3 has a delta function at 0.)

T 6. The proportional controller of figure 14.7 stabilizes the plant for $K > 1$. In this exercise, we try to achieve the same effect by the cascade compensator of figure 14.1.

(a) Assume that the plant $\hat{H}_2$ is as given in figure 14.7. Design $\hat{H}_1$ for the cascade composition of figure 14.1 so that $\hat{H}_2\hat{H}_1$ is the same as the closed-loop transfer function achieved in figure 14.7.

(b) Now suppose that the model of the plant is not perfect, and the plant's real transfer function is

$$\hat{H}_2(s) = \frac{1}{M(s - \epsilon)},$$

for some small value of $\epsilon > 0$. Using the same $\hat{H}_1$ that you designed in part (a), what is the transfer function of the cascade composition? Is it stable?

T 7. Consider a discrete-time causal plant with transfer function

$$\hat{H}_2(z) = \frac{z}{z - 2}.$$

(a) Where are the poles and zeros? Is the plant stable?

(b) Find the impulse response of the plant. Is it bounded?

(c) Give the closed-loop transfer function for the P controller for this plant.

(d) Sketch the root locus for the P controller for this plant.

(e) For what values of K is the closed-loop system stable?

(f) Find the step response of the closed-loop system. Identify the transient and steady-state responses. For $K = 10$, what is the steady-state tracking error?

(g) Suppose that the plant is instead defined by

$$\hat{H}_2 = \frac{z}{z - 2 - \epsilon},$$

for some real $\epsilon \geq 0$. For what values of K is the P controller robust for plants with $|\epsilon| < 0.5$?

T 8. Consider the feedback composition in figure 14.6. Suppose that $\hat{H}_1\hat{H}_2$ is causal and strictly proper, meaning that the order of the numerator is greater than the order of the denominator.

(a) Show that if $\hat{H}_1\hat{H}_2$ is causal and strictly proper, then so is $\hat{H}$, the transfer function of the feedback composition given by (14.1).

(b) For the discrete-time case, show that we can write

$$\hat{H}_1(z)\hat{H}_2(z) = z^{-1}\hat{G}(z), \qquad (14.14)$$

where $G(z)$ is proper, and is the transfer function of a causal system. Intuitively, this means that there must be a net unit delay in the feedback loop, because z^{-1} is the transfer function of a unit delay.

(c) Use the result of part (a) to argue that the system $\hat{H}_1\hat{H}_2$ has state-determined output.

(d) For the continuous-time case, show that we can write

$$\hat{H}_1(s)\hat{H}_2(s) = s^{-1}\hat{G}(s), \qquad (14.15)$$

where $G(s)$ is proper, and is the transfer function of a causal system. Intuitively, this means that there must be an integration in the feedback loop, because s^{-1} is the transfer function of an integrator.

(e) Use the result of part (c) to argue that the system $\hat{H}_1\hat{H}_2$ has state-determined output, assuming that the input is bounded and piecewise continuous.

E 9. Consider the mth order polynomial $s^N + a_{m-1}s^{m-1} + \cdots + a_1 s + a_0$. Suppose all its roots have negative real parts. Show that all coefficients of the polynomial must be positive—that is, $a_{m-1} > 0, \ldots, a_0 > 0$. Hint: Express the polynomial as $(s - p_1) \cdots (s - p_m)$ with $Re\{p_i\} > 0$. Note that complex roots must occur in complex-conjugate pairs. (The positiveness of all coefficients is a necessary condition. A sufficient condition is given by the **Routh-Hurwitz** criterion, described in control theory texts.)

T 10. Consider the feedback composition in figure 14.6. The plant's transfer function is $\hat{H}_2(s) = 1/s^2$.

(a) Show that no PI controller in the form $\hat{H}_1(s) = K_1 + K_2/s$ can stabilize the closed-loop system for any values of K_1, K_2. Hint: Use the result of problem 9.

(b) Show that by the proper choice of the coefficients K_1, K_2 of a PD controller in the form $\hat{H}_1(s) = K_1 + K_2 s$, you can place the closed-loop poles at any locations p_1, p_2 (these must be complex conjugate if they are complex).

T 11. Consider the feedback composition in figure 14.6. The plant's transfer function is $\hat{H}_2(s) = 1/(s^2 + 2s + 1)$. The PI controller is $\hat{H}_1(s) = K_1 + K_2/s$.

(a) Take $K_2 = 0$, and plot the root locus as K_1 varies. For what values of K_1 is the closed-loop system stable? What is the steady-state error to a step input as a function of K_1?

(b) Select K_1, K_2 such that the closed-loop system is stable and has zero steady-state error.

APPENDIX A
Sets and functions

This appendix establishes the notation of sets and functions as used in the text. We review the use of this mathematical language to describe sets in a variety of ways; to combine sets using the operations of union, intersection, and product; and to derive logical consequences. We also review how to formulate and understand predicates and define certain sets that occur frequently in the study of signals and systems. Finally, we review functions.

A.1 Sets

A **set** is a collection of **elements**. The set of **natural numbers**, for example, is the collection of all positive integers. This set is denoted (identified) by the name *Naturals*,

$$\boxed{Naturals = \{1, 2, 3, \dots\}.} \tag{A.1}$$

In (A.1) the left side is the name of the set and the right side is an enumeration or list of all the elements of the set. We read (A.1) as "*Naturals* is the set consisting of the numbers 1, 2, 3, and so on." The ellipsis . . . means "and so on." Because *Naturals* is an infinite set, we cannot enumerate all its elements, and so we have to use ellipsis. For a finite set, too, we may use ellipsis as a convenient shorthand, as in

$$A = \{1, 2, 3, \ldots, 100\}, \tag{A.2}$$

which defines A to be the set consisting of the first 100 natural numbers. *Naturals* and A are sets of numbers. The concept of sets is very general, as the following examples illustrate.

Students is the set of all students in a class, each element of which is referenced by a student's name:

$$\textit{Students} = \{\text{John Brown, Jane Doe, Jie Xin Zhou}, \ldots\}.$$

USCities consists of all cities in the United States, referenced by name:

$$\textit{USCities} = \{\text{Albuquerque, San Francisco, New York}, \ldots\}.$$

BooksInLib comprises all books in the U.C. Berkeley library, referenced by a 4-tuple (first author's last name, book title, publisher, year of publication):

$$\begin{aligned}\textit{BooksInLib} = \{&(\text{Lee}, \textit{Digital Communication}, \text{Kluwer, 1994}),\\ &(\text{Walrand}, \textit{Communication Networks}, \text{Morgan Kaufmann, 1996}), \ldots\}.\end{aligned}$$

BookFiles consists of all LaTeX documents for this book, referenced by their file name:

$$\textit{BookFiles} = \{\text{sets.tex, functions.tex}, \ldots\}.$$

We usually use either italicized, capitalized names for sets, such as *Reals* and *Integers*, or single capital letters, such as A, B, X, Y.

An element of a set is also said to be a **member** of the set. The number 10 is a member of the set A defined in (A.2), but the number 110 is not a member of A. We express these two facts by the two expressions:

$$10 \in A, \quad 110 \notin A.$$

The symbol "$\in$" is read "is a member of" or "belongs to" and the symbol "$\notin$" is read "is not a member of" or "does not belong to."

When we define a set by enumerating or listing all its elements, we enclose the list in curly braces $\{\ldots\}$. It is not always necessary to give the set a name. We can instead refer to it directly by enumerating its elements. Thus,

$$\{1, 2, 3, 4, 5\}$$

is the set consisting of the numbers $1, 2, \ldots, 5$.

The order in which the elements of the set appear in the list is not usually significant. When the order is significant, the set is called an **ordered set**.

An element is either a member of a set or it is not. It cannot be a member more than once. So, for example, $\{1, 2, 1\}$ is not a set.

Thus, a set is defined by an unordered collection of its elements, without repetition. Two sets are equal if and only if every element of the first set is also an element of the second set, and every element of the second set is a member of the first set. So if $B = \{1, 2, 3, 4, 5\}$ and $C = \{5, 3, 4, 1, 2\}$, then it is correct to state

$$B = C. \tag{A.3}$$

A.1.1 *Assignment and assertion*

Although the expressions (A.1) and (A.3) are both in the form of equations, the "=" in these two expressions have very different meanings. Expression (A.1) (as well as (A.2)) is an **assignment**: the set on the right side is assigned to the name *Naturals* on the left side. Expression (A.3) is an **assertion**, which is an expression that can be true or false. In other words, an assertion is an expression that has a truth value. Thus, (A.3) asserts that the two sides are equal. Since this is true, (A.3) is a true assertion. But the assertion

$$Naturals = A$$

is a false assertion. An assertion is true or false, while an assignment is a tautology (something that is trivially true because the definition makes it so). Some notation systems make a distinction between an assignment and an assertion by writing an assignment using the symbol ":=" instead of "=" as in

$$Naturals := \{1, 2, 3, \ldots\}.$$

Other notation systems use "=" for assignments and "==" for assertions. We do not make these notational distinctions (except in chapter 6) because it is clear from the context whether an expression is an assignment or an assertion.*

Context is essential in order to disambiguate whether an expression like "$MyNumbers = \{1, 3, 5\}$" is an assertion or an assignment. Thus, for example, in the context,

$$\text{Define the set } MyNumbers \text{ by } MyNumbers = \{1, 3, 5\},$$

*A symbol such as "=," which has more than one meaning depending on the context, is said to be **overloaded**. C++ uses overloaded symbols, while Java does not (Java does support overloaded methods, however). People often have difficulty reading other people's code written in a language that permits overloading.

the expression is clearly an assignment, as indicated by "Define the set . . .". However, in the following context,

If we define *MyNumbers* by *MyNumbers* $= \{1, 3, 5\}$, then *MyNumbers* $= \{3, 5, 1\}$,

the first "=" is an assignment, but the second is an assertion.

A.1.2 Sets of sets

Anything can be an element of a set, so of course sets can be elements of sets. Suppose, for example, that X is a set. Then we can construct the set of all subsets of X, which is written $P(X)$ and is called the **power set** of X.* Notice that since $\emptyset \subset X$, then $\emptyset \in P(X)$. ($\emptyset$ denotes the empty set.)

A.1.3 Variables and predicates

We can refer to a particular element of a set by using its name, for example, the element 55 in *Naturals*, or Berkeley in *USCities*. We often need to refer to a general element in a set. We do this by using a **variable**. We usually use lowercase letters for variable names, such as x, y, n, t. Thus, $n \in Naturals$ refers to any natural number. *Naturals* is the **range** of the variable n. We may also use a character string for a variable name such as $city \in USCities$. We say that "n is a variable over *Naturals*" and "*city* is a variable over *USCities*."

A variable can be assigned (or substituted by) any value in its range. Thus, the assignment $n = 5$ assigns the value 5 to n, and $city =$ Berkeley assigns the value Berkeley to *city*.

Once again the use of "=" in an expression like $n = 5$ is ambiguous, because the expression could be an assignment or an assertion, depending on the context. Thus, for example, in the context,

$$\text{Let } n = 5, m = 6, k = 2, \text{ then } m = k + n$$

the first three "=" are assignments, but the last is an assertion, which happens to be false.

*The power set of X is sometimes written 2^X.

PROBING FURTHER

Predicates in MATLAB®*

In MATLAB, "=" is always used as an assignment, while "==" is used to express an assertion. Thus the MATLAB program,

```
n = 5,   m = 6,   k = 2,   m = k + n
```

returns

```
n = 5,   m = 6,   k = 2,   m = 7
```

because the expression `m=k+n` assigns the value 7 to `m`. However, the MATLAB program

```
n = 5,   m = 6,   k = 2,   m == k + n
```

returns

```
n = 5,   m = 6,   k = 2,   ans = 0
```

where `ans = 0` means that the assertion `m == k+n` evaluates to false.

* MATLAB is a registered trademark of The MathWorks, Inc.

We can use variables to form expressions such as $n \leq 5$. Now, when we assign a particular value to n in this expression, the expression becomes an assertion that evaluates to true or false. (For instance, $n \leq 5$ evaluates to true for $n = 3$ and to false for $n = 6$.) An expression such as $x \leq 5$, which evaluates to true or false when we assign the variable a specific value, is called a **predicate** (in x).* Predicates are used to define new sets from old ones, as in this example:

$$B = \{x \in Naturals \mid x \leq 5\},$$

which reads "B is the set of all elements x in *Naturals* such that (the symbol '|' means 'such that') the predicate $x \leq 5$ is true." More generally, we use the following prototype expression to define a new set *NewSet* from an old set *Set*:

$$\boxed{NewSet = \{x \in Set \mid Pred(x)\}.} \tag{A.4}$$

* A specific value is said to **satisfy** a predicate if the predicate evaluates to true with that value, and **not to satisfy** the predicate if the predicate evaluates to false.

In this prototype expression for defining sets, $x \in Set$ means that x is a variable over the set *Set*, $Pred(x)$ is a predicate in x, and so *NewSet* is the set consisting of all elements x in *Set* for which $Pred(x)$ evaluates to true.

We illustrate the use of the prototype (A.4). In the following examples, note that the concept of predicate is very general. In essence, a predicate is a condition involving any attributes or properties of the elements of *Set*. Consider

$$TallStudents = \{name \in Students \mid name \text{ is more than 6 feet tall}\}.$$

Here the predicate is "*name* is more than 6 feet tall." If John Brown's height is 5′ 10″ and Jane Doe is 6′ 1″ tall, the predicate evaluates to true for *name* = Jane Doe, and to false for *name* = John Brown, so Jane Doe $\in$ *TallStudents* and John Brown $\notin$ *TallStudents*. Consider

$$NorthCities = \{city \in USCities \mid city \text{ is located north of Washington, DC}\}.$$

The predicate here evaluates to true for *city* = New York, and to false for *city* = Atlanta.

The variable name "x" used in (A.4) is not significant. The sets $\{x \in Naturals \mid x \geq 5\}$ and $\{n \in Naturals \mid n \geq 5\}$ are the *same* sets even though the variable names used in the predicates are different. This is because the predicates "$n \geq 5$" and "$x \geq 5$" both evaluate to true or both evaluate to false when the same value is assigned to n and x. We say that the variable name x used in (A.4) is a **dummy variable** since the meaning of the expression is unchanged if we substitute x with another variable name, say y. You are already familiar with the use of dummy variables in integrals. The two integrals below evaluate to the same number:

$$\int_0^1 x^2 dx = \int_0^1 y^2 dy.$$

A.1.4 *Quantification over sets*

Consider the set $A = \{1, 3, 4\}$. Suppose we want to make the assertion that every element of A is smaller than 7. We could do this by the three expressions

$$1 < 7, \quad 3 < 7, \quad 4 < 7,$$

which gets to be very clumsy if A has many more elements. So mathematicians have invented a shorthand. The idea is to be able to say that $x < 7$ for every value that the variable x takes in the set A. The precise expression is

$$\forall x \in A, \quad x < 7. \tag{A.5}$$

The symbol "$\forall$" reads "for all," so the expression reads, "for all values of x in A, $x < 7$." The phrase "$\forall\, x \in A$" is called **universal quantification**. Note that in the expression (A.5), x is again a dummy variable; the meaning of the expression is unchanged if we use another variable name. Note, also, that (A.5) is an assertion which, in this case, evaluates to true. However, the assertion

$$\forall\, x \in A, \quad x > 3$$

is false, since for at least one value of $x \in A$, namely $x = 1$, $x > 3$ is false.

Suppose we want to say that there is at least one element in A that is larger than 3. We can say this using **existential quantification** as in

$$\exists\, x \in A, \quad x > 3. \tag{A.6}$$

The symbol "$\exists$" reads "there exists," so the expression (A.6) reads "there exists a value of x in A, $x > 3$." Once again, any other variable name could be used in place of x, and the meaning of the assertion is unchanged.

In general, the expression,

$$\boxed{\forall\, x \in A, \quad Pred(x),} \tag{A.7}$$

is an assertion that evaluates to true if $Pred(x)$ evaluates to true for every value of $x \in A$, and

$$\boxed{\exists\, x \in A, \quad Pred(x),} \tag{A.8}$$

is an assertion that evaluates to true if $Pred(x)$ evaluates to true for at least one value of $x \in A$.

Conversely, the assertion (A.7) evaluates to false if $Pred(x)$ evaluates to false for at least one value of $x \in A$, and the assertion (A.8) evaluates to false if $Pred(x)$ evaluates to false for every value of $x \in A$.

We can use these two quantifiers to define sets using the prototype new set constructor (A.4). For example,

$$EvenNumbers = \{n \in Naturals \mid \exists\, k \in Naturals,\ n = 2k\}$$

is the set of all even numbers, since the predicate in the variable n,

$$\exists\, k \in Naturals, \quad n = 2k,$$

evaluates to true if and only if n is even.

A.1.5 Some useful sets

The following sets are frequently used in the book:

$Naturals = \{1, 2, \ldots\}$	natural numbers
$Naturals_0 = \{0, 1, 2, \ldots\}$	nonnegative integers
$Integers = \{\ldots, -2, -1, 0, 1, 2, \ldots\}$	integers
$Integers_+ = \{0, 1, 2, \ldots\}$	nonnegative integers, same as $Naturals_0$
$Reals = (-\infty, \infty)$	real numbers
$Reals_+ = [0, \infty)$	nonnegative real numbers
$Complex = \{x + jy \mid x, y \in Reals\}$	complex numbers, $j = \sqrt{-1}$

If α, β are real numbers, then

$$
\begin{aligned}
[\alpha, \beta] &= \{x \in Reals \mid \alpha \le x \le \beta\} \\
(\alpha, \beta) &= \{x \in Reals \mid \alpha < x < \beta\} \\
(\alpha, \beta] &= \{x \in Reals \mid \alpha < x \le \beta\} \\
(-\infty, \infty) &= Reals.
\end{aligned}
$$

Note the meaning of the difference in notation: both end-points α and β are included in $[\alpha, \beta]$, and we say that $[\alpha, \beta]$ is a **closed** interval; neither end-point is included in (α, β), and we call this an **open** interval; the interval $(\alpha, \beta]$ is said to be half-open, half-closed. Whether an end-point is included in an interval or not is indicated by the use of square brackets [,] or parentheses (,).

Other useful sets are:

$Binary = \{0, 1\}$, the binary values

$Binary^* = \{0, 1\}^*$, set of all finite binary strings

$Bools = \{true, false\}$, truth values

$Bools^* = \{true, false\}^*$, set of all finite sequences of truth values

$Char$ = set of all alphanumeric characters

$Char^*$ = set of all finite character strings

A.1.6 *Set operations: union, intersection, complement*

Let $A = \{1, 2, \ldots, 10\}$, and $B = \{1, \ldots, 5\}$. Clearly B is included in A, and A is included in *Naturals*. We express these assertions as:

$$B \subset A, \quad A \subset Naturals,$$

which we read "B is contained in or included in A" and "A is contained in or included in *Naturals*." For the sets above,

$$Naturals \subset Naturals_0 \subset Integers \subset Reals \subset Complex.$$

If A and B are sets, then $A \cap B$ is the set consisting of all elements that are in both A and B, and $A \cup B$ is the set consisting of all elements that are either in A or in B or in both A and B. $A \cap B$ is called the **intersection** of A and B, and $A \cup B$ is called the **union** of A and B. We can express these definitions using variables as:

$$\boxed{A \cap B = \{x \mid x \in A \wedge x \in B\}, \quad A \cup B = \{x \mid x \in A \vee x \in B\}}$$

where $\wedge$ is the notation for the logical **and** and $\vee$ is the symbol for the logical **or**. The predicate "$x \in A \wedge x \in B$" reads "x is a member of A and x is a member of B"; "$x \in A \vee x \in B$" reads "x is a member of A or x is a member of B." The logical and of two predicates is also called their **conjunction** and their logical or is also called their **disjunction**. The symbols $\wedge$ and $\vee$ are called **logical connectives**.

If A, X are sets, then $X \setminus A$ is the set consisting of all elements in X that are not in A (think of it as set subtraction, $X - A$). When $A \subset X$, $X \setminus A$ is called the **complement** of A in X. When X is understood, we can write A^c instead of $X \setminus A$.

We gain some intuitive understanding by depicting sets and set operations using pictures. Figure A.1 illustrates union, intersection, and complement.

A.1.7 *Predicate operations*

Given two predicates $P(x)$ and $Q(x)$, we can form their conjunction $P(x) \wedge Q(x)$ and their disjunction $P(x) \vee Q(x)$. These predicate operations correspond to the set operations of intersection and union:

$$\boxed{\begin{aligned} \{x \in X \mid P(x) \wedge Q(x)\} &= \{x \in X \mid P(x)\} \cap \{x \in X \mid Q(x)\} \\ \{x \in X \mid P(x) \vee Q(x)\} &= \{x \in X \mid P(x)\} \cup \{x \in X \mid Q(x)\}. \end{aligned}}$$

There is a helpful visual similarity between $\wedge$ and $\cap$, and between $\vee$ and $\cup$.

The counterpart of the complement of a set is the **negation** of a predicate. We denote by $\neg Pred(x)$ the predicate that evaluates to false for any value of x

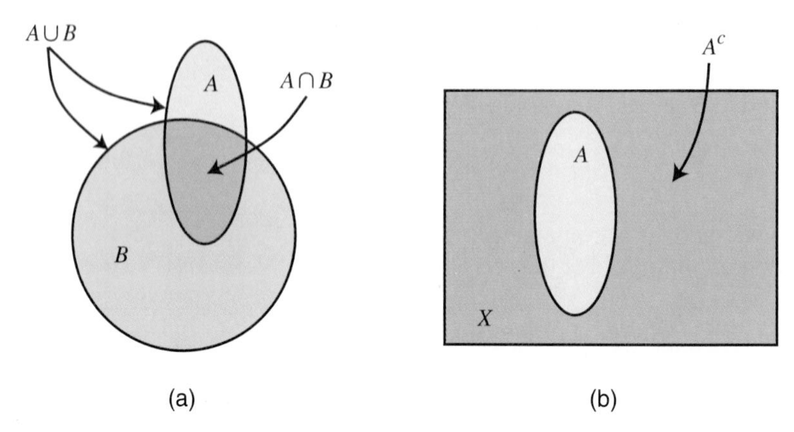

FIGURE A.1: (a) Union and intersection. (b) Set complement.

for which *Pred*(x) evaluates to true, and that evaluates to true for any value of x for which *Pred*(x) evaluates to false. We read "$\neg Pred(x)$" as "not *Pred*(x)" or the "negation of *Pred*(x)." For example,

$$\{n \in Naturals \mid \neg(n < 5)\} = \{5, 6, 7, \ldots\},$$

because $\neg(n < 5)$ evaluates to true if and only if $n < 5$ evaluates to false, which happens if and only if n is larger than or equal to 5.

In general we have the following correspondence between predicate negation and set complement:

$$\{x \in X \mid \neg Pred(x)\} = X \setminus \{x \in X \mid Pred(x)\}. \tag{A.9}$$

We can combine the operations to obtain more complex identities. If $P(x)$ and $Q(x)$ are predicates, then

$$\boxed{\begin{aligned} \{x \in X \mid \neg(P(x) \wedge Q(x))\} &= \{x \in X \mid \neg P(x) \vee \neg Q(x)\} \\ \{x \in X \mid \neg(P(x) \vee Q(x))\} &= \{x \in X \mid \neg P(x) \wedge \neg Q(x)\}. \end{aligned}}$$

These identities have counterparts for set operations. For some set X, if and $Y \subset X$ and $Z \subset X$, then

$$X \setminus (Y \cap Z) = (X \setminus Y) \cup (X \setminus Z)$$

$$X \setminus (Y \cup Z) = (X \setminus Y) \cap (X \setminus Z).$$

When the set X is understood, we can write these as

$$(Y \cap Z)^c = Y^c \cup Z^c$$
$$(Y \cup Z)^c = Y^c \cap Z^c.$$

These identities are called **de Morgan's rules**.

A.1.8 *Permutations and combinations*

Given a set X with a finite number n of elements, we sometimes wish to construct a subset with a fixed number $m \le n$ of elements. The number of such subsets is given by

$$\binom{n}{m} = \frac{n!}{m!(n-m)!}, \tag{A.10}$$

where the exclamation point denotes the factorial function. The notation $\binom{n}{m}$ is read "n **choose** m". It gives the number of **combinations** of m elements chosen from the set n.

A combination is a set, so order does not matter. Sometimes, however, order matters. Suppose, for example, that $X = \{a, b, c\}$. The number of subsets with two elements is

$$\binom{3}{2} = \frac{3!}{2!1!} = \frac{6}{2} = 3.$$

These subsets are $\{a, b\}$, $\{a, c\}$, and $\{b, c\}$. Suppose instead that we wish to construct *ordered* subsets of X with two elements. In other words, we wish to consider $[a, b]$ to be distinct from $[b, a]$ (note the temporary use of square brackets to avoid confusion with unordered sets). Such ordered subsets are called **permutations**. The number of m-element permutations of a set of size n is given by

$$\frac{n!}{(n-m)!}. \tag{A.11}$$

The number of permutations is a factor of $m!$ larger than the number of combinations in (A.10). For example, the number of 2-element permutations of $X = \{a, b, c\}$ is six. They are $[a, b]$, $[a, c]$, $[b, c]$, $[b, a]$, $[c, a]$, and $[c, b]$.

BASICS

Tuples, strings, and sequences

Given N sets, $X_1, X_2, \ldots, X_N$, which may be identical, an **N-tuple** is an ordered collection of one element from each set. It is written in parentheses, as in

$$(x_1, x_2, \ldots, x_N)$$

where

$$x_i \in X_i \text{ for each } i \in \{1, 2, \ldots, N\}.$$

The elements of an N-tuple are called its **components** or **coordinates**. Thus, x_i is the ith component or coordinate of $(x_1, \ldots, x_N)$. The order in which the components are given is important; it is part of the definition of the N-tuple. We can use a variable to refer to the entire N-tuple, as in $x = (x_1, \ldots, x_N)$.

Frequently, the sets from which the tuple components are drawn are all identical, as in

$$(x_1, x_2, \ldots, x_N) \in X^N.$$

This notation means simply that each component in the tuple is a member of the same set X. Of course, this means that a tuple may contain identical components. For example, if $X = \{a, b, c\}$, then (a, a) is a 2-tuple over X.

Recall that a permutation is ordered, like a tuple, but like a set and unlike a tuple, it does not allow duplicate elements. In other words, (a, a) is not a permutation of $\{a, b, c\}$. So a permutation is not the same as a tuple. Similarly, an ordered set does not allow duplicate elements, and thus is not the same as a tuple.

We define the set of finite **sequences** over X to be

$$\{(x_1, \ldots, x_N) \mid x_i \in X, 1 \le i \le N, N \in Naturals_0\},$$

where if $N = 0$ we call the sequence the **empty sequence**. This allows us to talk about tuples without specifying N. Finite sequences are also called **strings**, although by convention, strings are written differently, omitting the parentheses and commas, as in

$$x_1 x_2 \ldots x_N.$$

We may even wish to allow N to be infinite. We define the set of **infinite sequence** over a set X to be

$$\{(x_1, x_2, \ldots) \mid x_i \in X, i \in Naturals_0\}.$$

A.1.9 *Product sets*

The **product** $X \times Y$ of two sets X and Y consists of all pairs of elements (x, y) with $x \in X$ and $y \in Y$, that is,

$$X \times Y = \{(x, y) \mid x \in X, y \in Y\}.$$

The product of two sets may be visualized as in figure A.2. These pictures are informal, to be used only to reinforce intuition. In figure A.2(a), the set $X = [0, 6]$ is represented by a horizontal line segment, and the set $Y = [1, 8]$ is represented by the vertical line segment. The product set $X \times Y = [0, 6] \times [1, 8]$ is represented by the rectangle whose lower left corner is the pair (0, 1) and upper right corner is $(6, 8)$.

In figure A.2(b), the discrete set $X = \{1, 2, 3, 4, 5, 6\}$ is represented by six points, while $Y = [1, 8]$ is represented as before. The product set is depicted by six

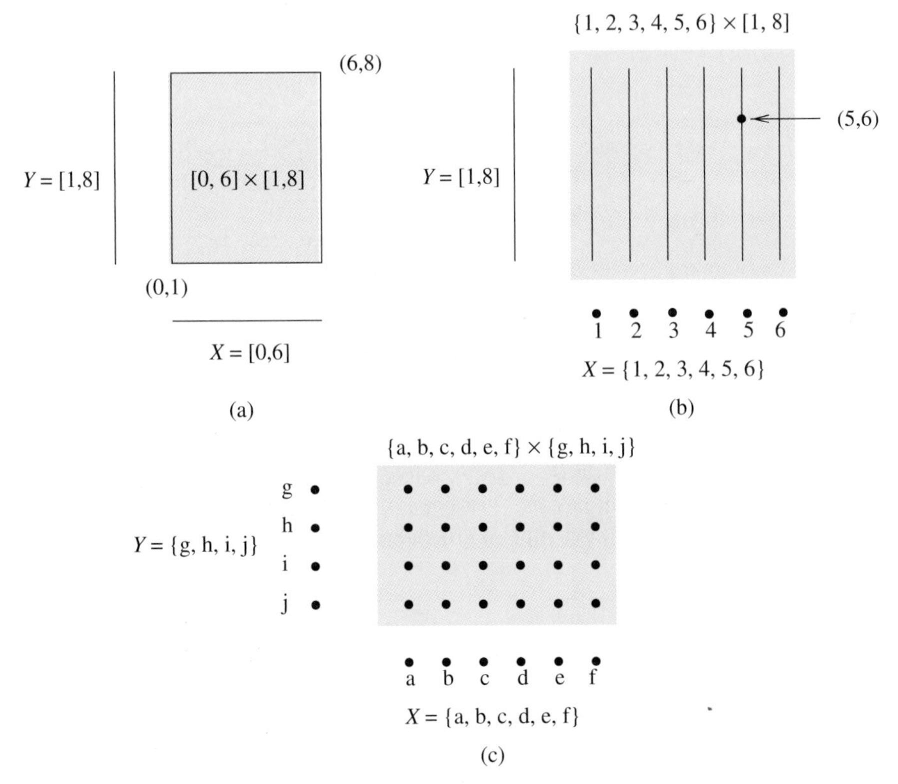

FIGURE A.2: Visualization of product sets. (a) The rectangle depicts the product set $[0, 6] \times [1, 8]$. (b) Together, the six vertical lines depict the product set $\{1, \ldots, 6\} \times [1, 8]$. (c) The array of dots depicts the set $\{a, b, c, d, e, f\} \times \{g, h, i, j\}$.

vertical line segments, one for each element of X. For example, the fifth segment from the left is the set $\{(5, y) \mid 1 \le y \le 8\}$. One point in that set is shown.

In figure A.2(c), the product of two discrete sets is shown as an array of points. Unless these are ordered sets, there is no significance to the left-to-right or top-to-bottom order in which the points are shown. In all three cases in the figure, there is no significance to the choice to depict the first set in the product $X \times Y$ on the horizontal axis, and the second set on the vertical axis. We could have done it the other way around. Although there is no significance to which is depicted on the vertical axis and which on the horizontal, there *is* significance to the order of X and Y in $X \times Y$. The set $X \times Y$ is not the same as $Y \times X$ unless $X = Y$.

We generalize the product notation to three or more sets. Thus, if X, Y, Z are sets, then $X \times Y \times Z$ is the set of all triples or 3-tuples,

$$X \times Y \times Z = \{(x, y, z) \mid x \in X, y \in Y, z \in Z\},$$

and if there are N sets, $X_1, X_2, \ldots, X_N$, their product is the set consisting of N-tuples,

$$\boxed{X_1 \times \cdots \times X_N = \{(x_1, \ldots, x_N) \mid x_i \in X_i, i = 1, \ldots, N\}.} \tag{A.12}$$

We can alternatively write (A.12) as

$$\boxed{\prod_{i=1}^{N} X_i.} \tag{A.13}$$

The large Π operator indicates a product of its arguments.

$X \times X$ is also written as X^2. The N-fold product of the same set X is also written as X^N. For example, $Reals^N$ is the set of all N-tuples of real numbers, and $Complex^N$ is the set of all N-tuples of complex numbers. In symbols,

$$\boxed{\begin{aligned} Reals^N &= \{x = (x_1, \ldots, x_N) \mid x_i \in Reals, i = 1, \ldots, N\} \\ Complex^N &= \{z = (z_1, \ldots, z_N) \mid z_i \in Complex, i = 1, \ldots, N\}. \end{aligned}}$$

Predicates on product sets

A variable over $X \times Y$ is denoted by a pair (x, y), with x as the variable over X and y as the variable over Y. We can use predicates in x and y to define subsets of $X \times Y$.

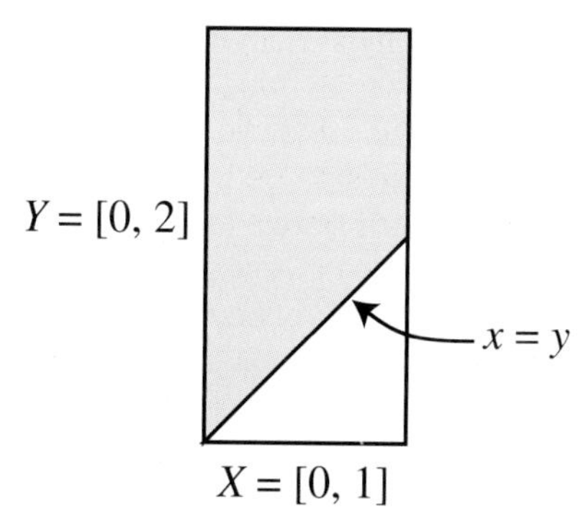

FIGURE A.3: The rectangle depicts the set $[0, 1] \times [0, 2]$, the shaded region depicts the set given by (A.14), and the unshaded triangle depicts the set given by (A.15).

Example A.1: The set

$$\{(x, y) \in [0, 1] \times [0, 2] \mid x \leq y\} \tag{A.14}$$

can be depicted by the shaded region in figure A.3. The unshaded triangle depicts the set

$$\{(x, y) \in [0, 1] \times [0, 2] \mid x \geq y\}. \; \square \tag{A.15}$$

Example A.2: The solid line in figure A.4(a) represents the set

$$\{(x, y) \in Reals^2 \mid x + y = 1\},$$

the shaded region (including the solid line) depicts

$$\{(x, y) \in Reals^2 \mid x + y \geq 1\},$$

and the unshaded region (excluding the solid line) depicts

$$\{(x, y) \in Reals^2 \mid x + y < 1\}.$$

Similarly, the shaded region in figure A.4(b) depicts the set

$$\{(x, y) \in Reals^2 \mid -x + y \geq 1\}.$$

The overlap region in figure A.4(c) depicts the intersection of the two shaded regions, and corresponds to the conjunction of two predicates:

$$\{(x, y) \in Reals^2 \mid [x + y \geq 1] \wedge [-x + y \geq 1]\}. \; \square$$

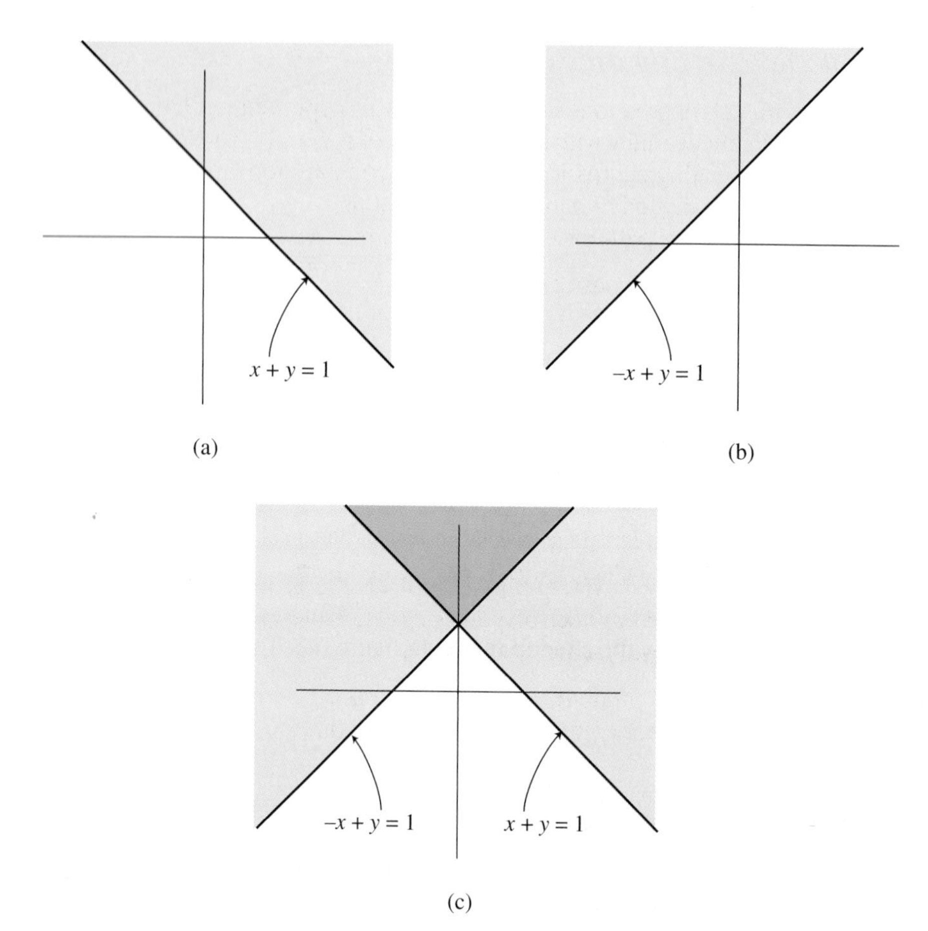

FIGURE A.4: (a) The solid line depicts the subset of $Reals^2$ satisfying the predicate $x + y = 1$. The shaded region satisfies $x + y \geq 1$. (b) The solid line satisfies $-x + y = 1$, and the shaded region satisfies $-x + y \geq 1$. (c) The overlap region satisfies $[x + y \geq 1] \wedge [-x + y \geq 1]$.

Example A.3: The set *TallerThan* consists of all pairs of students ($name_1$, $name_2$) such that $name_1$ is taller than $name_2$:

$$TallerThan = \{(name_1, name_2) \in Students^2 \mid name_1 \text{ is taller than } name_2\}.$$

NearbyCities consists of pairs of cities that are less than 50 miles apart:

$$NearbyCities = \{(city_1, city_2) \in USCities^2 \mid distance(city_1, city_2) \leq 50\}.$$

In the preceding predicate, $distance(city_1, city_2)$ is the distance in miles between $city_1$ and $city_2$. ❐

A.1.10 *Evaluating an expression*

We have evaluated expressions several times in this appendix, each time relying on our mathematical ability with simple expressions. We can develop systematic methods for evaluating expressions that rely less on intuition, and can therefore handle more complicated and intricate expressions.

Some examples of patterns of expressions are:

- $A, B, Naturals, \ldots$, names of sets
- $A = \{\text{list of elements}\}$
- $x \in A, x \notin A$, set membership
- $A = B, B \subset A$, and $A \supset B$, set inclusion
- $A \cap B, A \cup B, X \times Y, X \setminus A, A^c$, set operations
- $x, y, \ldots$, names of variables
- $P(x), Q(x), \ldots$, predicates in x
- $\forall x \in Set,\ P(x)$ where P is a predicate
- $NewSet = \{x \in Set \mid Pred(x)\}$, set definition
- $P(x) \wedge Q(x), P(x) \vee Q(x), \neg(P(x))$, predicate operations

The patterns define the rules of grammar of the notation. The notation itself consists of **expressions**, that are composed of

- **constants**, such as numbers and predefined sets,
- **variables**, which are names that are not predefined,
- **operators**, such as $\in$, $\cap$, or $=$ (as an assertion),
- **quantifiers**, such as $\forall$ and $\exists$, or
- **definitions**, name $=$ (as an assignment).

An expression is **well formed** if it conforms to the established patterns.* For example, if P, Q, and R are predicates, then the syntax implies that

$$\neg[[\neg(P(x)) \vee Q(x)] \wedge [P(x) \vee [R(x) \wedge \neg(P(x))]]] \quad (A.16)$$

is also a predicate. Just as in the case of a computer language, you learn the syntax of mathematical expressions through practice.†

In an expression, constants have a meaning. For example, "20" is a number, "*Berkeley*" is a city, and the constant "*true*" has a truth value. By contrast, a variable, such as x, has no meaning, unless it has been defined (which turns it into a constant). An expression on constants has a meaning. For example, "$10 + 3$" means "13." However, "$x + 3$" has no meaning, unless x has been defined. We

* In this text, we do not attempt to define precisely what these established patterns are. To do so, we would have to define a **grammar**, something that can be done formally, but is beyond our scope.

† The syntax of a language is the set of rules (or patterns) whereby words can be combined to form grammatical or well-formed sentences. Thus, the syntax of the "C" language is the set of rules that a sequence of characters (the code) must obey to be a C program. A C compiler checks whether the code conforms to the syntax. Of course, even if the code obeys the syntax, it may not be what you intend—that is, it may not execute the intended computation.

say that in the expression "$x + 3$," x is a **free variable**. It has no assigned value in the expression. An expression with a free variable acquires a meaning when that free variable is given a meaning, if it is well formed with that meaning. For example, "$x + 3$" has meaning 13 if x has meaning 10, but it is not well formed if x has meaning *Berkeley*. Quantifiers remove free variables from expressions.

> **Example A.4:** In the predicate "$x = 0$," x is free. In the expression "$\exists\ x \in Reals,\ \ x = 0$," x is no longer free. In fact, this expression has value *true*. In the expression "$\forall\ x \in Reals,\ \ x = 0$," again x is not free, but this expression has value *false*. The expression "$\exists\ y \in Reals,\ \ x + 1 = y$" still has free x. However, "$\forall\ x \in Reals, \exists\ y \in Reals,\ \ x + 1 = y$" has no free variables, and has value *true*. The expression "$\forall\ x \in Reals,\ \ x + 7$," however, is not well formed. It has neither free variables nor a value. ❐

A **predicate expression** is one that either has a meaning that is a truth value (*true* or *false*), or can have such a meaning if its free variables are appropriately defined. Expression (A.16) is an example of a predicate expression.

Expressions also contain **punctuation**, such as parentheses. These help define the relationships among the components of the expression. It is beyond the scope of this book to define completely the rules for constructing expressions,* but we can hint at the issues with a brief discussion of parsing.

Parsing

To show that (A.16) is indeed a well-formed predicate expression, we must show that it can be constructed using the syntax. We do this by **parsing** the expression (A.16) with the help of matching brackets and parentheses. Parsing the expression will also enable us to evaluate it in a systematic way.

The result is the parse tree shown in figure A.5. The leaves of this tree (the bottom-most nodes) are labeled by the elementary predicates P, Q, R, and the other nodes of the tree are labeled by one of the predicate operations $\wedge, \vee, \neg$. Each of the intermediate nodes corresponds to a sub-predicate of (A.16). Two such sub-predicates are shown in the figure. The last leaf on the right is labeled $P(x)$. Its parent node is labeled $\neg$, and so that parent node corresponds to the sub-predicate $\neg(P(x))$. Its parent node is labeled $\wedge$, and it has another child node labeled $R(x)$, so this node corresponds to the sub-predicate $R(x) \wedge \neg(P(x))$.

If we go up the tree in this way, we can see that the top of the tree (the root node) indeed corresponds to the predicate (A.16). Since at each intermediate node, the sub-predicate is constructed using the syntax, the root node is a well-formed predicate.

Evaluating

Suppose we know whether the predicates $P(x), Q(x), R(x)$ evaluate to true or false for some value of x. Then we can use the parse tree to figure out whether the

* A text on compilers for computer programming languages will typically do this.

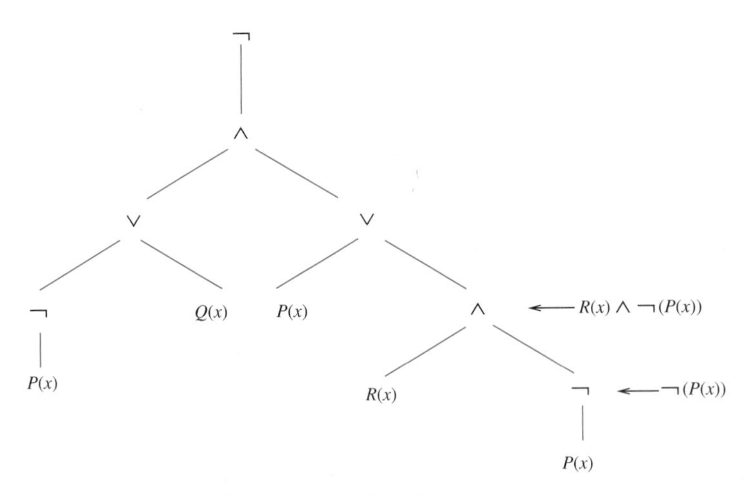

FIGURE A.5: Parse tree for the expression (A.16).

predicate (A.16) evaluates to true or false. To do this, we begin with the known truth values at the leaves of the parse tree, use the meaning of the predicate operations to figure out the truth values of the sub-predicates corresponding to the parents of the leaf nodes, and then the parents of those nodes, and work our way up the tree to figure out the truth value of the predicate (A.16) at the root node. In figure A.6, the parse tree is annotated with the truth values of each of

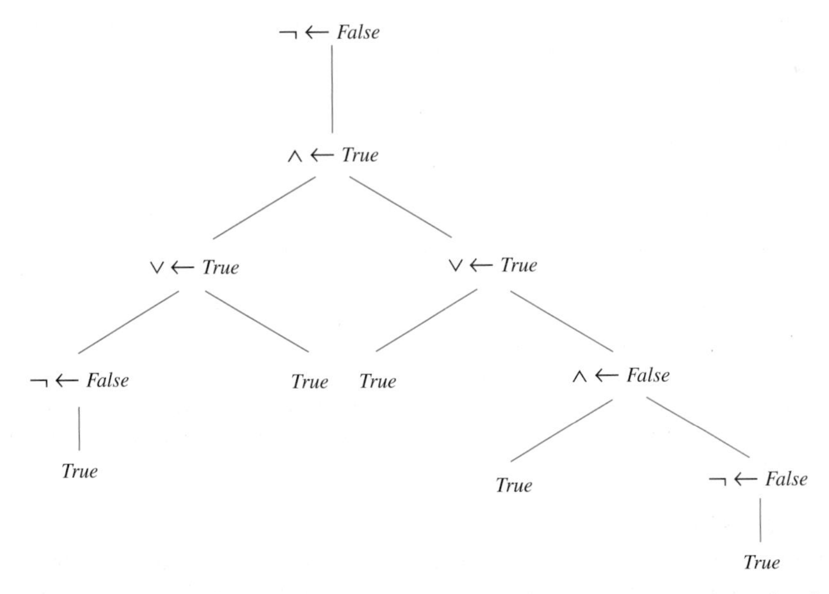

FIGURE A.6: Parse tree for the expression (A.16) annotated with the truth values of each of the nodes.

the nodes, where the values of $P(x)Q(x)$, $R(x)$ are all assumed to be true. Since the root node is annotated "false," we conclude that (A.16) evaluates to false.

Truth tables

The way in which the predicate operations transform the truth values is given in the following **truth table**:

P(x)	Q(x)	$\neg$ P(x)	P(x) $\wedge$ Q(x)	P(x) $\vee$ Q(x)
True	True	False	True	True
True	False	False	False	True
False	True	True	False	True
False	False	True	False	False

Consider a particular row of this table, say the first row. The first two entries specify that $P(x)$ is true and $Q(x)$ is true. The remaining three entries give the corresponding truth values for $\neg P(x)$, $P(x) \wedge Q(x)$, and $P(x) \vee Q(x)$, namely, $\neg P(x)$ is false, $P(x) \wedge Q(x)$ is true, and $P(x) \vee Q(x)$ is true. The four rows correspond to the four possible truth value assignments of $P(x)$ and $Q(x)$. This truth table can be used repeatedly to evaluate any well-formed expression given the truth value of $P(x)$ and $Q(x)$.

Thus, given the truth values of predicates $P_1(x), \ldots, P_n(x)$, the truth value of any well-formed expression involving these predicates can be obtained by a computer algorithm that constructs the parse tree and uses the truth table above. Such algorithms are used to evaluate logic circuits.

A.2 Functions

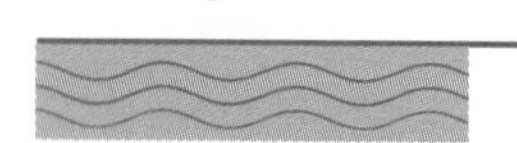

In the notation

$$f{:}X \to Y, \tag{A.17}$$

X and Y are sets, and f is the name of a function. The function is an assignment rule that assigns a value in Y to each element in X. If the element in X is x, then the value is written $f(x)$.

We read (A.17) as "f is (the name of) a function from X into (or to) Y." We also say that f maps X into Y. The set X is called the **domain** of f, written $X = domain(f)$, the set Y is called the **range** of f, written $Y = range(f)$.* When

* In some mathematics texts, the set Y, which we call the range, is called the **codomain**, and $range(f)$ is defined to be the set of all values that f takes, that is, $range(f) = \{f(x) \mid x \in X\}$. However, we do not use this terminology.

the domain and range are understood from the context, we write "f" by itself to represent the function or the map. If x is a variable over X, we also say "f is a function of x."

Example A.5: Recall the set *Students* of all the students in a class. Each element of *Students* is represented by a student's name,

$$\mathit{Students} = \{\text{John Brown, Jane Doe}, \ldots\}.$$

We assign to each name in *Students* the student's grades on the final examination, a number between 0 and 100. This name-to-marks assignment is an example of a function. Just as we give names to sets (e.g., *Students*), we give names to functions. In this example the function might be named *Score*. When we evaluate the function *Score* at any name, we get the marks assigned to that name. We write this as

$$\mathit{Score}(\text{John Brown}) = 90, \ \mathit{Score}(\text{Jane Doe}) = 91.2, \ldots$$

Figure A.7 illustrates the function *Score*. Three things are involved in defining *Score*: the set *Students*, the set $[0, 100]$ of possible marks, and the assignment of marks to each name. In the figure this assignment is depicted by the arrows: the tail of the arrow points to a name and the head of the arrow points to the marks assigned to that name. We denote these three things by

$$\mathit{Score}: \mathit{Students} \to [0, 100],$$

which we read as "*Score* is a function from *Students* into $[0, 100]$." The domain of the function *Score* is *Students*, and the range of *Score* is $[0, 100]$. ❐

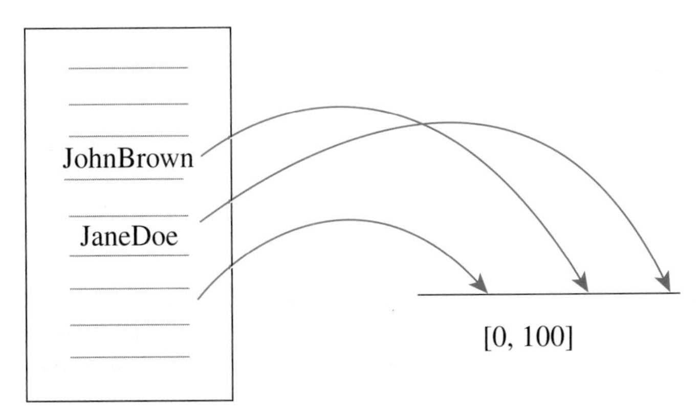

FIGURE A.7: Illustration of *Score*.

It is easy to imagine other functions with the same domain *Students*. For example, *Height* assigns to each student his or her height measured in cm, *SSN* assigns students their Social Security Number, and *Address* assigns students their address. The range of these functions is different. The range of *Height* might be defined to be $[0, 200]$ (200 cm is about 6.5 feet). Since a Social Security Number is a nine-digit number, we can take $\{0, 1, \ldots, 9\}^9$ to be the range of *SSN*. And we can take the range of *Address* to be $Char^{100}$, assuming that an address can be expressed as a string of 100 characters, including blank spaces.

We usually use lowercase letters for function names, such as f, g, h, or more descriptive names such as *Score*, *Voice*, *Video*, *SquareWave*, *AMSignal*.

> **Avoid a bad habit** : It is important to distinguish between a function f and its value $f(x)$ at a particular point $x \in domain(f)$. The function f is a *rule* that assigns a value in $range(f)$ to each $x \in domain(f)$, whereas $f(x)$ is a *point* or element in $range(f)$. Unfortunately, too many books encourage the bad habit by using "$f(x)$" as a shorthand for "f is a function of x." If you keep the distinction between f and $f(x)$, it will be easier to avoid confusion when we study systems.

A.2.1 Defining functions

To define a function, you must give the domain, the range, and the rule that produces an element in the range given an element in the domain. There are many ways to do this, as explored in greater depth in chapter 2. Here we mention only two. The first is enumeration. That is, in tabular form or some other form, each possible value in the domain is associated with a value in the range. This method would be appropriate for the *Score* function, for example. Alternatively, functions can be mathematically defined by the prototype: define $f: X \to Y$,

$$\forall x \in X, \quad f(x) = \text{expression in } x.$$

The "expression in x" may be specified by an algebraic expression, by giving the graph of f, by a table, or by a procedure.

A.2.2 Tuples and sequences as functions

An N-tuple $x = (x_1, \ldots, x_N) \in X^N$ can be viewed as a function

$$x: \{1, \ldots, N\} \to X.$$

For each integer in $\{1, \dots, N\}$, it assigns a value in X. An infinite sequence y over the set Y can also be viewed as a function

$$y\colon Naturals \to Y,$$

or

$$y\colon Naturals_0 \to Y,$$

depending on whether you wish to begin indexing the sequence at zero or one (both conventions are widely used). This view of sequences as functions is in fact our model for discrete-time signals and event traces, as developed in chapter 1.

A.2.3 *Function properties*

A function $f\colon X \to Y$ is **one-to-one** if

$$\forall\, x_1 \in X \text{ and } \forall\, x_2 \in X, \quad x_1 \neq x_2 \Rightarrow f(x_1) \neq f(x_2).$$

Here the logical symbol "$\Rightarrow$" means "implies" so the expression is read: If x_1, x_2 are two different elements in X, then $f(x_1), f(x_2)$ are different.

Example A.6: The function *Cube*: *Reals* → *Reals* given by

$$\forall\, x \quad Cube(x) = x^3$$

is one-to-one, because if $x_1 \neq x_2$, then $x_1^3 \neq x_2^3$. But *Square*: *Reals* → *Reals* is not one-to-one because, for example, $Square(1) = Square(-1)$. ❐

A function $f\colon X \to Y$ is **onto** if

$$\forall\, y \in Y,\ \exists\ x \in X, \text{ such that } f(x) = y.$$

The symbol "$\exists$" is the existential quantifier, which means "there exists" or "for some." So the preceding expression reads "For each y in Y, there exists x in X such that $f(x) = y$."

Accordingly, f is onto if for every y in its range there is some x in its domain such that $y = f(x)$.

Example A.7: The function *Cube*: *Reals* → *Reals* is onto, while *Square*: *Reals* → *Reals* is not onto. However, *Square*: *Reals* → $Reals_+$ *is* onto. ❐

PROBING FURTHER

Infinite sets

The size of a set A, denoted $|A|$, is the number of elements it contains. By counting, we immediately know that $\{1, 2, 3, 4\}$ and $\{a, b, c, d\}$ have the same number of elements, whereas $\{1, 2, 3\}$ has fewer elements. But we cannot count infinite sets. It is more difficult to compare the number of elements in the following infinite sets:

$$A = Naturals = \{1, 2, 3, 4, \ldots\}, \; B = \{2, 3, 4, 5 \ldots\}, \; C = [0, 1].$$

At first, we might say that A has one more element than B, since A includes B and has one additional element, $1 \in A$. In fact, these two sets have the same size.

The **cardinality** of a set is the number of elements in the set, but generalized to handle infinite sets. Comparing the cardinality of two sets is done by matching elements, using one-to-one functions. Consider two sets A and B, finite or infinite. We say that A has a smaller cardinality than B, written $|A| \leq |B|$, if there exists a one-to-one function mapping A into B. We say that A and B have the same cardinality, written $|A| = |B|$, if $|A| \leq |B|$ and $|B| \leq |A|$.

The cardinality of the infinite set $A = Naturals$ is denoted $\aleph_0$, read "aleph null" (aleph is the first letter of the Hebrew alphabet). It is quite easy to prove using the definition of cardinality that $n < \aleph_0$ for any finite number n.

We can now show that the cardinality of B is also $\aleph_0$. There is a one-to-one function $f: A \to B$, namely,

$$\forall n \in A, \quad f(n) = n + 1,$$

so that $|A| \leq |B|$, and there is a one-to-one function $g: B \to A$, namely,

$$\forall n \in B, \quad g(n) = n - 1,$$

so that $|B| \leq |A|$. A similar argument can be used to show that the set of even numbers and the set of odd numbers also have cardinality $\aleph_0$.

It is more difficult to show that the cardinality of $Naturals \times Naturals$ is also $\aleph_0$. To see this, we can define a one-to-one function $h: Naturals^2 \to Naturals$ as follows (see figure A.8).

$$h((1, 1)) = 1, \; h((2, 1)) = 2, \; h((2, 2)) = 3,$$
$$h((1, 2)) = 4, \; h((1, 3)) = 5, \; \ldots.$$

Observe that since a rational number m/n can be identified with the pair $(m, n) \in Naturals^2$, the argument shows that the cardinality of the set of all rational numbers is also $\aleph_0$.

PROBING FURTHER

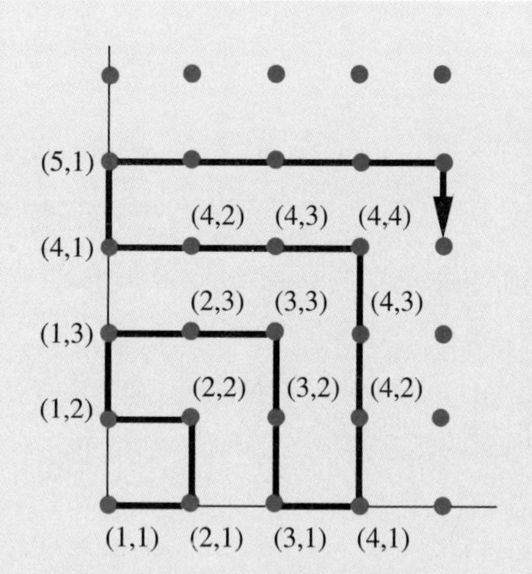

FIGURE A.8: A correspondence between $Naturals^2$ and *Naturals*.

PROBING FURTHER

Even bigger sets

We can show that the cardinality of [0, 1] is strictly larger than that of *Naturals* (i.e., $|[0, 1]| > |Naturals|$). Since the function $f: Naturals \to [0, 1]$ defined by

$$\forall\, n \in Naturals, \quad f(n) = 1/n$$

is one-to-one, we have $|Naturals| \leq |[0, 1]|$. However, we can show that there is no one-to-one function in the other direction. If there were such a function $g: [0, 1] \to Naturals$, then it would be possible to enumerate all the elements of [0, 1] in an ordered list,

$$[0, 1] = \{x^1, x^2, x^3, \ldots\}. \tag{A.18}$$

(The superscript here is not raising to a power, but just indexing.) We can show that this is not possible. If we express each element of [0, 1] by its decimal expansion (ignoring the element 1.0), this list looks like

$$x^1 = 0.x_1^1 x_2^1 x_3^1 \cdots$$
$$x^2 = 0.x_1^2 x_2^2 x_3^2 \cdots$$
$$x^3 = 0.x_1^3 x_2^3 x_3^3 \cdots$$
$$\cdots$$
$$x^n = 0.x_1^n x_2^n x_3^n \cdots$$
$$\cdots$$

continued on next page

PROBING FURTHER

Construct any number $y \in [0, 1]$ with the decimal expansion

$$y = 0.y_1 y_2 y_3 \ldots$$

such that for each i, $y_i \neq x_i^i$ where x_i^i is the ith term in the decimal expansion of x^i. Clearly, such a number exists and is in [0, 1]. But then for every i, $y \neq x^i$, so that y cannot be in the list $\{x^1, x^2, x^3, \ldots\}$. Thus, the list (A.18) is not complete in that it does not include all the elements of [0, 1].

The cardinality of [0, 1] is denoted $\aleph_1$, and is strictly greater than $\aleph_0$. In this sense we can say that the continuum [0, 1] has more elements than the denumerable set *Naturals*, even though both sets have infinite size.

The obvious question is whether there are cardinalities larger than $\aleph_1$. The answer is yes; in fact, there are sets of ever higher cardinality,

$$\aleph_0 < \aleph_1 < \aleph_2, \ldots$$

and sets with cardinality larger than all of these!

A.3 *Summary*

Sets are mathematical objects representing collections of elements. A variable is a representative for an element of a set. A predicate over a set is an expression involving a variable that evaluates to true or false when the variable is assigned a particular element of the set. Predicates are used to construct new sets from existing sets. If the variable in a predicate is quantified, the expression becomes an assertion.

Sets can be combined using the operations of union, intersection, and complement. The corresponding operations on predicates are disjunction, conjunction, and negation. New sets can also be obtained by the product of two sets.

There are precise rules that must be followed in using these operations. The collection of these rules is the syntax of sets and predicates. By parsing a complex expression, we can determine whether it is well formed. The truth value of a predicate constructed from elementary predicates using predicate operations can be calculated using the parse tree and the truth table.

Functions are mathematical objects representing a relationship between two sets, the domain and the range of the function. We have introduced the following patterns of expressions for functions

- $f, g, h, Score, \ldots$, names of functions,
- $f: X \to Y, X = domain(f), Y = range(f)$, a function from X to Y.

A function $f: X \to Y$ assigns to each value $x \in X$ a value $f(x) \in Y$.

KEY: E = mechanical T = requires plan of attack C = more than 1 answer

EXERCISES

E 1. In the spirit of figure A.2, give a picture of the following sets:

(a) $\{1, 2, 3\}$,

(b) $[0, 1] \times \{0, 1\}$,

(c) $[0, 1] \times [a, b]$.

(d) $\{1, 2, 3\} \times \{a, b\}$,

(e) $\{a, b\} \times [0, 1]$.

E 2. How many elements are there in the sets (a) $\{1, \ldots, 6\}$, (b) $\{-2, -1, \ldots, 10\}$, and (c) $\{0, 1, 2\} \times \{2, 3\}$?

T 3. Determine which of the following expressions are true and which are false:

(a) $\forall\, n \in Naturals, \quad n > 1$,

(b) $\exists\, n \in Naturals, \quad n < 10$,

(c) If $A = \{1, 2, 3\}$ and $B = \{2, 3, 4\}$, then $\forall\, x \in A, \forall\, y \in B, \quad x \leq y$,

(d) If $A = \{1, 2, 3\}$ and $B = \{2, 3, 4\}$, then $\forall\, x \in A, \exists\, y \in B, \quad x \leq y$,

(e) If $A = \{1, 2, 3\}$ and $B = \{2, 3, 4\}$, then $\exists\, x \in A, \forall\, y \in B, \quad x \leq y$.

T 4. In the following figure, $X = \{(x, y) \mid x^2 + y^2 = 1\}$ is depicted as a two-dimensional circle and $Z = [0, 2]$ is shown as a one-dimensional line segment. Explain why it is reasonable to show the product set as a three-dimensional cylinder.

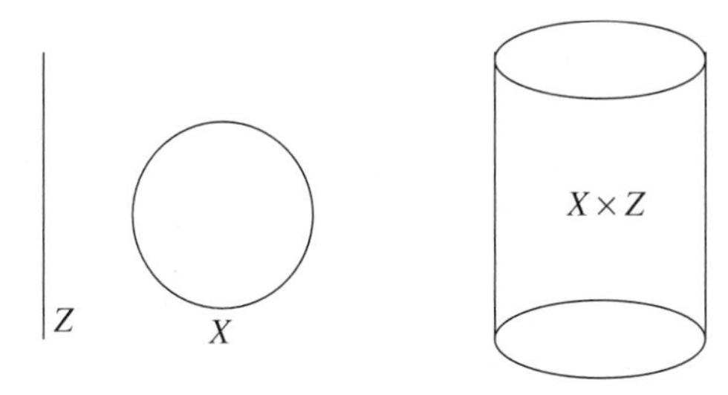

E 5. In the spirit of figure A.2, give a picture for the product set $\{M, Tu, W, Th, F\} \times [8.00, 17.00]$ and indicate on your drawing the lecture hours for this class.

E 6. In the spirit of figure A.2, give a picture for the set $A = \{(x, y) \mid x \in [1, 2], y \in [1, 2]\}$ and the set $B = \{(x, x) \mid x \in [1, 2]\}$. Explain why the two sets are different.

C 7. Give a precise expression for the predicate below so that *Triangle* is indeed a triangle:

$$Triangle = \{(x, y) \in Reals^2 \mid Pred(x, y)\}.$$

There are many ways of writing this predicate. One way is to express $Pred(x, y)$ as the conjunction of three linear inequality predicates. Hint: We used the conjunction of two linear inequality predicates in figure A.4.

T 8. If X has m elements and Y has n elements, how many elements are there in $X \times Y$? If X_i has m_i elements, for $i = 1, \ldots, I$, for some constant I, how many elements are there in

$$\prod_{i=1}^{i=I} X_i = X_1 \times \cdots \times X_I?$$

T 9. How many different 10-letter strings are there if each letter is drawn from the set *Alphabet* consisting of the 26 lowercase letters of the alphabet? How many such strings are there with *exactly* one occurrence of the letter a?

T 10. Recall that a set cannot contain duplicate elements. Now suppose X contains 10 elements.

(a) How many two-element combinations of elements from X are there?

(b) How many two-element permutations are there?

C 11. Construct predicates for use in the prototype (A.4) to define the following sets. Define them in terms of examples of sets introduced in this appendix.

(a) The set of U.S. cities with a population exceeding 1 million.

(b) The male students in a class.

(c) The old books in the library.

T 12. Which of the following expressions are well formed? For those that are well formed, state whether they are assertions. For those that are assertions, evaluate the assertion to true or false. For those that are not assertions, find an equivalent simpler expression.

(a) $2 \in \{1, 3, 4\}$,

(b) $3 \subset \{1, 3, 4\}$,

(c) $\{3\} \subset \{1, 2, 3\}$,

(d) $2 \cup \{1, 3, 4\}$,

(e) $\{2\} \cup \{1, 3, 4\}$,

(f) $[2.3, 3.4] = \{x \in Reals \mid 2.3 \le x \le 3.4\}$,

(g) $\{x \in Reals \mid x > 3 \wedge x < 4\}$,

(h) $[1,2] \cap [3,4] = \emptyset$.

E 13. Define the following sets in terms of the sets named in section A.1.5.

(a) The set of all 10-letter passwords.

(b) The set of all 5×6 matrices of real numbers.

(c) The set of all complex numbers with magnitude at most 1.

(d) The set of all two-dimensional vectors with magnitude exactly 1.

E 14. Give the set of all subsets $P(X)$ of the set $X = \{a, b, c\}$.

T 15. Suppose a set X has n elements. Let $P(X)$ be the power set of X. How many elements are there in $P(X)$?

T 16. Use MATLAB to depict the following sets using the plot command:

(a) $\{(t, x) \in Reals^2 \mid x = \sin(t),\ \text{and } t \in \{0, \frac{1}{20}2\pi, \frac{2}{20}2\pi, \ldots, \frac{20}{20}2\pi\}\}$,

(b) $\{(y, x) \in Reals^2 \mid y = e^x,\ \text{and } x \in \{-1, -1 + \frac{1}{20}, -1 + \frac{2}{20}, \ldots, 1\}\}$,

(c) $\{(y, x) \in Reals^2 \mid y = e^{-x},\ \text{and } x \in \{-1, -1 + \frac{1}{20}, -1 + \frac{2}{20}, \ldots, 1\}\}$.

T 17. Determine which of the following functions are onto, which are one-to-one, and which are neither, and give a short explanation for your answer.

(a) $License: CalVehicles \to Char*$ given by $\forall\, vehicle \in CalVehicles$, $License(vehicle)$ is the California license number of the vehicle

(b) $f: Reals \to [-2, 2]$, given by $\forall\, x \in Reals,\ f(x) = 2\sin(x)$

(c) $f: Reals \to Reals$, given by $\forall\, x \in Reals,\ f(x) = 2\sin(x)$

(d) $conj: Complex \to Complex$, the complex conjugate function

(e) $f: Complex \to Reals^2$, given by $\forall\, z \in Complex,\ f(z) = (Re(z), Im(z))$, where $Re(z)$ is the real part of z and $Im(z)$ is the imaginary part of z.

(f) $M: Reals^2 \to Reals^2$, $\forall\, (x_1, x_2) \in Reals^2$,

$$M(x_1, x_2) = (y_1, y_2),$$

where

$$\begin{bmatrix} x_1 \\ x_2 \end{bmatrix} \begin{bmatrix} 1 & 2 \\ 2 & 1 \end{bmatrix} = \begin{bmatrix} y_1 \\ y_2 \end{bmatrix}.$$

(g) $Zero: Reals^4 \to Reals^4, \forall\, x \in Reals^4,\ Zero(x) = (0, 0, 0, 0)$

T 18. Let A, B, and C be arbitrary sets. Let $f: A \to A$ and $g: A \to A$ be two functions with domain and range A. Which of the following assertions is true for all choices of A, B, C, f, and g? Note that $f \circ g$ is function composition, defined in section 2.1.5. $X \setminus Y$ is the complement of Y in X, as explained in section A.1.6. $P(X)$ is the power set of X, as explained in section A.1.2.

(a) $(A \cup B) \setminus C = (A \setminus C) \cup (B \setminus C)$.

(b) $P(A \cup B) = P(A) \cup P(B)$.

(c) $f \circ g = g \circ f$.

(d) If both f and g are one-to-one, then $f \circ g$ is one-to-one.

E 19. Each of the following expressions is intended to be a predicate expression. Determine whether it is *true* or *false*, or whether it is not well formed, or has a free variable. If it has a free variable, identify the free variable.

(a) $\forall$ sets x, $x \subset P(x)$, where $P(x)$ is the power set of x.

(b) $\exists\ x \in Reals,\ \{x \in Integers | x + 3 = 10\}$.

(c) $\forall\, n \in Naturals,\ n = 2 \Rightarrow (n, n+1) \in \{1, 2, 3\}^2$.

(d) $\exists\, x \in Naturals,\ x + y = 10$.

APPENDIX B
Complex numbers

Complex numbers are used extensively in the modeling of signals and systems for two reasons. The first reason is that complex numbers provide a compact and elegant way to talk simultaneously about the phase and amplitude of sinusoidal signals. Complex numbers are therefore heavily used in Fourier analysis, which represents arbitrary signals in terms of sinusoidal signals. The second reason is that a large class of systems, called linear time-invariant (LTI) systems, treat signals that can be described as complex exponential functions in an especially simple way. They simply scale the signals.

These uses of complex numbers are developed in detail in the main body of this text. This appendix summarizes essential properties of complex numbers themselves. We review complex number arithmetic, how to manipulate complex exponentials, Euler's formula, and the polar coordinate representation of complex numbers.

B.1 Imaginary numbers

The quadratic equation, where $x \in Reals$,

$$x^2 - 1 = 0,$$

has two solutions, $x = +1$ and $x = -1$. These solutions are said to be **roots** of the polynomial $x^2 - 1$. Thus, this polynomial has two roots, $+1$ and -1.

More generally, the roots of the nth degree polynomial,

$$x^n + a_1 x^{n-1} + \cdots + a_{n-1}x + a_n, \tag{B.1}$$

are defined to be the solutions to the polynomial equation

$$x^n + a_1 x^{n-1} + \cdots + a_{n-1}x + a_n = 0. \tag{B.2}$$

The roots of a polynomial provide a particularly useful factorization into first-degree polynomials. For example, we can factor the polynomial $x^2 - 1$ as

$$x^2 - 1 = (x - 1)(x + 1).$$

Notice the role of the roots, $+1$ and -1. In general, if (B.1) has roots $r_1, \ldots, r_n$, then we can factor the polynomial as follows

$$x^n + a_1 x^{n-1} + \cdots + a_{n-1}x + a_n = (x - r_1)(x - r_2) \cdots (x - r_n). \tag{B.3}$$

It is easy to see that if $x = r_i$ for any $i \in \{1, \ldots, n\}$, then the polynomial evaluates to zero, so (B.2) is satisfied.

This raises the question whether (B.2) always has a solution for x. In other words, can we always find roots for a polynomial?

The equation

$$x^2 + 1 = 0 \tag{B.4}$$

has no solution for x in the set of real numbers. Thus, it would appear that not all polynomials have roots. However, a surprisingly simple and clever mathematical device changes the picture dramatically. With the introduction of **imaginary numbers**, mathematicians ensure that all polynomials have roots. Moreover, they ensure that any polynomial of degree n has exactly n factors as in (B.3). The n values $r_1, \ldots, r_n$ (some of which may be repeated) are the **roots** of the polynomial.

If we try by simple algebra to solve (B.4) we discover that we need to find x such that

$$x^2 = -1.$$

This suggests that

$$x = \sqrt{-1}.$$

But -1 does not normally have a square root.

The clever device is to define an imaginary number, usually written i or j, that is equal to $\sqrt{-1}$. By definition,*

$$i \times i = \sqrt{-1} \times \sqrt{-1} = -1.$$

This imaginary number, thus, is a solution of the equation $x^2 + 1 = 0$.

For any real number y, iy is an imaginary number. Thus, we can define the set of imaginary numbers as

$$\boxed{ImaginaryNumbers = \{iy \mid y \in Reals, \text{ and } i = \sqrt{-1}\}.} \tag{B.5}$$

It is a profound result that this simple device is all we need to guarantee that every polynomial equation has a solution, and that every polynomial of degree n can be factored into n polynomials of degree one, as in (B.3).

B.2 Arithmetic of imaginary numbers

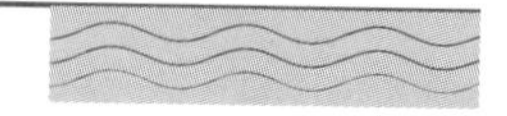

The sum of i and i is written $2i$ or $i2$. Sums and differences of imaginary numbers simplify like real numbers:

$$i3 + i2 = i5, \; i3 - i4 = -i.$$

If iy_1 and iy_2 are two imaginary numbers, then

$$\boxed{iy_1 + iy_2 = i(y_1 + y_2), \quad iy_1 - iy_2 = i(y_1 - y_2).} \tag{B.6}$$

The product of a real number x and an imaginary number iy is

$$x \times iy = iy \times x = ixy.$$

To take the product of two imaginary numbers, we must remember that $i^2 = -1$, and so for any two imaginary numbers, iy_1 and iy_2, we have

$$iy_1 \times iy_2 = -y_1 \times y_2. \tag{B.7}$$

The result is a real number. We can use rule (B.7) repeatedly to multiply as many imaginary numbers as we wish. For example,

$$i \times i = -1, \; i^3 = i \times i^2 = -i, \; i^4 = 1.$$

* Here, the operator, $\times$, is ordinary multiplication, not products of sets.

The ratio of two imaginary numbers iy_1 and iy_2 is a real number

$$\frac{iy_1}{iy_2} = \frac{y_1}{y_2}.$$

B.3 Complex numbers

The sum of a real number x and an imaginary number iy is called a **complex number**. This sum does not simplify as do the sums of two real numbers or two imaginary numbers, and it is written as $x + iy$ or $x + jy$.

Examples of complex numbers are

$$2 + i, \; -3 - i2, \; -\pi + i\sqrt{2}.$$

In general a complex number z is of the form

$$z = x + iy = x + \sqrt{-1}y,$$

where x, y are real numbers. The **real part** of z, written $Re\{z\}$, is x. The **imaginary part** of z, written $Im\{z\}$, is y. *Notice that, confusingly, the imaginary part is a real number.* The imaginary part times i is an imaginary number. So

$$\boxed{z = Re\{z\} + iIm\{z\}.}$$

The set of complex numbers, therefore, is defined by

$$\boxed{Complex = \{x + iy \mid x \in Reals, y \in Reals, \text{ and } i = \sqrt{-1}\}.} \tag{B.8}$$

Every real number x is in *Complex*, because $x = x + i0$; and every imaginary number iy is in *Complex*, because $iy = 0 + iy$.

Two complex numbers $z_1 = x_1 + iy_1$ and $z_2 = x_2 + iy_2$ are equal if and only if their real parts are equal and their imaginary parts are equal, that is, $z_1 = z_2$ if and only if

$$Re\{z_1\} = Re\{z_2\}, \text{ and } Im\{z_1\} = Im\{z_2\}.$$

B.4 Arithmetic of complex numbers

In order to add two complex numbers, we separately add their real and imaginary parts,

$$(x_1 + iy_1) + (x_2 + iy_2) = (x_1 + x_2) + i(y_1 + y_2).$$

The **complex conjugate** of $x + iy$ is defined to be $x - iy$. The complex conjugate of a complex number z is written z^*. Notice that

$$z + z^* = 2Re\{z\},\ z - z^* = 2iIm\{z\}.$$

Hence, the real and imaginary parts can be obtained using the complex conjugate,

$$Re\{z\} = \frac{z + z^*}{2}, \text{ and } Im\{z\} = \frac{z - z^*}{2i}.$$

The product of two complex numbers works as expected if you remember that $i^2 = -1$. So, for example,

$$(1 + 2i)(2 + 3i) = 2 + 3i + 4i + 6i^2 = 2 + 7i - 6 = -4 + 7i,$$

which seems strange, but follows mechanically from $i^2 = -1$. In general,

$$\boxed{(x_1 + iy_1)(x_2 + iy_2) = (x_1x_2 - y_1y_2) + i(x_1y_2 + x_2y_1).} \qquad \text{(B.9)}$$

If we multiply $z = x + iy$ by its complex conjugate z^* we get

$$zz^* = (x + iy)(x - iy) = x^2 + y^2,$$

which is a positive real number. Its positive square root is called the **modulus** or **magnitude** of z, and is written $|z|$,

$$|z| = \sqrt{zz^*} = \sqrt{x^2 + y^2}.$$

How to calculate the ratio of two complex numbers is less obvious, but it is equally mechanical. We convert the denominator into a real number by multiplying both numerator and denominator by the complex conjugate of the denominator,

$$\begin{aligned}\frac{2 + 3i}{1 + 2i} &= \frac{2 + 3i}{1 + 2i} \times \frac{1 - 2i}{1 - 2i} \\ &= \frac{(2 + 6) + (-4 + 3)i}{1 + 4} \\ &= \frac{8}{5} - \frac{1}{5}i.\end{aligned}$$

The general formula is

$$\frac{x_1 + iy_1}{x_2 + iy_2} = \frac{x_1x_2 + y_1y_2}{x_2^2 + y_2^2} + i\frac{-x_1y_2 + x_2y_1}{x_2^2 + y_2^2}. \tag{B.10}$$

In practice it is easier to calculate the ratio as in the example, rather than memorizing formula (B.10).

B.5 Exponentials

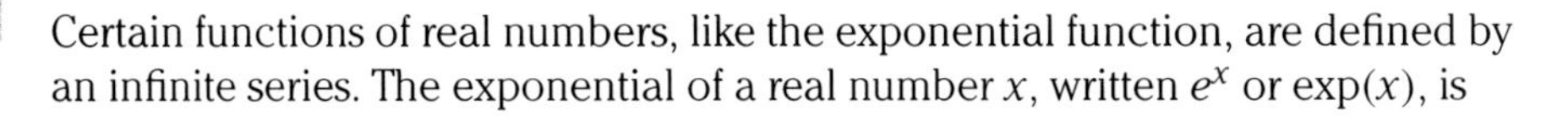

Certain functions of real numbers, like the exponential function, are defined by an infinite series. The exponential of a real number x, written e^x or $\exp(x)$, is

$$e^x = \sum_{k=0}^{\infty} \frac{x^k}{k!} = 1 + x + \frac{x^2}{2!} + \frac{x^3}{3!} + \cdots.$$

We also recall the infinite series expansion for cos and sin:

$$\cos(\theta) = 1 - \frac{\theta^2}{2} + \frac{\theta^4}{4!} - \cdots$$

$$\sin(\theta) = \theta - \frac{\theta^3}{3!} + \frac{\theta^5}{5!} - \cdots.$$

The exponential of a complex number z is written e^z or $\exp(z)$, and is defined in the same way as the exponential of a real number,

$$e^z = \sum_{k=0}^{\infty} \frac{z^k}{k!} = 1 + z + \frac{z^2}{2!} + \frac{z^3}{3!} + \cdots. \tag{B.11}$$

Note that $e^0 = 1$, as expected.

The exponential of an imaginary number $i\theta$ is very interesting,

$$\begin{aligned} e^{i\theta} &= 1 + (i\theta) + \frac{(i\theta)^2}{2!} + \frac{(i\theta)^3}{3!} + \cdots \\ &= [1 - \frac{\theta^2}{2} + \frac{\theta^4}{4!} - \cdots] + i[\theta - \frac{\theta^3}{3!} + \frac{\theta^5}{5!} - \cdots] \\ &= \cos(\theta) + i\sin(\theta). \end{aligned}$$

This identity is known as **Euler's formula**:

$$\boxed{e^{i\theta} = \cos(\theta) + i\sin(\theta).} \tag{B.12}$$

Euler's formula is used heavily in this text in the analysis of linear time invariant systems. It allows sinusoidal functions to be given as sums or differences of exponential functions,

$$\boxed{\cos(\theta) = (e^{i\theta} + e^{-i\theta})/2} \tag{B.13}$$

and

$$\boxed{\sin(\theta) = (e^{i\theta} - e^{-i\theta})/(2i).} \tag{B.14}$$

This proves useful because exponential functions turn out to be simpler mathematically (despite being complex valued) than sinusoidal functions.

An important property of the exponential function is the **product formula**:

$$\boxed{e^{z_1+z_2} = e^{z_1}e^{z_2}.} \tag{B.15}$$

We can obtain many trigonometric identities by combining (B.12) and (B.15). For example, since

$$e^{i\theta}e^{-i\theta} = e^{i\theta-i\theta} = e^0 = 1,$$

and

$$e^{i\theta}e^{-i\theta} = [\cos(\theta) + i\sin(\theta)][\cos(\theta) - i\sin(\theta)] = \cos^2(\theta) + \sin^2(\theta),$$

we have the identity

$$\boxed{\cos^2(\theta) + \sin^2(\theta) = 1.}$$

Here is another example. Using

$$e^{i(\alpha+\beta)} = e^{i\alpha}e^{i\beta}, \tag{B.16}$$

we get

$$\begin{aligned} \cos(\alpha+\beta) + i\sin(\alpha+\beta) &= [\cos(\alpha) + i\sin(\alpha)][\cos(\beta) + i\sin(\beta)] \\ &= [\cos(\alpha)\cos(\beta) - \sin(\alpha)\sin(\beta)] \\ &\quad + i[\sin(\alpha)\cos(\beta) + \cos(\alpha)\sin(\beta)]. \end{aligned}$$

Since the real part of the left side must equal the real part of the right side, we get the identity,

$$\boxed{\cos(\alpha+\beta)=\cos(\alpha)\cos(\beta)-\sin(\alpha)\sin(\beta).}$$

Since the imaginary part of the left side must equal the imaginary part of the right side, we get the identity

$$\boxed{\sin(\alpha+\beta)=\sin(\alpha)\cos(\beta)+\cos(\alpha)\sin(\beta).}$$

It is much easier to remember (B.16) than to remember these identities.

B.6 Polar coordinates

The representation of a complex number as a sum of a real and an imaginary number, $z = x + iy$, is called its **Cartesian representation**.

Recall from trigonometry that if x, y, r are real numbers and $r^2 = x^2 + y^2$, then there is a unique number θ with $0 \le \theta < 2\pi$ such that

$$\cos(\theta) = \frac{x}{r}, \ \sin(\theta) = \frac{y}{r}.$$

That number is

$$\theta = \cos^{-1}(x/r) = \sin^{-1}(y/r) = \tan^{-1}(y/x).$$

We can therefore express any complex number $z = x + iy$ as

$$z = |z|(\frac{x}{|z|} + i\frac{y}{|z|}) = |z|(\cos\theta + i\sin\theta) = |z|e^{i\theta},$$

where $\theta = \tan^{-1}(y/x)$. The **angle** or **argument** θ is measured in radians, and it is written as $\arg(z)$ or $\angle z$. So we have the **polar representation** of any complex number z as

$$\boxed{z = x + iy = re^{i\theta}.} \tag{B.17}$$

The two representations are related by

$$r = |z| = \sqrt{x^2 + y^2}$$

and

$$\theta = \arg(z) = \tan^{-1}(y/x).$$

The values x and y are called the **Cartesian coordinates** of z, while r and θ are its **polar coordinates**. Note that r is real and $r \geq 0$. Figure B.1 depicts the Cartesian and polar representations.

BASICS

From Cartesian to polar coordinates

The polar representation of a complex number z is

$$z = x + iy = re^{i\theta},$$

where

$$r = |z| = \sqrt{x^2 + y^2} \quad \text{and} \quad \theta = \arg(z) = \tan^{-1}(y/x).$$

However, you must be careful in calculating $\tan^{-1}(y/x)$. For any angle θ,

$$\tan(\theta) = \tan(\theta + \pi).$$

Thus, for any real number y/x, there are two possible values for $\theta = \tan^{-1}(y/x)$ that lie within the range $[0, 2\pi)$. You should select the one of these that yields a nonnegative value for r in

$$z = x + iy = re^{i\theta} = r(\cos(\theta) + i\sin(\theta)).$$

This choice makes it reasonable to interpret r as the magnitude of the complex number.

Note that for any integer K,

$$re^{i(2K\pi + \theta)} = re^{i\theta}.$$

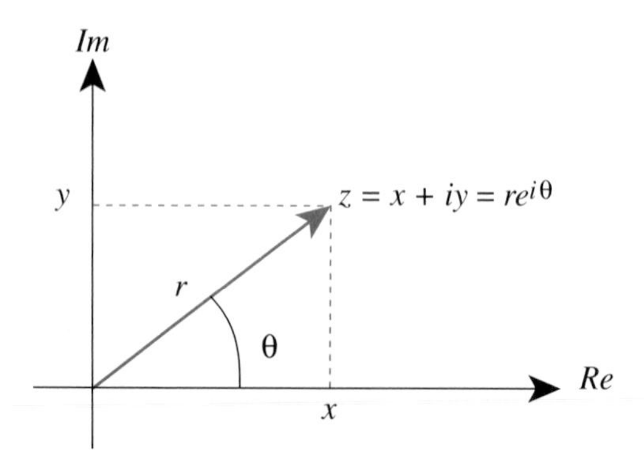

FIGURE B.1: A complex number z is represented in Cartesian coordinates as $z = x + iy$ and in polar coordinates as $z = re^{i\theta}$. The x axis is called the **real axis**, the y axis is called the **imaginary axis**. The **angle** θ in radians is measured counterclockwise from the real axis.

This is because

$$re^{i(2K\pi+\theta)} = re^{i2K\pi}e^{i\theta}$$

and

$$e^{i2K\pi} = \cos(2K\pi) + i\sin(2K\pi) = 1.$$

Thus, the polar coordinates (r, θ) and $(r, \theta + 2K\pi)$ for any integer K represent the same complex number. Thus, the polar representation is not unique; by convention, a unique polar representation can be obtained by requiring that the angle given by a value of θ satisfying $0 \leq \theta < 2\pi$ or $-\pi < \theta \leq \pi$. We normally require $0 \leq \theta < 2\pi$.

Example B.1: The polar representation of the number 1 is $1 = 1e^{i0}$. Notice that it is also true that $1 = 1e^{i2\pi}$, because the sine and cosine are periodic with period 2π. The polar representation of the number -1 is $-1 = 1e^{i\pi}$. Again, it is true that $-1 = 1e^{i3\pi}$, or, in fact, $-1 = 1e^{i\pi+K2\pi}$, for any integer K. ❐

Products of complex numbers represented in polar coordinates are easy to compute. If $z_i = r_i e^{i\theta_i}$, then

$$z_1 z_2 = r_1 r_2 e^{i(\theta_1+\theta_2)}.$$

Thus, the magnitude of a product is a product of the magnitudes, and the angle of a product is the sum of the angles,

$$\boxed{|z_1 z_2| = |z_1||z_2|, \quad \angle(z_1 z_2) = \angle(z_1) + \angle(z_2).}$$

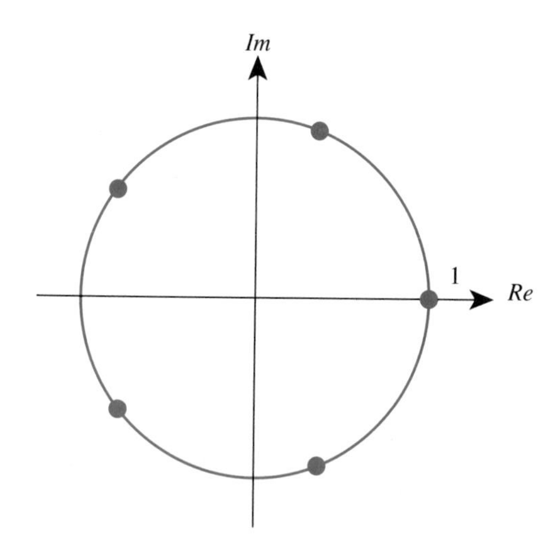

FIGURE B.2: The 5 roots of unity.

Example B.2: We can use the polar representation to find the n distinct roots of the equation $z^n = 1$. Write $z = re^{i\theta}$, and $1 = e^{i2k\pi}$, so

$$z^n = r^n e^{in\theta} = e^{i2k\pi},$$

which gives $r = 1$ and $\theta = 2k\pi/n$, $k = 0, 1, \ldots, n-1$. These are called the n **roots of unity**. Figure B.2 shows the 5 roots of unity. ❐

Whereas it is easy to solve the polynomial equation $z^n = 1$, solving a general polynomial equation is difficult.

Theorem. The polynomial equation

$$z^n + a_1 z^{n-1} + \cdots + a_{n-1} z + a_n = 0,$$

where $a_1, \ldots, a_n$ are complex constants, has exactly n factors of the form $(z - \alpha_i)$, where $\alpha_1, \ldots \alpha_n$ are called the n roots. In other words, we can always find the factorization,

$$z^n + a_1 z^{n-1} + \cdots + a_{n-1} z + a_n = \prod_{k=1}^{n} (z - \alpha_k).$$

Some of the roots may be identical.

Note that although this theorem ensures the existence of this factorization, it does not suggest a way to find the roots. Indeed, finding the roots can be difficult. Fortunately, software for finding roots is readily available, for example using the MATLAB®* `roots` function.

EXERCISES

KEY: E = mechanical T = requires plan of attack C = more than 1 answer

E 1. Simplify the following expressions:

(a)

$$\frac{3+4i}{5-6i} \times \frac{3+6i}{4-5i},$$

(b)

$$e^{2+\pi i}.$$

E 2. Express the following in polar coordinates:

$$2-2i,\ 2+2i,\ \frac{1}{2-2i},\ \frac{1}{2+2i},\ 2i,\ -2i.$$

E 3. Depict the following numbers graphically as in figure B.1:

$$i1,\ -2,\ -3-i,\ -1-i.$$

E 4. Find θ so that

$$Re\{(1+i)e^{i\theta}\} = -1.$$

E 5. Express the six distinct roots of unity, namely, the six solutions to

$$z^6 = 1$$

in Cartesian and polar coordinates.

T 6. Express the six roots of -1, namely, the six solutions to

$$z^6 = -1$$

in Cartesian and polar coordinates. Depict these roots as in figure B.2.

* MATLAB is a registered trademark of The MathWorks, Inc.

T 7. Figure out i^n for all positive and negative integers n. (For a negative integer n, $z^{-n} = 1/z^n$.)

T 8. Factor the polynomial $z^5 + 2$ as

$$z^5 + 2 = \prod_{k=1}^{5} (z - \alpha_k),$$

expressing the α_k in polar coordinates.

C 9. How would you define $\sqrt{1+i}$? More generally, how would you define $\sqrt{z}$ for any complex number z?

T 10. The logarithm of a complex number z is written $\log z$ or $\log(z)$. It can be defined as an infinite series, or as the inverse of the exponential (i.e., define $\log z = w$, if $e^w = z$). Using the latter definition, find the logarithm of the following complex numbers:

$$1, \ -1, \ i, \ -i, \ 1+i.$$

More generally, if $z \neq 0$ is expressed in polar coordinates, what is $\log z$? For which complex numbers z is $\log z$ not defined?

E 11. Use MATLAB to answer the following questions. Let $z_1 = 2 + 3i$ and $z_2 = 4 - 2i$. Hint: Consult MATLAB help on `i`, `j`, `exp`, `real`, `imag`, `abs`, `angle`, `conj`, and `complex`. Looking up "complex" in the help desk may also help.

(a) What is $z_1 + z_2$? What are the real and imaginary parts of the sum?

(b) Express the sum in polar coordinates.

(c) Draw by hand two rays in the complex plane, one from the origin to z_1 and the other from the origin to z_2. Now draw $z_1 + z_2$ and $z_1 - z_2$ on the same plane. Explain how you might systematically construct the sum and difference rays.

(d) Draw two rays in the complex plane to $z_3 = -2 - 3i$ and $z_4 = 3 - 3i$. Now draw $z_3 \times z_4$ and z_3/z_4.

(e) Consider $z_5 = 2e^{i\pi/6}$ and $z_6 = z_5^*$. Express z_6 in polar coordinates. What is $z_5 z_6$?

(f) Draw the ray to $z_0 = 1 + 1i$. Now draw rays to $z_n = z_0 e^{in\pi/4}$ for $n = 1, 2, 3, \ldots$. How many distinct z_n are there?

(g) Find all the solutions of the equation $z^7 = 1$. Hint: Express z in polar coordinates, $z = re^{i\theta}$ and solve for r, θ.

E 12. This problem explores how complex signals may be visualized and analyzed.

(a) Use MATLAB to plot the complex exponential function as follows:

```
plot(exp((-2+10i)*[0:0.01:1]))
```

The result is a spiraling curve corresponding to the signal $f\colon [0, 1] \to$ *Complex* where

$$\forall t \in [0, 1] \quad f(t) = e^{(-2+10i)t}.$$

In the plot window, under the Tools menu item, use "Axes properties" to turn on the grid. Print the plot and on it mark the points for which the function is purely imaginary. Is it evident what values of t yield purely imaginary $f(t)$?

(b) Find analytically the values of t that result in purely imaginary and purely real $f(t)$.

(c) Construct four plots, where the horizontal axis represents t and the vertical axis represents the real and imaginary parts of $f(t)$, and the magnitude and angle of $f(t)$. Give these as four subplots.

(d) Give the mathematical expressions for the four functions plotted in part (c).

T 13. Euler's formula is: for any real number θ,

$$e^{i\theta} = \cos\theta + i\sin\theta,$$

and the product formula is: for any complex numbers z_1, z_2,

$$e^{z_1+z_2} = e^{z_1}e^{z_2}.$$

The following problems show that these two formulas can be combined to obtain many useful identities.

(a) Express $\sin(2\theta)$ and $\cos(2\theta)$ as sums and products of $\sin\theta$ and $\cos\theta$. Hint: Write $e^{i2\theta} = e^{i\theta}e^{i\theta}$ (by the product formula) and then use Euler's formula.

(b) Express $\sin(3\theta)$ and $\cos(3\theta)$ also as sums and products of $\sin\theta$ and $\cos\theta$.

(c) The sum of several sinewaves of the *same* frequency ω but different phases is a sine wave of the same frequency, that is, given A_k, ϕ_k, $k = 1, \ldots, n$, we can find A, ϕ so that

$$A \cos(\omega t + \phi) = \sum_{k=1}^{n} A_k \cos(\omega t + \phi_k).$$

Express A, ϕ in terms of $\{A_k, \phi_k\}$.

Symbols

Index

Entries in *italics* are mathematical definitions, typically of sets or functions.
Page numbers in **bold** indicate where a key term is defined.